GENERAL MOTORS | ASTRO/SAFARI 1985- ___ ___ ___AL

CHILTON'S

**Covers U.S. and Canadian models of
Chevrolet Astro and GMC Safari**

Does not include information specific
to all-wheei drive models

**by Kevin M. G. Maher, A.S.E.
and Ken Freund**

CHILTON *Automotive Books*

PUBLISHED BY **HAYNES NORTH AMERICA. Inc.**

Manufactured in USA
©1997, 2005, 2007 Haynes North America, Inc.
ISBN 13: 978 1 56392 697 6
ISBN 10: 1 56392 697 0
Library of Congress Control Number 2007940563

Haynes Publishing Group
Sparkford Nr Yeovil
Somerset BA22 7JJ England

Haynes North America, Inc
861 Lawrence Drive
Newbury Park
California 91320 USA

ABCDE
F

11K4

Contents

Author, mechanic and photographer with 2003 Chevrolet Astro mini-van

ACKNOWLEDGEMENTS

We are grateful for the help and cooperation of Tomco Industries, 1435 Woodson Road, St. Louis, Missouri 63132, for their assistance with technical information and illustrations. Wiring diagrams and certain illustrations originated exclusively for Haynes North America, Inc. by Valley Forge Technical Information Services. Technical writers who contributed to this project include Larry Warren, Mike Stubblefield, Jeff Killingsworth and Marc Scribner.

About this manual

ITS PURPOSE

The purpose of this manual is to help you get the best value from your vehicle. It can do so in several ways. It can help you decide what work must be done, even if you choose to have it done by a dealer service department or a repair shop; it provides information and procedures for routine maintenance and servicing; and it offers diagnostic and repair procedures to follow when trouble occurs.

We hope you use the manual to tackle the work yourself. For many simpler jobs, doing it yourself may be quicker than arranging an appointment to get the vehicle into a shop and making the trips to leave it and pick it up. More importantly, a lot of money can be saved by avoiding the expense the shop must pass on to you to cover its labor and overhead costs. An added benefit is the sense of satisfaction and accomplishment that you feel after doing the job yourself.

USING THE MANUAL

The manual is divided into Chapters. Each Chapter is divided into numbered Sections. Each Section consists of consecutively numbered paragraphs.

At the beginning of each numbered Section you will be referred to any illustrations which apply to the procedures in that Section. The reference numbers used in illustration captions pinpoint the pertinent Section and the Step within that Section. That is, illustration 3.2 means the illustration refers to Section 3 and Step (or paragraph) 2 within that Section.

Procedures, once described in the text, are not normally repeated. When it's necessary to refer to another Chapter, the reference will be given as Chapter and Section number. Cross references given without use of the word "Chapter" apply to Sections and/or paragraphs in the same Chapter. For example, "see Section 8" means in the same Chapter.

References to the left or right side of the vehicle assume you are sitting in the driver's seat, facing forward.

Even though we have prepared this manual with extreme care, neither the publisher nor the author can accept responsibility for any errors in, or omissions from, the information given.

➡NOTE

A *Note* provides information necessary to properly complete a procedure or information which will make the procedure easier to understand.

※ CAUTION

A *Caution* provides a special procedure or special steps which must be taken while completing the procedure where the Caution is found. Not heeding a Caution can result in damage to the assembly being worked on.

※ WARNING

A *Warning* provides a special procedure or special steps which must be taken while completing the procedure where the Warning is found. Not heeding a Warning can result in personal injury.

Introduction to the Chevrolet Astro and GMC Safari

Chevrolet and GMC mini-vans are conventional front engine/rear wheel drive layout.

Inline four-cylinder and V6 engines are used for power, with fuel injection available on later models.

Power from the engine is transferred to either a four or five-speed manual or a four-speed automatic transmission.

Suspension is independent at the front with unequal length upper and lower control arms and coil springs. The rear suspension on all models consists of longitudinal composite fiberglass leaf springs.

The steering box is mounted to the left of the engine and is connected to the steering arms through a series of rods with power assist available as an option.

The brakes are vacuum assisted disc-type at the front and either self-adjusting drums or, on later models, discs at the rear.

Vehicle Identification numbers

Modifications are a continuing and unpublicized process in vehicle manufacturing. Since spare parts manuals and lists are compiled on a numerical basis, the individual vehicle numbers are essential to correctly identify the component required.

VEHICLE IDENTIFICATION NUMBER (VIN)

This very important identification number is located on a plate attached to the driver's side cowling just inside the windshield (see illustration). The VIN also appears on the Vehicle Certificate of Title and Registration. It contains information such as where and when the vehicle was manufactured, the model year and the body style.

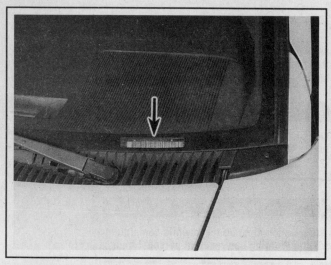

The Vehicle Identification Number (VIN) is on the driver's side cowl area - it is visible through the windshield

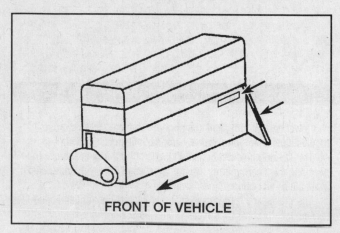

Engine ID number locations on four-cylinder engines

SERVICE PARTS IDENTIFICATION LABEL

This label should always be referred to when ordering parts. The vehicle Service parts identification label contains the VIN, wheelbase, paint and special equipment and option codes (see illustration).

ENGINE ID NUMBERS

The engine ID numbers can be found in a variety of locations, depending on engine type.

On the four-cylinder engine, the engine ID number is found on a pad located on the left side of the block, to the rear of the exhaust manifold.

On the V6 engine, the ID number is located on a pad at the front or at the left rear of the engine block (see illustrations).

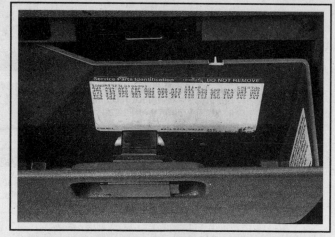

The Service parts ID label is located in one of three places, depending on the year: in the glove box, as shown here (1985 to 1993), in the engine compartment (1994 and 1995) or in the jamb of the right front door (1996 and later models)

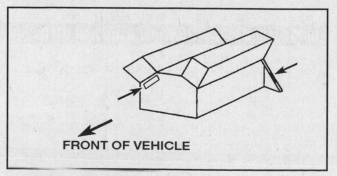

Engine ID number locations on V6 engines

Buying parts

Replacement parts are available from many sources, which generally fall into one of two categories - authorized dealer parts departments and independent retail auto parts stores. Our advice concerning these parts is as follows:

Retail auto parts stores: Good auto parts stores will stock frequently needed components which wear out relatively fast, such as clutch components, exhaust systems, brake parts, tune-up parts, etc. These stores often supply new or reconditioned parts on an exchange basis, which can save a considerable amount of money. Discount auto parts stores are often very good places to buy materials and parts needed for general vehicle maintenance such as oil, grease, filters, spark plugs, belts, touch-up paint, bulbs, etc. They also usually sell

tools and general accessories, have convenient hours, charge lower prices and can often be found not far from home.

Authorized dealer parts department: This is the best source for parts which are unique to the vehicle and not generally available elsewhere (such as major engine parts, transmission parts, trim pieces, etc.).

Warranty information: If the vehicle is still covered under warranty, be sure that any replacement parts purchased - regardless of the source - do not invalidate the warranty!

To be sure of obtaining the correct parts, have engine and chassis numbers available and, if possible, take the old parts along for positive identification.

Maintenance techniques, tools and working facilities

MAINTENANCE TECHNIQUES

There are a number of techniques involved in maintenance and repair that will be referred to throughout this manual. Application of these techniques will enable the home mechanic to be more efficient, better organized and capable of performing the various tasks properly, which will ensure that the repair job is thorough and complete.

Fasteners

Fasteners are nuts, bolts, studs and screws used to hold two or more parts together. There are a few things to keep in mind when working with fasteners. Almost all of them use a locking device of some type, either a lockwasher, locknut, locking tab or thread adhesive. All threaded fasteners should be clean and straight, with undamaged threads and undamaged corners on the hex head where the wrench fits. Develop the habit of replacing all damaged nuts and bolts with new ones. Special locknuts with nylon or fiber inserts can only be used once. If they are removed, they lose their locking ability and must be replaced with new ones.

Rusted nuts and bolts should be treated with a penetrating fluid to ease removal and prevent breakage. Some mechanics use turpentine in a spout-type oil can, which works quite well. After applying the rust penetrant, let it work for a few minutes before trying to loosen the nut or bolt. Badly rusted fasteners may have to be chiseled or sawed off or removed with a special nut breaker, available at tool stores.

If a bolt or stud breaks off in an assembly, it can be drilled and removed with a special tool commonly available for this purpose. Most automotive machine shops can perform this task, as well as other

repair procedures, such as the repair of threaded holes that have been stripped out.

Flat washers and lockwashers, when removed from an assembly, should always be replaced exactly as removed. Replace any damaged washers with new ones. Never use a lockwasher on any soft metal surface (such as aluminum), thin sheet metal or plastic.

Fastener sizes

For a number of reasons, automobile manufacturers are making wider and wider use of metric fasteners. Therefore, it is important to be able to tell the difference between standard (sometimes called U.S. or SAE) and metric hardware, since they cannot be interchanged.

All bolts, whether standard or metric, are sized according to diameter, thread pitch and length. For example, a standard 1/2 - 13 x 1 bolt is 1/2 inch in diameter, has 13 threads per inch and is 1 inch long. An M12 - 1.75 x 25 metric bolt is 12 mm in diameter, has a thread pitch of 1.75 mm (the distance between threads) and is 25 mm long. The two bolts are nearly identical, and easily confused, but they are not interchangeable.

In addition to the differences in diameter, thread pitch and length, metric and standard bolts can also be distinguished by examining the bolt heads. To begin with, the distance across the flats on a standard bolt head is measured in inches, while the same dimension on a metric bolt is sized in millimeters (the same is true for nuts). As a result, a standard wrench should not be used on a metric bolt and a metric wrench should not be used on a standard bolt. Also, most standard bolts have slashes radiating out from the center of the head to denote the grade or strength of the bolt, which is an indication of the amount of torque that

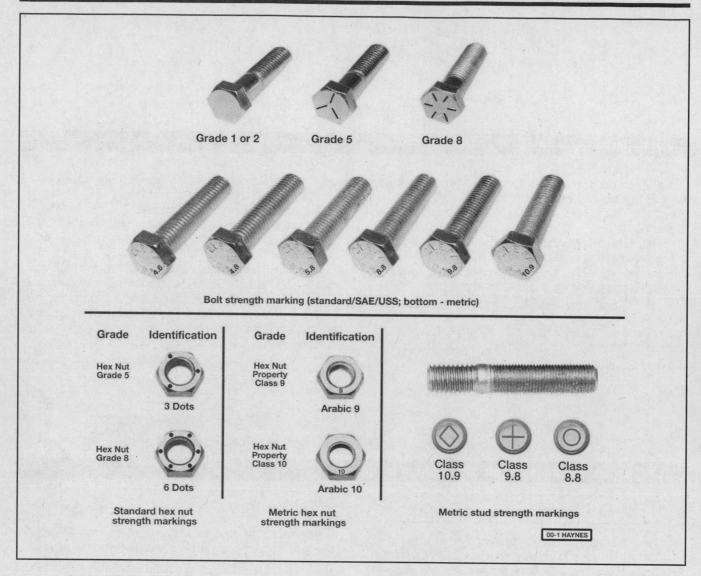

Grade 1 or 2   Grade 5   Grade 8

Bolt strength marking (standard/SAE/USS; bottom - metric)

Grade	Identification
Hex Nut Grade 5	3 Dots
Hex Nut Grade 8	6 Dots

Grade	Identification
Hex Nut Property Class 9	Arabic 9
Hex Nut Property Class 10	Arabic 10

Standard hex nut strength markings

Metric hex nut strength markings

Class 10.9   Class 9.8   Class 8.8

Metric stud strength markings

00-1 HAYNES

can be applied to it. The greater the number of slashes, the greater the strength of the bolt. Grades 0 through 5 are commonly used on automobiles. Metric bolts have a property class (grade) number, rather than a slash, molded into their heads to indicate bolt strength. In this case, the higher the number, the stronger the bolt. Property class numbers 8.8, 9.8 and 10.9 are commonly used on automobiles.

Strength markings can also be used to distinguish standard hex nuts from metric hex nuts. Many standard nuts have dots stamped into one side, while metric nuts are marked with a number. The greater the number of dots, or the higher the number, the greater the strength of the nut.

Metric studs are also marked on their ends according to property class (grade). Larger studs are numbered (the same as metric bolts), while smaller studs carry a geometric code to denote grade.

It should be noted that many fasteners, especially Grades 0 through 2, have no distinguishing marks on them. When such is the case, the only way to determine whether it is standard or metric is to measure the thread pitch or compare it to a known fastener of the same size.

Standard fasteners are often referred to as SAE, as opposed to metric. However, it should be noted that SAE technically refers to a non-metric fine thread fastener only. Coarse thread non-metric fasteners are referred to as USS sizes.

Since fasteners of the same size (both standard and metric) may

have different strength ratings, be sure to reinstall any bolts, studs or nuts removed from your vehicle in their original locations. Also, when replacing a fastener with a new one, make sure that the new one has a strength rating equal to or greater than the original.

Tightening sequences and procedures

Most threaded fasteners should be tightened to a specific torque value (torque is the twisting force applied to a threaded component such as a nut or bolt). Overtightening the fastener can weaken it and cause it to break, while undertightening can cause it to eventually come loose. Bolts, screws and studs, depending on the material they are made of and their thread diameters, have specific torque values, many of which are noted in the Specifications at the end of each Chapter. Be sure to follow the torque recommendations closely. For fasteners not assigned a specific torque, a general torque value chart is presented here as a guide. These torque values are for dry (unlubricated) fasteners threaded into steel or cast iron (not aluminum). As was previously mentioned, the size and grade of a fastener determine the amount of torque that can safely be applied to it. The figures listed here are approximate for Grade 2 and Grade 3 fasteners. Higher grades can tolerate higher torque values.

Fasteners laid out in a pattern, such as cylinder head bolts, oil pan

Metric thread sizes

	Ft-lbs	Nm
M-6	6 to 9	9 to 12
M-8	14 to 21	19 to 28
M-10	28 to 40	38 to 54
M-12	50 to 71	68 to 96
M-14	80 to 140	109 to 154

Pipe thread sizes

1/8	5 to 8	7 to 10
1/4	12 to 18	17 to 24
3/8	22 to 33	30 to 44
1/2	25 to 35	34 to 47

U.S. thread sizes

1/4 - 20	6 to 9	9 to 12
5/16 - 18	12 to 18	17 to 24
5/16 - 24	14 to 20	19 to 27
3/8 - 16	22 to 32	30 to 43
3/8 - 24	27 to 38	37 to 51
7/16 - 14	40 to 55	55 to 74
7/16 - 20	40 to 60	55 to 81
1/2 - 13	55 to 80	75 to 108

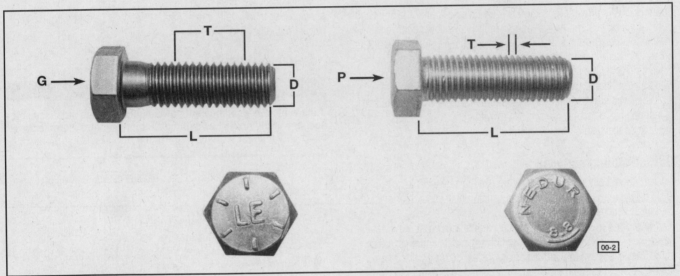

Standard (SAE and USS) bolt dimensions/grade marks

G   Grade marks (bolt strength)
L   Length (in inches)
T   Thread pitch (number of threads per inch)
D   Nominal diameter (in inches)

Metric bolt dimensions/grade marks

P   Property class (bolt strength)
L   Length (in millimeters)
T   Thread pitch (distance between threads in millimeters)
D   Diameter

bolts, differential cover bolts, etc., must be loosened or tightened in sequence to avoid warping the component. This sequence will normally be shown in the appropriate Chapter. If a specific pattern is not given, the following procedures can be used to prevent warping.

Initially, the bolts or nuts should be assembled finger-tight only. Next, they should be tightened one full turn each, in a criss-cross or diagonal pattern. After each one has been tightened one full turn, return to the first one and tighten them all one-half turn, following the same pattern. Finally, tighten each of them one-quarter turn at a time until each fastener has been tightened to the proper torque. To loosen and remove the fasteners, the procedure would be reversed.

Component disassembly

Component disassembly should be done with care and purpose to help ensure that the parts go back together properly. Always keep track of the sequence in which parts are removed. Make note of special characteristics or marks on parts that can be installed more than one way, such as a grooved thrust washer on a shaft. It is a good idea to lay the disassembled parts out on a clean surface in the order that they were removed. It may also be helpful to make sketches or take instant photos of components before removal.

When removing fasteners from a component, keep track of their locations. Sometimes threading a bolt back in a part, or putting the

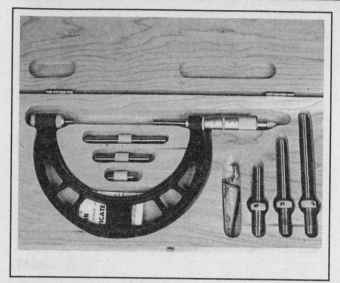

Micrometer set

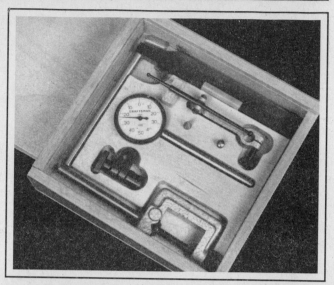

Dial indicator set

washers and nut back on a stud, can prevent mix-ups later. If nuts and bolts cannot be returned to their original locations, they should be kept in a compartmented box or a series of small boxes. A cupcake or muffin tin is ideal for this purpose, since each cavity can hold the bolts and nuts from a particular area (i.e. oil pan bolts, valve cover bolts, engine mount bolts, etc.). A pan of this type is especially helpful when working on assemblies with very small parts, such as the carburetor, alternator, valve train or interior dash and trim pieces. The cavities can be marked with paint or tape to identify the contents.

Whenever wiring looms, harnesses or connectors are separated, it is a good idea to identify the two halves with numbered pieces of masking tape so they can be easily reconnected.

Gasket sealing surfaces

Throughout any vehicle, gaskets are used to seal the mating surfaces between two parts and keep lubricants, fluids, vacuum or pressure contained in an assembly.

Many times these gaskets are coated with a liquid or paste-type gasket sealing compound before assembly. Age, heat and pressure can sometimes cause the two parts to stick together so tightly that they are very difficult to separate. Often, the assembly can be loosened by striking it with a soft-face hammer near the mating surfaces. A regular hammer can be used if a block of wood is placed between the hammer and the part. Do not hammer on cast parts or parts that could be easily damaged. With any particularly stubborn part, always recheck to make sure that every fastener has been removed.

Avoid using a screwdriver or bar to pry apart an assembly, as they can easily mar the gasket sealing surfaces of the parts, which must remain smooth. If prying is absolutely necessary, use an old broom handle, but keep in mind that extra clean up will be necessary if the wood splinters.

After the parts are separated, the old gasket must be carefully scraped off and the gasket surfaces cleaned. Stubborn gasket material can be soaked with rust penetrant or treated with a special chemical to soften it so it can be easily scraped off. A scraper can be fashioned from a piece of copper tubing by flattening and sharpening one end. Copper is recommended because it is usually softer than the surfaces to be scraped, which reduces the chance of gouging the part. Some gaskets can be removed with a wire brush, but regardless of the method

used, the mating surfaces must be left clean and smooth. If for some reason the gasket surface is gouged, then a gasket sealer thick enough to fill scratches will have to be used during reassembly of the components. For most applications, a non-drying (or semi-drying) gasket sealer should be used.

Hose removal tips

✷✷ WARNING:

If the vehicle is equipped with air conditioning, do not disconnect any of the A/C hoses without first having the system depressurized by a dealer service department or a service station.

Hose removal precautions closely parallel gasket removal precautions. Avoid scratching or gouging the surface that the hose mates against or the connection may leak. This is especially true for radiator hoses. Because of various chemical reactions, the rubber in hoses can bond itself to the metal spigot that the hose fits over. To remove a hose, first loosen the hose clamps that secure it to the spigot. Then, with slip-joint pliers, grab the hose at the clamp and rotate it around the spigot. Work it back and forth until it is completely free, then pull it off. Silicone or other lubricants will ease removal if they can be applied between the hose and the outside of the spigot. Apply the same lubricant to the inside of the hose and the outside of the spigot to simplify installation.

As a last resort (and if the hose is to be replaced with a new one anyway), the rubber can be slit with a knife and the hose peeled from the spigot. If this must be done, be careful that the metal connection is not damaged.

If a hose clamp is broken or damaged, do not reuse it. Wire-type clamps usually weaken with age, so it is a good idea to replace them with screw-type clamps whenever a hose is removed.

TOOLS

A selection of good tools is a basic requirement for anyone who plans to maintain and repair his or her own vehicle. For the owner who has few tools, the initial investment might seem high, but when

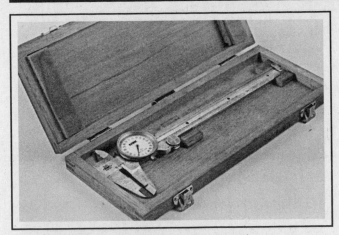

Dial caliper

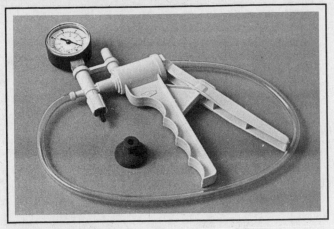

Hand-operated vacuum pump

Timing light

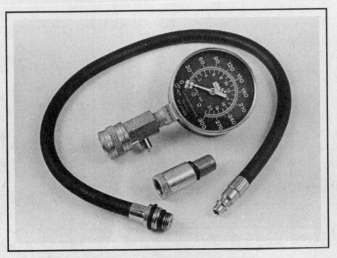

Compression gauge with spark plug hole adapter

compared to the spiraling costs of professional auto maintenance and repair, it is a wise one.

To help the owner decide which tools are needed to perform the tasks detailed in this manual, the following tool lists are offered: *Maintenance and minor repair, Repair/overhaul* and *Special*.

The newcomer to practical mechanics should start off with the *maintenance and minor repair* tool kit, which is adequate for the simpler jobs performed on a vehicle. Then, as confidence and experience grow, the owner can tackle more difficult tasks, buying additional tools as they are needed. Eventually the basic kit will be expanded into the *repair and overhaul* tool set. Over a period of time, the experienced do-it-yourselfer will assemble a tool set complete enough for most repair and overhaul procedures and will add tools from the special category when it is felt that the expense is justified by the frequency of use.

Maintenance and minor repair tool kit

The tools in this list should be considered the minimum required for performance of routine maintenance, servicing and minor repair work. We recommend the purchase of combination wrenches (box-end and open-end combined in one wrench). While more expensive than open end wrenches, they offer the advantages of both types of wrench.

Combination wrench set (1/4-inch to 1 inch or 6 mm to 19 mm)
Adjustable wrench, 8 inch
Spark plug wrench with rubber insert

Spark plug gap adjusting tool
Feeler gauge set
Brake bleeder wrench
Standard screwdriver (5/16-inch x 6 inch)
Phillips screwdriver (No. 2 x 6 inch)
Combination pliers - 6 inch
Hacksaw and assortment of blades
Tire pressure gauge
Grease gun
Oil can
Fine emery cloth
Wire brush
Battery post and cable cleaning tool
Oil filter wrench
Funnel (medium size)
Safety goggles
Jackstands (2)
Drain pan

➡**Note: If basic tune-ups are going to be part of routine maintenance, it will be necessary to purchase a good quality stroboscopic timing light and combination tachometer/dwell meter. Although they are included in the list of special tools, it is mentioned here because they are absolutely necessary for tuning most vehicles properly.**

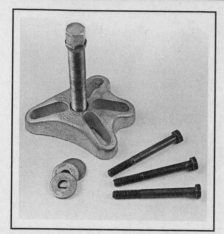

Damper/steering wheel puller

General purpose puller

Hydraulic lifter removal tool

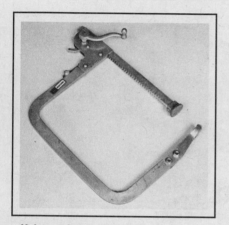

Valve spring compressor

Valve spring compressor

Ridge reamer

Repair and overhaul tool set

These tools are essential for anyone who plans to perform major repairs and are in addition to those in the maintenance and minor repair tool kit. Included is a comprehensive set of sockets which, though expensive, are invaluable because of their versatility, especially when various extensions and drives are available. We recommend the 1/2-inch drive over the 3/8-inch drive. Although the larger drive is bulky and more expensive, it has the capacity of accepting a very wide range of large sockets. Ideally, however, the mechanic should have a 3/8-inch drive set and a 1/2-inch drive set.

Socket set(s)
Reversible ratchet
Extension - 10 inch
Universal joint
Torque wrench (same size drive as sockets)
Ball peen hammer - 8 ounce
Soft-face hammer (plastic/rubber)
Standard screwdriver (1/4-inch x 6 inch)
Standard screwdriver (stubby - 5/16-inch)
Phillips screwdriver (No. 3 x 8 inch)
Phillips screwdriver (stubby - No. 2)
Pliers - vise grip
Pliers - lineman's
Pliers - needle nose
Pliers - snap-ring (internal and external)

Cold chisel - 1/2-inch
Scribe
Scraper (made from flattened copper tubing)
Centerpunch
Pin punches (1/16, 1/8, 3/16-inch)
Steel rule/straightedge - 12 inch
Allen wrench set (1/8 to 3/8-inch or 4 mm to 10 mm)
A selection of files
Wire brush (large)
Jackstands (second set)
Jack (scissor or hydraulic type)

➡ **Note: Another tool which is often useful is an electric drill with a chuck capacity of 3/8-inch and a set of good quality drill bits.**

Special tools

The tools in this list include those which are not used regularly, are expensive to buy, or which need to be used in accordance with their manufacturer's instructions. Unless these tools will be used frequently, it is not very economical to purchase many of them. A consideration would be to split the cost and use between yourself and a friend or friends. In addition, most of these tools can be obtained from a tool rental shop on a temporary basis.

This list primarily contains only those tools and instruments widely available to the public, and not those special tools produced by the vehicle manufacturer for distribution to dealer service depart-

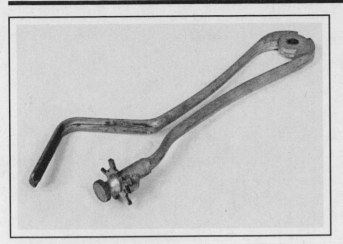

Piston ring groove cleaning tool

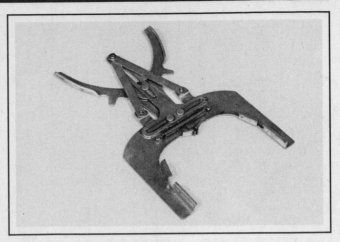

Ring removal/installation tool

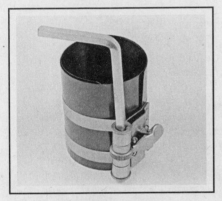

Ring compressor

Cylinder hone

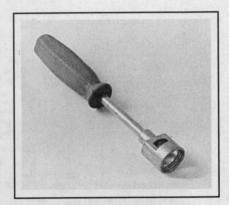

Brake hold-down spring tool

ments. Occasionally, references to the manufacturer's special tools are included in the text of this manual. Generally, an alternative method of doing the job without the special tool is offered. However, sometimes there is no alternative to their use. Where this is the case, and the tool cannot be purchased or borrowed, the work should be turned over to the dealer service department or an automotive repair shop.

Valve spring compressor
Piston ring groove cleaning tool
Piston ring compressor
Piston ring installation tool
Cylinder compression gauge
Cylinder ridge reamer
Cylinder surfacing hone
Cylinder bore gauge
Micrometers and/or dial calipers
Hydraulic lifter removal tool
Balljoint separator
Universal-type puller
Impact screwdriver
Dial indicator set
Stroboscopic timing light (inductive pick-up)
Hand operated vacuum/pressure pump
Tachometer/dwell meter
Universal electrical multimeter
Cable hoist
Brake spring removal and installation tools
Floor jack

Buying tools

For the do-it-yourselfer who is just starting to get involved in vehicle maintenance and repair, there are a number of options available when purchasing tools. If maintenance and minor repair is the extent of the work to be done, the purchase of individual tools is satisfactory. If, on the other hand, extensive work is planned, it would be a good idea to purchase a modest tool set from one of the large retail chain stores. A set can usually be bought at a substantial savings over the individual tool prices, and they often come with a tool box. As additional tools are needed, add-on sets, individual tools and a larger tool box can be purchased to expand the tool selection. Building a tool set gradually allows the cost of the tools to be spread over a longer period of time and gives the mechanic the freedom to choose only those tools that will actually be used.

Tool stores will often be the only source of some of the special tools that are needed, but regardless of where tools are bought, try to avoid cheap ones, especially when buying screwdrivers and sockets, because they won't last very long. The expense involved in replacing cheap tools will eventually be greater than the initial cost of quality tools.

Care and maintenance of tools

Good tools are expensive, so it makes sense to treat them with respect. Keep them clean and in usable condition and store them properly when not in use. Always wipe off any dirt, grease or metal chips before putting them away. Never leave tools lying around in the work area. Upon completion of a job, always check closely under the hood

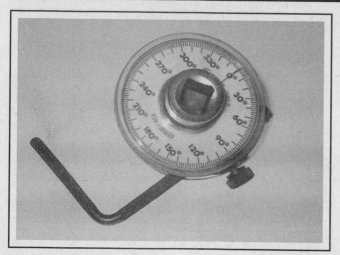

Torque angle gauge

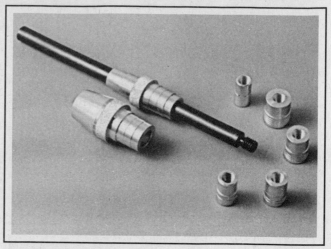

Clutch plate alignment tool

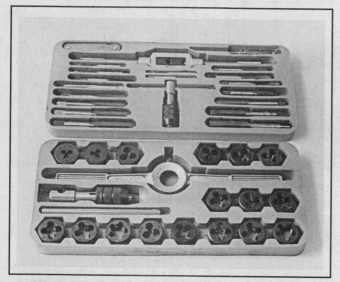

Tap and die set

for tools that may have been left there so they won't get lost during a test drive.

Some tools, such as screwdrivers, pliers, wrenches and sockets, can be hung on a panel mounted on the garage or workshop wall, while others should be kept in a tool box or tray. Measuring instruments, gauges, meters, etc. must be carefully stored where they cannot be damaged by weather or impact from other tools.

When tools are used with care and stored properly, they will last a very long time. Even with the best of care, though, tools will wear out if used frequently. When a tool is damaged or worn out, replace it. Subsequent jobs will be safer and more enjoyable if you do.

HOW TO REPAIR DAMAGED THREADS

Sometimes, the internal threads of a nut or bolt hole can become stripped, usually from overtightening. Stripping threads is an all-too-common occurrence, especially when working with aluminum parts, because aluminum is so soft that it easily strips out.

Usually, external or internal threads are only partially stripped. After they've been cleaned up with a tap or die, they'll still work. Sometimes, however, threads are badly damaged. When this happens, you've got three choices:

1) Drill and tap the hole to the next suitable oversize and install a larger diameter bolt, screw or stud.
2) Drill and tap the hole to accept a threaded plug, then drill and tap the plug to the original screw size. You can also buy a plug already threaded to the original size. Then you simply drill a hole to the specified size, then run the threaded plug into the hole with a bolt and jam nut. Once the plug is fully seated, remove the jam nut and bolt.
3) The third method uses a patented thread repair kit like Heli-Coil or Slimsert. These easy-to-use kits are designed to repair damaged threads in straight-through holes and blind holes. Both are available as kits which can handle a variety of sizes and thread patterns. Drill the hole, then tap it with the special included tap. Install the Heli-Coil and the hole is back to its original diameter and thread pitch.

Regardless of which method you use, be sure to proceed calmly and carefully. A little impatience or carelessness during one of these relatively simple procedures can ruin your whole day's work and cost you a bundle if you wreck an expensive part.

WORKING FACILITIES

Not to be overlooked when discussing tools is the workshop. If anything more than routine maintenance is to be carried out, some sort of suitable work area is essential.

It is understood, and appreciated, that many home mechanics do not have a good workshop or garage available, and end up removing an engine or doing major repairs outside. It is recommended, however, that the overhaul or repair be completed under the cover of a roof.

A clean, flat workbench or table of comfortable working height is an absolute necessity. The workbench should be equipped with a vise that has a jaw opening of at least four inches.

As mentioned previously, some clean, dry storage space is also required for tools, as well as the lubricants, fluids, cleaning solvents, etc. which soon become necessary.

Sometimes waste oil and fluids, drained from the engine or cooling system during normal maintenance or repairs, present a disposal problem. To avoid pouring them on the ground or into a sewage system, pour the used fluids into large containers, seal them with caps and take them to an authorized disposal site or recycling center. Plastic jugs, such as old antifreeze containers, are ideal for this purpose.

Always keep a supply of old newspapers and clean rags available. Old towels are excellent for mopping up spills. Many mechanics use rolls of paper towels for most work because they are readily available and disposable. To help keep the area under the vehicle clean, a large cardboard box can be cut open and flattened to protect the garage or shop floor.

Whenever working over a painted surface, such as when leaning over a fender to service something under the hood, always cover it with an old blanket or bedspread to protect the finish. Vinyl covered pads, made especially for this purpose, are available at auto parts stores.

Anti-theft audio systems

GENERAL INFORMATION

1   Some models are equipped with an anti-theft audio system which includes an anti-theft feature that will render the stereo inoperative if stolen. If the power source to the stereo is disconnected with the anti-theft feature activated, the stereo must be "unlocked" using the appropriate secret code to become operative again. Even if the power is immediately reconnected, the stereo will not function. If your vehicle is equipped with this anti-theft system, do not disconnect the battery, remove the stereo or disconnect related components unless you have disabled the anti-theft feature first. In order to deactivate this feature you must know the secret code.

2   Refer to your owner's manual for more complete information on your particular audio system and its anti-theft feature. If you lose or forget your code, contact your local dealer service department.

DISABLING THE ANTI-THEFT FEATURE

Delco LOC II system (1992 to 1995 models)

➡Note: This system uses a six digit code.

3   With the ignition ON and the stereo OFF, press the stereo's 1 and 4 buttons at the same time for five seconds. The display will show SEC, indicating the unit is in the secure mode (anti-theft feature activated).

4   Press the SET button. The display will show "000."

5   Press the SEEK right or SEEK left arrows as required to display the first digit of the secret code.

6   Press the SCAN button to make the next two digits of your code appear. On 1992 and 1993 models, rotate the TUNE knob to display your code.

7   Press the BAND knob, "000" will be displayed.

8   Repeat steps 5 and 6 to enter the last three digits of your code.

9   Press the BAND knob. If the display shows "——" the anti-theft feature has been deactivated. If SEC is displayed, the code entered was incorrect and the anti-theft feature is still enabled.

THEFTLOCK system (1996 and later models)

➡Note: This system uses a three or four digit code.

10   In order to successfully deactivate the anti-theft feature, pause no more than fifteen seconds between each step.

11   With the ignition ON and the stereo OFF, press the stereo's 1 and 4 buttons at the same time until the display shows SEC, indicating the unit is in the secure mode (anti-theft feature activated).

12   Press the MN button. The display will show "000."

13   Press the MN button again to display the LAST two digits of your code.

14   Next, press the HR button to display the FIRST one or two digits of your code.

15   Press the AM-FM button. If the display shows "——" the anti-theft feature has been deactivated. If SEC is displayed, the code entered was incorrect and the anti-theft feature is still enabled.

UNLOCKING THE STEREO AFTER A POWER LOSS

Delco LOC II system (1992 to 1995 models)

16   If the anti-theft feature was not deactivated before a power interruption occurred, when power is restored to the stereo LOC will appear on the display. To successfully unlock the anti-theft feature, pause no more than fifteen seconds between each step.

17   With the ignition ON and the stereo OFF, press the SET button. The display will show "000."

18   Press the SEEK right or SEEK left arrows as required to display the first digit of the secret code.

19   Press the SCAN button to make the next two digits of your code appear. On 1992 and 1993 models, rotate the TUNE knob to display your code.

20   Press the BAND knob, "000" will be displayed.

21   Repeat steps 18 and 19 to enter the last three digits of your code.

22   Press the BAND knob. If the display shows the time (hours and minutes) the anti-theft feature has been successfully deactivated and the stereo will function. If SEC is displayed, the code entered was incorrect and the anti-theft feature is still enabled.

THEFTLOCK system (1996 and later models)

23   If the anti-theft feature was not deactivated before a power interruption occurred, when power is restored to the stereo LOC will appear on the display. To successfully unlock the anti-theft feature, pause no more than fifteen seconds between each step.

24   With the ignition ON and the stereo OFF, press the MN button. The display will show "000."

25   Press the MN button again to display the LAST two digits of your code.

26   Next, press the HR button to display the FIRST one or two digits of your code.

27   Press the AM-FM button. If the display shows "SEC" the anti-theft feature has been successfully deactivated and the stereo will function. If REP is displayed, repeat Steps 24, 25 and 26. If you enter the wrong code eight times, the display will show "INOP." If this occurs, the ignition must remain in the ON position for one hour before you can try to unlock the stereo again. Then, on the next try, you will only get three chances to unlock the system before the display shows "INOP."

Jacking and towing

Four jacking points are provided - behind the front wheels and in front of the rear wheels. If your jack head is slotted, it engages the body seam at the notched area, as shown here. If your jack head is dished, it engages a large dimple on the frame. Raise the vehicle slowly and check to be sure the jack is steady and secure

JACKING

The jack supplied with the vehicle should only be used for raising the vehicle when changing a tire or placing jackstands under the frame.

✳✳ WARNING

Never work under the vehicle or start the engine while this jack is being used as the only means of support.

The vehicle should be on level ground with the wheels blocked and the transmission in Park (automatic) or Reverse (manual). If the wheel is being replaced, loosen the wheel lug nuts one-half turn and leave them in place until the wheel is raised off the ground.

Place the jack under the vehicle suspension in the indicated position (see illustration). Operate the jack with a slow, smooth motion until the wheel is raised off the ground. Remove the tire and install the spare. Tighten the lug nuts until they're snug, but wait until the vehicle is lowered to use the wrench.

Lower the vehicle, remove the jack and tighten the nuts (if loosened or removed) in a criss-cross pattern.

TOWING

The manufacturer does not recommend towing with all four wheels on the ground. Towing should therefore be left to a dealer or professional towing service.

Booster battery (jump) starting

Observe the following precautions when using a booster battery to start a vehicle:

a) *Before connecting the booster battery, make sure the ignition switch is in the Off position.*

b) *Turn off the lights, heater and other electrical loads.*

c) *Your eyes should be shielded. Safety goggles are a good idea.*

d) *Make sure the booster battery is the same voltage as the dead one in the vehicle.*

e) *The two vehicles MUST NOT TOUCH each other.*

f) *Make sure the transmission is in Park.*

g) *If the booster battery is not a maintenance-free type, remove the vent caps and lay a cloth over the vent holes.*

Connect the red jumper cable to the positive (+) terminals of each battery.

Connect one end of the black cable to the negative (–) terminal of the booster battery. The other end of this cable should be connected to a good ground on the engine block (see illustration). Make sure the cable will not come into contact with the fan, drivebelts or other moving parts of the engine.

Start the engine using the booster battery, then, with the engine running at idle speed, disconnect the jumper cables in the reverse order of connection

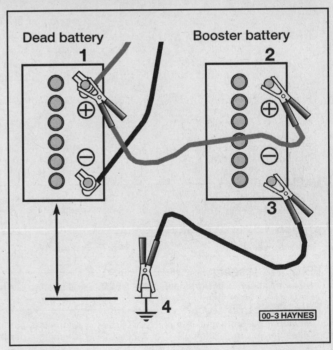

Make the booster battery cable connections in the numerical order shown (note that the negative cable of the booster battery is NOT attached to the negative terminal of the dead battery)

CONVERSION FACTORS

LENGTH (distance)

Inches (in)	X	25.4	= Millimeters (mm)	X	0.0394 = Inches (in)
Feet (ft)	X	0.305	= Meters (m)	X	3.281 = Feet (ft)
Miles	X	1.609	= Kilometers (km)	X	0.621 = Miles

VOLUME (capacity)

Cubic inches (cu in; in^3)	X	16.387	= Cubic centimeters (cc; cm^3)	X	0.061 = Cubic inches (cu in; in^3)
Imperial pints (Imp pt)	X	0.568	= Liters (l)	X	1.76 = Imperial pints (Imp pt)
Imperial quarts (Imp qt)	X	1.137	= Liters (l)	X	0.88 = Imperial quarts (Imp qt)
Imperial quarts (Imp qt)	X	1.201	= US quarts (US qt)	X	0.833 = Imperial quarts (Imp qt)
US quarts (US qt)	X	0.946	= Liters (l)	X	1.057 = US quarts (US qt)
Imperial gallons (Imp gal)	X	4.546	= Liters (l)	X	0.22 = Imperial gallons (Imp gal)
Imperial gallons (Imp gal)	X	1.201	= US gallons (US gal)	X	0.833 = Imperial gallons (Imp gal)
US gallons (US gal)	X	3.785	= Liters (l)	X	0.264 = US gallons (US gal)

MASS (weight)

Ounces (oz)	X	28.35	= Grams (g)	X	0.035 = Ounces (oz)
Pounds (lb)	X	0.454	= Kilograms (kg)	X	2.205 = Pounds (lb)

FORCE

Ounces-force (ozf; oz)	X	0.278	= Newtons (N)	X	3.6 = Ounces-force (ozf; oz)
Pounds-force (lbf; lb)	X	4.448	= Newtons (N)	X	0.225 = Pounds-force (lbf; lb)
Newtons (N)	X	0.1	= Kilograms-force (kgf; kg)	X	9.81 = Newtons (N)

PRESSURE

Pounds-force per square inch (psi; lbf/in^2; lb/in^2)	X	0.070	= Kilograms-force per square centimeter (kgf/cm^2; kg/cm^2)	X	14.223 = Pounds-force per square inch (psi; lbf/in^2; lb/in^2)
Pounds-force per square inch (psi; lbf/in^2; lb/in^2)	X	0.068	= Atmospheres (atm)	X	14.696 = Pounds-force per square inch (psi; lbf/in^2; lb/in^2)
Pounds-force per square inch (psi; lbf/in^2; lb/in^2)	X	0.069	= Bars	X	14.5 = Pounds-force per square inch (psi; lbf/in^2; lb/in^2)
Pounds-force per square inch (psi; lbf/in^2; lb/in^2)	X	6.895	= Kilopascals (kPa)	X	0.145 = Pounds-force per square inch (psi; lbf/in^2; lb/in^2)
Kilopascals (kPa)	X	0.01	= Kilograms-force per square centimeter (kgf/cm^2; kg/cm^2)	X	98.1 = Kilopascals (kPa)

TORQUE (moment of force)

Pounds-force inches (lbf in; lb in)	X	1.152	= Kilograms-force centimeter (kgf cm; kg cm)	X	0.868 = Pounds-force inches (lbf in; lb in)
Pounds-force inches (lbf in; lb in)	X	0.113	= Newton meters (Nm)	X	8.85 = Pounds-force inches (lbf in; lb in)
Pounds-force inches (lbf in; lb in)	X	0.083	= Pounds-force feet (lbf ft; lb ft)	X	12 = Pounds-force inches (lbf in; lb in)
Pounds-force feet (lbf ft; lb ft)	X	0.138	= Kilograms-force meters (kgf m; kg m)	X	7.233 = Pounds-force feet (lbf ft; lb ft)
Pounds-force feet (lbf ft; lb ft)	X	1.356	= Newton meters (Nm)	X	0.738 = Pounds-force feet (lbf ft; lb ft)
Newton meters (Nm)	X	0.102	= Kilograms-force meters (kgf m; kg m)	X	9.804 = Newton meters (Nm)

VACUUM

Inches mercury (in. Hg)	X	3.377	= Kilopascals (kPa)	X	0.2961 = Inches mercury
Inches mercury (in. Hg)	X	25.4	= Millimeters mercury (mm Hg)	X	0.0394 = Inches mercury

POWER

Horsepower (hp)	X	745.7	= Watts (W)	X	0.0013 = Horsepower (hp)

VELOCITY (speed)

Miles per hour (miles/hr; mph)	X	1.609	= Kilometers per hour (km/hr; kph)	X	0.621 = Miles per hour (miles/hr; mph)

FUEL CONSUMPTION *

Miles per gallon, Imperial (mpg)	X	0.354	= Kilometers per liter (km/l)	X	2.825 = Miles per gallon, Imperial (mpg)
Miles per gallon, US (mpg)	X	0.425	= Kilometers per liter (km/l)	X	2.352 = Miles per gallon, US (mpg)

TEMPERATURE

Degrees Fahrenheit = (°C x 1.8) + 32

Degrees Celsius (Degrees Centigrade; °C) = (°F - 32) x 0.56

*It is common practice to convert from miles per gallon (mpg) to liters/100 kilometers (l/100km), where mpg (Imperial) x l/100 km = 282 and mpg (US) x l/100 km = 235

FRACTION/DECIMAL/MILLIMETER EQUIVALENTS

DECIMALS TO MILLIMETERS

Decimal	mm	Decimal	mm
0.001	0.0254	0.500	12.7000
0.002	0.0508	0.510	12.9540
0.003	0.0762	0.520	13.2080
0.004	0.1016	0.530	13.4620
0.005	0.1270	0.540	13.7160
0.006	0.1524	0.550	13.9700
0.007	0.1778	0.560	14.2240
0.008	0.2032	0.570	14.4780
0.009	0.2286	0.580	14.7320
		0.590	14.9860
0.010	0.2540		
0.020	0.5080		
0.030	0.7620		
0.040	1.0160	0.600	15.2400
0.050	1.2700	0.610	15.4940
0.060	1.5240	0.620	15.7480
0.070	1.7780	0.630	16.0020
0.080	2.0320	0.640	16.2560
0.090	2.2860	0.650	16.5100
		0.660	16.7640
0.100	2.5400	0.670	17.0180
0.110	2.7940	0.680	17.2720
0.120	3.0480	0.690	17.5260
0.130	3.3020		
0.140	3.5560		
0.150	3.8100		
0.160	4.0640	0.700	17.7800
0.170	4.3180	0.710	18.0340
0.180	4.5720	0.720	18.2880
0.190	4.8260	0.730	18.5420
		0.740	18.7960
0.200	5.0800	0.750	19.0500
0.210	5.3340	0.760	19.3040
0.220	5.5880	0.770	19.5580
0.230	5.8420	0.780	19.8120
0.240	6.0960	0.790	20.0660
0.250	6.3500		
0.260	6.6040		
0.270	6.8580	0.800	20.3200
0.280	7.1120	0.810	20.5740
0.290	7.3660	0.820	20.8280
		0.830	21.0820
0.300	7.6200	0.840	21.3360
0.310	7.8740	0.850	21.5900
0.320	8.1280	0.860	21.8440
0.330	8.3820	0.870	22.0980
0.340	8.6360	0.880	22.3520
0.350	8.8900	0.890	22.6060
0.360	9.1440		
0.370	9.3980		
0.380	9.6520		
0.390	9.9060	0.900	22.8600
0.400	10.1600	0.910	23.1140
0.410	10.4140	0.920	23.3680
0.420	10.6680	0.930	23.6220
0.430	10.9220	0.940	23.8760
0.440	11.1760	0.950	24.1300
0.450	11.4300	0.960	24.3840
0.460	11.6840	0.970	24.6380
0.470	11.9380	0.980	24.8920
0.480	12.1920	0.990	25.1460
0.490	12.4460	1.000	25.4000

FRACTIONS TO DECIMALS TO MILLIMETERS

Fraction	Decimal	mm	Fraction	Decimal	mm
1/64	0.0156	0.3969	33/64	0.5156	13.0969
1/32	0.0312	0.7938	17/32	0.5312	13.4938
3/64	0.0469	1.1906	35/64	0.5469	13.8906
1/16	0.0625	1.5875	9/16	0.5625	14.2875
5/64	0.0781	1.9844	37/64	0.5781	14.6844
3/32	0.0938	2.3812	19/32	0.5938	15.0812
7/64	0.1094	2.7781	39/64	0.6094	15.4781
1/8	0.1250	3.1750	5/8	0.6250	15.8750
9/64	0.1406	3.5719	41/64	0.6406	16.2719
5/32	0.1562	3.9688	21/32	0.6562	16.6688
11/64	0.1719	4.3656	43/64	0.6719	17.0656
3/16	0.1875	4.7625	11/16	0.6875	17.4625
13/64	0.2031	5.1594	45/64	0.7031	17.8594
7/32	0.2188	5.5562	23/32	0.7188	18.2562
15/64	0.2344	5.9531	47/64	0.7344	18.6531
1/4	0.2500	6.3500	3/4	0.7500	19.0500
17/64	0.2656	6.7469	49/64	0.7656	19.4469
9/32	0.2812	7.1438	25/32	0.7812	19.8438
19/64	0.2969	7.5406	51/64	0.7969	20.2406
5/16	0.3125	7.9375	13/16	0.8125	20.6375
21/64	0.3281	8.3344	53/64	0.8281	21.0344
11/32	0.3438	8.7312	27/32	0.8438	21.4312
23/64	0.3594	9.1281	55/64	0.8594	21.8281
3/8	0.3750	9.5250	7/8	0.8750	22.2250
25/64	0.3906	9.9219	57/64	0.8906	22.6219
13/32	0.4062	10.3188	29/32	0.9062	23.0188
27/64	0.4219	10.7156	59/64	0.9219	23.4156
7/16	0.4375	11.1125	15/16	0.9375	23.8125
29/64	0.4531	11.5094	61/64	0.9531	24.2094
15/32	0.4688	11.9062	31/32	0.9688	24.6062
31/64	0.4844	12.3031	63/64	0.9844	25.0031
1/2	0.5000	12.7000	1	1.0000	25.4000

Automotive chemicals and lubricants

A number of automotive chemicals and lubricants are available for use during vehicle maintenance and repair. They include a wide variety of products ranging from cleaning solvents and degreasers to lubricants and protective sprays for rubber, plastic and vinyl.

CLEANERS

Carburetor cleaner and choke cleaner is a strong solvent for gum, varnish and carbon. Most carburetor cleaners leave a dry-type lubricant film which will not harden or gum up. Because of this film it is not recommended for use on electrical components.

Brake system cleaner is used to remove brake dust, grease and brake fluid from the brake system, where clean surfaces are absolutely necessary. It leaves no residue and often eliminates brake squeal caused by contaminants.

Electrical cleaner removes oxidation, corrosion and carbon deposits from electrical contacts, restoring full current flow. It can also be used to clean spark plugs, carburetor jets, voltage regulators and other parts where an oil-free surface is desired.

Demoisturants remove water and moisture from electrical components such as alternators, voltage regulators, electrical connectors and fuse blocks. They are non-conductive and non-corrosive.

Degreasers are heavy-duty solvents used to remove grease from the outside of the engine and from chassis components. They can be sprayed or brushed on and, depending on the type, are rinsed off either with water or solvent.

LUBRICANTS

Motor oil is the lubricant formulated for use in engines. It normally contains a wide variety of additives to prevent corrosion and reduce foaming and wear. Motor oil comes in various weights (viscosity ratings) from 0 to 50. The recommended weight of the oil depends on the season, temperature and the demands on the engine. Light oil is used in cold climates and under light load conditions. Heavy oil is used in hot climates and where high loads are encountered. Multi-viscosity oils are designed to have characteristics of both light and heavy oils and are available in a number of weights from 5W-20 to 20W-50.

Gear oil is designed to be used in differentials, manual transmissions and other areas where high-temperature lubrication is required.

Chassis and wheel bearing grease is a heavy grease used where increased loads and friction are encountered, such as for wheel bearings, balljoints, tie-rod ends and universal joints.

High-temperature wheel bearing grease is designed to withstand the extreme temperatures encountered by wheel bearings in disc brake equipped vehicles. It usually contains molybdenum disulfide (moly), which is a dry-type lubricant.

White grease is a heavy grease for metal-to-metal applications where water is a problem. White grease stays soft under both low and high temperatures (usually from -100 to +190-degrees F), and will not wash off or dilute in the presence of water.

Assembly lube is a special extreme pressure lubricant, usually containing moly, used to lubricate high-load parts (such as main and rod bearings and cam lobes) for initial start-up of a new engine. The assembly lube lubricates the parts without being squeezed out or washed away until the engine oiling system begins to function.

Silicone lubricants are used to protect rubber, plastic, vinyl and nylon parts.

Graphite lubricants are used where oils cannot be used due to contamination problems, such as in locks. The dry graphite will lubricate metal parts while remaining uncontaminated by dirt, water, oil or acids. It is electrically conductive and will not foul electrical contacts in locks such as the ignition switch.

Moly penetrants loosen and lubricate frozen, rusted and corroded fasteners and prevent future rusting or freezing.

Heat-sink grease is a special electrically non-conductive grease that is used for mounting electronic ignition modules where it is essential that heat is transferred away from the module.

SEALANTS

RTV sealant is one of the most widely used gasket compounds. Made from silicone, RTV is air curing, it seals, bonds, waterproofs, fills surface irregularities, remains flexible, doesn't shrink, is relatively easy to remove, and is used as a supplementary sealer with almost all low and medium temperature gaskets.

Anaerobic sealant is much like RTV in that it can be used either to seal gaskets or to form gaskets by itself. It remains flexible, is solvent resistant and fills surface imperfections. The difference between an anaerobic sealant and an RTV-type sealant is in the curing. RTV cures when exposed to air, while an anaerobic sealant cures only in the absence of air. This means that an anaerobic sealant cures only after the assembly of parts, sealing them together.

Thread and pipe sealant is used for sealing hydraulic and pneumatic fittings and vacuum lines. It is usually made from a Teflon compound, and comes in a spray, a paint-on liquid and as a wrap-around tape.

CHEMICALS

Anti-seize compound prevents seizing, galling, cold welding, rust and corrosion in fasteners. High-temperature anti-seize, usually made with copper and graphite lubricants, is used for exhaust system and exhaust manifold bolts.

Anaerobic locking compounds are used to keep fasteners from vibrating or working loose and cure only after installation, in the absence of air. Medium strength locking compound is used for small nuts, bolts and screws that may be removed later. High-strength locking compound is for large nuts, bolts and studs which aren't removed on a regular basis.

Oil additives range from viscosity index improvers to chemical treatments that claim to reduce internal engine friction. It should be noted that most oil manufacturers caution against using additives with their oils.

Gas additives perform several functions, depending on their chemical makeup. They usually contain solvents that help dissolve gum and varnish that build up on carburetor, fuel injection and intake parts. They also serve to break down carbon deposits that form on the inside surfaces of the combustion chambers. Some additives contain upper cylinder lubricants for valves and piston rings, and others contain chemicals to remove condensation from the gas tank.

MISCELLANEOUS

Brake fluid is specially formulated hydraulic fluid that can withstand the heat and pressure encountered in brake systems. Care must be taken so this fluid does not come in contact with painted surfaces or plastics. An opened container should always be resealed to prevent contamination by water or dirt.

Weatherstrip adhesive is used to bond weatherstripping around doors, windows and trunk lids. It is sometimes used to attach trim pieces.

Undercoating is a petroleum-based, tar-like substance that is designed to protect metal surfaces on the underside of the vehicle from corrosion. It also acts as a sound-deadening agent by insulating the bottom of the vehicle.

Waxes and polishes are used to help protect painted and plated surfaces from the weather. Different types of paint may require the use of different types of wax and polish. Some polishes utilize a chemical or abrasive cleaner to help remove the top layer of oxidized (dull) paint on older vehicles. In recent years many non-wax polishes that contain a wide variety of chemicals such as polymers and silicones have been introduced. These non-wax polishes are usually easier to apply and last longer than conventional waxes and polishes.

Regardless of how enthusiastic you may be about getting on with the job at hand, take the time to ensure that your safety is not jeopardized. A moment's lack of attention can result in an accident, as can failure to observe certain simple safety precautions. The possibility of an accident will always exist, and the following points should not be considered a comprehensive list of all dangers. Rather, they are intended to make you aware of the risks and to encourage a safety conscious approach to all work you carry out on your vehicle.

ESSENTIAL DOS AND DON'TS

DON'T rely on a jack when working under the vehicle. Always use approved jackstands to support the weight of the vehicle and place them under the recommended lift or support points.

DON'T attempt to loosen extremely tight fasteners (i.e. wheel lug nuts) while the vehicle is on a jack - it may fall.

DON'T start the engine without first making sure that the transmission is in Neutral (or Park where applicable) and the parking brake is set.

DON'T remove the radiator cap from a hot cooling system - let it cool or cover it with a cloth and release the pressure gradually.

DON'T attempt to drain the engine oil until you are sure it has cooled to the point that it will not burn you.

DON'T touch any part of the engine or exhaust system until it has cooled sufficiently to avoid burns.

DON'T siphon toxic liquids such as gasoline, antifreeze and brake fluid by mouth, or allow them to remain on your skin.

DON'T inhale brake lining dust - it is potentially hazardous (see *Asbestos* below).

DON'T allow spilled oil or grease to remain on the floor - wipe it up before someone slips on it.

DON'T use loose fitting wrenches or other tools which may slip and cause injury.

DON'T push on wrenches when loosening or tightening nuts or bolts. Always try to pull the wrench toward you. If the situation calls for pushing the wrench away, push with an open hand to avoid scraped knuckles if the wrench should slip.

DON'T attempt to lift a heavy component alone - get someone to help you.

DON'T rush or take unsafe shortcuts to finish a job.

DON'T allow children or animals in or around the vehicle while you are working on it.

DO wear eye protection when using power tools such as a drill, sander, bench grinder, etc. and when working under a vehicle.

DO keep loose clothing and long hair well out of the way of moving parts.

DO make sure that any hoist used has a safe working load rating adequate for the job.

DO get someone to check on you periodically when working alone on a vehicle.

DO carry out work in a logical sequence and make sure that everything is correctly assembled and tightened.

DO keep chemicals and fluids tightly capped and out of the reach of children and pets.

DO remember that your vehicle's safety affects that of yourself and others. If in doubt on any point, get professional advice.

ASBESTOS

Certain friction, insulating, sealing, and other products - such as brake linings, brake bands, clutch linings, torque converters, gaskets, etc. - may contain asbestos. Extreme care must be taken to avoid inhalation of dust from such products, since it is hazardous to health. If in doubt, assume that they do contain asbestos.

FIRE

Remember at all times that gasoline is highly flammable. Never smoke or have any kind of open flame around when working on a vehicle. But the risk does not end there. A spark caused by an electrical short circuit, by two metal surfaces contacting each other, or even by static electricity built up in your body under certain conditions, can ignite gasoline vapors, which in a confined space are highly explosive. Do not, under any circumstances, use gasoline for cleaning parts. Use an approved safety solvent.

Always disconnect the battery ground (-) cable at the battery before working on any part of the fuel system or electrical system. Never risk spilling fuel on a hot engine or exhaust component. It is strongly recommended that a fire extinguisher suitable for use on fuel and electrical fires be kept handy in the garage or workshop at all times. Never try to extinguish a fuel or electrical fire with water.

FUMES

Certain fumes are highly toxic and can quickly cause unconsciousness and even death if inhaled to any extent. Gasoline vapor falls into this category, as do the vapors from some cleaning solvents. Any draining or pouring of such volatile fluids should be done in a well ventilated area.

When using cleaning fluids and solvents, read the instructions on the container carefully. Never use materials from unmarked containers.

Never run the engine in an enclosed space, such as a garage. Exhaust fumes contain carbon monoxide, which is extremely poisonous. If you need to run the engine, always do so in the open air, or at least have the rear of the vehicle outside the work area.

If you are fortunate enough to have the use of an inspection pit, never drain or pour gasoline and never run the engine while the vehicle is over the pit. The fumes, being heavier than air, will concentrate in the pit with possibly lethal results.

THE BATTERY

Never create a spark or allow a bare light bulb near a battery. They normally give off a certain amount of hydrogen gas, which is highly explosive.

Always disconnect the battery ground (-) cable at the battery before working on the fuel or electrical systems.

If possible, loosen the filler caps or cover when charging the battery from an external source (this does not apply to sealed or maintenance-free batteries). Do not charge at an excessive rate or the battery may burst.

Take care when adding water to a non maintenance-free battery and when carrying a battery. The electrolyte, even when diluted, is very corrosive and should not be allowed to contact clothing or skin.

Always wear eye protection when cleaning the battery to prevent the caustic deposits from entering your eyes.

HOUSEHOLD CURRENT

When using an electric power tool, inspection light, etc., which operates on household current, always make sure that the tool is correctly connected to its plug and that, where necessary, it is properly grounded. Do not use such items in damp conditions and, again, do not create a spark or apply excessive heat in the vicinity of fuel or fuel vapor.

SECONDARY IGNITION SYSTEM VOLTAGE

A severe electric shock can result from touching certain parts of the ignition system (such as the spark plug wires) when the engine is running or being cranked, particularly if components are damp or the insulation is defective. In the case of an electronic ignition system, the secondary system voltage is much higher and could prove fatal.

Troubleshooting

CONTENTS

Section   Symptom

Engine

1   Engine will not rotate when attempting to start
2   Engine rotates but will not start
3   Starter motor operates without rotating engine
4   Engine hard to start when cold
5   Engine hard to start when hot
6   Starter motor noisy or excessively rough in engagement
7   Engine starts but stops immediately
8   Engine lopes while idling or idles erratically
9   Engine misses at idle speed
10 Engine misses throughout driving speed range
11 Engine stalls
12 Engine lacks power
13 Engine backfires
14 Pinging or knocking engine sounds during acceleration or uphill
15 Engine diesels (continues to run) after switching off

Engine electrical system

16 Battery will not hold a charge
17 Ignition light fails to go out
18 Ignition light fails to come on when key is turned on
19 "Service engine soon" light comes on

Fuel system

20 Excessive fuel consumption
21 Fuel leakage and/or fuel odor

Cooling system

22 Overheating
23 Overcooling
24 External coolant leakage
25 Internal coolant leakage
26 Clutch Coolant loss
27 Poor coolant circulation

Clutch

28 Fails to release (pedal pressed to the floor - shift lever does not move freely in and out of Reverse)
29 Clutch slips (engine speed increases with no increase in vehicle speed)
30 Grabbing (chattering) as clutch is engaged
31 Squeal or rumble with clutch fully engaged (pedal released)
32 Squeal or rumble with clutch fully disengaged (pedal depressed)
33 Clutch pedal stays on floor when disengaged

Manual transmission

34 Noisy in Neutral with engine running

Section   Symptom

35 Noisy in all gears
36 Noisy in one particular gear
37 Slips out of high gear
38 Difficulty in engaging gears
39 Oil leakage

Automatic transmission

40 General shift mechanism problems
41 Transmission will not downshift with accelerator pedal pressed to the floor
42 Transmission slips, shifts rough, is noisy or has no drive in forward or reverse gears
43 Fluid leakage

Driveshaft

44 Oil leak at front of driveshaft
45 Knock or clunk when the transmission is under initial load (just after transmission is put into gear)
46 Metallic grating sound consistent with vehicle speed
47 Vibration

Axles

48 Noise
49 Vibration
50 Oil leakage

Brakes

51 Vehicle pulls to one side during braking
52 Noise (high-pitched squeal with the brakes applied)
53 Excessive brake pedal travel
54 Brake pedal feels spongy when depressed
55 Excessive effort required to stop vehicle
56 Pedal travels to the floor with little resistance
57 Brake pedal pulsates during brake application

Suspension and steering systems

58 Vehicle pulls to one side
59 Shimmy, shake or vibration
60 Excessive pitching and/or rolling around corners or during braking
61 Excessively stiff steering
62 Excessive play in steering
63 Lack of power assistance
64 Excessive tire wear (not specific to one area)
65 Excessive tire wear on outside edge
66 Excessive tire wear on inside edge
67 Tire tread worn in one place

This section provides an easy reference guide to the more common problems which may occur during the operation of your vehicle. These problems and possible causes are grouped under various components or systems, such as Engine, Cooling system, etc., and also refer to the Chapter and/or Section which deals with the problem.

Remember that successful troubleshooting is not a mysterious black art practiced only by professional mechanics. It's simply the result of a bit of knowledge combined with an intelligent, systematic approach to the problem. Always work by a process of elimination, starting with the simplest solution and working through to the most complex - and never overlook the obvious. Anyone can forget to fill the gas tank or leave the lights on overnight, so don't assume that you are above such oversights.

Finally, always get clear in your mind why a problem has occurred and take steps to ensure that it doesn't happen again. If the electrical system fails because of a poor connection, check all other connections in the system to make sure that they don't fail as well. If a particular fuse continues to blow, find out why - don't just go on replacing fuses. Remember, failure of a small component can often be indicative of potential failure or incorrect functioning of a more important component or system.

ENGINE

1   Engine will not rotate when attempting to start

1   Battery terminal connections loose or corroded. Check the cable terminals at the battery. Tighten the cable or remove corrosion as necessary.

2   Battery discharged or faulty. If the cable connections are clean and tight on the battery posts, turn the key to the On position and switch on the headlights and/or windshield wipers. If they fail to function, the battery is discharged.

3   Automatic transmission not completely engaged in Park or clutch not completely depressed.

4   Broken, loose or disconnected wiring in the starting circuit. Inspect all wiring and connectors at the battery, starter solenoid and ignition switch.

5   Starter motor pinion jammed in flywheel ring gear. If equipped with a manual transmission, place the transmission in gear and rock the vehicle to manually turn the engine. Remove the starter and inspect the pinion and flywheel at earliest convenience.

6   Starter solenoid faulty (Chapter 5).

7   Starter motor faulty (Chapter 5).

8   Ignition switch faulty (Chapter 12).

2   Engine rotates but will not start

1   Fuel tank empty.

2   Battery discharged (engine rotates slowly). Check the operation of electrical components as described in previous Section.

3   Battery terminal connections loose or corroded. See previous Section.

4   Carburetor flooded and/or fuel level in carburetor incorrect. This will usually be accompanied by a strong fuel odor from under the engine cover. Wait a few minutes, depress the accelerator pedal all the way to the floor and attempt to start the engine.

5   Choke control inoperative (Chapter 4).

6   Fuel not reaching carburetor or fuel injector(s). With the ignition switch in the Off position, remove the engine cover, remove the top plate of the air cleaner assembly and observe the top of the carburetor

(manually move the choke plate back if necessary). Depress the accelerator pedal and check that fuel spurts into the carburetor. If not, check the fuel filter (Chapter 1), fuel lines and fuel pump (Chapter 4).

7   Fuel injector(s) or fuel pump faulty (fuel injected vehicles) (Chapter 4).

8   No power to fuel pump (Chapter 4).

9   Worn, faulty or incorrectly gapped spark plugs (Chapter 1).

10   Broken, loose or disconnected wiring in the starting circuit (see previous Section).

11   Distributor loose, causing ignition timing to change. Turn the distributor as necessary to start the engine, then set the ignition timing as soon as possible (Chapter 1).

12   Broken, loose or disconnected wires at the ignition coil or faulty coil (Chapter 5).

3   Starter motor operates without rotating engine

1   Starter pinion sticking. Remove the starter (Chapter 5) and inspect.

2   Starter pinion or flywheel teeth worn or broken. Remove the cover at the rear of the engine and inspect.

4   Engine hard to start when cold

1   Battery discharged or low. Check as described in Section 1.

2   Choke control inoperative or out of adjustment (Chapter 4).

3   Carburetor flooded (see Section 2).

4   Fuel supply not reaching the carburetor (see Section 2).

5   Carburetor/fuel injection system in need of overhaul (Chapter 4).

6   Distributor rotor carbon tracked and/or damaged (Chapter 1).

7   Fuel injection malfunction (Chapter 4).

5   Engine hard to start when hot

1   Air filter clogged (Chapter 1).

2   Fuel not reaching the injector(s) (see Section 2).

3   Corroded electrical leads at the battery (Chapter 1).

4   Bad engine ground (Chapter 12).

5   Starter worn (Chapter 5).

6   Corroded electrical leads at the fuel injector (Chapter 4).

6   Starter motor noisy or excessively rough in engagement

1   Pinion or flywheel gear teeth worn or broken. Remove the cover at the rear of the engine (if so equipped) and inspect.

2   Starter motor mounting bolts loose or missing.

7   Engine starts but stops immediately

1   Loose or faulty electrical connections at distributor, coil or alternator.

2   Insufficient fuel reaching the carburetor or fuel injector(s). Disconnect the fuel line. Place a container under the disconnected fuel line and observe the flow of fuel from the line. If little or none at all, check for blockage in the lines and/or replace the fuel pump (Chapter 4).

3   Vacuum leak at the gasket surfaces of the carburetor or fuel injection unit. Make sure that all mounting bolts/nuts are tightened securely and that all vacuum hoses connected to the carburetor or fuel injection unit and manifold are positioned properly and in good condition.

8   Engine lopes while idling or idles erratically

1   Vacuum leakage. Check mounting bolts/nuts at the carburetor/

fuel injection unit and intake manifold for tightness. Make sure that all vacuum hoses are connected and in good condition. Use a stethoscope or a length of fuel hose held against your ear to listen for vacuum leaks while the engine is running. A hissing sound will be heard. Check the carburetor/fuel injector and intake manifold gasket surfaces.

2 Leaking EGR valve or plugged PCV valve (see Chapters 1 and 6).
3 Air filter clogged (Chapter 1).
4 Fuel pump not delivering sufficient fuel to the carburetor/fuel injector (see Section 7).
5 Carburetor out of adjustment (Chapter 4).
6 Leaking head gasket. If this is suspected, take the vehicle to a repair shop or dealer where the engine can be pressure checked.
7 Timing chain and/or gears worn (Chapter 2).
8 Camshaft lobes worn (Chapter 2).

9 Engine misses at idle speed

1 Spark plugs worn or not gapped properly (Chapter 1).
2 Faulty spark plug wires (Chapter 1).
3 Choke not operating properly (Chapter 1).
4 Sticking or faulty emissions system components (Chapter 6).
5 Clogged fuel filter and/or foreign matter in fuel. Remove the fuel filter (Chapter 1) and inspect.
6 Vacuum leaks at the intake manifold or at hose connections. Check as described in Section 8.
7 Incorrect idle speed or idle mixture (Chapter 1).
8 Incorrect ignition timing (Chapter 1).
9 Uneven or low cylinder compression. Check compression as described in Chapter 2.

10 Engine misses throughout driving speed range

1 Fuel filter clogged and/or impurities in the fuel system (Chapter 1). Also check fuel output at the carburetor/fuel injector (see Section 7).
2 Faulty or incorrectly gapped spark plugs (Chapter 1).
3 Incorrect ignition timing (Chapter 1).
4 Check for cracked distributor cap, disconnected distributor wires and damaged distributor components (Chapter 1).
5 Faulty spark plug wires (Chapter 1).
6 Faulty emissions system components (Chapter 6).
7 Low or uneven cylinder compression pressures. Remove the spark plugs and test the compression with gauge (Chapter 2).
8 Weak or faulty ignition system (Chapter 5).
9 Vacuum leaks at the carburetor/fuel injection unit or vacuum hoses (see Section 8).

11 Engine stalls

1 Idle speed incorrect (Chapter 1).
2 Fuel filter clogged and/or water and impurities in the fuel system (Chapter 1).
3 Choke improperly adjusted or sticking (Chapter 4).
4 Distributor components damp or damaged (Chapter 5).
5 Faulty emissions system components (Chapter 6).
6 Faulty or incorrectly gapped spark plugs (Chapter 1). Also check spark plug wires (Chapter 1).
7 Vacuum leak at the carburetor/fuel injection unit or vacuum hoses. Check as described in Section 8.

12 Engine lacks power

1 Incorrect ignition timing (Chapter 1).
2 Excessive play in distributor shaft. At the same time, check for worn rotor, faulty distributor cap, wires, etc. (Chapters 1 and 5).
3 Faulty or incorrectly gapped spark plugs (Chapter 1).
4 Fuel injection unit not adjusted properly or excessively worn (Chapter 4).
5 Faulty coil (Chapter 5).
6 Brakes binding (Chapter 1).
7 Automatic transmission fluid level incorrect (Chapter 1).
8 Clutch slipping (Chapter 8).
9 Fuel filter clogged and/or impurities in the fuel system (Chapter 1).
10 Emissions control system not functioning properly (Chapter 6).
11 Use of substandard fuel. Fill tank with proper octane fuel.
12 Low or uneven cylinder compression pressures. Test with compression tester, which will detect leaking valves and/or blown head gasket (Chapter 2).

13 Engine backfires

1 Emissions system not functioning properly (Chapter 6).
2 Ignition timing incorrect (Chapter 1).
3 Faulty secondary ignition system (cracked spark plug insulator, faulty plug wires, distributor cap and/or rotor) (Chapters 1 and 5).
4 Carburetor/fuel injection unit in need of adjustment or worn excessively (Chapter 4).
5 Vacuum leak at the fuel injection unit or vacuum hoses. Check as described in Section 8.
6 Valves sticking (Chapter 2).
7 Crossed plug wires (Chapter 1).

14 Pinging or knocking engine sounds during acceleration or uphill

1 Incorrect grade of fuel. Fill tank with fuel of the proper octane rating.
2 Ignition timing incorrect (Chapter 1).
3 Carburetor/fuel injection unit in need of adjustment (Chapter 4).
4 Improper spark plugs. Check plug type against Emissions Control Information label located under hood. Also check plugs and wires for damage (Chapter 1).
5 Worn or damaged distributor components (Chapter 5).
6 Faulty emissions system (Chapter 6).
7 Vacuum leak. Check as described in Section 8.

15 Engine diesels (continues to run) after switching off

1 Idle speed too high (Chapter 1).
2 Electrical solenoid at side of carburetor not functioning properly (not all models, see Chapter 4).
3 Ignition timing incorrectly adjusted (Chapter 1).
4 Thermo-controlled air cleaner heat valve not operating properly (Chapter 1).
5 Excessive engine operating temperature. Probable causes of this are malfunctioning thermostat, clogged radiator, faulty water pump (Chapter 3).

ENGINE ELECTRICAL SYSTEM

16 Battery will not hold a charge

1 Alternator drivebelt defective or not adjusted properly (Chapter 1).
2 Electrolyte level low or battery discharged (Chapter 1).
3 Battery terminals loose or corroded (Chapter 1).
4 Alternator not charging properly (Chapter 5).
5 Loose, broken or faulty wiring in the charging circuit (Chapter 5).
6 Short in vehicle wiring causing a continual drain on battery.
7 Battery defective internally.

17 Ignition light fails to go out

1 Fault in alternator or charging circuit (Chapter 5).
2 Alternator drivebelt defective or not properly adjusted (Chapter 1).

18 Ignition light fails to come on when key is turned on

1 Warning light bulb defective (Chapter 12).
2 Alternator faulty (Chapter 5).
3 Fault in the printed circuit, dash wiring or bulb holder (Chapter 12).

19 'Service engine soon' light comes on

See Chapter 6.

FUEL SYSTEM

20 Excessive fuel consumption

1 Dirty or clogged air filter element (Chapter 1).
2 Incorrectly set ignition timing (Chapter 1).
3 Choke sticking or improperly adjusted (Chapter 1).
4 Emissions system not functioning properly (not all vehicles, see Chapter 6).
5 Carburetor idle speed and/or mixture not adjusted properly (Chapter 1).
6 Carburetor/fuel injection internal parts excessively worn or damaged (Chapter 4).
7 Low tire pressure or incorrect tire size (Chapter 1).

21 Fuel leakage and/or fuel odor

1 Leak in a fuel feed or vent line (Chapter 4).
2 Tank overfilled. Fill only to automatic shut-off.
3 Emissions system clogged or damaged (Chapter 6).
4 Vapor leaks from system lines (Chapter 4).
5 Carburetor/fuel injection internal parts excessively worn or out of adjustment (Chapter 4).

COOLING SYSTEM

22 Overheating

1 Insufficient coolant in system (Chapter 1).
2 Water pump drivebelt defective or not adjusted properly (Chapter 1).
3 Radiator core blocked or radiator grille dirty and restricted (Chapter 3).

4 Thermostat faulty (Chapter 3).
5 Fan blades broken or cracked (Chapter 3).
6 Radiator cap not maintaining proper pressure. Have cap pressure tested by gas station or repair shop.
7 Ignition timing incorrect (Chapter 1).

23 Overcooling

1 Thermostat faulty (Chapter 3).
2 Inaccurate temperature gauge (Chapter 12).

24 External coolant leakage

1 Deteriorated or damaged hoses or loose clamps. Replace hoses and/or tighten clamps at hose connections (Chapter 1).
2 Water pump seals defective. If this is the case, water will drip from the weep hole in the water pump body (Chapter 3).
3 Leakage from radiator core or header tank. This will require the radiator to be professionally repaired (see Chapter 3 for removal procedures).
4 Engine drain plugs or water jacket core plugs leaking (see Chapter 2).

25 Internal coolant leakage

➡Note: Internal coolant leaks can usually be detected by examining the oil. Check the dipstick and inside of the rocker arm cover(s) for water deposits and an oil consistency like that of a milkshake.

1 Leaking cylinder head gasket. Have the cooling system pressure tested.
2 Cracked cylinder bore or cylinder head. Dismantle engine and inspect (Chapter 2).

26 Coolant loss

1 Too much coolant in system (Chapter 1).
2 Coolant boiling away due to overheating (see Section 22).
3 Internal or external leakage (see Sections 24 and 25).
4 Faulty radiator cap. Have the cap pressure tested.

27 Poor coolant circulation

1 Inoperative water pump. A quick test is to pinch the top radiator hose closed with your hand while the engine is idling, then let it loose. You should feel the surge of coolant if the pump is working properly (Chapter 3).
2 Restriction in cooling system. Drain, flush and refill the system (Chapter 1). If necessary, remove the radiator (Chapter 3) and have it reverse flushed.
3 Water pump drivebelt defective or not adjusted properly (Chapter 1).
4 Thermostat sticking (Chapter 3).

CLUTCH

28 Fails to release (pedal pressed to the floor - shift lever does not move freely in and out of Reverse)

1 Clutch fork off ball stud. Look under the vehicle, on the left side of transmission.

2 Clutch plate warped or damaged (Chapter 8).
3 Clutch hydraulic system low or has air in system and needs to be bled (Chapter 8).

29 Clutch slips (engine speed increases with no increase in vehicle speed)

1 Clutch plate oil soaked or lining worn. Remove clutch (Chapter 8) and inspect.
2 Clutch plate not seated. It may take 30 or 40 normal starts for a new one to seat.
3 Pressure plate worn (Chapter 8).

30 Grabbing (chattering) as clutch is engaged

1 Oil on clutch plate lining. Remove (Chapter 8) and inspect. Correct any leakage source.
2 Worn or loose engine or transmission mounts. These units move slightly when clutch is released. Inspect mounts and bolts.
3 Worn splines on clutch plate hub. Remove clutch components (Chapter 8) and inspect.
4 Warped pressure plate or flywheel. Remove clutch components and inspect.

31 Squeal or rumble with clutch fully engaged (pedal released)

1 Release bearing binding on transmission bearing retainer. Remove clutch components (Chapter 8) and check bearing. Remove any burrs or nicks, clean and relubricate before reinstallation.
2 Weak linkage return spring. Replace the spring.

32 Squeal or rumble with clutch fully disengaged (pedal depressed)

1 Worn, defective or broken release bearing (Chapter 8).
2 Worn or broken pressure plate springs (or diaphragm fingers) (Chapter 8).
3 Air in hydraulic line (Chapter 8).

33 Clutch pedal stays on floor when disengaged

1 Bind in linkage or release bearing. Inspect linkage or remove clutch components as necessary.
2 Clutch hydraulic cylinder faulty or there is air in the system.

MANUAL TRANSMISSION

➡Note: All the following references are to Chapter 7, unless noted.

34 Noisy in Neutral with engine running

1 Input shaft bearing worn.
2 Damaged main drive gear bearing.
3 Worn countershaft bearings.
4 Worn or damaged countershaft end play shims.

35 Noisy in all gears

1 Any of the above causes, and/or:
2 Insufficient lubricant (see checking procedures in Chapter 1).

36 Noisy in one particular gear

1 Worn, damaged or chipped gear teeth for that particular gear.
2 Worn or damaged synchronizer for that particular gear.

37 Slips out of high gear

1 Transmission mounting bolts loose.
2 Shift rods not working freely.
3 Damaged mainshaft pilot bushing.
4 Dirt between transmission case and engine or misalignment of transmission.

38 Difficulty in engaging gears

1 Loose, damaged or out-of-adjustment shift linkage. Make a thorough inspection, replacing parts as necessary.
2 Air in hydraulic system (Chapter 8).

39 Oil leakage

1 Excessive amount of lubricant in transmission (see Chapter 1 for correct checking procedures). Drain lubricant as required.
2 Side cover loose or gasket damaged.
3 Rear oil seal or speedometer oil seal in need of replacement.
4 Clutch hydraulic system leaking (Chapter 8).

AUTOMATIC TRANSMISSION

➡Note: Due to the complexity of the automatic transmission, it is difficult for the home mechanic to properly diagnose and service this component. For problems other than the following, the vehicle should be taken to a dealer or reputable repair shop.

40 General shift mechanism problems

1 Chapter 7 deals with checking and adjusting the shift linkage on automatic transmissions. Common problems which may be attributed to poorly adjusted linkage are:

Engine starting in gears other than Park or Neutral
Indicator on shifter pointing to a gear other than the one actually being used
Vehicle moves when in Park

2 Refer to Chapter 7 to adjust the linkage.

41 Transmission will not downshift with accelerator pedal pressed to the floor

Chapter 7 deals with adjusting the TV cable to enable the transmission to downshift properly.

42 Transmission slips, shifts rough, is noisy or has no drive in forward or reverse gears

1 There are many probable causes for the above problems, but the home mechanic should be concerned with only one possibility - fluid level.
2 Before taking the vehicle to a repair shop, check the level and condition of the fluid as described in Chapter 1. Correct fluid level as necessary or change the fluid and filter if needed. If the problem persists, have a professional diagnose the probable cause.

43 Fluid leakage

1 Automatic transmission fluid is a deep red color. Fluid leaks should not be confused with engine oil, which can easily be blown by air flow to the transmission.

2 To pinpoint a leak, first remove all built-up dirt and grime from around the transmission. Degreasing agents and/or steam cleaning will achieve this. With the underside clean, drive the vehicle at low speeds so air flow will not blow the leak far from its source. Raise the vehicle and determine where the leak is coming from. Common areas of leakage are:

a) *Pan: Tighten mounting bolts and/or replace pan gasket as necessary (see Chapter 1).*

b) *Filler pipe: Replace the rubber seal where pipe enters transmission case.*

c) *Transmission oil lines: Tighten connectors where lines enter transmission case and/or replace lines.*

d) *Vent pipe: Transmission overfilled and/or water in fluid (see checking procedures, Chapter 1).*

e) *Speedometer connector: Replace the O-ring where speedometer cable enters transmission case (Chapter 7).*

DRIVESHAFT

44 Oil leak at front of driveshaft

Defective transmission rear oil seal. See Chapter 7 for replacement procedures. While this is done, check the splined yoke for burrs or a rough condition which may be damaging the seal. Burrs can be removed with crocus cloth or a fine whetstone.

45 Knock or clunk when the transmission is under initial load (just after transmission is put into gear)

1 Loose or disconnected rear suspension components. Check all mounting bolts, nuts and bushings (Chapter 10).

2 Loose driveshaft bolts. Inspect all bolts and nuts and tighten them to the specified torque.

3 Worn or damaged universal joint bearings. Check for wear (Chapter 8).

46 Metallic grating sound consistent with vehicle speed

Pronounced wear in the universal joint bearings. Check as described in Chapter 8.

47 Vibration

➡Note: Before assuming that the driveshaft is at fault, make sure the tires are perfectly balanced and perform the following test.

1 Install a tachometer inside the vehicle to monitor engine speed as the vehicle is driven. Drive the vehicle and note the engine speed at which the vibration (roughness) is most pronounced. Now shift the transmission to a different gear and bring the engine speed to the same point.

2 If the vibration occurs at the same engine speed (rpm) regardless of which gear the transmission is in, the driveshaft is NOT at fault since the driveshaft speed varies.

3 If the vibration decreases or is eliminated when the transmission is in a different gear at the same engine speed, refer to the following probable causes.

4 Bent or dented driveshaft. Inspect and replace as necessary (Chapter 8).

5 Undercoating or built-up dirt, etc. on the driveshaft. Clean the shaft thoroughly and recheck.

6 Worn universal joint bearings. Remove and inspect (Chapter 8).

7 Driveshaft and/or companion flange out-of-balance. Check for missing weights on the shaft. Remove the driveshaft (Chapter 8) and reinstall 180° from original position, then retest. Have the driveshaft professionally balanced if the problem persists.

AXLES

48 Noise

1 Road noise. No corrective procedures available.

2 Tire noise. Inspect the tires and check tire pressures (Chapter 1).

3 Rear wheel bearings loose, worn or damaged (Chapter 8).

49 Vibration

See probable causes under Driveshaft. Proceed under the guidelines listed for the driveshaft. If the problem persists, check the rear wheel bearings by raising the rear of the vehicle and spinning the wheels by hand. Listen for evidence of rough (noisy) bearings. Remove and inspect (Chapter 8).

50 Oil leakage

1 Pinion seal damaged (Chapter 8).

2 Axleshaft oil seals damaged (Chapter 8).

3 Differential inspection cover leaking. Tighten the bolts or replace the gasket as required (Chapters 1 and 8).

BRAKES

➡Note: Before assuming that a brake problem exists, make sure that the tires are in good condition and inflated properly (see Chapter 1), that the front end alignment is correct and that the vehicle is not loaded with weight in an unequal manner.

51 Vehicle pulls to one side during braking

1 Defective, damaged or oil contaminated brake pads or shoes on one side. Inspect as described in Chapter 9.

2 Excessive wear of brake shoe or pad material or drum/disc on one side. Inspect and correct as necessary.

3 Loose or disconnected front suspension components. Inspect and tighten all bolts to the specified torque (Chapter 10).

4 Defective drum brake or caliper assembly. Remove the drum or caliper and inspect for a stuck piston or other damage (Chapter 9).

52 Noise (high-pitched squeal with the brakes applied)

Disc brake pads worn out. The noise comes from the wear sensor rubbing against the disc. Replace the pads with new ones immediately (Chapter 9).

53 Excessive brake pedal travel

1 Partial brake system failure. Inspect the entire system (Chapter 9) and correct as required.
2 Insufficient fluid in the master cylinder. Check (Chapter 1), add fluid and bleed the system if necessary (Chapter 9).
3 Brakes not adjusting properly. Make a series of starts and stops with the vehicle is in Reverse. If this does not correct the situation, remove the drums and inspect the self-adjusters (Chapter 9).

54 Brake pedal feels spongy when depressed

1 Air in the hydraulic lines. Bleed the brake system (Chapter 9).
2 Faulty flexible hoses. Inspect all system hoses and lines. Replace parts as necessary.
3 Master cylinder mounting bolts/nuts loose.
4 Master cylinder defective (Chapter 9).

55 Excessive effort required to stop vehicle

1 Power brake booster not operating properly (Chapter 9).
2 Excessively worn linings or pads. Inspect and replace if necessary (Chapters 1 and 9).
3 One or more caliper pistons or wheel cylinders seized or sticking. Inspect and rebuild as required (Chapter 9).
4 Brake linings or pads contaminated with oil or grease. Inspect and replace as required (Chapters 1 and 9).
5 New pads or shoes installed and not yet seated. It will take a while for the new material to seat against the drum (or rotor).

56 Pedal travels to the floor with little resistance

Little or no fluid in the master cylinder reservoir caused by leaking wheel cylinder(s), leaking caliper piston(s), loose, damaged or disconnected brake lines. Inspect the entire system and correct as necessary.

57 Brake pedal pulsates during brake application

1 Wheel bearings not adjusted properly or in need of replacement (Chapter 1).
2 Caliper not sliding properly due to improper installation or obstructions. Remove and inspect (Chapter 9).
3 Rotor or drum defective. Remove the rotor or drum (Chapter 9) and check for excessive lateral runout, out-of-round and parallelism. Have the drum or rotor resurfaced or replace it with a new one.

SUSPENSION AND STEERING SYSTEMS

58 Vehicle pulls to one side

1 Tire pressures uneven (Chapter 1).
2 Defective tire (Chapter 1).
3 Excessive wear in suspension or steering components (Chapter 10).
4 Front end in need of alignment.
5 Front brakes dragging. Inspect the brakes as described in Chapter 9.

59 Shimmy, shake or vibration

1 Tire or wheel out-of-balance or out-of-round. Have professionally balanced.

2 Loose, worn or out-of-adjustment wheel bearings (Chapters 1 and 8).
3 Shock absorbers and/or suspension components worn or damaged (Chapter 10).

60 Excessive pitching and/or rolling around corners or during braking

1 Defective shock absorbers. Replace as a set (Chapter 10).
2 Broken or weak springs and/or suspension components. Inspect as described in Chapter 10.

61 Excessively stiff steering

1 Lack of fluid in power steering fluid reservoir (Chapter 1).
2 Incorrect tire pressures (Chapter 1).
3 Lack of lubrication at steering joints (Chapter 1).
4 Front end out of alignment.
5 See Section 63.

62 Excessive play in steering

1 Loose front wheel bearings (Chapter 1).
2 Excessive wear in suspension or steering components (Chapter 10).
3 Steering gearbox out of adjustment (Chapter 10).

63 Lack of power assistance

1 Steering pump drivebelt faulty or not adjusted properly (Chapter 1).
2 Fluid level low (Chapter 1).
3 Hoses or lines restricted. Inspect and replace parts as necessary.
4 Air in power steering system. Bleed the system (Chapter 10).

64 Excessive tire wear (not specific to one area)

1 Incorrect tire pressures (Chapter 1).
2 Tires out-of-balance. Have professionally balanced.
3 Wheels damaged. Inspect and replace as necessary.
4 Suspension or steering components excessively worn (Chapter 10).

65 Excessive tire wear on outside edge

1 Inflation pressures incorrect (Chapter 1).
2 Excessive speed in turns.
3 Front end alignment incorrect (excessive toe-in). Have professionally aligned.
4 Suspension arm bent or twisted (Chapter 10).

66 Excessive tire wear on inside edge

1 Inflation pressures incorrect (Chapter 1).
2 Front end alignment incorrect (toe-out). Have professionally aligned.
3 Loose or damaged steering components (Chapter 10).

67 Tire tread worn in one place

1 Tires out-of-balance.
2 Damaged or buckled wheel. Inspect and replace if necessary.
3 Defective tire (Chapter 1).

Section

Reference to other Chapters

1

TUNE-UP AND ROUTINE MAINTENANCE

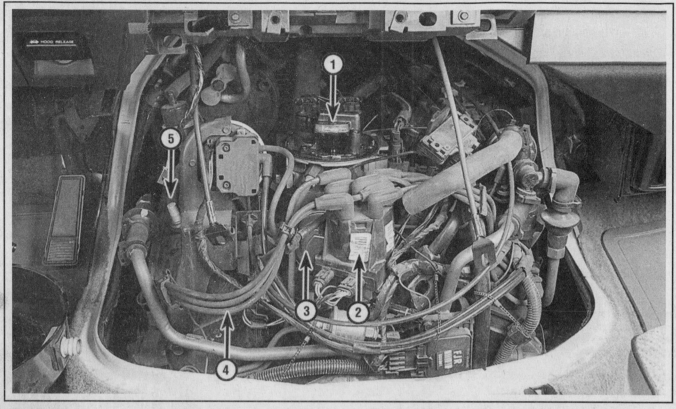

Typical V6 engine compartment component layout (viewed from the passenger compartment with the engine cover removed)

1	Throttle Body Injection (TBI) (air cleaner assembly removed)	2	Distributor (HEI)	4	Spark plug wires
		3	EGR valve	5	PCV valve

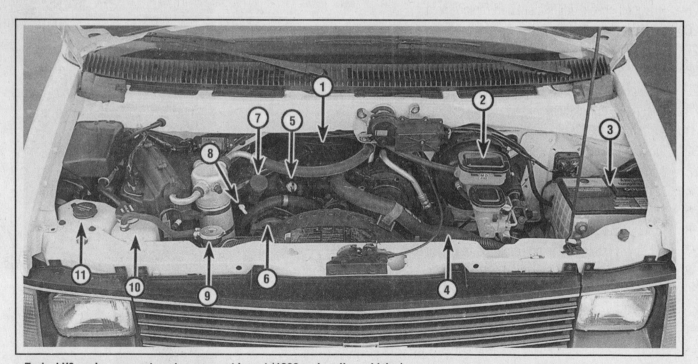

Typical V6 engine compartment component layout (1992 and earlier vehicles)

1	Air cleaner assembly	5	Automatic transmission fluid dipstick	9	Radiator cap
2	Brake fluid reservoir	6	Alternator	10	Coolant reservoir
3	Battery	7	Engine oil filler cap	11	Windshield washer fluid reservoir
4	Radiator hose	8	Engine oil dipstick		

Typical V6 engine compartment layout (1996 and later models)

1	Windshield washer fluid reservoir	5	Radiator pressure cap	9	Brake fluid reservoir
2	Coolant reservoir	6	Air cleaner assembly	10	Underhood electrical center
3	Automatic transmission fluid dipstick	7	Engine oil filler cap		(relays and fuses)
4	Engine oil dipstick	8	Power steering fluid reservoir	11	Battery

Underside view of engine/transmission (V6 engine shown)

1	Radiator hose	4	Balljoint grease fitting	6	Exhaust pipe
2	Drivebelt	5	Automatic transmission pan	7	Shock absorber
3	Steering linkage grease fitting				

Typical rear underside component layout

1 Spring
2 Shock absorber
3 Muffler

4 Driveshaft universal joint
5 Rear axle/differential
6 Exhaust pipe

7 Brake line
8 Parking brake cable

1   Maintenance schedule

The following maintenance intervals are based on the assumption that the vehicle owner will be doing the maintenance or service work, as opposed to having a dealer service department do the work. Although the time/mileage intervals are loosely based on factory recommendations, most have been shortened to ensure, for example, that such items as lubricants and fluids are checked/changed at intervals that promote maximum engine/driveline service life. Also, subject to the preference of the individual owner interested in keeping his or her vehicle in peak condition at all times, and with the vehicle's ultimate resale in mind, many of the maintenance procedures may be performed more often than recommended in the following schedule. We encourage such owner initiative.

When the vehicle is new it should be serviced initially by a factory authorized dealer service department to protect the factory warranty. In many cases the initial maintenance check is done at no cost to the owner (check with your dealer service department for more information).

EVERY 250 MILES OR WEEKLY, WHICHEVER COMES FIRST

Check the engine oil level (Section 4)
Check the engine coolant level (Section 4)
Check the windshield washer fluid level (Section 4)
Check the brake and clutch fluid levels (Section 4)
Check the tires and tire pressures (Section 5)

EVERY 3000 MILES OR 3 MONTHS, WHICHEVER COMES FIRST

All items listed above plus:
Check the automatic transmission fluid level (Section 6)
Check the power steering fluid level (Section 7)
Check and service the battery (Section 8)
Check the cooling system (Section 9)
Inspect and replace, if necessary, all underhood hoses (Section 10)
Inspect and replace, if necessary, the windshield wiper blades (Section 11)

EVERY 7500 MILES OR 12 MONTHS, WHICHEVER COMES FIRST

All items listed above plus:
Change the engine oil and filter (Section 12)*
Lubricate the chassis components (Section 13)
Inspect the suspension and steering components (Section 14)*
Inspect the exhaust system (Section 15)*
Check the manual transmission lubricant level (Section 16)*
Check the differential (rear axle) oil level (Section 17)*
Rotate the tires (Section 18)
Check the brakes (Section 19)*
Inspect the fuel system (Section 20)
Check the carburetor choke operation (Section 21)

Check the carburetor/throttle body mounting nut torque (Section 22)
Check the throttle linkage (Section 23)
Check the thermostatically-controlled air cleaner (Section 24)
Check the engine drivebelts (Section 25)
Check the seatbelts (Section 26)
Check the starter safety switch (Section 27)
Check the spare tire and jack (Section 28)

EVERY 30,000 MILES OR 24 MONTHS, WHICHEVER COMES FIRST

All items listed above plus:
Check and adjust, if necessary, the engine idle speed (Section 29)
Replace the fuel filter (Section 30)
Replace the air and PCV filters (Section 31)
Check and adjust, if necessary, the ignition timing (Section 32)
Change the automatic transmission fluid (Section 33)**
Change the manual transmission lubricant (Section 34)
Change the differential (rear axle) oil (Section 35)
Check and repack the front wheel bearings (2002 and earlier models only) (Section 36)
Service the cooling system (drain, flush and refill) (1985 to 1995 models) (Section 37)
Inspect and replace, if necessary, the PCV valve (Section 38)
Inspect the evaporative emissions control system (Section 39)
Check the EGR system (Section 40)
Replace the spark plugs (1985 to 1995 models) (Section 41)
Inspect the spark plug wires, distributor cap and rotor (Sections 42 and 43)

EVERY 100,000 MILES OR 48 MONTHS, WHICHEVER COMES FIRST

Replace spark plugs (1996 and later models) (Section 41)
Service the cooling system (drain, flush and refill) (1996 and later models) (Section 37)

*This item is affected by "severe" operating conditions as described below. If your vehicle is operated under severe conditions, perform all maintenance indicated with an asterisk (*) at 3000 mile/3 month intervals. Severe conditions are indicated if you mainly operate your vehicle under one or more of the following:

Operating in dusty areas
Towing a trailer
Idling for extended periods and/or low speed operation
Operating when outside temperatures remain below freezing and when most trips are less than four miles

** If operated under one or more of the following conditions, change the automatic transmission fluid every 12,000 miles:

In heavy city traffic where the outside temperature regularly reaches 90° F (32° C) or higher
In hilly or mountainous terrain
Frequent trailer pulling

2 Introduction

This Chapter is designed to help the home mechanic maintain the Chevrolet Astro/GMC Safari with the goals of maximum performance, economy, safety and reliability in mind.

Included is a master maintenance schedule, followed by procedures dealing specifically with each item on the schedule. Visual checks, adjustments, component replacement and other helpful items are included. Refer to the accompanying illustrations of the engine compartment and the underside of the vehicle for the locations of various components.

Servicing your vehicle in accordance with the mileage/time maintenance schedule and the step-by-step procedures will result in a planned maintenance program that should produce a long and reliable service life. Keep in mind that it is a comprehensive plan, so maintaining some items but not others at the specified intervals will not produce the same results.

As you service your vehicle, you will discover that many of the procedures can - and should - be grouped together because of the nature of the particular procedure you're performing or because of the close proximity of two otherwise unrelated components to one another.

For example, if the vehicle is raised for chassis lubrication, you should inspect the exhaust, suspension, steering and fuel systems while you're under the vehicle. When you're rotating the tires, it makes good sense to check the brakes since the wheels are already removed. Finally, let's suppose you have to borrow or rent a torque wrench. Even if you only need it to tighten the spark plugs, you might as well check the torque of as many critical fasteners as time allows.

The first step in this maintenance program is to prepare yourself before the actual work begins. Read through all the procedures you're planning to do, then gather up all the parts and tools needed. If it looks like you might run into problems during a particular job, seek advice from a mechanic or an experienced do-it-yourselfer.

3 Tune-up general information

The term tune-up is used in this manual to represent a combination of individual operations rather than one specific procedure.

If, from the time the vehicle is new, the routine maintenance schedule is followed closely and frequent checks are made of fluid levels and high wear items, as suggested throughout this manual, the engine will be kept in relatively good running condition and the need for additional work will be minimized.

More likely than not, however, there will be times when the engine is running poorly due to lack of regular maintenance. This is even more likely if a used vehicle, which has not received regular and frequent maintenance checks, is purchased. In such cases, an engine tune-up will be needed outside of the regular routine maintenance intervals.

The first step in any tune-up or diagnostic procedure to help correct a poor running engine is a cylinder compression check. A compression check (see Chapter 2 Part C) will help determine the condition of internal engine components and should be used as a guide for tune-up and repair procedures. If, for instance, a compression check indicates serious internal engine wear, a conventional tune-up will not improve the performance of the engine and would be a waste of time and money. Because of its importance, the compression check should be done by someone with the right equipment and the knowledge to use it properly.

The following procedures are those most often needed to bring a generally poor running engine back into a proper state of tune.

MINOR TUNE-UP

Check all engine related fluids (Section 4)
Clean, inspect and test the battery (Section 8)
Check and adjust the drivebelts (Section 25)
Replace the spark plugs (Section 41)
Inspect the distributor cap and rotor (Section 42)
Inspect the spark plug and coil wires (Section 42)
Check and adjust the ignition timing (Section 32)
Check the PCV valve (Section 38)
Check the air and PCV filters (Section 31)
Check the cooling system (Section 9)
Check all underhood hoses (Section 10)

MAJOR TUNE-UP

All items listed under Minor tune-up plus . . .
Check the EGR system (Section 40)
Check the ignition system (Chapter 5)
Check the charging system (Chapter 5)
Check the fuel system (Section 20)
Replace the air and PCV filters (Section 31)
Replace the distributor cap and rotor (Section 43)
Replace the spark plug wires (Section 42)

4 Fluid level checks

▶ **Refer to illustrations 4.4, 4.6, 4.8 and 4.19**

➡**Note: The following are fluid level checks to be done on a 250 mile or weekly basis. Additional fluid level checks can be found in specific maintenance procedures which follow. Regardless of intervals, be alert to fluid leaks under the vehicle which would indicate a fault to be corrected immediately.**

1   Fluids are an essential part of the lubrication, cooling, brake, clutch and windshield washer systems. Because the fluids gradually become depleted and/or contaminated during normal operation of the vehicle, they must be periodically replenished. See *Recommended lubricants and fluids* at the end of this Chapter before adding fluid to any of the following components.

➡**Note: The vehicle must be on level ground when fluid levels are checked.**

ENGINE OIL

2 The engine oil level is checked with a dipstick that extends through a tube and into the oil pan at the bottom of the engine (see illustration 4.6).

3 The oil level should be checked before the vehicle has been driven, or about 5 minutes after the engine has been shut off. If the oil is checked immediately after driving the vehicle, some of the oil will remain in the upper engine components, resulting in an inaccurate reading on the dipstick.

4 Pull the dipstick from the tube and wipe all the oil from the end with a clean rag or paper towel. Insert the clean dipstick all the way back into the tube, then pull it out again. Note the oil at the end of the dipstick. Add oil as necessary to keep the level between the ADD mark and the FULL mark on the dipstick (see illustration).

5 Do not overfill the engine by adding too much oil since this may result in oil fouled spark plugs, oil leaks or oil seal failures.

6 Oil is added to the engine after removing a pull off cap (see illustration). An oil can spout or funnel may help to reduce spills.

7 Checking the oil level is an important preventive maintenance step. A consistently low oil level indicates oil leakage through damaged seals, defective gaskets or past worn rings or valve guides. If the oil looks milky in color or has water droplets in it, the cylinder head gasket(s) may be blown or the head(s) or block may be cracked. The engine should be checked immediately. The condition of the oil should also be checked. Whenever you check the oil level, slide your thumb and index finger up the dipstick before wiping off the oil. If you see small dirt or metal particles clinging to the dipstick, the oil should be changed (Section 12).

ENGINE COOLANT

⁂ WARNING:

Do not allow antifreeze to come in contact with your skin or painted surfaces of the vehicle. Flush contaminated areas immediately with plenty of water. Don't store new coolant or leave old coolant lying around where it's accessible to children or pets - they're attracted by its sweet smell. Ingestion of even a small amount of coolant can be fatal! Wipe up garage floor and drip pan coolant spills immediately. Keep antifreeze containers covered and repair leaks in your cooling system immediately.

⁂ CAUTION:

1996 and later models use a special coolant called DEX-COOL. It has a much longer service life than standard coolant (see Maintenance schedule). DEX-COOL is orange in color and should not be mixed with ethylene glycol type (green colored) coolants. Mixing coolant types will result in premature corrosion of engine cooling system components. If only a small amount of coolant is needed to bring the level back to full, adding distilled water is acceptable.

8 All vehicles covered by this manual are equipped with a pressurized coolant recovery system. A white plastic coolant reservoir located in the engine compartment is connected by a hose to the radiator filler neck (see illustration). If the engine overheats, coolant escapes through a valve in the radiator cap and travels through the hose into the reservoir. As the engine cools, the coolant is automatically drawn back into the cooling system to maintain the correct level.

9 The coolant level in the reservoir should be checked regularly.

⁂ WARNING:

Do not remove the radiator cap to check the coolant level when the engine is warm. The level in the reservoir varies with the temperature of the engine. When the engine is cold, the coolant level should be at or slightly above the FULL COLD mark on the reservoir. Once the engine has warmed up, the level should be at or near the FULL HOT mark. If it isn't, allow the engine to cool, then remove the cap from the reservoir and add a 50/50 mixture of the appropriate coolant and water. See Recommended lubricants and fluids in this Chapter.

10 Drive the vehicle and recheck the coolant level. If only a small amount of coolant is required to bring the system up to the proper level, water can be used. However, repeated additions of water will dilute the antifreeze and water solution. In order to maintain the proper ratio of antifreeze and water, always top up the coolant level with the correct mixture. An empty plastic milk jug or bleach bottle makes an excellent container for mixing coolant. Do not use rust inhibitors or additives.

11 If the coolant level drops consistently, there may be a leak in the system. Inspect the radiator, hoses, filler cap, drain plugs and water

![dipstick illustration]

4.4 The engine oil level must be maintained between the marks at all times - it takes one quart of oil to raise the level from the ADD mark to the FULL mark

4.6 Oil is added to the engine after removing the pull-off cap from the filler tube. Engine oil dipstick is indicated by the arrow

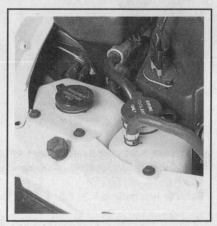

4.8 The engine coolant and windshield washer fluid reservoirs are located next to each other - DO NOT mix them up when adding fluids!

pump (see Section 9). If no leaks are noted, have the radiator cap pressure tested by a service station.

12 If you have to remove the radiator cap, wait until the engine has cooled, then wrap a thick cloth around the cap and turn it to the first stop. If coolant or steam escapes, let the engine cool down longer, then remove the cap.

13 Check the condition of the coolant as well. It should be relatively clear. If it's brown or rust colored, the system should be drained, flushed and refilled. Even if the coolant appears to be normal, the corrosion inhibitors wear out, so it must be replaced at the specified intervals.

WINDSHIELD WASHER FLUID

14 Fluid for the windshield washer system is located in a plastic reservoir in the engine compartment (see illustration 4.8).

15 In milder climates, plain water can be used in the reservoir, but it should be kept no more than 2/3 full to allow for expansion if the water freezes. In colder climates, use windshield washer system antifreeze, available at any auto parts store, to lower the freezing point of the fluid. Mix the antifreeze with water in accordance with the manufacturer's directions on the container.

❄❄ CAUTION:

Don't use cooling system antifreeze - it will damage the vehicle's paint.

16 To help prevent icing in cold weather, warm the windshield with the defroster before using the washer.

BATTERY ELECTROLYTE

17 All vehicles with which this manual is concerned are equipped with a battery which is permanently sealed (except for vent holes) and has no filler caps. Water doesn't have to be added to these batteries at any time. If a maintenance-type battery is installed, the caps on the top of the battery should be removed periodically to check for a low water level. This check is most critical during the warm summer months.

BRAKE AND CLUTCH FLUID

18 The brake master cylinder is mounted on the front of the power booster unit in the engine compartment. The clutch cylinder used on manual transmissions is mounted adjacent to it on the firewall.

19 The fluid inside is readily visible. The level should be above the MIN marks on the reservoirs (see illustration). If a low level is indicated, be sure to wipe the top of the reservoir cover with a clean rag to prevent contamination of the brake and/or clutch system before remov-

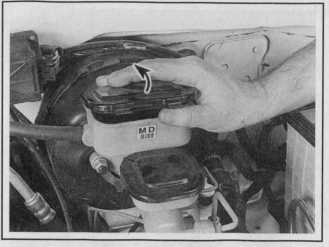

4.19 The brake fluid level is easily checked by looking through the clear reservoir - when adding fluid, grasp the tabs and rotate the cover up as shown

ing the cover.

20 When adding fluid, pour it carefully into the reservoir to avoid spilling it onto surrounding painted surfaces. Be sure the specified fluid is used, since mixing different types of brake fluid can cause damage to the system. See *Recommended lubricants and fluids* at the end of this Chapter or your owner's manual.

❄❄ WARNING:

Brake fluid can harm your eyes and damage painted surfaces, so use extreme caution when handling or pouring it. Do not use brake fluid that has been standing open or is more than one year old. Brake fluid absorbs moisture from the air. Excess moisture can cause a dangerous loss of braking effectiveness.

21 At this time the fluid and master cylinder can be inspected for contamination. The system should be drained and refilled if deposits, dirt particles or water droplets are seen in the fluid.

22 After filling the reservoir to the proper level, make sure the cover is on tight to prevent fluid leakage.

23 The brake fluid level in the master cylinder will drop slightly as the pads and the brake shoes at each wheel wear down during normal operation. If the master cylinder requires repeated additions to keep it at the proper level, it's an indication of leakage in the brake system, which should be corrected immediately. Check all brake lines and connections (see Section 19 for more information).

24 If, upon checking the master cylinder fluid level, you discover one or both reservoirs empty or nearly empty, the brake system should be bled (Chapter 9).

5   Tire and tire pressure checks

▸ **Refer to illustrations 5.2, 5.3, 5.4a, 5.4b and 5.8**

1   Periodic inspection of the tires may spare you the inconvenience of being stranded with a flat tire. It can also provide you with vital information regarding possible problems in the steering and suspension systems before major damage occurs.

2   The original tires on this vehicle are equipped with 1/2-inch wide bands that will appear when tread depth reaches 1/16-inch, at which

point the tires can be considered worn out. Tread wear can be monitored with a simple, inexpensive device known as a tread depth indicator (see illustration).

3   Note any abnormal tread wear (see illustration). Tread pattern irregularities such as cupping, flat spots and more wear on one side than the other are indications of front end alignment and/or balance problems. If any of these conditions are noted, take the vehicle to a tire

shop or service station to correct the problem.

4 Look closely for cuts, punctures and embedded nails or tacks. Sometimes a tire will hold air pressure for a short time or leak down very slowly after a nail has embedded itself in the tread. If a slow leak persists, check the valve stem core to make sure it's tight (see illustration). Examine the tread for an object that may have embedded itself in the tire or for a "plug" that may have begun to leak (radial tire punctures are repaired with a plug that's installed in a puncture). If a puncture is suspected, it can be easily verified by spraying a solution of soapy water onto the puncture area (see illustration). The soapy solution will bubble if there's a leak. Unless the puncture is unusually large, a tire shop or service station can usually repair the tire.

5 Carefully inspect the inner sidewall of each tire for evidence of brake fluid leakage. If you see any, inspect the brakes immediately.

6 Correct air pressure adds miles to the lifespan of the tires, improves mileage and enhances overall ride quality. Tire pressure cannot be accurately estimated by looking at a tire, especially if it's a radial. A tire pressure gauge is essential. Keep an accurate gauge in the vehicle. The pressure gauges attached to the nozzles of air hoses at gas stations are often inaccurate.

7 Always check tire pressure when the tires are cold. Cold, in this case, means the vehicle has not been driven over a mile in the three hours preceding a tire pressure check. A pressure rise of four to eight pounds is not uncommon once the tires are warm.

8 Unscrew the valve cap protruding from the wheel or hubcap and push the gauge firmly onto the valve stem (see illustration). Note the reading on the gauge and compare the figure to the recommended tire pressure shown on the placard on the driver's side door pillar. Be sure

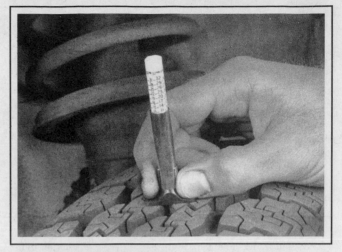

5.2 A tire tread depth indicator should be used to monitor tire wear - they are available at auto parts stores and service stations and cost very little

to reinstall the valve cap to keep dirt and moisture out of the valve stem mechanism. Check all four tires and, if necessary, add enough air to bring them up to the recommended pressure.

9 Don't forget to keep the spare tire inflated to the specified pressure (refer to your owner's manual or the tire sidewall). Note that the pressure recommended for the compact spare is higher than for the tires on the vehicle.

UNDERINFLATION

CUPPING

OVERINFLATION

Cupping may be caused by:
- **Underinflation and/or mechanical irreguarities such as out-of-balance condition of wheel and/or tire, and bent or damaged wheel.**
- **Loose or worn steering tie-rod or steering idler arm.**
- **Loose, damaged or worn front suspension parts.**

INCORRECT TOE-IN OR EXTREME CAMBER

FEATHERING DUE TO MISALIGNMENT

5.3 This chart will help you determine the condition of your tires, the probable cause(s) of abnormal wear and the corrective action necessary

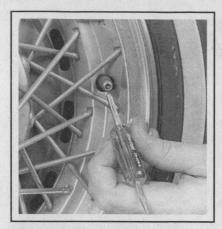

5.4a If a tire loses air on a steady basis, check the valve core first to make sure it's snug (special inexpensive wrenches are commonly available at auto parts stores)

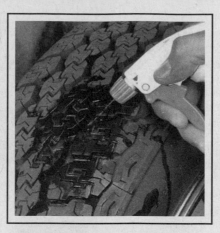

5.4b If the valve core is tight, raise the corner of the vehicle with the low tire and spray a soapy water solution onto the tread as the tire is turned slowly - slow leaks will cause small bubbles to appear

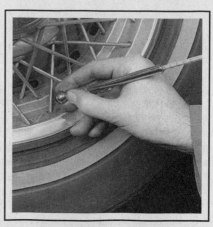

5.8 To extend the life of your tires, check the air pressure at least once a week with an accurate gauge (don't forget the spare!)

6 Automatic transmission fluid level check

▶ **Refer to illustration 6.6**

1 The automatic transmission fluid level should be carefully maintained. Low fluid level can lead to slipping or loss of drive, while overfilling can cause foaming and loss of fluid.

2 With the parking brake set, start the engine, then move the shift lever through all the gear ranges, ending in Park. The fluid level must be checked with the vehicle level and the engine running at idle.

➡**Note: Incorrect fluid level readings will result if the vehicle has just been driven at high speeds for an extended period, in hot weather in city traffic, or if it has been pulling a trailer. If any of these conditions apply, wait until the fluid has cooled (about 30 minutes).**

3 With the transmission at normal operating temperature, remove the dipstick from the filler tube. The dipstick is located at the front of the engine compartment on the passenger's side.

4 Carefully touch the fluid at the end of the dipstick to determine if it is cool, warm or hot. Wipe the fluid from the dipstick with a clean rag and push it back into the filler tube until the cap seats.

5 Pull the dipstick out again and note the fluid level.

6 If the fluid felt cool, the level should be about 1/8 to 3/8-inch above the ADD mark (see illustration). If it felt warm, the level should be near the lower part of the operating range. If the fluid was hot, the level should be near the FULL HOT mark. If additional fluid is required, add it directly into the tube using a funnel. It takes about one pint to

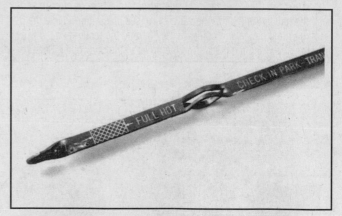

6.6 When checking the automatic transmission fluid level be sure to note the fluid temperature

raise the level from the ADD mark to the FULL HOT mark with a hot transmission, so add the fluid a little at a time and keep checking the level until it's correct.

7 The condition of the fluid should also be checked along with the level. If the fluid at the end of the dipstick is a dark reddish-brown color, or if it smells burned, it should be changed. If you are in doubt about the condition of the fluid, purchase some new fluid and compare the two for color and smell.

7 Power steering fluid level check

▶ **Refer to illustrations 7.2 and 7.6**

1 Unlike manual steering, the power steering system relies on fluid which may, over a period of time, require replenishing.

2 The fluid reservoir for the power steering pump is located on the pump body at the front of the engine (see illustration).

3 For the check, the front wheels should be pointed straight ahead and the engine should be off.

4 Use a clean rag to wipe off the reservoir cap and the area around

the cap. This will help prevent any foreign matter from entering the reservoir during the check.

5 Twist off the cap and check the temperature of the fluid at the end of the dipstick with your finger.

6 Wipe off the fluid with a clean rag, reinsert the dipstick, then withdraw it and read the fluid level. The level should be at the HOT mark if the fluid was hot to the touch (see illustration). It should be at the COLD mark if the fluid was cool to the touch. Never allow the fluid

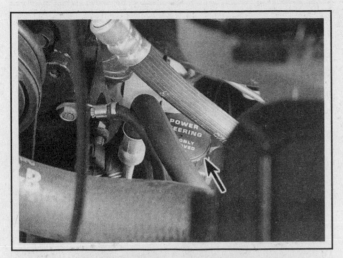

7.2 The power steering fluid reservoir on 1985 to 1992 models is located near the front of the engine below the brake master cylinder (arrow). On 1993 and later models, it's mounted on the firewall

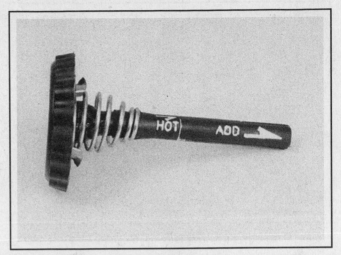

7.6 The marks on the power steering fluid dipstick indicate the safe range

level to drop below the ADD mark.

7 If additional fluid is required, pour the specified type directly into the reservoir, using a funnel to prevent spills.

8 If the reservoir requires frequent fluid additions, all power steering hoses, hose connections and the power steering pump should be carefully checked for leaks.

8 Battery check and maintenance

▶ Refer to illustrations 8.1, 8.6a and 8.6b

✳✳ WARNING:

Certain precautions must be followed when checking and servicing the battery. Hydrogen gas, which is highly flammable, is always present in the battery cells, so keep lighted tobacco and all other open flames and sparks away from the battery. The electrolyte inside the battery is actually dilute sulfuric acid, which will cause injury if splashed on your skin or in your eyes. It will also ruin clothes and painted surfaces. When removing the battery cables, always detach the negative cable first and hook it up last!

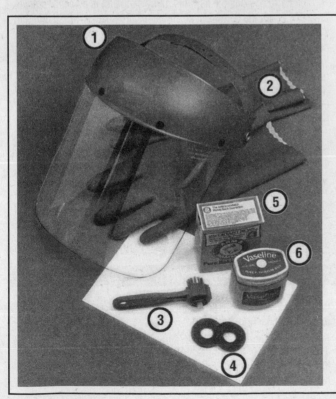

8.1 Tools and materials required for battery maintenance

1 **Face shield/safety goggles** - *When removing corrosion with a brush, the acidic particles can easily fly up into your eyes*

2 **Rubber gloves** - *Another safety item to consider when servicing the battery - remember that's acid inside the battery!*

3 **Battery terminal/cable cleaner** - *This wire brush cleaning tool will remove all traces of corrosion from the battery and cable*

4 **Treated felt washers** - *Placing one of these on each terminal, directly under the cable end, will help prevent corrosion (be sure to get the correct type for side-terminal batteries)*

5 **Baking soda** - *A solution of baking soda and water can be used to neutralize corrosion*

6 **Petroleum jelly** - *A layer of this on the battery terminal bolts will help prevent corrosion*

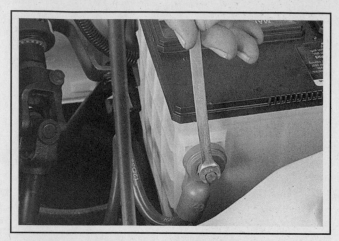

8.6a Make sure the battery terminal bolts are tight

✳✳ CAUTION:

On models equipped with an anti-theft audio system, disable the anti-theft feature before performing any operation that requires disconnecting the battery or disrupting power to the stereo (see the Anti-theft audio system procedures at the front of this manual).

1 Battery maintenance is an important procedure which will help ensure that you are not stranded because of a dead battery. Several tools are required for this procedure (see illustration).

2 When checking/servicing the battery, always turn the engine and all accessories off.

3 A sealed (sometimes called maintenance-free), side-terminal battery is standard equipment on these vehicles. The cell caps cannot be removed, no electrolyte checks are required and water cannot be added to the cells. However, if a standard top-terminal aftermarket battery has been installed, the following maintenance procedure can be used.

4 Remove the caps and check the electrolyte level in each of the battery cells. It must be above the plates. There's usually a split-ring indicator in each cell to indicate the correct level. If the level is low, add distilled water only, then reinstall the cell caps.

✳✳ CAUTION:

Overfilling the cells may cause electrolyte to spill over during periods of heavy charging, causing corrosion and damage to nearby components.

5 The external condition of the battery should be checked periodically. Look for damage such as a cracked case.

6 Check the tightness of the battery cable bolts (see illustration) to ensure good electrical connections. Inspect the entire length of each

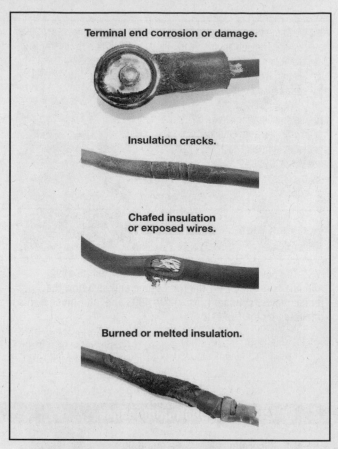

8.6b If you find any of these conditions as you're inspecting a battery cable, replace it immediately

cable, looking for cracked or abraded insulation and frayed conductors (see illustration).

7 If corrosion (visible as white, fluffy deposits) is evident, remove the cables from the terminals, clean them with a battery brush and reinstall them. Corrosion can be kept to a minimum by applying a layer of petroleum jelly or grease to the bolt threads.

8 Make sure the battery carrier is in good condition and the hold-down clamp is tight. If the battery is removed (see Chapter 5 for the removal and installation procedure), make sure that no parts remain in the bottom of the carrier when it's reinstalled. When reinstalling the hold-down clamp, don't overtighten the bolt.

9 Corrosion on the carrier, battery case and surrounding areas can be removed with a solution of water and baking soda. Apply the mixture with a small brush, let it work, then rinse it off with plenty of clean water.

10 Any metal parts of the vehicle damaged by corrosion should be coated with a zinc-based primer, then painted.

11 Additional information on the battery, charging and jump starting can be found in the front of this manual and in Chapter 5.

9 Cooling system check

▸ **Refer to illustration 9.4**

1 Many major engine failures can be attributed to a faulty cooling system. If the vehicle is equipped with an automatic transmission, the cooling system also cools the transmission fluid and thus plays an important role in prolonging transmission life.

2 The cooling system should be checked with the engine cold. Do

this before the vehicle is driven for the day or after it has been shut off for at least three hours.

3 Remove the radiator cap by turning it to the left until it reaches a stop. If you hear a hissing sound (indicating there is still pressure in the system), wait until this stops. Now press down on the cap with the palm of your hand and continue turning to the left until the cap can be

removed. Thoroughly clean the cap, inside and out, with clean water. Also clean the filler neck on the radiator. All traces of corrosion should be removed. The coolant inside the radiator should be relatively transparent. If it is rust colored, the system should be drained and refilled (see Section 37). If the coolant level is not up to the top, add additional antifreeze/coolant mixture (see Section 4).

4   Carefully check the large upper and lower radiator hoses along with the smaller diameter heater hoses which run from the engine to the firewall. On some models the heater return hose runs directly to the radiator. Inspect each hose along its entire length, replacing any hose which is cracked, swollen or shows signs of deterioration. Cracks may become more apparent if the hose is squeezed (see illustration). Regardless of condition, it's a good idea to replace hoses with new

ones every two years.

5   Make sure that all hose connections are tight. A leak in the cooling system will usually show up as white or rust colored deposits on the areas adjoining the leak. If wire-type clamps are used at the ends of the hoses, it may be a good idea to replace them with more secure screw-type clamps.

6   Use compressed air or a soft brush to remove bugs, leaves, etc. from the front of the radiator or air conditioning condenser. Be careful not to damage the delicate cooling fins or cut yourself on them.

7   Every other inspection, or at the first indication of cooling system problems, have the cap and system pressure tested. If you don't have a pressure tester, most gas stations and repair shops will do this for a minimal charge.

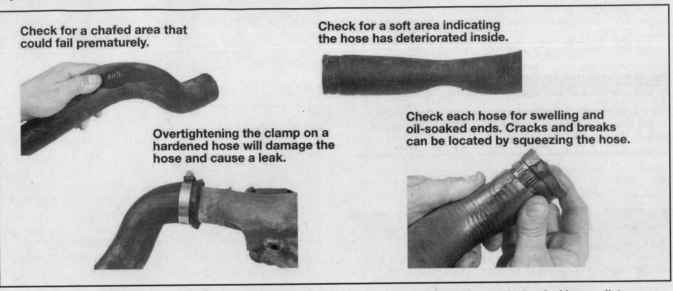

Check for a chafed area that could fail prematurely.

Check for a soft area indicating the hose has deteriorated inside.

Overtightening the clamp on a hardened hose will damage the hose and cause a leak.

Check each hose for swelling and oil-soaked ends. Cracks and breaks can be located by squeezing the hose.

9.4 Hoses, like drivebelts, have a habit of failing at the worst possible time - to prevent the inconvenience of a blown radiator or heater hose, inspect them carefully as shown here

10 Underhood hose check and replacement

▶ Refer to illustration 10.1

GENERAL

✳ CAUTION:

Replacement of air conditioning hoses must be left to a dealer service department or air conditioning shop that has the equipment to depressurize the system safely. Never remove air conditioning components or hoses (see illustration) until the system has been de pressurized.

1   High temperatures in the engine compartment can cause the deterioration of the rubber and plastic hoses used for engine, accessory and emission systems operation. Periodic inspection should be made for cracks, loose clamps, material hardening and leaks. Information specific to the cooling system hoses can be found in Section 9.

2   Some, but not all, hoses are secured to the fittings with clamps. Where clamps are used, check to be sure they haven't lost their tension, allowing the hose to leak. If clamps aren't used, make sure the hose has not expanded and/or hardened where it slips over the fitting, allowing it to leak.

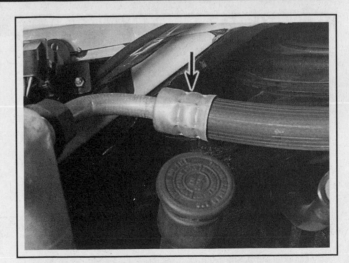

10.1 Air conditioning hoses are best identified by the metal tubes used at all bends (arrow) - DO NOT disconnect or accidentally damage the air conditioning hoses as the system is under high pressure

VACUUM HOSES

3   It's quite common for vacuum hoses, especially those in the emissions system, to be color coded or identified by colored stripes molded into them. Various systems require hoses with different wall thicknesses, collapse resistance and temperature resistance. When replacing hoses, be sure the new ones are made of the same material.

4   Often the only effective way to check a hose is to remove it completely from the vehicle. If more than one hose is removed, be sure to label the hoses and fittings to ensure correct installation.

5   When checking vacuum hoses, be sure to include any plastic T-fittings in the check. Inspect the fittings for cracks and the hose where it fits over the fitting for distortion, which could cause leakage.

6   A small piece of vacuum hose (1/4-inch inside diameter) can be used as a stethoscope to detect vacuum leaks. Hold one end of the hose to your ear and probe around vacuum hoses and fittings, listening for the "hissing" sound characteristic of a vacuum leak.

✳✳ WARNING:

When probing with the vacuum hose stethoscope, be very careful not to come into contact with moving engine components such as the drivebelt, cooling fan, etc.

FUEL HOSE

✳✳ WARNING:

There are certain precautions which must be taken when inspecting or servicing fuel system components. Work in a well ventilated area and do not allow open flames (cigarettes, appliance pilot lights, etc.) or bare light bulbs near the work area. Mop up any spills immediately and do not store fuel soaked rags where they could ignite. On vehicles equipped with fuel injection, the fuel system is under pressure, so if any fuel lines are to be disconnected, the pressure in the system must be relieved first (see Chapter 4 for more information).

7   Check all rubber fuel lines for deterioration and chafing. Check especially for cracks in areas where the hose bends and just before fittings, such as where a hose attaches to the fuel filter.

8   High quality fuel line, usually identified by the word Fluroelastomer printed on the hose, should be used for fuel line replacement. Never, under any circumstances, use unreinforced vacuum line, clear plastic tubing or water hose for fuel lines.

9   Spring-type clamps are commonly used on fuel lines. These clamps often lose their tension over a period of time, and can be "sprung" during removal. Replace all spring-type clamps with screw clamps whenever a hose is replaced.

METAL LINES

10   Sections of metal line are often used for fuel line between the fuel pump and carburetor or fuel injection unit. Check carefully to be sure the line has not been bent or crimped and that cracks have not started in the line.

11   If a section of metal fuel line must be replaced, only seamless steel tubing should be used, since copper and aluminum tubing don't have the strength necessary to withstand normal engine vibration.

12   Check the metal brake lines where they enter the master cylinder and brake proportioning unit (if used) for cracks in the lines or loose fittings. Any sign of brake fluid leakage calls for an immediate thorough inspection of the brake system.

11   Wiper blade inspection and replacement

▶ **Refer to illustration 11.6**

1   The windshield wiper and blade assembly should be inspected periodically for damage, loose components and cracked or worn blade elements.

2   Road film can build up on the wiper blades and affect their efficiency, so they should be washed regularly with a mild detergent solution.

3   The action of the wiping mechanism can loosen the bolts, nuts and fasteners, so they should be checked and tightened, as necessary, at the same time the wiper blades are checked.

4   If the wiper blade elements (sometimes called inserts) are cracked, worn or warped, they should be replaced with new ones.

5   Pull the wiper blade/arm assembly away from the glass.

6   Depress the blade-to-arm connector and slide the blade assembly off the wiper arm and over the retaining stud (see illustration).

7   Pinch the tabs at the end, then slide the element out of the blade assembly.

8   Compare the new element with the old for length, design, etc.

9   Slide the new element into place. It will automatically lock at the

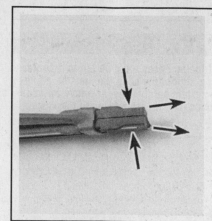

11.6 Use needle-nose pliers to push in and pull out the metal clip, then slide the old element out - slide the new element in and lock in place with the metal clip

correct location.

10   Reinstall the blade assembly on the arm, wet the windshield and check for proper operation.

12 Engine oil and filter change

▶ **Refer to illustrations 12.3, 12.9, 12.14 and 12.18**

1 Frequent oil changes are the most important preventive maintenance procedures that can be done by the home mechanic. As engine oil ages, it becomes diluted and contaminated, which leads to premature engine wear.

2 Although some sources recommend oil filter changes every other oil change, we feel that the minimal cost of an oil filter and the relative ease with which it is installed dictate that a new filter be installed every time the oil is changed.

3 Gather together all necessary tools and materials before beginning this procedure (see illustration).

4 You should have plenty of clean rags and newspapers handy to mop up any spills. Access to the underside of the vehicle is greatly improved if the vehicle can be lifted on a hoist, driven onto ramps or supported by jackstands.

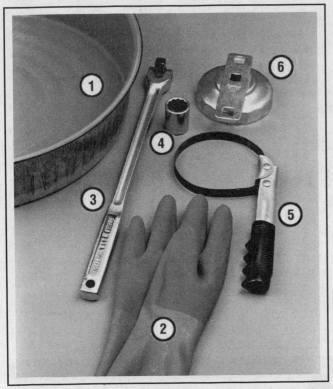

12.3 These tools are required when changing the engine oil and filter

1 **Drain pan** - *It should be fairly shallow in depth, but wide to prevent spills*

2 **Rubber gloves** - *When removing the drain plug and filter, you will get oil on your hands (the gloves will prevent burns)*

3 **Breaker bar** - *Sometimes the oil drain plug is tight, and a long breaker bar is needed to loosen it*

4 **Socket** - *To be used with the breaker bar or a ratchet (must be the correct size to fit the drain plug - six-point preferred)*

5 **Filter wrench** - *This is a metal band-type wrench, which requires clearance around the filter to be effective*

6 **Filter wrench** - *This type fits on the bottom of the filter and can be turned with a ratchet or breaker bar (different-size wrenches are available for different types of filters)*

✳✳ WARNING:

Do not work under a vehicle which is supported only by a bumper, hydraulic or scissors-type jack.

---

5 If this is your first oil change, get under the vehicle and familiarize yourself with the locations of the oil drain plug and the oil filter. The engine and exhaust components will be warm during the actual work, so note how they are situated to avoid touching them when working under the vehicle.

6 Warm the engine to normal operating temperature. If the new oil or any tools are needed, use this warm-up time to gather everything necessary for the job. The correct type of oil for your application can be found in *Recommended lubricants and fluids* at the end of this Chapter.

7 With the engine oil warm (warm engine oil will drain better and more built-up sludge will be removed with it), raise and support the vehicle. Make sure it's safely supported!

8 Move all necessary tools, rags and newspapers under the vehicle. Set the drain pan under the drain plug. Keep in mind that the oil will initially flow from the pan with some force; position the pan accordingly.

9 Being careful not to touch any of the hot exhaust components, use a wrench to remove the drain plug near the bottom of the oil pan (see illustration). Depending on how hot the oil is, you may want to wear gloves while unscrewing the plug the final few turns.

10 Allow the old oil to drain into the pan. It may be necessary to move the pan as the oil flow slows to a trickle.

11 After all the oil has drained, wipe off the drain plug with a clean rag. Small metal particles may cling to the plug and would immediately contaminate the new oil.

12 Clean the area around the drain plug opening and reinstall the plug. Tighten the plug securely with the wrench. If a torque wrench is available, use it to tighten the plug.

13 Move the drain pan into position under the oil filter.

14 Use the filter wrench to loosen the oil filter (see illustration). Chain or metal band filter wrenches may distort the filter canister, but it doesn't matter since the filter will be discarded anyway.

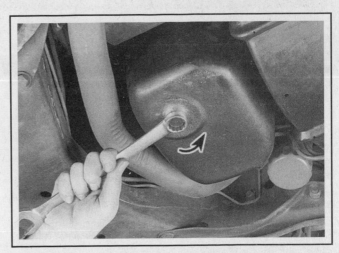

12.9 The oil drain plug is located at the bottom of the pan and should be removed with a socket or box-end wrench - DO NOT use an open-end wrench, as the corners on the bolt can be easily rounded off

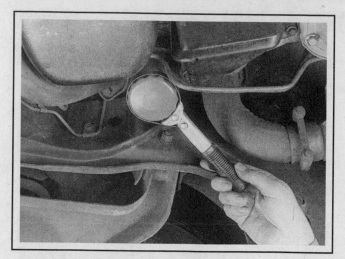

12.14 Use a strap-type oil filter wrench to loosen the filter - if access makes removal difficult, other types of filter wrenches are available

12.18 Lubricate the oil filter gasket with clean engine oil before installing the filter on the engine

15 Completely unscrew the old filter. Be careful; it's full of oil. Empty the oil inside the filter into the drain pan.

16 Compare the old filter with the new one to make sure they're the same type.

17 Use a clean rag to remove all oil, dirt and sludge from the area where the oil filter mounts to the engine. Check the old filter to make sure the rubber gasket isn't stuck to the engine. If the gasket is stuck to the engine (use a flashlight if necessary), remove it.

18 Apply a light coat of clean oil to the rubber gasket on the new oil filter (see illustration).

19 Attach the new filter to the engine, following the tightening directions printed on the filter canister or packing box. Most filter manufacturers recommend against using a filter wrench due to the possibility of overtightening and damage to the seal.

20 Remove all tools, rags, etc. from under the vehicle, being careful not to spill the oil in the drain pan, then lower the vehicle.

21 Move to the engine compartment and locate the oil filler cap.

22 If an oil can spout is used, push the spout into the top of the oil can and pour the fresh oil through the filler opening. A funnel may also be used.

23 Pour four quarts of fresh oil into the engine. Wait a few minutes to allow the oil to drain into the pan, then check the level on the oil dipstick (see Section 4 if necessary). If the oil level is above the ADD mark, start the engine and allow the new oil to circulate.

24 Run the engine for only about a minute and then shut it off. Immediately look under the vehicle and check for leaks at the oil pan drain plug and around the oil filter. If either is leaking, tighten with a bit more force.

25 With the new oil circulated and the filter now completely full, recheck the level on the dipstick and add more oil as necessary.

26 During the first few trips after an oil change, make it a point to check frequently for leaks and proper oil level.

27 The old oil drained from the engine cannot be reused in its present state and should be disposed of. Oil reclamation centers, auto repair shops and gas stations will normally accept the oil, which can be refined and used again. After the oil has cooled it can be drained into a suitable container (capped plastic jugs, topped bottles, milk cartons, etc.) for transport to one of these disposal sites.

13 Chassis lubrication

▶ **Refer to illustrations 13.1, 13.2 and 13.6**

1 Refer to *Recommended lubricants and fluids* at the end of this Chapter to obtain the necessary grease, etc. You'll also need a grease gun (see illustration). Occasionally plugs will be installed rather than grease fittings. If so, grease fittings will have to be purchased and installed.

2 Look under the vehicle and see if grease fittings or plugs are installed (see illustration). If there are plugs, remove them and buy grease fittings, which will thread into the component. A dealer or auto parts store will be able to supply the correct fittings. Straight, as well as angled, fittings are available.

3 For easier access under the vehicle, raise it with a jack and place jackstands under the frame. Make sure it's safely supported by the stands. If the wheels are to be removed at this interval for tire rotation or brake inspection, loosen the lug nuts slightly while the vehicle is still on the ground.

4 Before beginning, force a little grease out of the nozzle to remove any dirt from the end of the gun. Wipe the nozzle clean with a rag.

5 With the grease gun and plenty of clean rags, crawl under the vehicle and begin lubricating the components.

6 Wipe the balljoint grease fitting nipple clean and push the nozzle firmly over it (see illustration). Squeeze the trigger on the grease gun to force grease into the component. The balljoints should be lubricated until the rubber seal is firm to the touch. Do not pump too much grease into the fittings as it could rupture the seal. For all other suspension and steering components, continue pumping grease into the fitting until it oozes out of the joint between the two components. If it escapes around the grease gun nozzle, the nipple is clogged or the nozzle is not completely seated on the fitting. Resecure the gun nozzle to the fitting and try again. If necessary, replace the fitting with a new one.

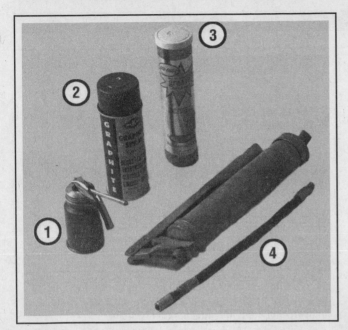

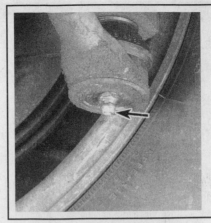

13.2 Look for all the grease fittings (arrow). If there are plugs, replace them with fittings, which are available from auto parts stores

13.1 Materials required for chassis and body lubrication

1 *Engine oil* - Light engine oil in a can like this can be used for door and hood hinges
2 *Graphite spray* - Used to lubricate lock cylinders
3 *Grease* - Grease, in a variety of types and weights, is available for use in a grease gun. Check the Specifications for your requirements
4 *Grease gun* - A common grease gun, shown here with a detachable hose and nozzle, is needed for chassis lubrication. After use, clean it thoroughly

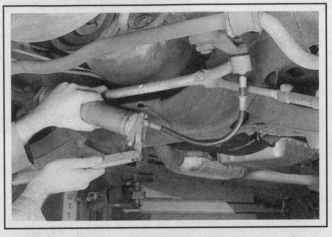

13.6 After wiping the grease fitting clean, push the nozzle firmly into place and pump the grease into the component - usually about two pumps of the gun will be sufficient

7 Wipe the excess grease from the components and the grease fitting. Repeat the procedure for the remaining fittings.

8 If equipped with a manual transmission, lubricate the shift linkage with a little multi-purpose grease. Later 4-speed transmissions also have a grease fitting so the grease gun can be used.

9 On manual transmission equipped models, lubricate the clutch linkage pivot points with clean engine oil. Lubricate the pushrod-to-fork contact points with chassis grease.

10 While you are under the vehicle, clean and lubricate the parking brake cable, along with the cable guides and levers. This can be done by smearing some of the chassis grease onto the cable and its related parts with your fingers.

11 The steering gear seldom requires the addition of lubricant, but if there is obvious leakage of grease at the seals, remove the plug or

cover and check the lubricant level. If the level is low, add the specified lubricant.

12 Open the hood and smear a little chassis grease on the hood latch mechanism. Have an assistant pull the hood release lever from inside the vehicle as you lubricate the cable at the latch.

13 Lubricate all the hinges (door, hood, etc.) with engine oil to keep them in proper working order.

14 The key lock cylinders can be lubricated with spray graphite or silicone lubricant, which is available at auto parts stores.

15 Lubricate the door weatherstripping with silicone spray. This will reduce chafing and retard wear.

14 Suspension and steering check

1 Indications of a fault in these systems are excessive play in the steering wheel before the front wheels react, excessive sway around corners, body movement over rough roads or binding at some point as the steering wheel is turned.

2 Raise the front of the vehicle periodically and visually check the suspension and steering components for wear. Because of the work to be done, make sure the vehicle cannot fall from the stands.

3 Check the wheel bearings. Do this by spinning the front wheels. Listen for any abnormal noises and watch to make sure the wheel spins true (doesn't wobble). Grab the top and bottom of the tire and pull in-and-out on it. Notice any movement which would indicate a loose

wheel bearing assembly. If the bearings are suspect, refer to Section 36 and Chapter 10 for more information.

4 From under the vehicle check for loose bolts, broken or disconnected parts and deteriorated rubber bushings on all suspension and steering components. Look for grease or fluid leaking from the steering assembly. Check the power steering hoses and connections for leaks.

5 Have an assistant turn the steering wheel from side-to-side and check the steering components for free movement, chafing and binding. If the steering doesn't react with the movement of the steering wheel, try to determine where the slack is located.

15 Exhaust system check

♦ **Refer to illustration 15.2**

1   With the engine cold (at least three hours after the vehicle has been driven), check the complete exhaust system from the manifold to the end of the tailpipe. Be careful around the catalytic converter, which may be hot even after three hours. The inspection should be done with the vehicle on a hoist to permit unrestricted access. If a hoist isn't available, raise the vehicle and support it securely on jackstands.

2   Check the exhaust pipes and connections for signs of leakage and/or corrosion indicating a potential failure (see illustration). Make sure that all brackets and hangers are in good condition and tight.

3   Inspect the underside of the body for holes, corrosion, open seams, etc. which may allow exhaust gases to enter the passenger compartment. Seal all body openings with silicone or body putty.

4   Rattles and other noises can often be traced to the exhaust system, especially the hangers, mounts and heat shields. Try to move the pipes, mufflers and catalytic converter. If the components can come in contact with the body or suspension parts, secure the exhaust system with new brackets and hangers.

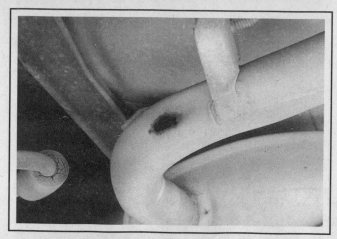

15.2 On light-colored exhaust pipes, leaks usually show up as brown or black stains - this stain around a small hole is indicative of a tailpipe needing replacement

16 Manual transmission lubricant level check

1   The manual transmission has an inspection and fill plug which must be removed to check the oil level. If the vehicle is raised to gain access to the plug, be sure to support it safely on jackstands - DO NOT crawl under a vehicle which is supported only by a jack!

2   Remove the plug from the transmission and use your little finger to reach inside the housing to feel the oil level. The level should be at or near the bottom of the plug hole.

3   If it isn't, add the recommended oil through the plug hole with a syringe or squeeze bottle.

4   Install and tighten the plug and check for leaks after the first few miles of driving.

17 Differential oil level check

♦ **Refer to illustrations 17.2 and 17.3**

1   The differential has a check/fill plug which must be removed to check the oil level. If the vehicle is raised to gain access to the plug, be sure to support it safely on jackstands - DO NOT crawl under the vehicle when it's supported only by the jack.

2   Remove the oil check/fill plug from the side of the differential (see illustration).

3   The oil level should be at the bottom of the plug opening (see illustration). If not, use a syringe to add the recommended lubricant until it just starts to run out of the opening. On some models a tag is located in the area of the plug which gives information regarding lubricant type, particularly on models equipped with a limited slip differential.

4   Install the plug and tighten it securely.

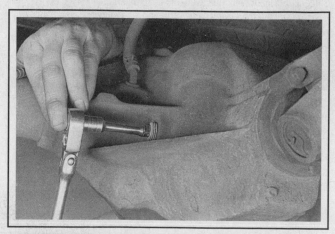

17.2 Use a ratchet or breaker bar and a 3/8-inch drive extension to remove the differential check/fill plug

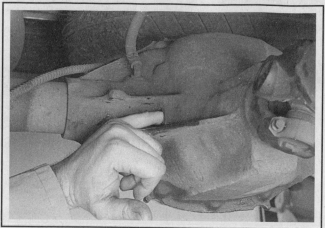

17.3 Use your little finger as a dipstick to make sure the differential oil level is even with the bottom of the opening

18 Tire rotation

⯈ **Refer to illustration 18.2**

1   The tires should be rotated at the specified intervals and whenever uneven wear is noticed.

2   Refer to the accompanying illustration for the preferred tire rotation pattern.

3   Refer to the information in *Jacking and towing* at the front of this manual for the proper procedures to follow when raising the vehicle and changing a tire. If the brakes are to be checked, don't apply the parking brake as stated. Make sure the tires are blocked to prevent the vehicle from rolling as it's raised.

4   Preferably, the entire vehicle should be raised at the same time. This can be done on a hoist or by jacking up each corner and then lowering the vehicle onto jackstands placed under the frame rails. Always use four jackstands and make sure the vehicle is safely supported.

5   After rotation, check and adjust the tire pressures as necessary and be sure to check the lug nut tightness.

6   For additional information on the wheels and tires, refer to Chapter 10.

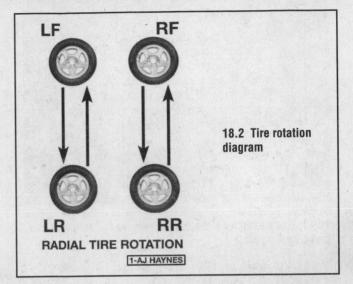

18.2 Tire rotation diagram

19 Brake check

⯈ **Refer to illustrations 19.4, 19.6, 19.11, 19.12 and 19.14**

❋ WARNING:

Brake system dust is hazardous to your health. DO NOT blow it out with compressed air and DO NOT inhale it. DO NOT use gasoline or solvents to remove the dust. Use brake system cleaner or denatured alcohol only.

➡**Note: For detailed photographs of the brake system, refer to Chapter 9.**

1   In addition to the specified intervals, the brakes should be inspected every time the wheels are removed or whenever a defect is suspected.

2   To check the brakes, raise the vehicle and place it securely on jackstands. Remove the wheels (see *Jacking and towing* at the front of the manual, if necessary).

DISC BRAKES

3   Disc brakes are used on the front wheels. Extensive rotor damage can occur if the pads are not replaced when needed.

4   These vehicles are equipped with a wear sensor attached to the inner pad. This is a small, bent piece of metal which is visible from the inner side of the brake caliper. When the pad wears to the specified limit, the metal sensor rubs against the rotor and makes a squealing sound (see illustration).

5   The disc brake calipers, which contain the pads, are visible with the wheels removed. There is an outer pad and an inner pad in each caliper. All pads should be inspected.

6   Each caliper has a "window" to inspect the pads. Check the thickness of the pad lining by looking into the caliper at each end and down through the inspection window at the top of the housing (see illustration). If the wear sensor is very close to the rotor or the pad material has worn to about 1/8-inch or less, the pads should be replaced.

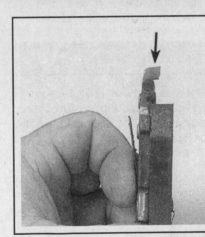

19.4 The disc brake pads have built-in wear indicators that contact the rotor and emit a squealing sound when the pads have worn to their limit

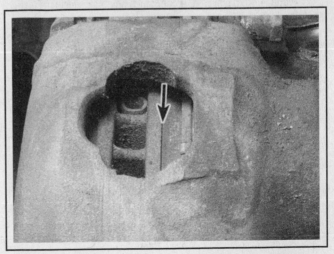

19.6 The front disc brake pads can be checked easily by looking through the inspection window in each caliper

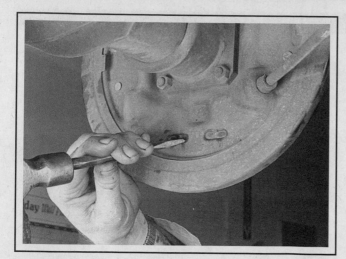

19.11 Use a hammer and chisel to remove the plug from the brake backing plate

7 If you're unsure about the exact thickness of the remaining lining material, remove the pads for further inspection or replacement (refer to Chapter 9).

8 Before installing the wheels, check for leakage and/or damage (cracks, splitting, etc.) around the brake hose connections. Replace the hose or fittings as necessary, referring to Chapter 9.

9 Check the condition of the rotor. Look for score marks, deep scratches and burned spots. If these conditions exist, the hub/rotor assembly should be removed for servicing (see Section 36).

DRUM BRAKES

10 On rear brakes, remove the drum by pulling it off the axle and brake assembly. If this proves difficult, make sure the parking brake is released, then squirt penetrating oil around the center hub areas. Allow the oil to soak in and try to pull the drum off again.

11 If the drum still cannot be pulled off, the brake shoes will have to be adjusted. This is done by first removing the plug from the backing plate with a hammer and chisel (see illustration).

12 With the plug removed, push the lever off the star wheel and then use a small screwdriver to turn the star wheel, which will move the brake shoes away from the drum (see illustration).

13 With the drum removed, do not touch any brake dust (see the **Warning** at the beginning of this Section).

14 Note the thickness of the lining material on both the front and rear brake shoes. If the material has worn away to within 1/16-inch of the recessed rivets or metal backing, the shoes should be replaced (see illustration). The shoes should also be replaced if they're cracked, glazed (shiny surface) or contaminated with brake fluid.

15 Make sure that all the brake assembly springs are connected and in good condition.

16 Check the brake components for any signs of fluid leakage. With your finger, carefully pry back the rubber cups on the wheel cylinders located at the top of the brake shoes. Any leakage is an indication that the wheel cylinders should be overhauled immediately (see Chapter 9). Also check brake hoses and connections for signs of leakage.

17 Wipe the inside of the drum with a clean rag and brake cleaner or denatured alcohol. Again, be careful not to breath the dangerous brake dust.

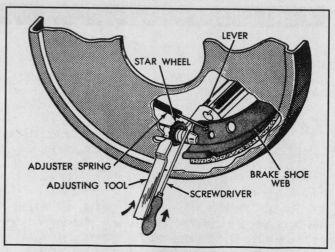

19.12 Use a screwdriver and adjusting tool to back off the rear brake shoes if necessary so the brake drum can be removed

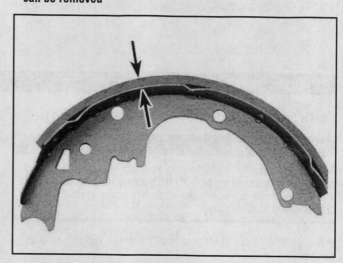

19.14 The brake shoe lining thickness is measured from the outer surface of the lining to the metal shoe

18 Check the inside of the drum for cracks, score marks, deep scratches and hard spots, which will appear as small discolorations. If these imperfections cannot be removed with fine emery cloth, the drum must be taken to a machine shop equipped to turn the drums.

19 If after the inspection process all parts are in good working condition, reinstall the brake drum (using a metal or rubber plug if the knockout was removed).

20 Install the wheels and lower the vehicle. Tighten the lug nuts to the torque listed in this Chapter's Specifications.

PARKING BRAKE

21 The parking brake operates from a foot pedal and locks the rear brake system. The easiest, and perhaps most obvious method of periodically checking the operation of the parking brake assembly is to park the vehicle on a steep hill with the parking brake set and the transmission in Neutral. If the parking brake cannot prevent the vehicle from rolling, it's in need of adjustment (see Chapter 9).

20 Fuel system check

✳✳ WARNING:

There are certain precautions to take when inspecting or servicing the fuel system components. Work in a well ventilated area and don't allow open flames (cigarettes, appliance pilot lights, etc.) in the work area. Mop up spills immediately and don't store fuel soaked rags where they could ignite. On fuel injection equipped models the fuel system is under pressure. No components should be disconnected until the pressure has been relieved (see Chapter 4).

1   On most models the main fuel tank is located under the left side of the vehicle.

2   The fuel system is most easily checked with the vehicle raised on a hoist so the components underneath the vehicle are readily visible and accessible.

3   If the smell of gasoline is noticed while driving or after the vehicle has been in the sun, the system should be thoroughly inspected immediately.

4   Remove the gas tank cap and check for damage, corrosion and an unbroken sealing imprint on the gasket. Replace the cap with a new one if necessary.

5   With the vehicle raised, check the gas tank and filler neck for punctures, cracks and other damage. The connection between the filler neck and the tank is especially critical. Sometimes a rubber filler neck will leak due to loose clamps or deteriorated rubber, problems a home mechanic can usually rectify.

✳✳ WARNING:

Do not, under any circumstances, try to repair a fuel tank yourself (except rubber components). A welding torch or any open flame can easily cause the fuel vapors to explode if the proper precautions are not taken!

6   Carefully check all rubber hoses and metal lines leading away from the fuel tank. Look for loose connections, deteriorated hoses, crimped lines and other damage. Follow the lines to the front of the vehicle, carefully inspecting them all the way. Repair or replace damaged sections as necessary.

7   If a fuel odor is still evident after the inspection, refer to Section 39.

21 Carburetor choke check (1985 models)

▶ **Refer to illustration 21.3**

1   The choke operates only when the engine is cold, so this check should be performed before the engine has been started for the day.

2   Remove the engine cover (see Chapter 11) and take off the top plate of the air cleaner assembly (see Chapter 4). It's usually held in place by a wing nut at the center. If any vacuum hoses must be disconnected, make sure you tag the hoses for reinstallation in their original positions. Place the top plate and wing nut aside, out of the way of moving engine components.

3   Look at the center of the air cleaner housing. You will notice a flat plate at the carburetor opening (see illustration).

4   Press the accelerator pedal to the floor. The plate should close completely. Start the engine while you watch the plate at the carburetor. Don't position your face near the carburetor, as the engine could backfire, causing serious burns. When the engine starts, the choke plate should open slightly.

5   Allow the engine to continue running at an idle speed. As the engine warms up to operating temperature, the plate should slowly open, allowing more air to enter through the top of the carburetor.

6   After a few minutes, the choke plate should be fully open to the vertical position. Blip the throttle to make sure the fast idle cam disengages.

7   You'll notice that the engine speed corresponds with the plate opening. With the plate fully closed, the engine should run at a fast idle

21.3 The carburetor choke plate is visible after removing the air cleaner top plate

speed. As the plate opens and the throttle is moved to disengage the fast idle cam, the engine speed will decrease.

8   Refer to Chapter 4 for specific information on adjusting and servicing the choke components.

22 Carburetor/throttle body mounting nut torque check (1986 to 1995 VIN Z models)

1 The carburetor or TBI unit is attached to the top of the intake manifold by several bolts or nuts. These fasteners can sometimes work loose from vibration and temperature changes during normal engine operation and cause a vacuum leak.

2 Remove the engine cover (Chapter 11).

3 If you suspect that a vacuum leak exists at the bottom of the carburetor or throttle body, obtain a length of hose. Start the engine and place one end of the hose next to your ear as you probe around the base with the other end. You will hear a hissing sound if a leak exists (be careful of hot or moving engine components).

4 Remove the air cleaner assembly, tagging each hose to be dis-connected with a piece of numbered tape to make reassembly easier.

5 Locate the mounting nuts or bolts at the base of the carburetor or throttle body. Decide what special tools or adapters will be necessary, if any, to tighten the fasteners.

6 Tighten the nuts or bolts to the specified torque. Don't over-tighten them, as the threads could strip.

7 If, after the nuts or bolts are properly tightened, a vacuum leak still exists, the carburetor or throttle body must be removed and a new gasket installed. See Chapter 4 for more information.

8 After tightening the fasteners, reinstall the air cleaner and return all hoses to their original positions. Install the engine cover.

23 Accelerator linkage inspection

1 Inspect the accelerator linkage for damage and missing parts and for binding and interference when the accelerator pedal is depressed.

2 Lubricate the various linkage pivot points with engine oil.

24 Thermostatic air cleaner check (carbureted and TBI models)

◆ **Refer to illustrations 24.5 and 24.6**

1 Some engines are equipped with a thermostatically controlled air cleaner which draws air to the carburetor from different locations, depending on engine temperature.

2 This is a visual check. If access is limited, a small mirror may have to be used.

3 Remove the engine cover and locate the damper door inside the air cleaner assembly. It's inside the long snorkel of the metal air cleaner housing.

4 If there is a flexible air duct attached to the end of the snorkel, leading to an area behind the grille, disconnect it at the snorkel. This will enable you to look through the end of the snorkel and see the damper inside.

5 The check should be done when the engine is cold. Start the engine and look through the snorkel at the damper, which should move to a closed position. With the damper closed, air cannot enter through the end of the snorkel, but instead enters the air cleaner through the flexible duct attached to the exhaust manifold and the heat stove pas-sage (see illustration).

6 As the engine warms up to operating temperature, the damper should open to allow air through the snorkel end (see illustration). Depending on outside temperature, this may take 10-to-15 minutes. To speed up this check you can reconnect the snorkel air duct, drive the vehicle, then check to see if the damper is completely open.

7 If the thermo-controlled air cleaner isn't operating properly see Chapter 6 for more information.

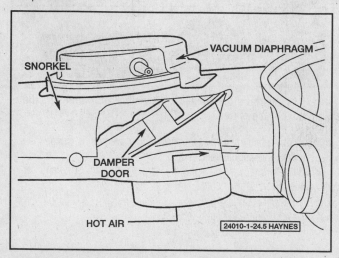

24.5 When the engine is cold, the damper door closes off the snorkel passage, allowing air warmed by the exhaust manifold to enter the carburetor

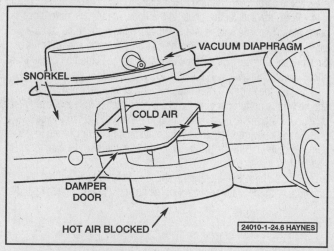

24.6 As the engine warms up, the damper door moves down to close off the heat stove passage and open the snorkel passage so outside air can enter the carburetor

25 Drivebelt check, adjustment and replacement

▶ Refer to illustrations 25.3a, 25.3b and 25.4

➡Note: 1987 and later models are equipped with one serpentine drivebelt that runs all engine accessories. The belt tension is automatically controlled - no check or adjustment is required. On later models, a raised pointer on the drivebelt tensioner casting will align with an indentation in the drivebelt tensioner. Replace the belt if the pointer is past the raised index mark to the left of the indentation.

1   The drivebelts, or V-belts as they are often called, are located at the front of the engine and play an important role in the overall operation of the engine and accessories. Due to their function and material makeup, the belts are prone to failure after a period of time and should be inspected and adjusted periodically to prevent major engine damage.

2   The number of belts used on a particular vehicle depends on the accessories installed. Drivebelts are used to turn the alternator, power steering pump, water pump and air conditioning compressor. Depending on the pulley arrangement, more than one of these components may be driven by a single belt.

CHECK

3   With the engine off, locate the drivebelt(s) at the front of the engine. Using your fingers (and a flashlight, if necessary), move along the belts checking for cracks and separation of the belt plies. Also check for fraying and glazing, which gives the belt a shiny appearance (see illustrations). Both sides of each belt should be inspected, which means you will have to twist the belt to check the underside. Check the pulleys for nicks, cracks, distortion and corrosion.

4   On 1985 and 1986 models, the tension of each belt is checked by pushing on it at a distance halfway between the pulleys. Push firmly with your thumb and see how much the belt moves (deflects) (see illustration). A rule of thumb is that if the distance from pulley center-to-pulley center is between 7 and 11-inches, the belt should deflect 1/4-inch. If the belt travels between pulleys spaced 12-to-16 inches apart, the belt should deflect 1/2-inch.

5   If adjustment is needed, either to make the belt tighter or looser, it's done by moving the belt-driven accessory on the bracket.

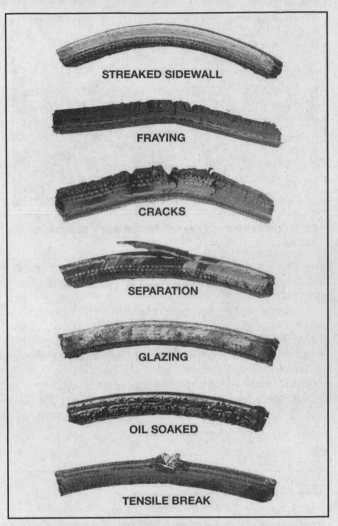

25.3a Here are some of the more common problems associated with drivebelts (check the belts very carefully to prevent an untimely breakdown)

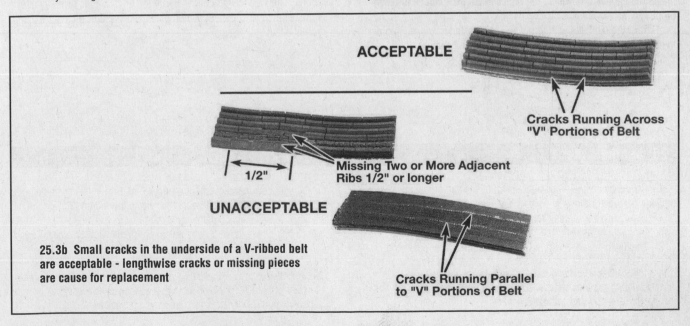

25.3b Small cracks in the underside of a V-ribbed belt are acceptable - lengthwise cracks or missing pieces are cause for replacement

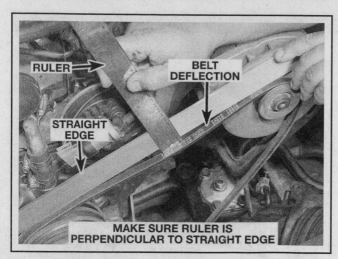

25.4 Drivebelt tension can be checked with a straightedge and ruler

6   For each component there will be an adjusting bolt and a pivot bolt. Both bolts must be loosened slightly to enable you to move the component.

7   After the two bolts have been loosened, move the component away from the engine to tighten the belt or toward the engine to loosen the belt. Hold the accessory in position and check the belt tension. If it's correct, tighten the two bolts until just snug, then recheck the tension. If the tension is all right, tighten the bolts.

8   It will often be necessary to use some sort of pry bar to move the accessory while the belt is adjusted. If this must be done to gain the proper leverage, be very careful not to damage the component being moved or the part being pried against.

REPLACEMENT

1985 and 1986 models

9   To replace a belt, follow the above procedures for drivebelt adjustment but slip the belt off the pulleys and remove it. Since belts tend to wear out more or less at the same time, it's a good idea to replace all of them at the same time. Mark each belt and the corresponding pulley grooves so the replacement belts can be installed properly.

10   Take the old belts with you when purchasing new ones in order to make a direct comparison for length, width and design.

11   Adjust the belts as described earlier in this Section.

25.13 Using a ratchet to relieve the drivebelt tension (fan shroud, fan and fan clutch removed for clarity) 1998 CSFI engine shown

1987 and later models

▸ **Refer to illustration 25.13**

12   Remove the air cleaner and the engine air intake duct assembly (see Chapter 4).

13   The serpentine drivebelt tension is maintained by a spring loaded tensioner. To replace the drivebelt, the tension must be relieved. To accomplish this a 1/2 inch or 3/8 inch square hole is provided in the tensioner arm so a breaker bar or ratchet can be used to move the tensioner away from the drivebelt (see illustration).

➡ **Note: Some tensioners do not have a square hole provided in the arm. If this is the case, use a socket on the tensioner pulley bolt to move the tensioner.**

14   Using the appropriate breaker bar or ratchet, relieve the tension on the drivebelt and carefully remove it from the pulleys and then release the tensioner. Withdraw the belt from the vehicle.

15   Route the new drivebelt onto the various pulleys according to the drivebelt routing schematic (usually found on the power steering pump reservoir), again rotating the tensioner to allow the belt to be installed, then releasing the tensioner. Make sure the drivebelt fits properly into the pulley grooves - it must be completely engaged or premature damage will occur.

26 Seatbelt check

1   Check the seatbelts, buckles, latch plates and guide loops for any obvious damage or signs of wear.

2   Make sure the seatbelt reminder light comes on when the key is turned on.

3   The seatbelts are designed to lock up during a sudden stop or impact, yet allow free movement during normal driving. The retractors should hold the belt against your chest while driving and rewind the belt when the buckle is unlatched.

4   If any of the above checks reveal problems with the seatbelt system, replace parts as necessary.

27 Park/Neutral switch check

❋❋ WARNING:

During the following checks there is a chance that the vehicle could lunge forward, possibly causing damage or injuries. Allow plenty of room around the vehicle, apply the parking brake firmly and hold down the regular brake pedal during the checks.

1   These models are equipped with a starter/Neutral safety switch which prevents the engine from starting unless the clutch pedal is depressed (manual transmission) or the shift lever is in Neutral or Park (automatic transmission).

2   On automatic transmission vehicles, try to start the vehicle in each gear. The engine should crank only in Park or Neutral. For replacement procedures, refer to Chapter 7B.

3   If equipped with a manual transmission, place the shift lever in Neutral. The engine should crank only with the clutch pedal depressed. For replacement procedures, refer to Chapter 8.

4   Make sure the steering column lock allows the key to go into the Lock position only when the shift lever is in Park (automatic transmission) or Reverse (manual transmission).

5   The ignition key should come out only in the Lock position.

28 Spare tire and jack check

1   Check the spare tire to make sure it's securely fastened so it cannot come loose when the vehicle is in motion.

2   Make sure the jack and components are secured in place. Lubricate the jack threads with engine oil after each use.

29 Idle speed check and adjustment

1   The idle speed on these models is controlled by the on-board computer via the Idle Air Control valve (IAC) and is not adjustable by conventional methods. If adjustment is required, take the vehicle to your local dealer service department or other qualified repair shop.

2   For more information on the IAC, refer to Chapter 4.

30 Fuel filter replacement

❋❋ WARNING:

Gasoline is extremely flammable, so take extra precautions when you work on any part of the fuel system. Don't smoke or allow open flames or bare light bulbs near the work area, and don't work in a garage where a gas-type appliance (such as a water heater or clothes dryer) is present. Since gasoline is carcinogenic, wear latex gloves when there is a possibility of being exposed to fuel, and if you spill any on your skin, rinse it off immediately with soap and water. Mop up any spills immediately and do not store fuel-soaked rags where they could ignite. When you perform any kind of work on the fuel system, wear safety glasses and have a Class B type fire extinguisher on hand.

CARBURETOR EQUIPPED VEHICLES (1985 ONLY)

▶ **Refer to illustrations 30.6 and 30.8**

1   On these models the fuel filter is located inside the fuel inlet nut at the carburetor. It's made of either pleated paper or porous bronze and cannot be cleaned or reused.

2   The job should be done with the engine cold (after sitting at least three hours) and the engine cover removed (see Chapter 11). The necessary tools include open-end wrenches to fit the fuel line nuts. Flare nut wrenches (which wrap around the nut) should be used if available. In addition, you have to obtain the replacement filter (make sure it's for your specific vehicle and engine) and some clean rags.

3   Remove the air cleaner assembly. If vacuum hoses must be disconnected, be sure to note their positions and/or tag them to ensure that they are reinstalled correctly.

30.6 Hold the large fuel inlet nut with a wrench to prevent it from turning while loosening the fuel line fitting with a flare-nut wrench

4   Follow the fuel line from the fuel pump to the point where it enters the carburetor. In most cases the fuel line will be metal all the way from the fuel pump to the carburetor.

5   Place some rags under the fuel inlet fittings to catch spilled fuel as the fittings are disconnected.

6   With the proper size wrench, hold the fuel inlet nut immediately next to the carburetor body. Now loosen the fitting at the end of the metal fuel line. Make sure the fuel inlet nut next to the carburetor is held securely while the fuel line is disconnected (see illustration).

30.8 Carburetor-mounted fuel filter component layout

7   After the fuel line is disconnected, move it aside for better access to the inlet nut. Don't kink the fuel line.

8   Unscrew the fuel inlet nut, which was previously held steady. As this fitting is drawn away from the carburetor body, be careful not to lose the thin washer-type gasket on the nut or the spring, located behind the fuel filter. Also pay close attention to how the filter is installed (see illustration).

9   Compare the old filter with the new one to make sure they're the same length and design.

10 Reinstall the spring in the carburetor body.

11 Place the filter in position (a gasket is usually supplied with the new filter) and tighten the nut. Make sure it's not cross-threaded. Tighten it securely, but be careful not to overtighten it as the threads can strip easily, causing fuel leaks. Reconnect the fuel line to the fuel inlet nut, again using caution to avoid cross-threading the nut. Use a back-up wrench on the fuel inlet nut while tightening the fuel line fitting.

12 Start the engine and check carefully for leaks. If the fuel line fitting leaks, disconnect it and check for stripped or damaged threads. If the fuel line fitting has stripped threads, remove the entire line and have a repair shop install a new fitting. If the threads look all right, purchase

30.17 Use two wrenches to detach the fuel lines from the filter

some thread sealing tape and wrap the threads with it. Inlet nut repair kits are available at most auto parts stores to overcome leaking at the fuel inlet nut.

FUEL INJECTED VEHICLES

▶ **Refer to illustration 30.17**

13 The fuel filter on EFI models is an in-line type and located under the vehicle along the left side frame rail near the engine.

14 Perform the fuel pressure relief procedure (see Chapter 4).

15 Raise the vehicle and support it securely on jackstands.

16 Place a container, newspapers or rags under the fuel filter.

17 Using two wrenches, one to steady the filter and the other to unscrew the fuel line nut, disconnect the fuel lines from the filter (see illustration). If available, it's a good idea to use a flare nut wrench to avoid rounding off the wrenching flats on the fuel line nuts.

18 Detach the filter from the bracket.

19 Installation is the reverse of removal; however, make sure the flow indication arrow on the filter is pointing towards the engine, not the fuel tank. After installation, start the engine and check for leaks.

31 Air filter and PCV filter replacement

▶ **Refer to illustrations 31.3 and 31.5**

1   At the specified intervals, the air filter and (if equipped) PCV filter should be replaced with new ones. The engine air cleaner also supplies filtered air to the PCV system.

CARBURETED AND TBI MODELS

➡**Note: These models can be identified by a Z in the eighth position of the Vehicle Identification Number (VIN) number.**

2   Remove the engine cover (Chapter 11).

3   The filter is located on top of the carburetor or Throttle Body Injection (TBI) unit and is replaced by unscrewing the wing nut from the top of the filter housing and lifting off the cover (see illustration).

4   While the top plate is off, be careful not to drop anything down into the carburetor, TBI or air cleaner assembly.

5   Lift the air filter element out of the housing (see illustration) and wipe out the inside of the air cleaner housing with a clean rag.

6   Place the new filter in the air cleaner housing. Make sure it seats

properly in the bottom of the housing.

7   The PCV filter is also located inside the air cleaner housing on some models. Remove the top plate and air filter as previously described, then locate the PCV filter on the inside of the housing.

8   Remove the old filter.

9   Install the new PCV filter and the new air filter.

10 Install the top plate and any hoses which were disconnected.

11 Install the engine cover.

CENTRAL MULTIPORT FUEL INJECTION (CMFI) AND CENTRAL SEQUENTIAL FUEL INJECTION (CSFI) MODELS

➡**Note: These models can be identified by a W or X in the eighth position of the Vehicle Identification Number (VIN) number.**

12 Loosen the clamp screw on the intake air duct. Disconnect the intake air duct from the air filter cover.

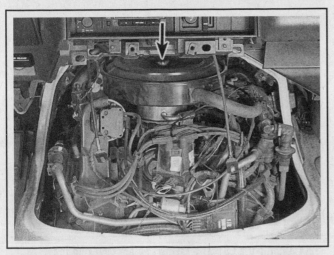

31.3 With the engine cover removed, the air cleaner wing nut (arrow) is visible on top of the air cleaner assembly

13 Detach the clips securing the air filter cover and remove the cover (see illustration).

14 Lift the filter out of the housing, noting the direction in which it's installed (see illustration).

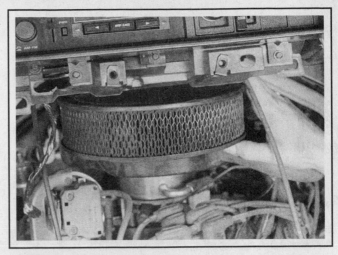

31.5 Lift the air cleaner element out of the housing; on V6 models it's a tight fit

15 Installation is the reverse of removal. Note that many replacement filters are marked with an arrow to show the direction of airflow through the filter.

31.13 Detach the cover clips (VIN W models)

31.14 Lift the filter out and note how it's installed (VIN W models)

32 Ignition timing check and adjustment - 1985 to 1994 vehicles only

▶ Refer to illustrations 32.4 and 32.5

➡ **Note:** *If the information in this Section differs from the Vehicle Emission Control Information label in the engine compartment of your vehicle, the label should be considered correct.*

1 The engine must be at normal operating temperature and the air conditioner must be Off. Make sure the idle speed is correct.

2 Apply the parking brake and block the wheels to prevent movement of the vehicle. The transmission must be in Park (automatic) or Neutral (manual).

3 If the SERVICE ENGINE SOON light is on, don't proceed with the ignition timing check (see Chapter 6 for more information).

4 The Electronic Spark Timing (EST) system must be bypassed prior to checking the ignition timing. Remove the engine cover (Chapter 11). Locate the single tan wire with a black stripe that's connected to the distributor and unplug the connector (see illustration). Don't unplug the 4-wire harness connector at the distributor.

5 Locate the timing marks at the front of the engine (they should be visible from above after the hood is opened) (see illustration). The crankshaft pulley or vibration damper has a notch or groove in it and a

small metal plate with notches and numbers is attached to the timing cover. Clean the plate with solvent so the numbers are visible.

6   Use chalk or white paint to mark the notch or groove in the pulley/vibration damper.

7   Highlight the notch or point on the timing plate that corresponds to the ignition timing specification on the Emission Control Information label (0-degrees or TDC will most likely be specified).

8   Hook up the timing light by following the manufacturer's instructions (an inductive pick-up timing light is preferred). Generally, the power leads are attached to the battery terminals and the pick-up lead is attached to the number one spark plug wire. On the four-cylinder engine, the number one spark plug is the very front one. On the V6 engine, the number one spark plug is the front one on the driver's side of the engine.

❊❊ CAUTION:

If an inductive pick-up timing light isn't available, don't puncture the spark plug wire to attach the timing light pick-up lead. Instead, use an adapter between the spark plug and plug wire. If the insulation on the plug wire is damaged, the secondary voltage will jump to ground at the damaged point and the engine will misfire.

9   Make sure the timing light wires are routed away from the drivebelts and fan, then start the engine.

10   Allow the idle speed to stabilize, then point the flashing timing light at the timing marks - be very careful of moving engine components!

11   The mark on the pulley/vibration damper will appear stationary. If

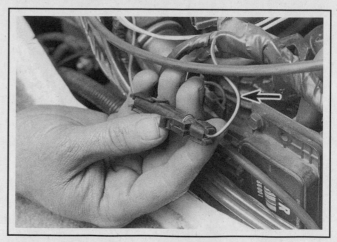

32.4   Unplug the EST wire (arrow) before checking the ignition timing

it's aligned with the specified point on the timing plate, the ignition timing is correct.

12   If the marks aren't aligned, adjustment is required. Loosen the distributor hold-down bolt and turn the distributor very slowly until the marks are aligned. Since access to the bolt is tight, a special distributor wrench may be needed.

13   Tighten the bolt and recheck the timing.

14   Turn off the engine and remove the timing light (and adapter, if used).

15   Reconnect the EST wire harness connector, then clear any PCM trouble codes set during the ignition timing procedure (see Chapter 6).

32.5   The ignition timing marks are located at the front of the engine

33   Automatic transmission fluid and filter change

▸ Refer to illustrations 33.7, 33.10a and 33.10b

➡Note: **Lubricants and fluids used on your vehicle are hazardous waste and must be disposed of properly. When disposal is required, place the waste into a sealable container and deliver it to a service station, auto parts store or other facility which accepts hazardous waste for recycling.**

1   At the specified time intervals, the transmission fluid should be drained and replaced. Since the fluid will remain hot long after driving, perform this procedure only after the engine has cooled down completely.

2   Before beginning work, purchase the specified transmission fluid (see *Recommended lubricants and fluids* at the end of this Chapter) and a new filter.

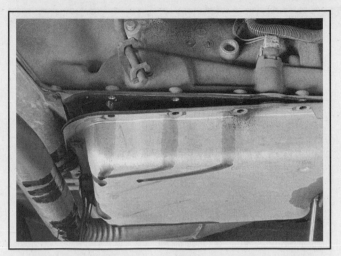

33.7 With the rear bolts in place but loose, pull the front of the pan down to let the transmission fluid drain

33.10a Rotate the filter out of the retaining clip and then lower it from the transmission

3   Other tools necessary for this job include jackstands to support the vehicle in a raised position, a drain pan capable of holding at least eight pints, newspapers and clean rags.

4   Raise the vehicle and support it securely on jackstands.

5   With a drain pan in place, remove the front and side pan mounting bolts.

6   Loosen the rear pan bolts approximately four turns.

7   Carefully pry the transmission pan loose with a screwdriver, allowing the fluid to drain (see illustration).

8   Remove the remaining bolts, pan and gasket. Carefully clean the gasket surface of the transmission to remove all traces of the old gasket and sealant.

9   Drain the fluid from the transmission pan, clean it with solvent and dry it with compressed air. If there is a magnet inside the oil pan, be sure to clean it and place it back in the same location.

10 Remove the filter and O-ring from the mount inside the transmission (see illustrations).

11 Install a new filter and O-ring.

12 Make sure the gasket surface on the transmission pan is clean, then install a new gasket. Put the pan in place against the transmission and, working around the pan, tighten each bolt a little at a time until the final torque listed in this Chapter's Specifications is reached.

13 Lower the vehicle and add the specified amount of automatic transmission fluid through the filler tube (see Section 6).

33.10b Reach up into the filter opening with your finger to retrieve the O-ring if it doesn't come out with the filter

14 With the transmission in Park and the parking brake set, run the engine at a fast idle, but don't race it.

15 Move the gear selector through each range and back to Park. Check the fluid level.

16 Check under the vehicle for leaks during the first few trips.

34   Manual transmission lubricant change

➡**Note: Lubricants and fluids used on your vehicle are hazardous waste and must be disposed of properly. When disposal is required, place the waste into a sealable container and deliver it to a service station, auto parts store or other facility which accepts hazardous waste for recycling.**

1   Raise the vehicle and support it securely on jackstands.

2   Move a drain pan, rags, newspapers and wrenches under the transmission.

3   Remove the fill plug from the side of the transmission case. Remove the transmission drain plug at the bottom of the case (see illustration 16.1) and allow the oil to drain into the pan.

4   After the oil has drained completely, reinstall the plug and tighten it securely.

5   Using a hand pump, syringe or funnel, fill the transmission with the correct amount of the specified lubricant. Reinstall the fill plug and tighten it securely.

6   Lower the vehicle.

7   Drive the vehicle for a short distance then check the drain and fill plugs for leakage.

35 Differential oil change

♦ **Refer to illustrations 35.6a, 35.6b, 35.6c and 35.8**

➡ **Note: Lubricants and fluids used on your vehicle are hazardous waste and must be disposed of properly. When disposal is required, place the waste into a sealable container and deliver it to a service station, auto parts store or other facility which accepts hazardous waste for recycling.**

1   Some differentials can be drained by removing the drain plug, while on others it's necessary to remove the cover plate on the differential housing. As an alternative, a hand suction pump can be used to remove the differential lubricant through the filler hole. If there is no drain plug and a suction pump isn't available, be sure to obtain a new gasket at the same time the gear lubricant is purchased.

2   Raise the vehicle and support it securely on jackstands. Move a drain pan, rags, newspapers and wrenches under the vehicle.

3   Remove the fill plug from the differential.

4   If equipped with a drain plug, remove the plug and allow the differential oil to drain completely. After the oil has drained, install the plug and tighten it securely.

5   If a suction pump is being used, insert the flexible hose. Work the hose down to the bottom of the differential housing and pump the oil out.

6   If the differential is being drained by removing the cover plate, remove the bolts on the lower half of the plate (see illustration). Loosen the bolts on the upper half and use them to keep the cover loosely attached (see illustration). Allow the oil to drain into the pan, then completely remove the cover (see illustration).

7   Using a lint-free rag, clean the inside of the cover and the accessible areas of the differential housing. As this is done, check for chipped gears and metal particles in the lubricant, indicating that the differential should be more thoroughly inspected and/or repaired.

8   Thoroughly clean the gasket mating surfaces of the differential housing and the cover plate. Use a gasket scraper or putty knife to remove all traces of the old gasket (see illustration).

9   Apply a thin layer of RTV sealant to the cover flange and then press a new gasket into position on the cover. Make sure the bolt holes align properly.

10   Place the cover on the differential housing and install the bolts. Tighten the bolts securely.

11   On all models, use a hand pump, syringe or funnel to fill the differential housing with the specified lubricant until it's level with the bottom of the plug hole.

12   Install the filler plug and tighten it securely.

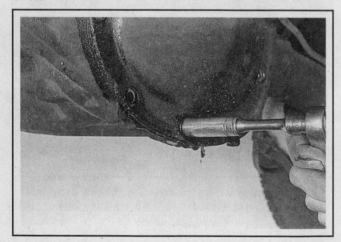

35.6a Remove the bolts from the lower edge of the cover . . .

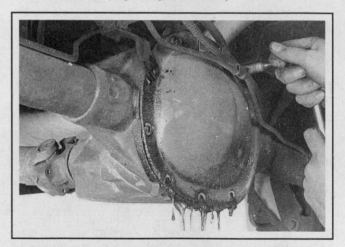

35.6b . . . then loosen the top bolts and let the oil drain

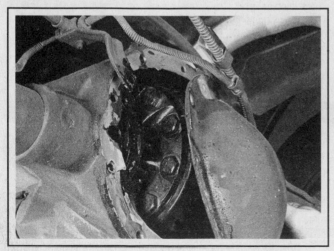

35.6c After the oil has drained, remove the cover

35.8 Carefully scrape the old gasket material off to ensure a leak-free seal with the new gasket

36 Front wheel bearing check, repack and adjustment

▶ Refer to illustrations 36.1, 36.6, 36.7, 36.8, 36.11 and 36.15

➡Note: This procedure applies to 2002 and earlier models only.

1 In most cases the front wheel bearings will not need servicing until the brake pads are changed. However, the bearings should be checked whenever the front of the vehicle is raised for any reason. Several items, including a torque wrench and special grease, are required for this procedure (see illustration).

2 With the vehicle securely supported on jackstands, spin each wheel and check for noise, rolling resistance and free play.

3 Grasp the top of each tire with one hand and the bottom with the other. Move the wheel in-and-out on the spindle. If there's any noticeable movement, the bearings should be checked and then repacked with grease or replaced if necessary.

4 Remove the wheel.

5 Remove the brake caliper (see Chapter 9) and hang it out of the way on a piece of wire.

6 Pry the dust cap out of the hub using a screwdriver or hammer and chisel (see illustration).

7 Straighten the bent ends of the cotter pin, then pull the cotter pin out of the locking nut (see illustration). Discard the cotter pin and use a new one during reassembly.

8 Remove the spindle nut and washer from the end of the spindle (see illustration).

9 Pull the hub assembly out slightly, then push it back into its original position. This should force the outer bearing off the spindle enough so it can be removed.

10 Pull the hub off the spindle.

11 Use a screwdriver to pry the seal out of the rear of the hub (see illustration). As this is done, note how the seal is installed.

12 Remove the inner wheel bearing from the hub.

13 Use solvent to remove all traces of the old grease from the bearings, hub and spindle. A small brush may prove helpful; however make sure no bristles from the brush embed themselves inside the bearing rollers. Allow the parts to air dry.

14 Carefully inspect the bearings for cracks, heat discoloration, worn rollers, etc. Check the bearing races inside the hub for wear and damage. If the bearing races are defective, use a soft metal drift to drive out the old races from the back side - tap them out slowly and evenly - remove and install one race at a time. Once the old race is removed, clean any remaining grease from inside the hub. Do not discard the old race, you'll need it for installation of the new race. Start the new race

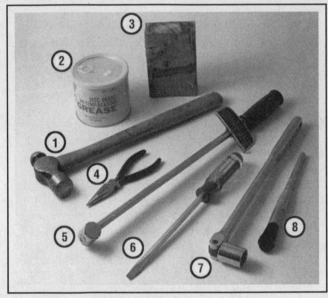

36.1 Tools and materials needed for front wheel bearing maintenance

1 **Hammer** - A common hammer will do just fine
2 **Grease** - High-temperature grease that is formulated specially for front wheel bearings should be used
3 **Wood block** - If you have a scrap piece of 2x4, it can be used to drive the new seal into the hub
4 **Needle-nose pliers** - Used to straighten and remove the cotter pin in the spindle
5 **Torque wrench** - This is very important in this procedure; if the bearing is too tight, the wheel won't turn freely - if it's too loose, the wheel will "wobble" on the spindle. Either way, it could mean extensive damage
6 **Screwdriver** - Used to remove the seal from the hub (a long screwdriver is preferred)
7 **Socket/breaker bar** - Needed to loosen the nut on the spindle if it's extremely tight
8 **Brush** - Together with some clean solvent, this will be used to remove old grease from the hub and spindle

36.6 Dislodge the dust cap by working around the outer circumference with a hammer and chisel

36.7 Remove the cotter pin and discard it - use a new one when the hub is reinstalled

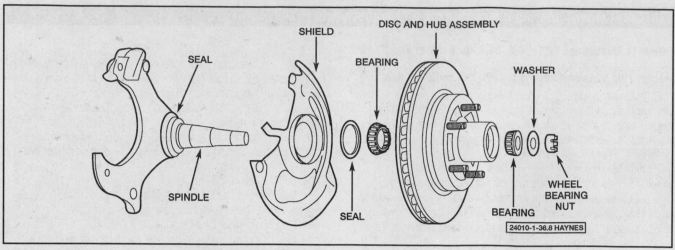

36.8 Front wheel hub and bearing components (2002 and earlier models) - exploded view

squarely in its bore. Tap it in slightly with a soft-faced hammer or a standard hammer and piece of wood. Once the race is squarely engaged in the bore, place the old race on top of the new race (facing the same direction as the race being installed) and, using a hammer, drive it in until it's bottomed in the bore.

➡**Note: Wear safety goggles!**

If the old race gets stuck in the bore, turn the hub over and drive it out with a soft metal drift, being careful not to damage the new race. Note that the bearings and races come as matched sets and old bearings should never be installed on new races.

15 Use high-temperature front wheel bearing grease to pack the bearings. Work the grease completely into the bearings, forcing it between the rollers, cone and cage from the back side (see illustration).

16 Apply a thin coat of grease to the spindle at the outer bearing seat, inner bearing seat, shoulder and seal seat.

17 Put a small quantity of grease inboard of each bearing race inside the hub. Using your finger, form a dam at these points to provide extra grease availability and to keep thinned grease from flowing out of the bearing.

18 Place the grease-packed inner bearing into the rear of the hub and put a little more grease outboard of the bearing.

19 Place a new seal over the inner bearing and tap the seal evenly into place with a hammer and block of wood until it's flush with the hub.

20 Carefully place the hub assembly onto the spindle and push the grease-packed outer bearing into position.

21 Install the washer and spindle nut. Tighten the nut only slightly (no more than 12 ft-lbs of torque).

22 Spin the hub in a forward direction to seat the bearings and remove any grease or burrs which could cause excessive bearing play later.

23 Check to see that the tightness of the spindle nut is still approximately 12 ft-lbs.

24 Loosen the spindle nut until it's just loose, no more.

25 Using your hand (not a wrench of any kind), tighten the nut until it's snug. Install a new cotter pin through the hole in the spindle and spindle nut. If the nut slots don't line up, loosen the nut slightly until they do. From the hand-tight position, the nut should not be loosened more than one-half flat to install the cotter pin.

26 Bend the ends of the cotter pin until they're flat against the nut. Cut off any extra length which could interfere with the dust cap.

27 Install the dust cap, tapping it into place with a hammer.

28 Place the brake caliper near the rotor and carefully remove the wood spacer. Install the caliper (see Chapter 9).

29 Install the tire/wheel assembly on the hub and tighten the lug nuts.

30 Grasp the top and bottom of the tire and check the bearings in the manner described earlier in this Section.

31 Lower the vehicle and tighten the wheel lug nuts to the torque listed in this Chapter's Specifications.

36.11 Use a screwdriver to pry the grease seal from the rear of the hub

36.15 Work the grease into each bearing until it's full

37 Cooling system servicing (draining, flushing and refilling)

▶ **Refer to illustration 37.4**

✳✳ WARNING:

Antifreeze is a corrosive and poisonous solution, so be careful not to spill any of the coolant mixture on the vehicle's paint or your skin. If this happens, rinse immediately with plenty of clean water. Consult local authorities regarding proper disposal procedures for antifreeze before draining the cooling system. In many areas, reclamation centers have been established to collect used oil and coolant mixtures.

✳✳ CAUTION:

1996 and later models use a special coolant called DEX-COOL. It has a much longer service life than standard coolant (see Maintenance schedule). DEX-COOL is orange in color and should not be mixed with ethylene glycol type (green colored) coolants. Mixing coolant types will result in premature corrosion of engine cooling system components.

➡**Note 1:** Whenever the entire cooling system is to be drained and refilled with new coolant, install GM engine coolant supplement sealer, or equivalent, in the radiator along with the new coolant to prevent premature water pump leakage.

➡**Note 2:** A back flush kit, available at auto parts stores, may be a preferable alternative to the following method. It involves installing a fitting in the heater inlet hose to which a garden hose can be attached. This permits back flushing the entire cooling system.

1 Periodically, the cooling system should be drained, flushed and refilled to replenish the antifreeze mixture and prevent formation of rust and corrosion, which can impair the performance of the cooling system and cause engine damage. When the cooling system is serviced, all hoses and the radiator cap should be checked and replaced if necessary.

2 Apply the parking brake and block the wheels. If the vehicle has just been driven, wait several hours to allow the engine to cool down before beginning this procedure.

3 Once the engine is completely cool, remove the radiator cap.

4 Move a large container under the radiator drain to catch the coolant. Attach a 3/8-inch diameter hose to the drain fitting to direct the coolant into the container, then open the drain fitting (see illustration) (a pair of pliers may be required to turn it).

5 After the coolant stops flowing out of the radiator, move the container under the engine block drain plug (if so equipped). Remove the plug and allow the coolant in the block to drain.

6 While the coolant is draining, check the condition of the radiator hoses, heater hoses and clamps (refer to Section 9 if necessary).

7 Replace any damaged clamps or hoses.

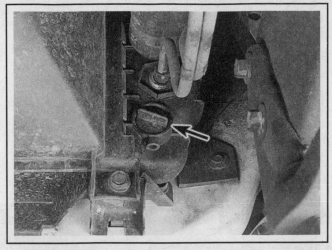

37.4 The drain fitting (arrow) is located at the lower right corner of the radiator

8 Once the system is completely drained, flush the radiator with fresh water from a garden hose until it runs clear at the drain. The flushing action of the water will remove sediments from the radiator but will not remove rust and scale from the engine and cooling tube surfaces.

9 These deposits can be removed with a chemical cleaner. Follow the procedure outlined in the manufacturer's instructions. If the radiator is severely corroded, damaged or leaking, it should be removed (Chapter 3) and taken to a radiator repair shop.

10 Remove the overflow hose from the coolant recovery reservoir. Drain the reservoir and flush it with clean water, then reconnect the hose.

11 Close and tighten the radiator drain. Install and tighten the block drain plug.

12 Place the heater temperature control in the maximum heat position.

13 Slowly add new coolant (a 50/50 mixture of water and antifreeze) to the radiator until it's full. Add coolant to the reservoir up to the lower mark.

14 Leave the radiator cap off and run the engine in a well-ventilated area until the thermostat opens (coolant will begin flowing through the radiator and the upper radiator hose will become hot).

15 Turn the engine off and let it cool. Add more coolant mixture to bring the level back up to the lip on the radiator filler neck.

16 Squeeze the upper radiator hose to expel air, then add more coolant mixture if necessary. Replace the radiator cap.

17 Start the engine, allow it to reach normal operating temperature and check for leaks.

38 Positive Crankcase Ventilation (PCV) valve check and replacement

▶ **Refer to illustration 38.2**

1 The PCV valve is usually located in the rocker arm cover.

2 With the engine idling at normal operating temperature, pull the valve (with hose attached) from the rubber grommet in the cover (see illustration).

3 Place your finger over the valve opening. If there's no vacuum at the valve, check for a plugged hose, manifold port, or the valve itself. Replace any plugged or deteriorated hoses.

4 Turn off the engine and shake the PCV valve, listening for a rattle. If the valve doesn't rattle, replace it with a new one.

5   To replace the valve, pull it from the end of the hose, noting its installed position and direction.

6   When purchasing a replacement PCV valve, make sure it's for your particular vehicle and engine size. Compare the old valve with the new one to make sure they're the same.

7   Push the valve into the end of the hose until it's seated.

8   Inspect the rubber grommet for damage and replace it with a new one if necessary.

9   Push the PCV valve and hose securely into position.

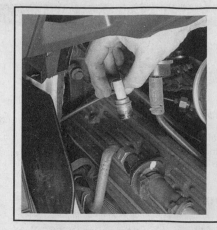

38.2 The PCV valve fits into the rocker arm cover - place your finger over the opening to feel for suction and shake the valve, listening for a rattling sound

39 Evaporative emissions control system check

▶ **Refer to illustration 39.2**

1   The function of the evaporative emissions control system is to draw fuel vapors from the gas tank and fuel system, store them in a charcoal canister and route them to the intake manifold during normal engine operation.

2   The most common symptom of a fault in the evaporative emissions system is a strong fuel odor in the engine compartment. If a fuel odor is detected, inspect the charcoal canister and all hoses for damage. On 1985 to 1995 models the canister is located in the engine compartment (see illustration), on 1996 and later models it's mounted to the frame rail underneath the vehicle.

3   The evaporative emissions control system is explained in more detail in Chapter 6.

39.2 The evaporative emissions control system canister on 1985 to 1995 vehicles is located at the left front corner of the engine compartment (arrow) - inspect the various hoses attached to it and the canister itself for damage

40 Exhaust Gas Recirculation (EGR) system check

CARBURETED AND TBI MODELS

▶ **Refer to illustration 40.2**

1   The EGR valve is usually located on the intake manifold, adjacent to the carburetor, TBI unit or throttle body. Most of the time when a problem develops in this emissions system, it's due to a stuck or corroded EGR valve.

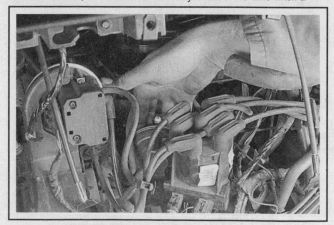

40.2 The EGR diaphragm should move easily with finger pressure (TBI engine)

2   With the engine cold to prevent burns, push on the EGR valve diaphragm. Using moderate pressure, you should be able to press the diaphragm in-and-out within the housing (see illustration).

3   If the diaphragm doesn't move or moves only with much effort, replace the EGR valve with a new one. If in doubt about the condition of the valve, compare the free movement of your EGR valve with a new valve. Next, check the EGR vacuum diaphragm by attaching a hand-held vacuum pump to the vacuum fitting on the valve. Pull 12 to 15 in-hg of vacuum on the valve. The pressure should hold steady. If vacuum will not develop or the pressure drops off, replace the EGR valve.

4   Refer to Chapter 6 for more information on the EGR system.

CENTRAL MULTIPORT FUEL INJECTION (CMFI) AND CENTRAL SEQUENTIAL FUEL INJECTION (CSFI) MODELS

5   These models incorporate an electronically actuated linear EGR valve which is controlled by the Powertrain Control Module (PCM).

6   Refer to Chapter 6 for information on this EGR system and check/ replacement procedures.

41 Spark plug replacement

▶ Refer to illustrations 41.2, 41.5a, 41.5b, 41.6, 41.10a, 41.10b and 41.11

➡Note: 1996 and later models with V6 engines are equipped with long-life platinum tipped spark plugs which have a service life of 100,000 miles. These plugs should not be cleaned or re-gapped; always replace them if their condition is questionable.

1　Open the hood and remove the engine cover (see Chapter 11).

2　In most cases, the tools necessary for spark plug replacement include a spark plug socket which fits onto a ratchet (spark plug sockets are padded inside to prevent damage to the porcelain insulators on the new plugs), various extensions and a gap gauge to check and adjust the gaps on the new plugs (see illustration). A special plug wire removal tool is available for separating the wire boots from the spark plugs, but it isn't absolutely necessary. A torque wrench should be used to tighten the new plugs.

3　The best approach when replacing the spark plugs is to purchase the new ones in advance, adjust them to the proper gap and replace them one at a time. When buying the new spark plugs, be sure to obtain the correct plug type for your particular engine. This information can be found on the Emission Control Information label located under the hood and in the factory owner's manual. If differences exist between the plug specified on the emissions label and in the owner's manual, assume that the emissions label is correct.

4　Allow the engine to cool completely before attempting to remove any of the plugs. While you're waiting for the engine to cool, check the new plugs for defects and adjust the gaps.

5　The gap is checked by inserting the proper thickness gauge between the electrodes at the tip of the plug (see illustration). The gap between the electrodes should be the same as the one specified on the Emissions Control Information label. The gauge should just slide between the electrodes with a slight amount of drag. If the gap is incorrect, use the adjuster on the gauge body to bend the curved side electrode slightly until the proper gap is obtained (see illustration). If the side electrode is not exactly over the center electrode, bend it with the adjuster until it is. Check for cracks in the porcelain insulator (if any are found, the plug should not be used).

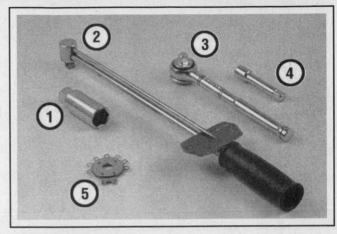

41.2 Tools required for changing spark plugs

1　*Spark plug socket* - This will have special padding inside to protect the spark plug's porcelain insulator
2　*Torque wrench* - Although not mandatory, using this tool is the best way to ensure the plugs are tightened properly
3　*Ratchet* - Standard hand tool to fit the spark plug socket
4　*Extension* - Depending on model and accessories, you may need special extensions and universal joints to reach one or more of the plugs
5　*Spark plug gap gauge* - This gauge for checking the gap comes in a variety of styles. Make sure the gap for your engine is included

➡Note: When checking the gap on platinum-type spark plugs, be careful not to scrape the thin platinum coating from the electrodes (this will dramatically shorten the life of the plugs).

6　With the engine cool, remove the spark plug wire from one spark plug. Pull only on the boot at the end of the wire - do not pull on the wire. A plug wire removal tool should be used if available (see illustration).

7　If compressed air is available, use it to blow any dirt or foreign

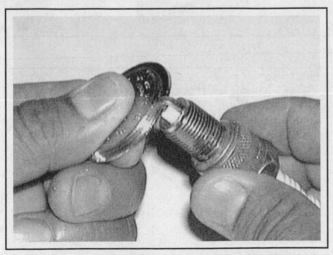

41.5a Spark plug manufacturers recommend using a tapered thickness gauge when checking the gap - slide the thin side into the gap and turn until the gauge just fills the gap, then read the thickness on the gauge - do not force the tool into the gap or use the tapered portion to widen a gap

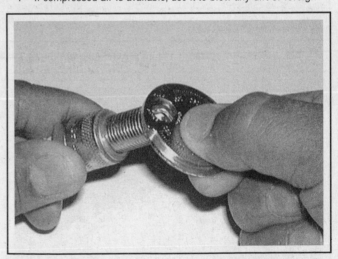

41.5b To change the gap, bend the side electrode only, using the adjuster hole in the tool, and be very careful not to crack or chip the porcelain insulator surrounding the center electrode

material away from the spark plug hole. A common bicycle pump will also work. The idea here is to eliminate the possibility of debris falling into the cylinder as the spark plug is removed.

8   The spark plugs on these models are, for the most part, difficult to reach. To gain access to the rear plugs, remove the engine cover (see Chapter 11). Also, access to the sides of the engine can be obtained through the wheel wells. If necessary, raise the front of the vehicle, support it securely on jackstands and remove the wheels. Remove the splash shield to access the side of the engine.

9   Place the spark plug socket over the plug and remove it from the engine by turning it in a counterclockwise direction.

10 Compare the spark plug with the chart shown **(see illustration)** to get an indication of the general running condition of the engine. Also, before installing the new spark plugs, it is a good idea to apply a thin coat of anti-seize compound to the threads (see illustration).

11 Thread one of the new plugs into the hole until you can no longer

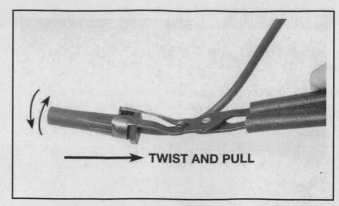

TWIST AND PULL

41.6 When removing the spark plug wires, pull only on the boot and use a twisting/pulling motion

A normally worn spark plug should have light tan or gray deposits on the firing tip.

A carbon fouled plug, identified by soft, sooty, black deposits, may indicate an improperly tuned vehicle. Check the air cleaner, ignition components and engine control system.

An oil fouled spark plug indicates an engine with worn piston rings and/or bad valve seals allowing excessive oil to enter the chamber.

This spark plug has been left in the engine too long, as evidenced by the extreme gap- Plugs with such an extreme gap can cause misfiring and stumbling accompanied by a noticeable lack of power.

A physically damaged spark plug may be evidence of severe detonation in that cylinder. Watch that cylinder carefully between services, as a continued detonation will not only damage the plug, but could also damage the engine.

A bridged or almost bridged spark plug, identified by a build-up between the electrodes caused by excessive carbon or oil build-up on the plug.

41.10a Inspect the spark plug to determine engine running conditions

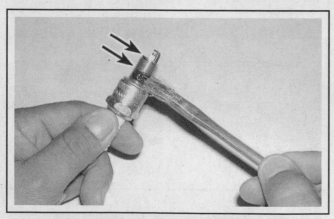

41.10b Apply a thin coat of anti-seize compound to the spark plug threads, being careful not to get any near the lower threads (arrows)

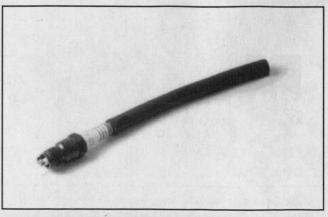

41.11 A length of snug-fitting rubber hose will save time and prevent damaged threads when installing the spark plugs

turn it with your fingers, then tighten it with a torque wrench (if available) or the ratchet. It might be a good idea to slip a short length of rubber hose over the end of the plug to use as a tool to thread it into place (see illustration). The hose will grip the plug well enough to turn it, but will start to slip if the plug begins to cross-thread in the hole - this will prevent damaged threads and the accompanying repair costs. Tighten the spark plug to the torque listed in this Chapter's specifications.

12 Before pushing the spark plug wire onto the end of the plug, inspect it following the procedures outlined in Section 42.

13 Attach the plug wire to the new spark plug, again using a twisting motion on the boot until it's seated on the spark plug.

14 Repeat the procedure for the remaining spark plugs, replacing them one at a time to prevent mixing up the spark plug wires.

2 Spark plug wire check and replacement

1   The spark plug wires should be checked at the recommended intervals and whenever new spark plugs are installed in the engine.

2   The wires should be inspected one at a time to prevent mixing up the order, which is essential for proper engine operation.

3   Disconnect the plug wire from one spark plug. To do this, grab the rubber boot, twist slightly and pull the wire free. Do not pull on the wire itself, only on the rubber boot (see illustration 41.6).

4   Check inside the boot for corrosion, which will look like a white crusty powder. Push the wire and boot back onto the end of the spark plug. It should be a tight fit on the plug. If it isn't, remove the wire and use a pair of pliers to carefully crimp the metal connector inside the boot until it fits securely on the end of the spark plug.

5   Using a clean rag, wipe the entire length of the wire to remove any built-up dirt and grease. Once the wire is clean, check for holes, burned areas, cracks and other damage. Don't bend the wire excessively or the conductor inside might break.

6   Disconnect the wire from the distributor cap. A retaining ring at the top of the distributor may have to be removed to free the wires. Again, pull only on the rubber boot. Check for corrosion and a tight fit in the same manner as the spark plug end. Using an ohmmeter, measure the resistance of the spark plug wire and compare it with the value listed in this Chapter's specifications. Reattach the wire to the distributor cap.

7   Check the remaining spark plug wires one at a time, making sure they are securely fastened at the distributor and the spark plug when the check is complete.

8   If new spark plug wires are required, purchase a new set for your specific engine model. Wire sets are available pre-cut, with the rubber boots already installed. Remove and replace the wires one at a time to avoid mix-ups in the firing order. The wire routing is extremely important, so be sure to note exactly how each wire is situated before removing it.

3 Distributor cap and rotor check and replacement

▶ **Refer to illustrations 43.4a, 43.4b, 43.5, 43.7a, 43.7b and 43.13**

➡**Note: It's common practice to install a new distributor cap and rotor whenever new spark plug wires are installed. On models that have the ignition coil mounted in the cap, the coil will have to be transferred to the new cap.**

1   Remove the engine cover (see Chapter 11).

CHECK

2   To gain access to the distributor cap, remove the air cleaner assembly (see Chapter 4).

3   Loosen the distributor cap mounting screws (note that the screws have a shoulder so they don't come completely out). On some models, the cap is held in place with latches that look like screws - to release them, push down with a screwdriver and turn them about 1/2-turn. Pull up on the cap, with the wires attached, to separate it from the distributor, then position it to one side.

4   The rotor is now visible on the end of the distributor shaft. Check it carefully for cracks and carbon tracks. Make sure the center terminal spring tension is adequate and look for corrosion and wear on the rotor tip (see illustrations). If in doubt about its condition, replace it with a

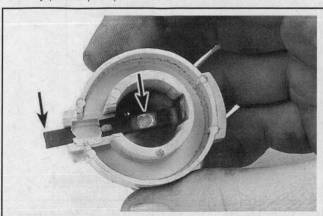

43.4a Check the distributor rotor contacts (arrows) for wear and burn marks (remote coil-type distributor)

43.4b Carbon tracking on the rotor is caused by a leaking arc seal between the distributor cap and the ignition coil (always replace both the seal and rotor when this condition is present) (coil in cap-type distributor)

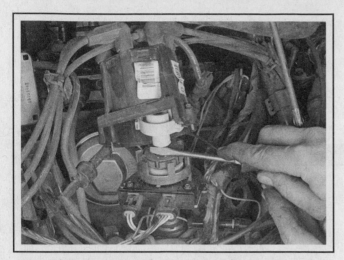

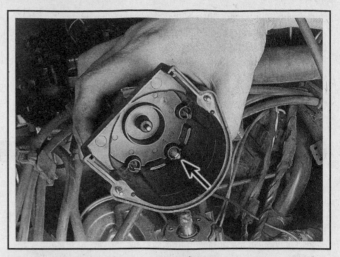

43.5 This type rotor can be pried off the distributor shaft with a screwdriver - other types have two screws which must be removed (be careful! not to drop anything down into the distributor)

43.7a Inspect the inside of the cap, especially the metal terminals (arrow) for corrosion and wear

new one.

5   If replacement is required, detach the rotor from the shaft and install a new one. On some models, the rotor is press fit on the shaft and can be pried or pulled off (see illustration). On other models, the rotor is attached to the distributor shaft with two screws.

6   The rotor is indexed to the shaft so it can only be installed one

way. Press fit rotors have an internal key that must line up with a slot in the end of the shaft (or vice versa). Rotors held in place with screws have one square and one round peg on the underside that must fit into holes with the same shape.

7   Check the distributor cap for carbon tracks, cracks and other damage. Closely examine the terminals on the inside of the cap for excessive corrosion and damage (see illustrations). Slight deposits are normal. Again, if in doubt about the condition of the cap, replace it with

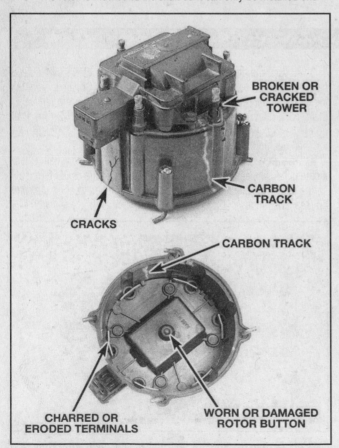

43.7b Shown here are some of the common defects to look for when inspecting the distributor cap (if in doubt about its condition, install a new one)

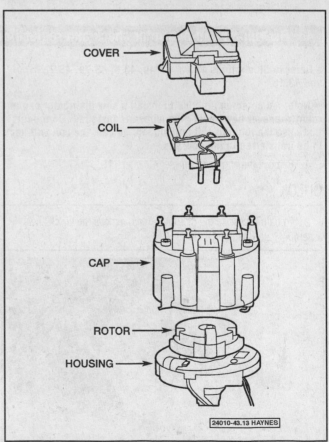

43.13 Coil-in-cap distributor components - exploded view (HEI system)

a new one. Be sure to apply a small dab of silicone lubricant to each terminal before installing the cap. Also, make sure the carbon brush (center terminal) is correctly installed in the cap - a wide gap between the brush and rotor will result in rotor burn-through and/or damage to the distributor cap.

REPLACEMENT

Conventional distributor

8   On models with a separately mounted ignition coil, simply separate the cap from the distributor and transfer the spark plug wires, one at a time, to the new cap. Be very careful not to mix up the wires!

9   Reattach the cap to the distributor, then tighten the screws or reposition the latches to hold it in place.

Coil-in-cap distributor

10   Use your thumbs to push the spark plug wire retainer latches away from the coil cover.

11   Lift the retainer ring away from the distributor cap with the spark plug wires attached to the ring. It may be necessary to work the wires off the distributor cap towers so they remain with the ring.

12   Disconnect the battery/tachometer/coil electrical connector from the distributor cap.

13   Remove the two coil cover screws and lift off the coil cover (see illustration).

14   There are three small spade connectors on wires extending from the coil into the electrical connector hood at the side of the distributor cap. Note which terminals the wires are attached to, then use a small screwdriver to push them free.

15   Remove the four coil mounting screws and lift the coil out of the cap.

16   When installing the coil in the new cap, be sure to install a new rubber arc seal in the cap.

17   Install the coil screws, the wires in the connector hood, and the coil cover.

18   Install the cap on the distributor.

19   Plug in the coil electrical connector to the distributor cap.

20   Install the spark plug wire retaining ring on the distributor cap.

Specifications

Recommended lubricants and fluids

→**Note: Listed here are manufacturer recommendations at the time this manual was written. Manufacturers occasionally upgrade their fluid and lubricant specifications, so check with your auto parts store for current recommendations.**

Engine oil type	API "certified for gasoline engines"
Engine oil viscosity	See accompanying chart
Automatic transmission fluid	Dexron III automatic transmission fluid (H-Specification))
Manual transmission lubricant	Dexron III automatic transmission fluid

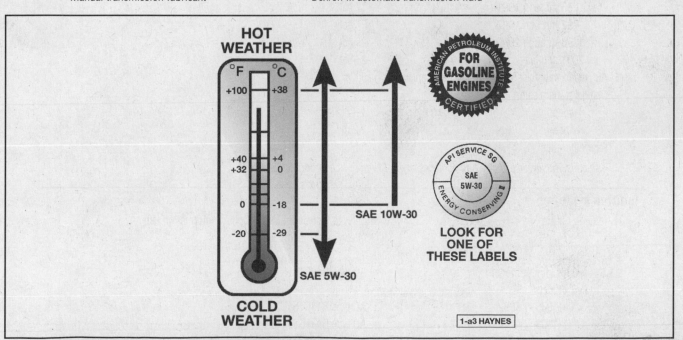

Recommended SAE viscosity grade engine oils. For best fuel economy and cold starting, select the lowest SAE viscosity grade oil for the expected temperature range

Recommended lubricants and fluids (continued)

Differential lubricant
  US
    1985 through 1999 — SAE 80W or SAE 80W-90 GL-5 gear lubricant
    2000 and later — SAE 75W-90 synthetic axle lubricant
  Canada — SAE 80W GL-5 gear lubricant
  Limited slip differential (all) — Add GM limited-slip additive to the specified lubricant
Chassis grease — NLGI No. 2 chassis grease
Engine coolant
  1985 to 1995 models — 50/50 mixture of water and ethylene glycol-base antifreeze (green colored)
  1996 and later models — 50/50 mixture of water and DEX-COOL silicate-free antifreeze (orange colored)
Brake fluid — DOT-3 brake fluid
Clutch fluid — DOT-3 brake fluid
Power steering fluid — GM power steering fluid or equivalent
Manual steering box lubricant — EP chassis grease
Wheel bearing grease — NLGI No. 2 moly-base hi-temp wheel bearing grease

Capacities*

Engine oil (with filter change, approximate)
  Four-cylinder engine — 3 qts
  V6 engine — 4.5 qts
Cooling system
  Four-cylinder engine — 10 qts
    Without rear heater — 10 qts
    With rear heater — 13 qts
  V6 engine — 13.5 qts
    Without rear heater — 14.3 qts
    With rear heater — 16.5 qts
Fuel tank
  1985 through 1991 — 17 gal
  1992 through 1997 — 27 gal
  1998 through 2000 — 25 gal
  2001 and later — 27 gal
Automatic transmission (fluid and filter replacement) — 5 qts
Manual transmission
  4-speed — 1.3 qts
  5-speed — 2.2 qts
Rear differential — 3.5 pts

All capacities approximate. Add as necessary to bring up to appropriate level.

Ignition system

Ignition timing — Refer to Vehicle Emission Control Information label in engine compartment

Spark plug type
  Four-cylinder models — AC type R43CTS-6
  V6 models
    1985 to 1992 — AC type R43CTS
    1993 — AC type R43TS
    1994 and 1995 — AC type RCR43TSM
    1996 and later (platinum tipped) — AC type 41-932

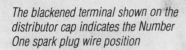

The blackened terminal shown on the distributor cap indicates the Number One spark plug wire position

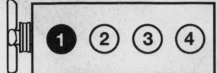

FIRING ORDER
1-3-4-2

24010-1-specs HAYNES

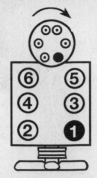

FIRING ORDER
1-6-5-4-3-2

with HEI ignition system

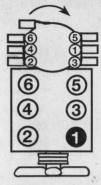

FIRING ORDER
1-6-5-4-3-2

with Enhanced Distributor Ignition (EDI) system

Cylinder location and distributor rotation

Spark plug gap	
Four-cylinder models	0.060 inch
V6 models	
Through 1991	0.035 inch
1992 to 1995	
With TBI (VIN Z)	0.035 inch
With CPI (VIN W)	0.045 inch
1996 and later	0.060 inch
Spark plug wire resistance	
1985 through 1990	
0 to 15 inches long	3K to 10K ohms
15 to 25 inches long	4K to 15K ohms
25 to 35 inches long	6K to 20K ohms
Over 35 inches long	5K to 10K per foot
1991 and later	1000 ohms per foot
Firing order	
Four-cylinder engine	1-3-4-2
V6 engine	1-6-5-4-3-2

General

Engine idle speed	Refer to Vehicle Emission Control Information label in engine compartment
Radiator pressure cap rating	15 psi

Drivebelt tension (conventional V-belts only - 1985 and 1986 vehicles)

Four-cylinder engine	
Alternator	
Vehicles with C60 A/C compressor	
New	169 lbs
Used	90 lbs
Vehicles with C41 heater	
New	146 lbs
Used	67 lbs

Drivebelt tension (conventional V-belts only - 1985 and 1986 vehicles) (continued)

Power steering pump		
	New	146 lbs
	Used	67 lbs
Air conditioning compressor		
	New	169 lbs
	Used	90 lbs
V6 engine		
Alternator		
	New	135 lbs
	Used	67 lbs
Power steering pump		
	New	146 lbs
	Used	67 lbs
Air conditioning compressor		
	New	169 lbs
	Used	90 lbs
AIR pump		
	New	146 lbs
	Used	67 lbs

Filters — AC part no.

	AC part no.
Oil filter type	
Four-cylinder engine	AC PF47
V6 engine	
1985 to 1995	AC PF51
1996 through 2000	AC PF52
2001 and later	AC PF47
Air filter type	
Four cylinder engine	AC 785C
V6 engine	
1985 through 1992	AC 773C
1993 VIN Z	AC 773C
1993 VIN W	AC A1163C
1994 and later	AC A1163C
PCV valve	
Four-cylinder engine	
1985 and 1986 models	AC CV881C
1987 models	AC CV845C
1988 models	AC CV895C
V6 engine	
1985 to 1995 VIN Z (carburetor and TBI models)	AC CV789C
1992 to 1995 VIN W (CMFI models)	AC CV892C
1996 and 1997	AC CV774C
1998	AC CV746C
1999 and later	AC CV769C
PCV filter	
Four-cylinder engine	AC FB73
V6 engine	AC FB59
Fuel filter (EFI equipped engine only)	AC GF481

Brakes

Brake pad wear limit (front)　　　　　1/8 inch
Brake shoe wear limit (rear)　　　　　1/16 inch

Torque specifications	Ft-lbs (unless otherwise indicated)
Differential (axle) fill plug	20 to 27
Engine oil drain plug	18
Wheel lug nuts	
1985 through 1992	90
1993 through 2002	100
2003	125
2004 and later	140
Manual transmission check/fill plug	15 to 25
Manual transmission drain plug	15 to 25
Automatic transmission oil pan bolts	84 to 120 in-lbs
Carburetor bolts	
Long	84 in-lbs
Short	132 in-lbs
Throttle body nuts	144 in-lbs
Throttle body studs	80 in-lbs
Carburetor-mounted fuel filter nut	18
Spark plugs	
Four cylinder engine	84 to 180 in-lbs
V6 engine	
New cylinder head	22
Subsequent installation	132 in-lbs

Notes

2A
FOUR-CYLINDER ENGINE

Section

Reference to other Chapters

1 General information

This Part of Chapter 2 is devoted to in-vehicle repair procedures for the 2.5 liter four-cylinder engine. Information concerning engine removal and installation, as well as engine block and cylinder head overhaul, is in Part C of this Chapter.

The following repair procedures are based on the assumption that the engine is installed in the vehicle. If the engine has been removed from the vehicle and mounted on a stand, many of the steps included in this Part of Chapter 2 will not apply.

The Specifications included in this Part of Chapter 2 apply only to the engine and procedures in this Part. The Specifications necessary for rebuilding the block and cylinder head are found in Part C.

❈❈ CAUTION:

On models equipped with an anti-theft audio system, disable the anti-theft feature before performing any operation that requires disconnecting the battery or disrupting power to the stereo (see the Anti-theft audio system procedures at the front of this manual).

2 Repair operations possible with the engine in the vehicle

Many major repair operations can be accomplished without removing the engine from the vehicle.

Clean the engine compartment and the exterior of the engine with some type of pressure washer before any work is done. A clean engine will make the job easier and will help keep dirt out of the internal areas of the engine.

Depending on the components involved, remove the engine cover and, if necessary, the hood to improve access to the engine as repairs are performed (refer to Chapter 11 if necessary).

If vacuum, exhaust, oil or coolant leaks develop, indicating a need for gasket or seal replacement, the repairs can generally be made with the engine in the vehicle. The intake and exhaust manifold gaskets, oil pan gasket and cylinder head gasket are all accessible with the engine in place.

Exterior engine components such as the intake and exhaust manifolds, the oil pan (and the oil pump), the water pump, the starter motor, the alternator, the distributor and the fuel injection system can be removed for repair with the engine in place.

Since the cylinder head can be removed without pulling the engine, valve component servicing can also be accomplished with the engine in the vehicle.

In extreme cases caused by a lack of necessary equipment, repair or replacement of piston rings, pistons, connecting rods and rod bearings is possible with the engine in the vehicle. However, this practice is not recommended because of the cleaning and preparation work that must be done to the components involved.

3 Rocker arm cover - removal and installation

▶ **Refer to illustrations 3.5, 3.11, 3.12 and 3.13**

1 Disconnect the cable from the negative terminal of the battery.

❈❈ CAUTION:

On models equipped with an anti-theft audio system, disable the anti-theft feature before performing any operation that requires disconnecting the battery or disrupting power to the stereo (see the Anti-theft audio system procedures at the front of this manual).

2 Remove the engine cover (see Chapter 11).
3 Remove the air cleaner assembly (see Chapter 4).
4 Unbolt the dipstick and oil filler tubes at the water outlet.
5 Remove the PCV valve from the rocker arm cover (see illustration).
6 Label each spark plug wire before removal to ensure that all wires are reinstalled correctly, then remove the wires from the plugs (refer to Chapter 1). Detach the wires and retaining clips from the rocker arm cover.
7 Remove the EGR valve (see Chapter 6).
8 Detach the vacuum rail at the intake manifold and water outlet.
9 Remove the rocker arm cover bolts.

10 Starting with 1987 models, a special tool, available at auto parts stores, is recommended for rocker arm cover removal. But if you pry carefully, you may get by without it.

11 Remove the rocker arm cover. If it sticks to the head, use a soft-face hammer or a block of wood and a hammer to dislodge it. If the cover still won't come loose, pry on it carefully at several points until

3.5 The PCV valve can be pulled out of the rubber grommet in the rocker arm cover (leave the hose attached to the valve)

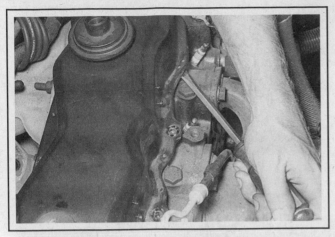

3.11 The rocker arm cover is sealed with RTV - if you have to pry it off the head, try to avoid bending the flange

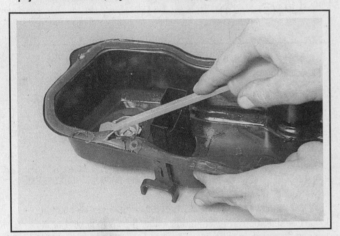

3.12 Remove the old sealant from the rocker arm cover flange and the cylinder head with a gasket scraper, then clean the mating surfaces with lacquer thinner or acetone

the seal is broken, but don't distort the cover flange (see illustration).

➡Note: If you bend the cover, straighten it with a block of wood and a hammer.

12 Prior to reinstallation, remove all dirt, oil and old gasket material from the cover and cylinder head with a scraper (see illustration). Clean the mating surfaces with lacquer thinner or acetone.

13 Apply a continuous 3/16-inch (5 mm) diameter bead of RTV sealant to the flange on the cover. Be sure the sealant is applied to the inside of the bolt holes (see illustration).

➡Note: Don't get the sealant in the bolt holes in the head or damage to the head may occur.

14 Place the rocker arm cover on the cylinder head while the sealant is still wet and install the mounting bolts. Tighten the bolts a little at a time until the specified torque is reached.

15 Complete the installation by reversing the removal procedure.

16 Start the engine and check for oil leaks at the rocker arm cover-to-head joint.

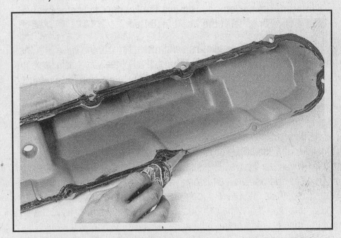

3.13 Make sure the sealant is applied to the INSIDE of the bolt holes or oil will leak out around the bolt threads

4   Valve springs, retainers and seals - replacement

♦ Refer to illustrations 4.4, 4.8a, 4.8b, 4.16 and 4.17

➡Note: Broken valve springs and defective valve stem seals can be replaced without removing the cylinder head. Two special tools and a compressed air source are normally required to perform this operation, so read through this Section carefully and rent or buy the tools before beginning the job. If compressed air isn't available, a length of nylon rope can be used to keep the valves from falling into the cylinder during this procedure.

1 Refer to Section 3 and remove the rocker arm cover.

2 Remove the spark plug from the cylinder which has the defective component. If all of the valve stem seals are being replaced, all of the spark plugs should be removed.

3 Turn the crankshaft until the piston in the affected cylinder is at top dead center on the compression stroke (refer to Section 13 for instructions). If you're replacing all of the valve stem seals, begin with cylinder number one and work on the valves for one cylinder at a time. Move from cylinder-to-cylinder following the firing order sequence (1-3-4-2).

4 Thread an adapter into the spark plug hole and connect an air hose from a compressed air source to it (see illustration). Most auto

4.4 This is what the air hose adapter that threads into the spark plug hole looks like - they're commonly available from auto parts stores

parts stores can supply the air hose adapter.

➡Note: Many cylinder compression gauges utilize a screw-in fitting that may work with your air hose quick-disconnect fitting.

4.8a A lever-type valve spring compressor is used to compress the spring and remove the keepers to replace valve seals or springs with the head installed

5   Remove the bolt, pivot ball and rocker arm for the valve with the defective part and pull out the pushrod. If all of the valve stem seals are being replaced, all of the rocker arms and pushrods should be removed (refer to Section 6).

6   Apply compressed air to the cylinder. The valves should be held in place by the air pressure. If the valve faces or seats are in poor condition, leaks may prevent the air pressure from retaining the valves - refer to the alternative procedure below.

7   If you don't have access to compressed air, an alternative method can be used. Position the piston at a point just before TDC on the compression stroke, then feed a long piece of nylon rope through the spark plug hole until it fills the combustion chamber. Be sure to leave the end of the rope hanging out of the engine so it can be removed easily. Use a large breaker bar and socket to rotate the crankshaft in the normal direction of rotation until slight resistance is felt.

8   Stuff shop rags into the cylinder head holes above and below the valves to prevent parts and tools from falling into the engine, then use a valve spring compressor to compress the spring (see illustrations). Remove the keepers with small needle-nose pliers or a magnet.

➡Note: A couple of different types of tools are available for compressing the valve springs with the head in place. One type grips the lower spring coils and presses on the retainer as the knob is turned, while the other type, shown here, utilizes the rocker arm bolt for leverage. Both types work very well, although the lever type is usually less expensive.

9   Remove the valve stem O-ring seal, spring retainer, shield and valve spring, then remove the umbrella type guide seal, if equipped (the O-ring seal will most likely be hardened and will probably break when removed, so plan on installing a new one each time the original is removed).

➡Note: If air pressure fails to hold the valve in the closed position during this operation, the valve face or seat is probably damaged. If so, the cylinder head will have to be removed for additional repair operations.

10 Wrap a rubber band or tape around the top of the valve stem so the valve will not fall into the combustion chamber, then release the air pressure.

➡Note: If a rope was used instead of air pressure, turn the crankshaft slightly in the direction opposite normal rotation.

11 Inspect the valve stem for damage. Rotate the valve in the guide

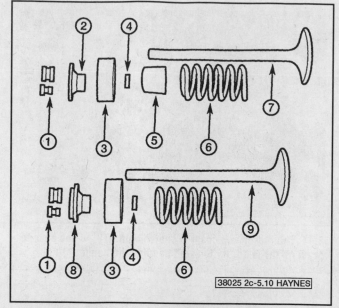

4.8b Typical engine valves and related components - exploded view

1	Keepers	6	Spring
2	Retainer	7	Intake valve
3	Oil shield	8	Retainer/rotator
4	O-ring oil seal	9	Exhaust valve
5	Umbrella seal (if equipped)		

and check the end for eccentric movement, which would indicate that the valve is bent.

12 Move the valve up-and-down in the guide and make sure it doesn't bind. If the valve stem binds, either the valve is bent or the guide is damaged. In either case, the head will have to be removed for repair.

13 Reapply air pressure to the cylinder to retain the valve in the closed position, then remove the tape or rubber band from the valve stem. If a rope was used instead of air pressure, rotate the crankshaft in the normal direction of rotation until slight resistance is felt.

14 Lubricate the valve stem with engine oil and install a new umbrella type guide seal, if used.

15 Install the spring and shield in position over the valve.

16 Install the valve spring retainer. Compress the valve spring and carefully install the new O-ring seal in the lower groove of the valve stem. Make sure the seal isn't twisted - it must lie perfectly flat in the groove (see illustration).

4.16 Make sure the O-ring seal under the retainer is seated in the groove and not twisted before installing the keepers

4.17 Keepers don't always stay in place, so apply a small dab of grease to each one as shown here before installation – it'll hold them in place on the valve stem as the spring is released

17 Position the keepers in the upper groove. Apply a small dab of grease to the inside of each keeper to hold it in place if necessary (see illustration). Remove the pressure from the spring tool and make sure the keepers are seated. Refer to Chapter 2, Part C, and check the seals with a vacuum pump.

18 Disconnect the air hose and remove the adapter from the spark plug hole. If a rope was used in place of air pressure, pull it out of the cylinder.

19 Refer to Section 6 and install the rocker arm(s) and pushrod(s).

20 Install the spark plug(s) and hook up the wire(s).

21 Refer to Section 3 and install the rocker arm cover.

22 Start and run the engine, then check for oil leaks and unusual sounds coming from the rocker arm cover area.

5 Pushrod cover - removal and installation

▶ Refer to illustrations 5.9, 5.12 and 5.13

1 Disconnect the negative cable from the battery.

❉❉ CAUTION:

On models equipped with an anti-theft audio system, disable the anti-theft feature before performing any operation that requires disconnecting the battery or disrupting power to the stereo (see the Anti-theft audio system procedures at the front of this manual).

2 Remove the engine cover (see Chapter 11).

3 Remove the alternator and brackets (see Chapter 5).

4 Remove the intake manifold-to-block brace.

5 Drain the cooling system (see Chapter 1).

6 Detach the lower radiator and heater hoses.

7 Remove the oil pressure sending unit.

8 Detach the wiring harness brackets from the cover.

9 Remove the four pushrod cover nuts (see illustration). Flip one nut over so the washer faces out and reinstall it on the long inner stud. Put a second nut on the stud with the washer facing in.

5.9 The pushrod cover is held in place with four nuts

Put two 6 mm nuts on the shorter stud. Jam them together with two wrenches. Unscrew the studs by turning the inner nuts until the cover breaks free.

10 Remove the pushrod cover.

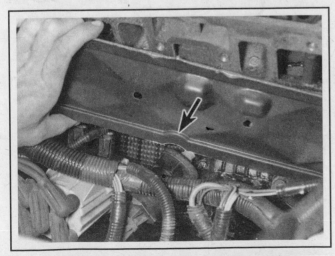

5.12 Install the pushrod cover while the sealant is still tacky – be sure the semi-circular cutout (arrow) is facing down

5.13 Don't forget to install new rubber sealing washers around the pushrod cover mounting studs or oil will leak past the studs

✳✳ CAUTION:

Careless prying may damage the sealing surface of the cover. If you bend the cover during removal, place it on a flat surface and straighten it with a soft-face hammer.

11 Remove all traces of old gasket material with a scraper, then clean the mating surfaces with lacquer thinner or acetone.

12 Apply a continuous 3/16-inch diameter bead of RTV sealant to the mating surface of the pushrod cover (see illustration).

13 Install new rubber pushrod cover seals (see illustration).

14 Install the cover while the sealant is still wet. Make sure the semicircular cutout in the edge of the pushrod cover is facing down.

15 Tighten the nuts gradually until they're snug, then tighten them to the torque listed in this Chapter's Specifications.

16 The remaining installation steps are the reverse of removal.

17 Start and run the engine, then check for oil and coolant leaks.

---

6 Rocker arms and pushrods - removal, inspection and installation

▶ Refer to illustrations 6.4a, 6.4b and 6.5

REMOVAL

1 Refer to Section 3 and detach the rocker arm cover from the cylinder head.

2 Beginning at the front of the cylinder head, loosen the rocker arm bolts.

➡ **Note: If the pushrods are the only items being removed, rotate the rocker arms to one side so the pushrods can be lifted out.**

3 Remove the bolts, the rocker arms and the pivot balls and store them in marked containers (they must be reinstalled in their original locations).

4 Remove the pushrods and store them separately to make sure they don't get mixed up during installation (see illustrations).

5 If the pushrod guides must be removed for any reason, make sure they're marked so they can be reinstalled in their original locations (see illustration).

INSPECTION

6 Check each rocker arm for wear, cracks and other damage, especially where the pushrods and valve stems contact the rocker arm faces.

7 Make sure the hole at the pushrod end of each rocker arm is open.

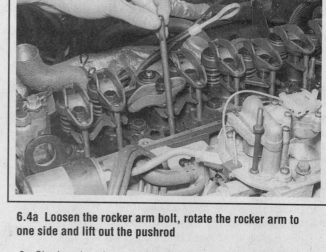

6.4a Loosen the rocker arm bolt, rotate the rocker arm to one side and lift out the pushrod

8 Check each rocker arm pivot area for wear, cracks and galling. If the rocker arms are worn or damaged, replace them with new ones and use new pivot balls as well.

9 Inspect the pushrods for cracks and excessive wear at the ends. Roll each pushrod across a piece of plate glass to see if it's bent (if it wobbles, it's bent).

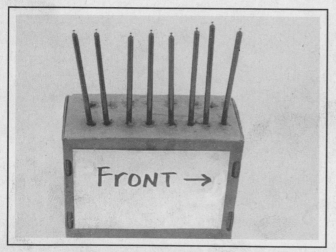

6.4b If more than one pushrod is being removed, store them in a perforated cardboard box to prevent mixups during installation- note the label indicating the front of the engine

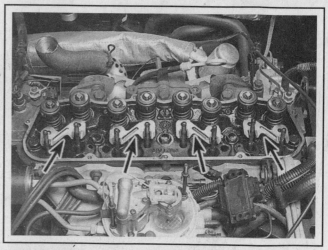

6.5 If they're removed, make sure the pushrod guides (arrows) are kept in order also

INSTALLATION

10 Lubricate the lower ends of the pushrods with clean engine oil or moly-base grease and install them in their original locations. Make sure each pushrod seats completely in the lifter socket.

11 Apply moly-base grease to the ends of the valve stems and the upper ends of the pushrods before positioning the rocker arms and installing the bolts.

12 Set the rocker arms in place, then install the pivot balls and bolts. Apply moly-base grease to the pivot balls to prevent damage to the mating surfaces before engine oil pressure builds up. Tighten the bolts to the specified torque.

7 Hydraulic lifters - removal, inspection and installation

REMOVAL

♦ Refer to illustrations 7.7a, 7.7b, 7.8 and 7.9

1 A noisy valve lifter can be isolated when the engine is idling. Place a length of hose or tubing near the position of each valve while listening at the other end. Or remove the rocker arm cover and, with the engine idling, place a finger on each of the valve spring retainers, one at a time. If a valve lifter is defective, it'll be evident from the shock felt at the retainer as the valve seats.

2 The most likely cause of a noisy valve lifter is a piece of dirt trapped between the plunger and the lifter body.

3 Remove the rocker arm cover (see Section 3).

4 Loosen both rocker arm bolts at the cylinder with the noisy lifter and rotate the rocker arms away from the pushrods.

5 Remove the pushrod guides and pushrods (see Section 6).

6 Remove the pushrod cover (see Section 5).

7 Remove the lifter guide retainer by unscrewing the locknuts on the pushrod cover studs. Remove the lifter guide (see illustrations).

8 There are several ways to extract a lifter from its bore. A special removal tool is available, but isn't always necessary. On newer engines without a lot of varnish buildup, lifters can often be removed with a small magnet or even with your fingers (see illustration). A scribe can also be used to pull the lifter out of the bore.

> ❋❋ **CAUTION:**
>
> **Don't use pliers of any type to remove a lifter unless you intend to replace it with a new one - they will damage the precision machined and hardened surface of the lifter, rendering it useless.**

9 Store the lifters in a clearly labeled box to insure their reinstallation in the same lifter bores (see illustration).

7.7a Pushrod cover stud locknuts (arrows) - manifold removed for clarity

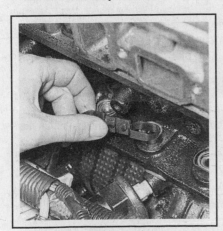

7.7b Remove the lifter guide - if you're removing more than one guide, keep them in order to prevent mixups during installation

7.8 On engines that haven't become sticky with sludge and varnish, the lifters can usually be removed by hand

7.9 If you're removing more than one lifter, keep them in order in a clearly labeled box

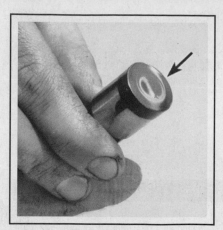

7.11a If the bottom (foot) of any lifter is worn concave, scratched or pitted, replace the entire set with new lifters

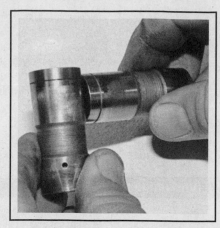

7.11b The bottom (foot) of each lifter should be slightly convex - the side of another lifter can be used as a straightedge to check it (if it appears flat, it's worn and should be discarded)

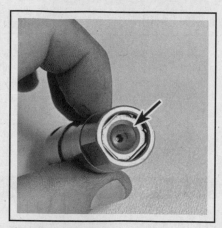

7.11c Check the pushrod seat (arrow) in the top of each lifter for wear

INSPECTION

▶ Refer to illustrations 7.11a, 7.11b, 7.11c and 7.13

Conventional lifters

10 Clean the lifters with solvent and dry them thoroughly without mixing them up.

11 Check each lifter wall, pushrod seat and foot for scuffing, score marks and uneven wear. Each lifter foot (the surface that rides on the cam lobe) must be slightly convex, although this can be difficult to determine by eye. If the base of the lifter is concave (see illustrations), the lifters and camshaft must be replaced. If the lifter walls are damaged or worn (which isn't very likely), inspect the lifter bores in the engine block as well. If the pushrod seats (see illustration) are worn, check the pushrod ends.

12 If new lifters are being installed, a new camshaft must also be installed. If a new camshaft is installed, then use new lifters as well. Never install used lifters unless the original camshaft is used and the lifters can be installed in their original location!

Roller lifters

13 Check the rollers carefully for wear and damage and make sure they turn freely without excessive play (see illustration). The inspection procedure for conventional lifters also applies to roller lifters.

14 Used roller lifters can be reinstalled with a new camshaft and the original camshaft can be used if new lifters are installed.

INSTALLATION

15 The used lifters must be installed in their original bores. Coat them with moly-base grease or engine assembly lube.

16 Lubricate the bearing surfaces of the lifter bores with engine oil.

17 Install the lifter(s) in the lifter bore(s).

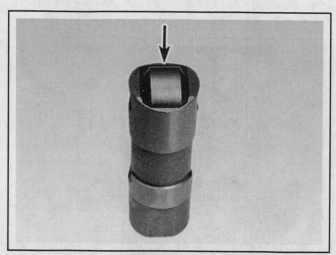

7.13 The roller on roller lifters must turn freely - check for wear and excessive play as well

➡ Note: Make sure that the oil orifice is facing toward the front of the engine.

18 Install the lifter guide(s) and retainer(s).

19 Install the pushrods, pushrod guides, rocker arms and rocker arm retaining bolts (see Section 6).

❋❋ CAUTION:

Make sure that each pair of lifters is on the base circle of the camshaft; that is, with both valves closed, before tightening the rocker arm bolts.

20 Tighten the rocker arm bolts to the specified torque.

21 Install the pushrod cover (see Section 5).

22 Install the rocker arm cover (see Section 3).

8 Intake manifold - removal and installation

▶ **Refer to illustrations 8.10, 8.15a and 8.15b**

1 Disconnect the cable from the negative battery terminal, then remove the engine cover (see Chapter 11).

✳✳ CAUTION:

On models equipped with an anti-theft audio system, disable the anti-theft feature before performing any operation that requires disconnecting the battery or disrupting power to the stereo (see the Anti-theft audio system procedures at the front of this manual).

2 Remove the air cleaner assembly (see Chapter 4).
3 Remove the PCV valve and hose.
4 Drain the cooling system (see Chapter 1).
5 Refer to Chapter 4 and detach the fuel lines, vacuum lines and wire leads from the fuel injection assembly. When disconnecting the fuel line be prepared to catch some fuel, then plug the line to prevent contamination.
6 After noting how it's installed, disconnect the fuel injection throttle linkage. Disconnect the cruise control and TV linkage (if so equipped).
7 Tag and disconnect all remaining vacuum lines and wires hooked to components on the manifold. Remove the coolant hoses from the manifold.
8 Remove the ignition coil. Unbolt the alternator and bracket and set it aside (see Chapter 5).
9 Remove the mounting bolts and separate the manifold from the cylinder head. Don't pry between the manifold and head, as damage to the gasket sealing surfaces may result.
10 Remove the old gasket (see illustration).

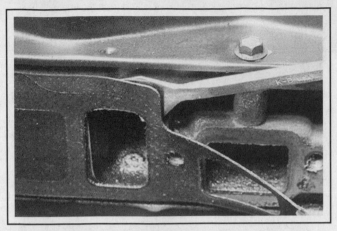

8.10 Remove the old intake manifold gasket with a scraper - don't leave any material on the mating surfaces

11 If a new manifold is being installed, transfer all components still attached to the old manifold to the new one.
12 Before installing the manifold, clean the cylinder head and manifold gasket surfaces with lacquer thinner or acetone. All gasket material and sealing compound must be removed prior to installation.
13 Apply a thin coat of RTV sealant to the intake manifold and cylinder head mating surfaces. Make certain that the sealant will not spread into the air or coolant passages when the manifold is installed.
14 Place a new gasket on the manifold, hold the manifold in position against the cylinder head and install the mounting bolts finger tight.
15 Tighten the mounting bolts a little at a time in the correct sequence (see illustrations) until they're all at the specified torque.
16 Install the remaining components in the reverse order of removal.
17 Fill the radiator with coolant, start the engine and check for leaks.

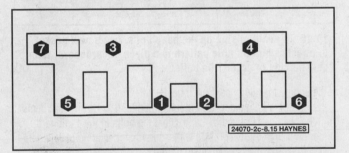

8.15a 1990 and earlier model intake manifold mounting bolt tightening sequence

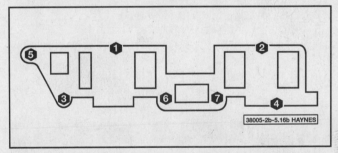

8.15b Use this intake manifold bolt tightening sequence on 1991 and later models

9 Exhaust manifold - removal and installation

▶ **Refer to illustration 9.13**

1 Remove the cable from the negative battery terminal.

✳✳ CAUTION:

On models equipped with an anti-theft audio system, disable the anti-theft feature before performing any operation that requires disconnecting the battery or disrupting power to the stereo (see the Anti-theft audio system procedures at the front of this manual).

2 Remove the engine cover (see Chapter 11).
3 Remove the heat stove pipe at the exhaust manifold.
4 Raise the vehicle and support it securely on jackstands.
5 Disconnect the oxygen sensor.
6 Label the four spark plug wires, then disconnect them and secure them out of the way.
7 Disconnect the exhaust pipe from the exhaust manifold. You may have to apply penetrating oil to the fastener threads, as they are usually corroded. The exhaust pipe can be hung from the frame with a piece of wire.

8   Remove the rear air conditioning compressor bracket.

9   Remove the exhaust manifold end bolts first, then remove the center bolts and separate the exhaust manifold from the engine.

10 Remove the exhaust manifold gasket.

11 Before installing the manifold, clean the gasket mating surfaces on the cylinder head and manifold. All old gasket material and carbon deposits must be removed. Check the bolt threads for damage.

12 Place a new exhaust manifold gasket in position on the cylinder head, then place the manifold in position and install the mounting bolts finger tight.

13 Tighten the mounting bolts a little at a time in the correct sequence (see illustration) until all of the bolts are at the specified torque.

14 Reconnect the exhaust pipe and reinstall the air conditioning compressor bracket.

15 Lower the vehicle.

16 Install the remaining components in the reverse order of removal.

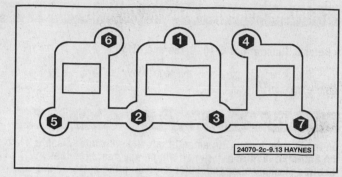

9.13 Recommended tightening sequence for the exhaust manifold bolts

17 Start the engine and check for exhaust leaks between the manifold and cylinder head and between the manifold and exhaust pipe.

10 Cylinder head - removal and installation

REMOVAL

▶ **Refer to illustrations 10.15, 10.16 and 10.17**

1   Disconnect the negative cable from the battery and remove the engine cover (see Chapter 11).

❄❄ CAUTION:

On models equipped with an anti-theft audio system, disable the anti-theft feature before performing any operation that requires disconnecting the battery or disrupting power to the stereo (see the Anti-theft audio system procedures at the front of this manual).

2   Drain the cooling system (see Chapter 1).

3   Remove the air cleaner assembly (see Chapter 4).

4   Remove the throttle, cruise control and TV cables (as equipped).

5   Remove the dipstick tube and thermostat housing.

6   Remove the alternator and brackets (see Chapter 5).

7   Unbolt the air conditioner compressor and swing it out of the way for clearance.

❄❄ CAUTION:

Don't disconnect any of the air conditioning lines unless the system has been depressurized by a dealer service department or repair shop.

8   Disconnect all wires and vacuum hoses from the cylinder head and manifold. Be sure to label them to simplify reinstallation. Refer to Chapter 4 and detach the fuel lines from the TBI unit.

9   Remove the upper radiator, water pump bypass and heater hoses.

10 Disconnect the spark plug wires and remove the spark plugs. Be sure to label the plug wires to simplify reinstallation.

11 Remove the rocker arm cover (see Section 3).

12 Remove the pushrods (see Section 6).

13 Remove the ignition coil (see Chapter 5).

14 Raise the vehicle and support it securely on jackstands, then unbolt the exhaust pipe from the manifold.

15 Using a new head gasket, outline the cylinders and bolt pattern on a piece of cardboard (see illustration). Be sure to indicate the front

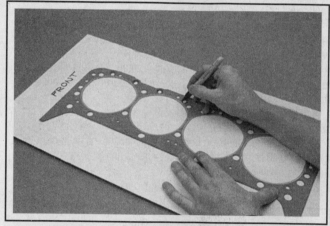

10.15 To avoid mixing up the head bolts, use a new gasket to transfer the bolt hole pattern to a piece of cardboard, then punch holes to accept the bolts . . .

of the engine for reference. Punch holes at the bolt locations.

16 Loosen the cylinder head mounting bolts in 1/4-turn increments until they can be removed by hand. Store the bolts in the cardboard holder as they're removed - this will ensure that they are reinstalled in their original locations (see illustration).

17 Lift the head off the engine. If it's stuck, pry on it only at the overhang on the thermostat end of the head (see illustration).

❄❄ CAUTION:

If you pry on the head anywhere else, damage to the gasket surface may result.

18 Place the head on a block of wood to prevent damage to the gasket surface. Refer to Part C for cylinder head disassembly and valve service procedures.

INSTALLATION

▶ **Refer to illustrations 10.23, 10.25, 10.26 and 10.28**

19 If a new cylinder head is being installed, transfer all external parts

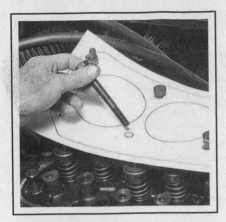

10.16 . . . and push each head bolt through the matching hole in the cardboard

10.17 If the head is stuck, pry it up at the overhang just behind and below the thermostat housing

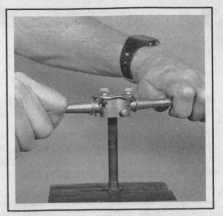

10.23 A die should be used to remove sealant and corrosion from the head bolt threads prior to installation

from the old cylinder head to the new one.

20 The mating surfaces of the cylinder head and block must be perfectly clean when the head is installed. It's also a good idea to have the head checked for distortion (warpage) and cracks by an automotive machine shop.

21 Use a gasket scraper to remove all traces of carbon and old gasket material, then clean the mating surfaces with lacquer thinner or acetone. If there's oil on the mating surfaces when the head is installed, the gasket may not seal correctly and leaks may develop. Use a vacuum cleaner to remove any debris that falls into the cylinders.

22 Check the block and head mating surfaces for nicks, deep scratches and other damage. If damage is slight, it can be removed with a file; if it's excessive, machining may be the only alternative.

23 Use a tap of the correct size to chase the threads in the head bolt holes. Mount each bolt in a vise and run a die down the threads to remove corrosion and restore the threads (see illustration). Dirt, corrosion, sealant and damaged threads will affect critical head bolt torque readings.

24 Position the new gasket over the dowel pins in the block, then carefully position the head on the block without disturbing the gasket.

25 Coat the threads and the undersides of the heads of cylinder head bolt numbers 9 and 10 with sealing compound and install the bolts finger tight (see illustration).

26 Tighten each of the bolts a little at a time in sequence until all bolts are at 18 ft-lbs (see illustration).

27 On 1985 models only, continue tightening the bolts, in sequence, to 40, then to 70 and finally to 92 ft-lbs. On 1986 models the final torque is 90 ft-lbs.

28 On 1987 and later models, tighten all bolts EXCEPT the left front bolt (number 9 in the sequence) to 22 ft-lbs. Tighten number 9 to 30 ft-lbs. Following the sequence one more time, tighten all bolts EXCEPT number 9 an additional 120-degrees (1/3-turn). Tighten number 9 an additional 90-degrees (1/4-turn) (see illustration).

29 The remaining installation steps are the reverse of removal.

30 Change the oil and filter, run the engine and check for leaks.

10.25 The cylinder head mounting bolts should be coated with sealant at the locations shown (arrows) before installation

10.28 The final step in tightening the cylinder head bolts is known as "angle torquing" - all bolts except number 9 should be turned an additional 120-degrees, while bolt 9 should be turned an additional 90-degrees

10	6	2	3	9
8	5	1	4	7

24070-2c-10.25 HAYNES

10.26 Cylinder head bolt positions and tightening sequence

11 Crankshaft pulley and hub - removal and installation

▶ Refer to illustrations 11.6 and 11.9

1 Remove the cable from the negative battery terminal.

❋❋❋ CAUTION:

On models equipped with an anti-theft audio system, disable the anti-theft feature before performing any operation that requires disconnecting the battery or disrupting power to the stereo (see the Anti-theft audio system procedures at the front of this manual).

2 Remove the drivebelts (see Chapter 1). Tag each belt as it's removed to simplify reinstallation.

3 Raise the vehicle and place it securely on jackstands.

4 If your vehicle is equipped with a manual transmission, apply the parking brake and put the transmission in gear to prevent the engine from turning over, then remove the crankshaft pulley bolts. If your vehicle is equipped with an automatic transmission, it may be necessary to remove the starter motor (see Chapter 5) and immobilize the starter ring gear with a large screwdriver while an assistant loosens the pulley bolts.

5 To loosen the crankshaft hub retaining bolt, install a bolt in one of the pulley bolt holes. Attach a breaker bar, extension and socket to the hub retaining bolt and immobilize the hub by wedging a large screwdriver between the bolt and the socket. Remove the hub retaining bolt.

6 Remove the crankshaft hub. Use a puller if necessary (see illustration).

7 Refer to Section 12 for the front oil seal replacement procedure.

8 Apply a thin layer of moly-base grease to the seal contact surface of the hub.

9 Slide the pulley hub onto the crankshaft until it bottoms against the crankshaft timing gear. Note that the slot in the hub must be aligned with the Woodruff key in the end of the crankshaft. The hub retaining bolt can also be used to press the hub into position (see illustration).

10 Tighten the hub-to-crankshaft bolt to the specified torque.

11 Install the crankshaft pulley on the hub. Use Locktite on the bolt threads.

12 Install the drivebelts (see Chapter 1).

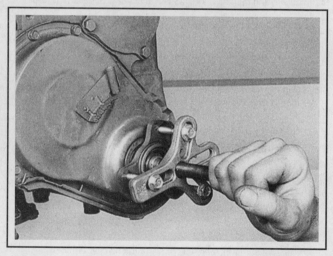

11.6 Use a puller to remove the hub from the crankshaft

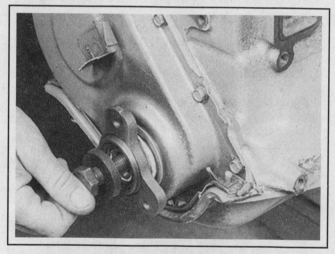

11.9 Use the pulley hub bolt to press the hub onto the crankshaft

12 Front crankshaft oil seal - replacement

➡Note: The front crankshaft oil seal can be replaced with the timing gear cover in place. However, due to the limited amount of room available, you may conclude that the procedure would be easier if the cover were removed from the engine first. If so, refer to Section 14 for the cover removal and installation procedure.

TIMING GEAR COVER IN PLACE

▶ Refer to illustration 12.2

1 Disconnect the negative battery cable from the battery, then remove the crankshaft pulley hub (see Section 11).

❋❋❋ CAUTION:

On models equipped with an anti-theft audio system, disable the anti-theft feature before performing any operation that requires disconnecting the battery or disrupting power to the stereo (see the Anti-theft audio system procedures at the front of this manual).

2 Note how the seal is installed - the new one must face the same direction! Carefully pry the oil seal out of the cover with a seal puller or a large screwdriver (see illustration). Be very careful not to distort the cover or scratch the crankshaft!

3 Apply clean engine oil or multi-purpose grease to the outer edge

of the new seal, then install it in the cover with the lip (open end) facing IN. Drive the seal into place with a large socket and a hammer (if a large socket isn't available, a piece of pipe will also work). Make sure the seal enters the bore squarely and stop when the front face is flush with the cover.

4   Install the pulley hub (see Section 11).

TIMING GEAR COVER REMOVED

▶ **Refer to illustrations 12.6, 12.8a and 12.8b**

5   Remove the timing gear cover as described in Section 14.

6   Using a large screwdriver, pry the old seal out of the cover (see illustration). Be careful not to distort the cover or scratch the wall of the seal bore. If the engine has accumulated a lot of miles, apply penetrating oil to the seal-to-cover joint and allow it to soak in before attempting to remove the seal.

7   Clean the bore to remove any old seal material and corrosion. Support the cover on a block of wood and position the new seal in the bore with the lip (open end) of the seal facing IN. A small amount of oil applied to the outer edge of the new seal will make installation easier - don't overdo it!

8   Drive the seal into the bore with a large socket and hammer until it's completely seated (see illustration). Select a socket that's the same

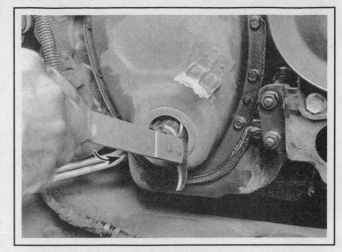

12.2 The front crankshaft seal can be removed in the vehicle with a seal removal tool or a large screwdriver (V6 engine shown - four-cylinder similar)

outside diameter as the seal. A section of pipe or even a block of wood can be used if a socket isn't available) (see illustration).

9   Reinstall the timing gear cover.

12.6 Once the timing gear cover is removed, place it on a flat surface and gently pry the old seal out with a large screwdriver

12.8a Clean the bore, then apply a small amount of oil to the outer edge of the new seal and drive it squarely into the opening with a large socket . . .

12.8b . . . or a block of wood and a hammer - don't damage the seal in the process!

13 Top Dead Center (TDC) for number 1 piston - locating

▶ **Refer to illustration 13.5**

1   Top Dead Center (TDC) is the highest point in the cylinder that each piston reaches as it travels up-and-down when the crankshaft turns. Each piston reaches TDC on the compression stroke and again on the exhaust stroke, but TDC generally refers to piston position on the compression stroke. The timing marks on the pulley installed on the front of the crankshaft are referenced to the number one piston at TDC on the compression stroke.

2   Positioning the piston(s) at TDC is an essential part of many procedures such as rocker arm removal, camshaft and timing gear removal

and distributor removal.

3   In order to bring any piston to TDC, the crankshaft must be turned using one of the methods outlined below. When looking at the front of the engine, normal crankshaft rotation is clockwise.

❉❉ **WARNING:**

Before beginning this procedure, be sure to place the transmission in Neutral and unplug the distributor wire harness connector to disable the ignition system.

a) The preferred method is to turn the crankshaft with a large socket and breaker bar attached to the crankshaft pulley hub bolt threaded into the front of the crankshaft.

b) A remote starter switch, which may save some time, can also be used. Attach the switch leads to the S (switch) and B (battery) terminals on the starter motor. Once the piston is close to TDC, use a socket and breaker bar as described in the previous paragraph.

c) If an assistant is available to turn the ignition switch to the Start position in short bursts, you can get the piston close to TDC without a remote starter switch. Use a socket and breaker bar as described in Paragraph a) to complete the procedure.

4   Note the position of the terminal for the number one spark plug wire on the distributor cap. Use a felt-tip pen or chalk to make a mark on the distributor body directly under the terminal. Remove the screws, detach the cap from the distributor and set it aside.

5   Turn the crankshaft (see Paragraph 3 above) until the notch in the crankshaft pulley is aligned with the 0 on the timing plate (located at the front of the engine) (see illustration).

6   Look at the distributor rotor - it should be pointing directly at the mark you made on the distributor body. If the rotor is pointing at the terminal for the number four spark plug, the number one piston is at TDC on the exhaust stroke.

7   To get the piston to TDC on the compression stroke, turn the crankshaft one complete turn (360°) clockwise. The rotor should now be pointing at the mark on the distributor. When the rotor is pointing at the number one spark plug wire terminal in the distributor cap and the

13.5 Turn the crankshaft until the notch in the drivebelt pulley (lower arrow) is directly opposite the zero mark on the timing plate (upper arrow)

ignition timing marks are aligned, the number one piston is at TDC on the compression stroke.

8   After the number one piston has been positioned at TDC on the compression stroke, TDC for any of the remaining pistons can be located by turning the crankshaft 180° at a time and following the firing order (1-3-4-2).

14 Timing gear cover - removal and installation

▶ **Refer to illustrations 14.9 and 14.13**

1   Detach the cable from the negative terminal of the battery.

❊❊ CAUTION:

On models equipped with an anti-theft audio system, disable the anti-theft feature before performing any operation that requires disconnecting the battery or disrupting power to the stereo (see the Anti-theft audio system procedures at the front of this manual).

2   Remove the power steering reservoir (if equipped). Be prepared to catch the fluid in a drain pan.

3   Remove the upper fan shroud (see Chapter 3).

4   Remove the drivebelts (see Chapter 1).

5   Detach the fan and drivebelt pulley.

6   Disconnect and unbolt the alternator and brackets (see Chapter 5).

14.9 Timing gear cover bolt locations

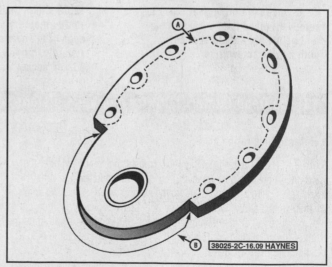

14.13 Timing gear cover sealant application details

A   1/4-inch by 1/8-inch bead of RTV sealant
B   3/8-inch by 3/16-inch bead of RTV sealant

7   Raise the front of the vehicle and support it on jackstands. Apply the parking brake. Drain the coolant (see Chapter 1), then detach the lower radiator hose from the water pump.

8   Remove the crankshaft pulley and hub (see Section 11).

9   Remove the timing gear cover-to-block bolts (see illustration).

10 Remove the cover by carefully prying it off. The cover is sealed with RTV, so it may be stuck to the block. The flange between the cover and oil pan may be bent during removal. Try to minimize damage to the flange or it may be too distorted to be straightened.

11 Use a scraper to remove all old sealant from the cover, oil pan and block, then clean the mating surfaces with lacquer thinner or acetone.

12 Check the cover flanges for distortion, particularly around the bolt holes. If necessary, place the cover on a block of wood and use a hammer to flatten and restore the mating surfaces.

13 Apply a 3/8-inch wide by 3/16-inch thick bead of RTV sealant to the timing gear cover flange that mates with the oil pan. Apply a 1/4-inch wide by 1/8-inch thick bead to the cover-to-block flange (see illustration).

14 Apply a dab of sealant to the joints between the oil pan and the engine block.

15 Place the cover in position and loosely install a couple of mounting bolts.

16 Install the crankshaft hub to center the cover (see Section 11). Be sure to lubricate the seal contact surface of the hub.

17 Install the remaining mounting bolts and tighten them to the specified torque.

18 Install the crankshaft pulley, then hook up the lower radiator hose.

19 Lower the vehicle.

20 Refill the cooling system (see Chapter 1).

21 Install the components removed to gain access to the cover.

22 Reattach the cable to the negative terminal of the battery.

23 Start the engine and check for oil leaks at the seal.

---

15 Camshaft, timing gears and bearings - removal, inspection and installation

CAMSHAFT LOBE LIFT CHECK

▶ **Refer to illustration 15.3**

1   In order to determine the extent of cam lobe wear, the lobe lift should be checked prior to camshaft removal. Refer to Section 3 and remove the rocker arm cover.

2   Position the number one piston at TDC on the compression stroke (see Section 13).

3   Beginning with the number one cylinder valves, mount a dial indicator on the engine and position the plunger against the top surface of the first rocker arm. The plunger should be directly above and in line with the pushrod (see illustration).

4   Zero the dial indicator, then very slowly turn the crankshaft in the normal direction of rotation (clockwise when looking at the front of the engine) until the indicator needle stops and begins to move in the opposite direction. The point at which it stops indicates maximum cam lobe lift.

5   Record this figure for future reference, then reposition the piston at TDC on the compression stroke.

6   Move the dial indicator to the remaining number one cylinder rocker arm and repeat the check. Be sure to record the results for each valve.

7   Repeat the check for the remaining valves. Since each piston must be at TDC on the compression stroke for this procedure, work from cylinder-to-cylinder following the firing order sequence. Turn the crankshaft 180-degrees when moving from one cylinder to the next.

8   After the check is complete, compare the results to the Specifications. If camshaft lobe lift is less than specified, cam lobe wear has occurred and a new camshaft should be installed.

REMOVAL

▶ **Refer to illustration 15.24**

9   Disconnect the negative battery cable from the battery.

10 Remove the engine cover (see Chapter 11).

✳✳ CAUTION:

On models equipped with an anti-theft audio system, disable the anti-theft feature before performing any operation that requires disconnecting the battery or disrupting power to the stereo (see the Anti-theft audio system procedures at the front of this manual).

11 Detach the power steering reservoir from the fan shroud. Be prepared to catch the fluid in a drain pan.

12 Remove the radiator (see Chapter 3).

13 Remove the drivebelts (see Chapter 1).

14 Detach the fan and drivebelt pulley (see Chapter 3).

15 Remove the air cleaner assembly.

16 Remove the pushrod cover (see Section 5).

17 Remove the EGR valve (see Chapter 6).

18 Tag and disconnect the vacuum hoses at the intake manifold and thermostat housing.

19 Remove the pushrods and lifters.

15.3 When checking the camshaft lobe lift, the dial indicator plunger must be positioned directly above and in line with the pushrod

20 Refer to Section 14 and detach the timing gear cover.
21 Remove the distributor (see Chapter 5) and the oil pump driveshaft (see Section 16).
22 Remove the headlight bezels, the grille and the bumper filler panel (see Chapter 11).
23 Remove the air conditioning condenser baffles. Unbolt the condenser and support it out of the way - don't disconnect the refrigerant lines!
24 Turn the crankshaft until the holes in the camshaft gear are aligned with the thrust plate bolts, then remove the bolts (see illustration).
25 Carefully pull the camshaft and gear assembly out of the block.

✳✳ CAUTION:

To avoid damage to the camshaft bearings as the lobes pass over them, support the camshaft near the block as it's withdrawn.

26 The crankshaft gear should slide off the crankshaft without a great deal of resistance.

INSPECTION

▶ **Refer to illustration 15.28**

Camshaft

27 After the camshaft has been removed from the engine, cleaned with solvent and dried, inspect the bearing journals for uneven wear, pitting and evidence of seizure. If the journals are damaged, the bearing inserts in the block are probably damaged as well. Both the camshaft and bearings will have to be replaced.
28 If the journals are in good condition, measure the bearing journals with a micrometer (see illustration) to determine their sizes and whether or not they're out-of-round. The inside diameter of each bearing can be measured with a telescoping gauge and micrometer. Subtract each cam journal diameter from the corresponding bearing inside diameter to obtain the bearing oil clearance.
29 Compare the clearance for each bearing to the Specifications. If it's excessive for any of the bearings, have new bearings installed by an automotive machine shop.
30 Check the camshaft lobes for heat discoloration, score marks,

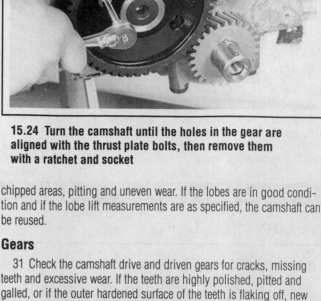

15.24 Turn the camshaft until the holes in the gear are aligned with the thrust plate bolts, then remove them with a ratchet and socket

chipped areas, pitting and uneven wear. If the lobes are in good condition and if the lobe lift measurements are as specified, the camshaft can be reused.

Gears

31 Check the camshaft drive and driven gears for cracks, missing teeth and excessive wear. If the teeth are highly polished, pitted and galled, or if the outer hardened surface of the teeth is flaking off, new parts will be required. If one gear is worn or damaged, replace both gears as a set. Never install one new and one used gear.
32 Check the gear end clearance with a feeler gauge between the thrust plate and camshaft journal face and compare it to the Specifications. If it's less than the minimum specified, the spacer ring should be replaced. If it's excessive, the thrust plate must be replaced. In either case, the gear will have to be pressed off the camshaft, so take the parts to an automotive machine shop.

Bearing replacement

33 Camshaft bearing replacement requires special tools and expertise that place it outside the scope of the home mechanic. Take the block to an automotive machine shop to ensure that the job is done correctly.

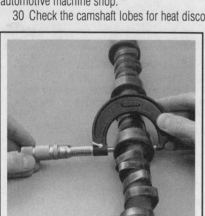

15.28 The camshaft bearing journal diameters are checked to pinpoint excessive wear and out-of-round conditions

15.34 Be sure to prelube the bearing journals and lobes prior to camshaft installation

15.36 Align the timing marks as shown here when installing the camshaft

INSTALLATION

▶ **Refer to illustrations 15.34 and 15.36**

34 Lubricate the camshaft bearing journals and cam lobes with moly-base grease, engine assembly lube or camshaft and lifter prelube (see illustration).

35 Slide the camshaft into the engine. Support the cam near the block and be careful not to scrape or nick the bearings.

36 Install the gear on the end of the crankshaft (if not already done). Don't forget the Woodruff key and don't hammer the gear onto the shaft. Align the timing marks on the gears as the gears mesh (see illustration).

37 Line up the access holes in the gear with the thrust plate holes and the bolt holes in the block. Apply Locktite to the threads, then install the thrust plate bolts and tighten them to the specified torque.

38 The remaining installation steps are the reverse of removal.

16 Oil pump driveshaft - removal and installation

▶ **Refer to illustrations 16.3, 16.6, 16.7 and 16.8**

1 Disconnect the cable from the negative terminal of the battery, then remove the engine cover (see Chapter 11).

❋❋ CAUTION:

On models equipped with an anti-theft audio system, disable the anti-theft feature before performing any operation that requires disconnecting the battery or disrupting power to the stereo (see the Anti-theft audio system procedures at the front of this manual).

2 Raise the vehicle and support it on jackstands.

3 Remove the oil pump driveshaft retainer plate bolts (see illustration).

4 Remove the oil pump driveshaft and bushing with a magnet.

5 Clean the mating surfaces of the block and retainer plate with lacquer thinner or acetone.

6 Check the bushing and driveshaft for wear (see illustration). Replace them if they're worn or damaged.

7 Install the bushing and oil pump driveshaft in the block. The shaft driven gear must mesh with the camshaft drive gear and the slot in the lower end of the shaft must mate with the oil pump gear tang (see illustration).

8 Apply a 1/16-inch bead of RTV sealant to the retainer plate so it completely seals around the oil pump driveshaft hole in the block (see illustration).

9 Install the retainer plate and tighten the mounting bolts securely.

16.3 The oil pump driveshaft retainer plate is located on the engine block, just below the pushrod cover and just above the oil filter

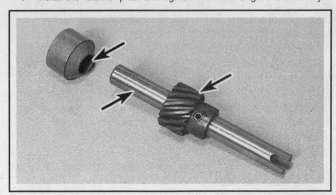

16.6 Inspect the oil pump driveshaft, gear and bushing for wear and damage

16.7 If the slotted oil pump driveshaft is properly mated with the oil pump gear tang, the top of the bushing will be flush with the retainer plate mounting surface

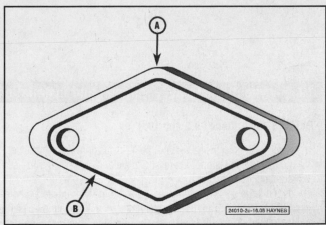

16.8 Apply a 1/16-inch bead of RTV sealant (B) to the oil pump driveshaft retainer plate (A) in the area shown

17 Oil pan - removal and installation

1   Disconnect the cable from the negative battery terminal.

※ CAUTION:

On models equipped with an anti-theft audio system, disable the anti-theft feature before performing any operation that requires disconnecting the battery or disrupting power to the stereo (see the Anti-theft audio system procedures at the front of this manual).

2   Raise the vehicle and support it securely on jackstands.
3   Drain the engine oil and remove the oil filter (see Chapter 1).
4   Disconnect the exhaust pipe at the manifold and hangers and tie the system aside.
5   Remove the starter (see Chapter 5) and the bellhousing dust cover.
6   Remove the bolts and detach the oil pan. Don't pry between the block and pan or damage to the sealing surfaces may result and oil

leaks could develop. If the pan is stuck, dislodge it with a block of wood and a hammer.
7   Use a scraper to remove all traces of sealant from the pan and block, then clean the mating surfaces with lacquer thinner or acetone.
8   Apply a 3/16-inch wide by 1/8-inch thick bead of RTV sealant to the oil pan flange. Make the bead 3/8-inch wide by 3/16-inch thick between the bolt holes at the rear end of the pan. Apply a 1/8-inch bead of sealant to the block at the rear main bearing cap joints and the timing gear cover joints.
9   Install the oil pan and tighten the mounting bolts to the specified torque. Start at the center of the pan and work out toward the ends in a spiral pattern.
10   Install the bellhousing dust cover and the starter, then reconnect the exhaust pipe to the manifold and hanger brackets.
11   Lower the vehicle.
12   Install a new filter and add oil to the engine.
13   Reconnect the negative battery cable.
14   Start the engine and check for leaks.

18 Oil pump - removal and installation

▶ **Refer to illustration 18.2**

1   Remove the oil pan (see Section 17).
2   Remove the two oil pump mounting bolts and the pick-up tube bracket nut from the main bearing cap bolt (see illustration).
3   Detach the oil pump and pick-up assembly from the block.
4   If the pump is defective, replace it with a new one. If the engine is being completely overhauled, install a new oil pump - don't reuse the original or attempt to rebuild it.
5   To install the pump, turn the shaft so the gear tang mates with the slot on the lower end of the oil pump driveshaft. The oil pump should slide easily into place over the oil pump driveshaft lower bushing. If it doesn't, pull it off and turn the tang until it's aligned with the pump driveshaft slot.
6   Install the pump mounting bolts and the tube bracket nut. Tighten them to the specified torque.
7   Reinstall the oil pan (see Section 17).
8   Add oil, run the engine and check for leaks.

18.2 Remove the oil pump flange mounting bolts and the pick-up tube bracket nut

19 Flywheel/driveplate - removal and installation

▶ **Refer to illustrations 19.2 and 19.3**

1   Refer to Chapter 7 and remove the transmission. If your vehicle has a manual transmission, the pressure plate and clutch will also have to be removed (see Chapter 8).
2   Jam a large screwdriver in the starter ring gear or driveplate hole to keep the crankshaft from turning, then remove the mounting bolts (see illustration). Since it's fairly heavy, support the flywheel as the last bolt is removed.
3   Pull straight back on the flywheel/driveplate to detach it from the

crankshaft. Note the shim installed between the driveplate and crankshaft on vehicles equipped with an automatic transmission (see illustration).
4   On manual transmission equipped vehicles, check the pilot bushing and replace it if necessary (see Chapter 8).
5   Installation is the reverse of removal. Be sure to align the hole in the flywheel/driveplate with the dowel pin in the crankshaft. Use Locktite on the bolt threads and tighten them to the specified torque in a criss-cross pattern.

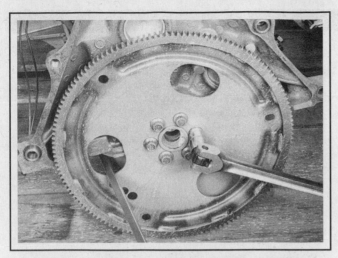

19.2 A large screwdriver wedged in the starter ring gear teeth or one of the holes in the driveplate can be used to keep the flywheel/driveplate from turning as the mounting bolts are removed

19.3 Don't lose the shim used on automatic transmission equipped vehicles

20 Rear main oil seal - replacement

▶ **Refer to illustrations 20.5 and 20.8**

1   The rear main bearing oil seal can be replaced without removing the oil pan or crankshaft.

2   Remove the transmission (see Chapter 7).

3   If equipped with a manual transmission, remove the pressure plate and clutch disc (see Chapter 8).

4   Remove the flywheel or driveplate (see Section 19).

5   Using a seal removal tool or a large screwdriver, carefully pry the seal out of the block (see illustration). Don't scratch or nick the crankshaft in the process.

6   Clean the bore in the block and the seal contact surface on the crankshaft. Check the crankshaft surface for scratches and nicks that could damage the new seal lip and cause oil leaks. If the crankshaft is damaged, the only alternative is a new or different crankshaft.

7   Apply a light coat of engine oil or multi-purpose grease to the outer edge of the new seal. Lubricate the seal lip with moly-base grease.

8   Press the new seal into place with a special tool, available at most auto parts stores. The seal lip must face toward the front of the engine. If the special tool isn't available, carefully work the seal lip over the end of the crankshaft and tap the seal in with a hammer and punch until it's seated in the bore (see illustration).

9   Install the flywheel or driveplate.

10 If equipped with a manual transmission, reinstall the clutch disc and pressure plate.

11 Reinstall the transmission as described in Chapter 7.

20.5 Carefully pry the oil seal out with a screwdriver - don't nick or scratch the crankshaft or the new seal will be damaged and leaks will develop

20.8 Tap around the outer edge of the new seal with a hammer and punch to seat it squarely in the bore

21 Engine mounts - replacement

⁂ WARNING:

Improper lifting methods or devices are hazardous and could result in severe injury or death. DO NOT place any part of your body under the engine/transmission when it's supported only by a jack. Failure of the lifting device could result in serious injury or death.

1 If the rubber mounts have hardened, cracked or separated from the metal backing plates, they must be replaced. This operation may be carried out with the engine/transmission still in the vehicle.

2 Disconnect the negative cable from the battery.

⁂ CAUTION:

On models equipped with an anti-theft audio system, disable the anti-theft feature before performing any operation that requires disconnecting the battery or disrupting power to the stereo (see the Anti-theft audio system procedures at the front of this manual).

3 Raise the front of the vehicle and support it securely on jackstands.

4 Support the engine with a jack. Position a block of wood between the jack head and the oil pan.

5 Remove the engine mount-to-chassis bolts.

6 Remove the mount-to-engine support bracket bolts. It's not necessary to detach the support bracket from the engine.

7 Raise the engine just enough to clear the bracket, then remove the engine mount.

8 Place the new mount in position.

9 Install the mount-to-engine support bracket bolts and tighten them securely.

10 Tighten the mount-to-chassis bolts securely.

11 Remove the jack.

12 Remove the jackstands and lower the vehicle.

Specifications

General

Cylinder numbers (front-to-rear)	1-2-3-4
Firing order	1-3-4-2
Oil pressure	36 to 41 @ 2000 rpm

Camshaft

Lobe lift (intake and exhaust)	0.398 inch
Bearing journal diameter	1.869 inches
Bearing oil clearance	0.0007 to 0.0027 inch
Gear/thrust plate end clearance	0.0015 to 0.0050 inch

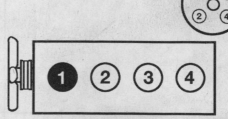

Cylinder location and distributor rotation

The blackened terminal shown on the distributor cap indicates the Number One spark plug wire position

FIRING ORDER
1-3-4-2

24010-1-specs HAYNES

Torque specifications   Ft-lbs (unless otherwise noted)

Cylinder head bolts	
1985	92
1986	90
1987 on (see illustration 10.26)	
Step 1 - all bolts, in sequence	18
Step 2 - all bolts, except 9, in sequence	22
Step 2 - bolt 9	30
Step 3 - all bolts, except 9, in sequence	Tighten an additional 120°
Step 3 - bolt 9	Tighten an additional 90°
Intake manifold-to-cylinder head bolts	
1985	29
1986 (see illustration 8.15a or 8.15b)	
Bolts 1, 2 and 6	38
Bolts 3, 4 and 5	25
Bolt 7	37
1987 on	25
Exhaust manifold bolts	
1985	44
1986 on (see illustration 9.13)	
Bolts 1, 2 and 7	36
Bolts 3, 4, 5 and 6	32
Flywheel-to-crankshaft bolts	
1985	44
1986	55
1987 on	65
Driveplate-to-crankshaft bolts	
1985	44
1986 on	55
Crankshaft pulley hub-to-crankshaft bolt	160
Lifter guide retainer-to-block stud	90 in-lbs
Oil pan bolts	
1985	75 in-lbs
1986 on	90 in-lbs
Oil pan drain plug	25
Oil pick-up tube bracket nut	37
Oil pump-to-block bolts	22
Pushrod cover nuts	90 in-lbs
Rocker arm bolts	
1985 and 1986	20
1987 on	24
Rocker arm cover bolts	
1985 and 1986	72 in-lbs
1987 on	48 in-lbs
Timing gear cover bolts	90 in-lbs
Camshaft thrust plate bolts	90 in-lbs

Notes

Section

Reference to other Chapters

2B

V6 ENGINE

1 General information

This Part of Chapter 2 is devoted to in-vehicle repair procedures for the V6 engine. All information concerning engine removal and installation and engine block and cylinder head overhaul can be found in Part C of this Chapter.

The VIN W V6 engine introduced in 1992 is very similar to the other V6 engines in this Chapter, except it incorporates a balancer shaft and the Central Multiport Fuel Injection (CMFI) system. In 1994, the balancer shaft became standard on all V6 engines. An improved fuel delivery system called Central Sequential Fuel Injection (CSFI) was introduced in 1996 and is installed on subsequent V6 engines. Refer to Chapter 4 for more information on the different fuel systems used on these engines. If there's any doubt as to which engine you have, refer to the Vehicle Identification Number (VIN) located on the forward edge of the dashboard on the driver's side. The VIN is visible from outside the vehicle, through the windshield. The eighth position in the alphanumeric code indicates the engine type.

The following repair procedures are based on the assumption that the engine is installed in the vehicle. If the engine has been removed from the vehicle and mounted on a stand, many of the steps outlined in this Part of Chapter 2 will not apply.

The Specifications included in this Part of Chapter 2 apply only to the procedures contained in this Part. Part C of Chapter 2 contains the Specifications necessary for cylinder head and engine block rebuilding.

✳✳ CAUTION:

On models equipped with an anti-theft audio system, disable the anti-theft feature before performing any operation that requires disconnecting the battery or disrupting power to the stereo (see the Anti-theft audio system procedures at the front of this manual).

2 Repair operations possible with the engine in the vehicle

Many major repair operations can be accomplished without removing the engine from the vehicle.

Clean the engine compartment and the exterior of the engine with some type of pressure washer before any work is done. It will make the job easier and help keep dirt out of the internal areas of the engine.

Remove the engine cover and the hood, if necessary, to improve access to the engine as repairs are performed (refer to Chapter 11 if necessary).

If vacuum, exhaust, oil or coolant leaks develop, indicating a need for gasket or seal replacement, the repairs can generally be made with the engine in the vehicle. The intake and exhaust manifold gaskets, timing cover gasket, oil pan gasket, crankshaft oil seals and cylinder head gaskets are all accessible with the engine in place.

Exterior engine components, such as the intake and exhaust manifolds, the oil pan (and the oil pump), the water pump, the starter motor, the alternator, the distributor and the fuel system components can be removed for repair with the engine in place.

Since the cylinder heads can be removed without pulling the engine, valve component servicing can also be accomplished with the engine in the vehicle. Replacement of the timing chain and sprockets is also possible with the engine in the vehicle.

In extreme cases caused by a lack of necessary equipment, repair or replacement of piston rings, pistons, connecting rods and rod bearings is possible with the engine in the vehicle. However, this practice is not recommended because of the cleaning and preparation work that must be done to the components involved.

3 Rocker arm covers - removal and installation

REMOVAL

1 Disconnect the cable from the negative terminal of the battery.

✳✳ CAUTION:

On models equipped with an anti-theft audio system, disable the anti-theft feature before performing any operation that requires disconnecting the battery or disrupting power to the stereo (see the Anti-theft audio system procedures at the front of this manual).

2 Remove the engine cover (see Chapter 11).
3 On carbureted and TBI engines, remove the air cleaner assembly and heat stove tube (see Chapter 4).

Right side

▶ **Refer to illustrations 3.4a, 3.4b, 3.5 and 3.13**

4 On carbureted and TBI engines, unbolt the oil filler tube, engine

3.4a Remove the dipstick tube mounting bolt and the oil filler tube bolts (arrows) (some have already been removed in this photo)

oil dipstick and transmission dipstick tube (if equipped) from the alternator bracket (see illustration). Detach the wiring harness from the oil filler tube (see illustration), then remove the tube from the rocker arm cover.

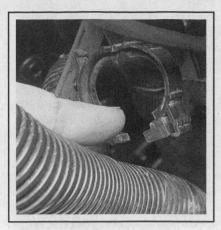

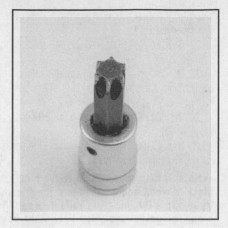

3.4b Unclip the wiring harness to make room for rocker arm cover removal

3.5 The diverter valve (arrow) is mounted rear of the right rocker arm cover

3.13 The valve cover bolts require a T-30 Torx bit for removal

5   On carbureted and TBI engines, remove the air injection diverter valve, bracket and hoses (see illustration) (see Chapter 6 if necessary).

6   On carbureted and TBI engines, unbolt the dipstick tube bracket from the cylinder head and move it aside.

7   Disconnect the spark plug wires from the brackets. If any wire will inhibit rocker arm cover removal, label the wire and disconnect it from the distributor.

8   If equipped, remove the spark plug wire bracket from the side of the cylinder head.

9   Remove the PCV valve or vent hose (as applicable) from the rocker arm cover and place it out of the way.

10 On CMFI engines, remove the ignition coil and bracket and place it out of the way.

11 On CSFI engines, remove the EVAP canister purge solenoid/bracket (see Chapter 6), air cleaner assembly (see Chapter 4) and unbolt the engine oil dipstick tube from the front of the cylinder head.

12 Label and disconnect any wiring harnesses and vacuum hoses that will interfere with rocker cover removal.

13 Use a T-30 Torx bit to remove the rocker arm cover bolts, then detach the cover from the head (see illustration).

➡Note: If the cover is stuck to the head, bump one end with a block of wood and a hammer to jar it loose. If that doesn't work, try to slip a flexible putty knife between the head and cover to break the gasket seal. Don't pry at the cover-to-head joint or damage to the sealing surfaces may occur (leading to oil leaks in the future).

Left side

▶ Refer to illustrations 3.18 and 3.19

14 Perform Steps 1, 2 and 3 above.

15 On CSFI engines, remove the air cleaner assembly and intake duct (see Chapter 4).

16 If equipped, detach the cruise control servo/bracket and set it aside (see illustration 5.8).

17 If equipped, remove the A/C compressor mounting bolts and then move the compressor forward to provide clearance for rocker arm cover removal (see illustration 5.6) (see Chapter 3, if necessary).

❋❋ **CAUTION:**

Do not disconnect the A/C refrigerant lines!

18 On models equipped with a vacuum power brake booster, disconnect the vacuum hose from the intake manifold (see illustration).

19 Remove the PCV valve from the rocker arm cover and position it out of the way (see illustration).

20 On late model models, remove the engine oil filler tube.

21 Disconnect the spark plug wires from the brackets. If any wire

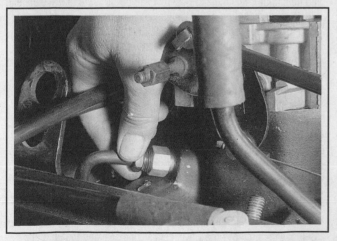

3.18 Unscrewing the power brake booster vacuum line fitting

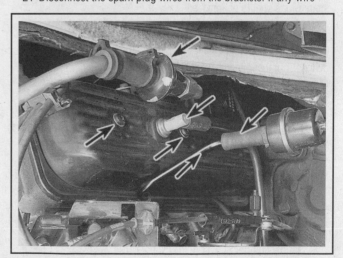

3.19 Rear view of the left rocker arm cover showing the vacuum brake booster line, the PCV valve, the AIR check valve and tube and the rocker arm cover mounting bolts (arrows)

will inhibit rocker arm cover removal, label the wire and disconnect it from the distributor.

22 On CMFI engines, remove the spark plug wire bracket from the cylinder head.

23 On carbureted and TBI engines, disconnect the accelerator/TV cables from the throttle lever. Unbolt the bracket from the intake manifold and place them out of the way.

24 On CSFI engines, remove the EGR tube (see Chapter 6, if necessary).

25 Use a T-30 TORX bit to remove the rocker arm cover bolts, then detach the cover from the head.

➡Note: If the cover is stuck to the head, bump one end with a block of wood and a hammer to jar it loose. If that doesn't work, try to slip a flexible putty knife between the head and cover to break the gasket seal. Don't pry at the cover-to-head joint or damage to the sealing surfaces may occur (leading to oil leaks in the future).

INSTALLATION

26 The mating surfaces of each cylinder head and rocker arm cover must be perfectly clean when the covers are installed. Use a gasket scraper to remove all traces of sealant and old gasket material, then clean the mating surfaces with lacquer thinner or acetone. If there's sealant or oil on the mating surfaces when the cover is installed, oil leaks may develop.

27 Clean the mounting bolt threads with a die to remove any corrosion and restore damaged threads. Make sure the threaded holes in the head are clean - run a tap into them to remove corrosion and restore damaged threads.

28 The gaskets should be mated to the covers before the covers are installed. Apply a thin coat of RTV sealant to the cover flange, then position the gasket inside the cover lip and allow the sealant to set up so the gasket adheres to the cover. If the sealant isn't allowed to set, the gasket may fall out of the cover as it's installed on the engine.

29 Carefully position the cover on the head and install the bolts.

30 Tighten the bolts in three or four steps to the specified torque.

31 The remaining installation steps are the reverse of removal.

32 Start the engine and check carefully for oil leaks as the engine warms up.

---

4   Valve spring, retainer and seals - replacement

▶ **Refer to illustrations 4.4, 4.8a, 4.8b, 4.9, 4.16 and 4.17**

➡Note: **Broken valve springs and defective valve stem seals can be replaced without removing the cylinder heads. Two special tools and a compressed air source are normally required to perform this operation, so read through this Section carefully and rent or buy the tools before beginning the job. If compressed air isn't available, a length of nylon rope can be used to keep the valves from falling into the cylinder during this procedure.**

1   Refer to Section 3 and remove the rocker arm cover from the affected cylinder head. If all of the valve stem seals are being replaced, remove both rocker arm covers.

2   Remove the spark plug from the cylinder which has the defective component. If all of the valve stem seals are being replaced, all of the spark plugs should be removed.

3   Turn the crankshaft until the piston in the affected cylinder is at top dead center on the compression stroke (refer to Section 9 for instructions). If you are replacing all the valve stem seals, begin with cylinder number one and work on the valves one cylinder at a time. Move from cylinder-to-cylinder following the firing order sequence (1-6-5-4-3-2).

4   Thread an adapter into the spark plug hole (see illustration) and connect an air hose from a compressed air source to it. Most auto parts stores can supply the air hose adapter.

➡Note: **Many cylinder compression gauges utilize a screw-in fitting that may work with your air hose quick-disconnect fitting.**

5   Remove the nut, pivot ball and rocker arm for the valve with the defective part and pull out the pushrod. If all of the valve stem seals are being replaced, all of the rockers arms and pushrods should be removed (refer to Section 7).

6   Apply compressed air to the cylinder. The valves should be held in place by the air pressure. If the valve faces or seats are in poor condition, leaks may prevent air pressure from retaining the valves - refer to the alternative procedure below.

7   If you don't have access to compressed air, an alternative method

4.4 Use compressed air, if available, to hold the valves closed when the springs are removed - the air hose adapter (arrow) threads into the spark plug hole and accepts the hose from the compressor

can be used. Position the piston at a point just before TDC on the compression stroke, then feed a long piece of nylon rope through the spark plug hole until it fills the combustion chamber. Be sure to leave the end of the rope hanging out of the engine so it can be removed easily. Use a large breaker bar and socket to rotate the crankshaft in the normal direction of rotation until slight resistance is felt.

8   Stuff shop rags into the cylinder head holes above and below the valves to prevent parts and tools from falling into the engine, then use a valve spring compressor to compress the spring/damper assembly. Remove the keepers with small needle-nose pliers or a magnet (see illustration).

➡Note: **A couple of different types of tools are available for compressing the valve springs with the head in place. One type grips the lower spring coils and presses on the retainer as the knob is turned, while the other type utilizes the rocker arm stud and nut for leverage (see illustration). Both types work very well, although the lever type is usually less expensive.**

9   Remove the spring retainer or rotator, oil shield and valve spring assembly, then remove the valve stem O-ring seal and the umbrella-

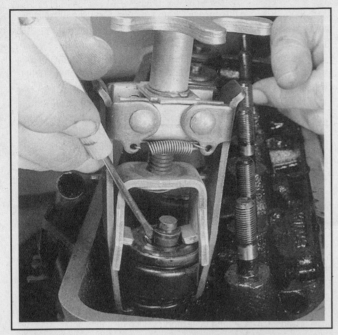

4.8a Once the spring is depressed, the keepers can be removed with a small magnet or needle-nose pliers (a magnet is preferred to prevent dropping the keepers)

4.8b The stamped steel lever-type valve spring compressor is usually less expensive than the type that grips the spring coils

type guide seal (the O-ring seal will most likely be hardened and will probably break when removed, so plan on installing a new one each time the original is removed) (see illustration).

➡**Note: If air pressure fails to hold the valve in the closed position during this operation, the valve face and/or seat is probably damaged. If so, the cylinder head will have to be removed for additional repair operations.**

10 Wrap a rubber band or tape around the top of the valve stem so the valve won't fall into the combustion chamber, then release the air pressure.

➡**Note: If a rope was used instead of air pressure, turn the crankshaft slightly in the direction opposite normal rotation.**

11 Inspect the valve stem for damage. Rotate the valve in the guide and check the end for eccentric movement, which would indicate that

the valve is bent.

12 Move the valve up-and-down in the guide and make sure it doesn't bind. If the valve stem binds, either the valve is bent or the guide is damaged. In either case, the head will have to be removed for repair.

13 Reapply air pressure to the cylinder to retain the valve in the closed position, then remove the tape or rubber band from the valve stem. If a rope was used instead of air pressure, rotate the crankshaft in the normal direction of rotation until slight resistance is felt.

14 Lubricate the valve stem, guide and new seal with clean engine oil.

➡**Note: On 1998 and later models, a new exhaust valve oil seal was introduced. It has the letters "EX" molded into the top of the seal and is either brown with a white stripe around its circumference or solid red. They are not interchangeable with the intake valve seals, which are black.**

Place the valve seal onto the guide and, using an appropriate seal installation tool and hammer (a deep socket will also work), drive the seal onto the valve guide. On all engines except 1998 and later models, drive the seal until it's bottomed on the guide. On 1998 and later

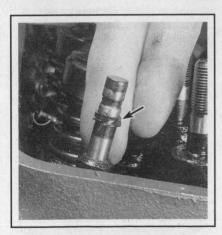

4.9 The O-ring seal (arrow) should be replaced with a new one each time the keepers and retainer are removed

4.16 Make sure the O-ring seal under the retainer is seated in the groove and not twisted before installing the keepers

4.17 Apply a small dab of grease to each keeper as shown here before installation - it will hold them in place on the valve stem as the spring is released

engines, drive the seal onto the guide until the gap between the bottom of the seal and the start of the chamfer on the guide is 0.040 to 0.078 inch.

15 Install the damper spring (if used), valve spring and oil shield (if used) onto the valve/guide.

16 Install the valve spring retainer or rotator. Compress the valve spring assembly and carefully install the new O-ring seal in the lower groove of the valve stem. Make sure the seal isn't twisted - it must lie perfectly flat in the groove (see illustration).

17 Position the keepers in the upper groove. Apply a small dab of grease to the inside of each keeper to hold it in place if necessary (see illustration). Remove the pressure from the spring tool and make sure the keepers are seated. Refer to Chapter 2, Part C, Section 11 and check the seals with a vacuum pump.

18 Disconnect the air hose and remove the adapter from the spark plug hole. If a rope was used in place of air pressure, pull it out of the cylinder.

19 Refer to Section 7 and install the rocker arm(s) and pushrod(s).

20 Install the spark plug(s) and hook up the wire(s).

21 Refer to Section 3 and install the rocker arm cover(s).

22 Start and run the engine, then check for oil leaks and unusual sounds coming from the rocker arm cover area.

5 Intake manifold - removal and installation

REMOVAL

♦ **Refer to illustrations 5.8, 5.13, 5.15, 5.17, 5.18, 5.20, 5.24 and 5.27**

➥**Note: On 1996 and later models, it is not necessary to remove the upper intake manifold to remove the lower manifold.**

1 On 1986 and later models, perform the fuel pressure relief procedure (see Chapter 4).

2 Disconnect the negative cable from the battery.

❊❊ CAUTION:

On models equipped with an anti-theft audio system, disable the anti-theft feature before performing any operation that requires disconnecting the battery or disrupting power to the stereo (see the Anti-theft audio system procedures at the front of this manual).

3 Drain the radiator (see Chapter 1).

4 Remove the air cleaner assembly and adapter or air intake duct (see Chapter 4).

5 Remove the upper fan shroud (see Chapter 3, Section 5). This is unnecessary on later models.

6 Remove the engine cover (see Chapter 11).

7 Disconnect the upper radiator and heater hoses from the intake manifold.

8 If equipped, detach the cruise control servo/bracket (see illustration) and position it out of the way.

9 Disconnect the PCV valve hose from the intake manifold and place it aside. On 1999 and later models, also disconnect the EVAP canister purge solenoid valve electrical harness.

10 Locate the engine on TDC for the number one piston (see Section 9).

11 Label and disconnect the spark plug wires from the distributor cap (see Chapter 1 if necessary).

12 Remove the distributor (see Chapter 5).

13 Detach the accelerator cable, cruise control cable (if equipped) and TV cable (if equipped) from the throttle lever and intake manifold (see illustration) (see Chapter 4 if necessary).

14 Label and disconnect the fuel lines (see Chapter 4 if necessary).

➥**Note: On TBI engines, unbolt the fuel line from the rear of the cylinder head. It may be necessary to use locking pliers to keep the stud from turning.**

15 On 1985 through 1993 models, detach the power brake vacuum booster line from the intake manifold (see illustration).

➥**Note: Some 1993 models are equipped with a hydraulic type brake booster, in which case this step can be ignored.**

16 Remove the ignition coil and bracket (see Chapter 5).

➥**Note: This step does not apply to coil-in-cap HEI systems.**

5.8 The arrow points to the optional cruise control servo bracket bolt location - the socket wrench is on the lower bracket mounting bolt

5.13 Pinch the cable tabs together to get them out of the bracket (cable removed for clarity)

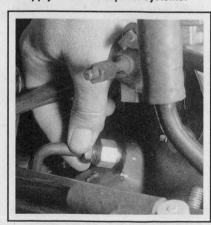

5.15 The vacuum brake booster line-to-intake manifold fitting is to the left of the throttle body

5.17 Remove the ESC module bracket from the side of the intake manifold (TBI engine)

5.18 Location of the bolts holding the dipstick and oil filler tubes to the alternator bracket (arrows)

5.20 Disconnect the coolant tube at the manifold and set it aside

17 On TBI engines, unbolt the ESC module/bracket from the intake manifold (see illustration) and position it out of the way (see Chapter 6 if necessary).

18 Detach the engine oil filler tube and transmission dipstick from the alternator bracket and remove the alternator bracket from the intake manifold (see illustration).

19 On carbureted and TBI engines, detach any Air Management system components that may interfere with intake manifold removal (see Chapter 6 if necessary).

20 On carbureted and TBI engines, remove the coolant tube from the right side of the intake manifold (see illustration).

21 On 2002 and earlier CSFI engines, remove the EGR pipe connecting the left exhaust manifold to the intake manifold (see Chapter 6, if necessary). If difficulty is encountered, apply penetrating oil to the fittings to ease removal.

22 On CMFI engines, remove the upper intake manifold (see Chapter 4).

➡**Note: On CSFI engines it is not necessary to remove the upper intake manifold to remove the lower manifold.**

23 On CSFI engines, remove the coolant bypass hose connecting the water pump to the intake manifold.

24 If equipped, remove the A/C compressor from its mounting (without detaching the refrigerant lines) and set it aside (see illustration).

For later models, see Step 25. On some models, the A/C compressor and mounting bracket may have to be removed (see Chapter 3 for more information).

❊❊❊ **WARNING:**

The A/C system is under high pressure. DO NOT disassemble any part of the system unless it has been properly discharged.

---

25 Removing the front manifold bolt on later models will require removing the drivebelt and the air conditioning compressor side brace (if equipped). Then loosen the nuts and bolts for the power steering pump mounting bracket and rear bracket. Leaving the power steering pump and air conditioning compressor (if equipped) on the bracket, slide the bracket forward to access the front intake manifold bolt (see Step 27).

26 Label and disconnect all vacuum and electrical connections attached to the intake manifold, carburetor, throttle body, or upper intake plenum.

27 Make note of the location of each manifold bolt or stud for reassembly. Loosen the manifold mounting bolts (12 bolts on 1985 to 1995 models, 8 bolts on 1996 and later models) in 1/4 turn increments until they can be removed by hand (see illustration).

5.24 Rear of the A/C compressor showing the brace attached to the intake manifold stud (arrow) (throttle body removed for clarity)

5.27 On CMFI engines, don't forget the four bolts (arrows) located inside the lower intake manifold

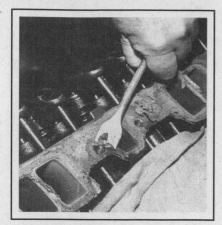

5.29 After covering the lifter valley, use a gasket scraper to remove all traces of sealant and old gasket material from the head and manifold mating surfaces

5.30a The bolt hole threads must be cleaned and dry to ensure accurate torque readings when the manifold mounting bolts are installed

5.30b Clean the bolt holes with compressed air, but be careful - wear safety goggles!

28 The manifold will probably be stuck to the cylinder heads and force may be required to break the gasket seal. A large pry bar can be positioned under the cast-in lug near the left front mounting bolt to pry up the front of the manifold, but make sure all bolts have been removed first!

✳✳ CAUTION:

Don't pry between the block and manifold or the heads and manifold or damage to the gasket sealing surfaces may occur, leading to vacuum leaks.

INSTALLATION

▶ Refer to illustrations 5.29, 5.30a, 5.30b, 5.31, 5.32, 5.33, 5.34, 5.36a, 5.36b, 5.36c and 5.36d

➡Note: The mating surfaces of the cylinder heads, block and manifold must be perfectly clean when the manifold is installed. Gasket removal solvents in aerosol cans are available at most auto parts stores and may be helpful when removing old gasket

material that's stuck to the heads and manifold (since the manifold is made of aluminum, aggressive scraping can cause damage). Be sure to follow the directions printed on the container.

29 Use a gasket scraper to remove all traces of sealant and old gasket material, then clean the mating surfaces with lacquer thinner or acetone. If there's old sealant or oil on the mating surfaces when the manifold is installed, oil or vacuum leaks may develop. When working on the heads and block, cover the lifter valley with shop rags to keep debris out of the engine (see illustration). Use a vacuum cleaner to remove any gasket material that falls into the intake ports in the heads.

30 Use a tap of the correct size to chase the threads in the bolt holes, then use compressed air (if available) to remove the debris from the holes (see illustrations).

✳✳ WARNING:

Wear safety glasses or a face shield to protect your eyes when using compressed air! Remove excessive carbon deposits and corrosion from the exhaust, EGR and coolant passages in the heads and manifold.

31 Apply a 3/16-inch wide bead of RTV sealant to the front and rear

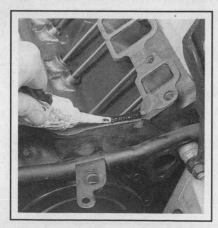

5.31 Apply a 3/16-inch bead of RTV sealant to the front and rear manifold mating surface of the block

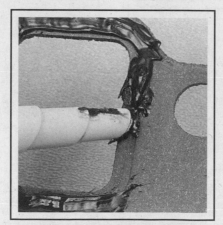

5.32 RTV sealant should be used around the coolant passage holes in the new intake manifold gaskets

5.33 Be sure to install the gaskets with the marks UP!

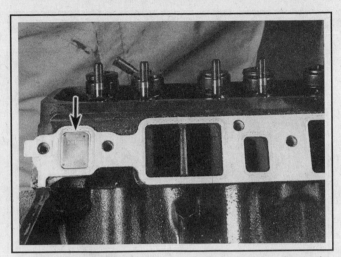

5.34 The rear coolant passages on some models are blocked off - make sure the gasket is installed with the blocked off hole at the rear!

manifold mating surfaces of the block (see illustration). Make sure the beads extend up the heads 1/2-inch on each side.

32 Apply a thin coat of RTV sealant around the coolant passage holes on the cylinder head side of the new intake manifold gaskets (see illustration).

33 Position the gaskets on the cylinder heads, with the ears at each end overlapping the bead of RTV sealant on the head. The upper side of each gasket will have a THIS SIDE UP label stamped into it to ensure correct installation (see illustration).

34 Make sure all intake port openings, coolant passage holes and bolt holes are aligned correctly.

→Note: The gaskets used on 1986 and later models have the rear coolant passages blocked off (see illustration). Be sure the gaskets are installed with the blocked off passages at the rear of the engine. Some gaskets may have small tabs which must be bent over until they're flush with the rear surface of each head.

35 Carefully set the manifold in place while the sealant is still wet.

※ CAUTION:

Don't disturb the gaskets and don't move the manifold fore and aft after it contacts the sealant on the block.

36 Install the bolts/studs in their proper locations. Following the recommended sequence, tighten the bolts to the torque listed in this

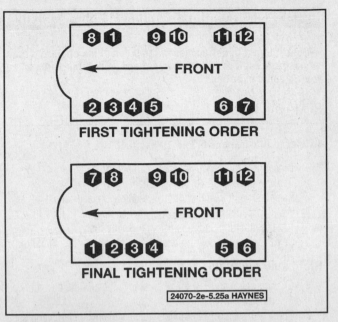

5.36a Intake manifold bolt tightening sequence - note that the initial sequence differs from the final sequence! (Carburetor and TBI models)

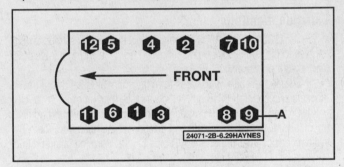

5.36b 1991 through 1995 VIN Z engine intake manifold bolt tightening sequence - bolt "A" must be tightened an additional amount (see this Chapter's Specifications)

Chapter's Specifications (see illustrations). Notice there are different procedures for different models.

37 The remaining installation steps are the reverse of removal. Start the engine and check carefully for oil and coolant leaks at the intake manifold joints.

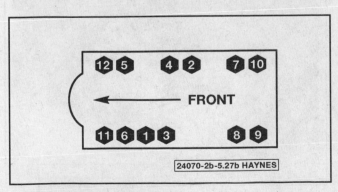

5.36c Intake manifold bolt tightening sequence (1992 through 1995 VIN W models)

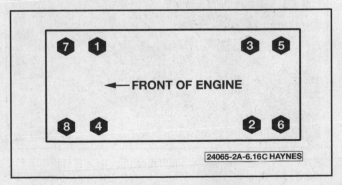

5.36d 1996 and later V6 engine intake manifold bolt tightening sequence (CSFI engine)

6 Exhaust manifolds - removal and installation

➡ Note: To gain better access to the sides of the engine, it may be easier to remove the front wheel and work through the wheel well in some situations.

REMOVAL

▶ Refer to illustrations 6.7, 6.10 and 6.13

1 Disconnect the negative battery cable from the battery.
2 Remove the engine cover (see Chapter 11).
3 Raise the vehicle and support it securely on jackstands.
4 Working under the vehicle, apply penetrating oil to the exhaust pipe-to-manifold studs and nuts (they're usually rusty).
5 Remove the nuts holding the exhaust pipe(s) to the manifold(s).

Right side manifold

6 On carbureted and TBI models, remove the air filter assembly and heat stove pipe (see Chapter 4).
7 On carbureted and TBI models, release the hose clamp and separate the hose from the AIR system check valve (see illustration).
8 Detach the spark plug wires from the plugs and brackets and position them out of the way (see Chapter 1).

Left side manifold

9 On 1986 through 1995 models, follow the oxygen sensor wiring harness to the connector and disconnect it from the wiring harness (see Chapter 6 if necessary).
10 Detach the spark plug wires from the spark plugs and routing brackets and position them out of the way (see illustration).
11 On 1985 through 1990 models, unscrew the AIR pipe fitting at the manifold, release the clamp and disconnect the hose from the AIR system check valve. Unbolt the AIR pipe bracket from the manifold and remove it from the engine.
12 On 1996 through 2002 models, unscrew the EGR pipe fittings at the intake and exhaust manifolds, remove the clamp bolt securing it to the intake manifold and remove the EGR tube from the engine.

➡ Note: Apply some penetrating oil to the pipe fittings to ease removal.

6.10 Left side exhaust manifold mounting details showing the spark plug wire bracket, heat shield, AIR system check valve, AIR tube-to-manifold fitting and AIR tube bracket bolt (arrows) (engine removed for clarity)

6.7 Right side exhaust manifold mounting details showing the AIR system check valve, spark plug wire bracket and heat stove pipe mount (arrows) (engine removed for clarity)

Both manifolds

13 Bend the lock tabs back (see illustration), then remove the mounting bolts and separate the manifold from the head. The heat shields (if equipped) will come off after the bolts are removed.

INSTALLATION

14 Check the manifold for cracks and make sure the bolt threads are clean and undamaged. The manifold and cylinder head mating surfaces must be clean before the manifolds are reinstalled - use a gasket scraper to remove all carbon deposits.
15 Before installing the exhaust manifold, apply some anti-seize compound to the threads of all fasteners.
16 Position the manifold, tab washers and heat shields (if equipped) on the head and install the mounting bolts.
17 When tightening the mounting bolts, work from the center to the ends and be sure to use a torque wrench. Tighten the bolts in three equal steps until the specified torque is reached.
18 The remaining installation steps are the reverse of removal.
19 Start the engine and check for exhaust leaks.

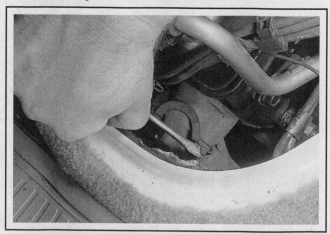

6.13 Flatten the tabs on the lock washers before removing the exhaust manifold mounting bolts

7 Rocker arms and pushrods- removal, inspection, installation and valve adjustment

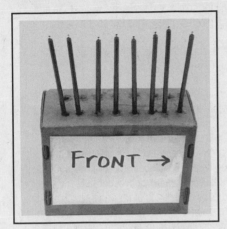

7.4 A perforated cardboard box can be used to store the pushrods to ensure that they are reinstalled in their original locations - note the label indicating the front of the engine

7.10 The ends of the pushrods and the valve stems should be lubricated with moly-base grease prior to installation of the rocker arms

7.11 Moly-base grease applied to the pivot balls will ensure adequate lubrication until oil pressure builds up when the engine is started

1985 THROUGH 1999 MODELS

Removal

♦ **Refer to illustration 7.4**

1 Refer to Section 3 and detach the rocker arm cover(s) from the cylinder head(s).

2 Beginning at the front of one cylinder head, loosen and remove the rocker arm stud nuts. Store them separately in marked containers to ensure that they will be reinstalled in their original locations.

➡**Note: If the pushrods are the only items being removed, loosen each nut just enough to allow the rocker arms to be rotated to the side so the pushrods can be lifted out.**

3 Lift off the rocker arms and pivot balls and store them in the marked containers with the nuts (they must be reinstalled in their original locations).

4 Remove the pushrods and store them separately to make sure they don't get mixed up during installation (see illustration).

Inspection

5 Check each rocker arm for wear, cracks and other damage, especially where the pushrods and valve stems contact the rocker arm faces.

6 Make sure the hole at the pushrod end of each rocker arm is open.

7 Check each rocker arm pivot area for wear, cracks and galling. If the rocker arms are worn or damaged, replace them with new ones and use new pivot balls as well.

8 Inspect the pushrods for cracks and excessive wear at the ends. Roll each pushrod across a piece of plate glass to see if it's bent (if it wobbles, it's bent).

Installation

♦ **Refer to illustrations 7.10, 7.11 and 7.13**

9 Lubricate the lower end of each pushrod with clean engine oil or moly-base grease and install them in their original locations. Make sure each pushrod seats completely in the lifter.

10 Apply moly-base grease to the ends of the valve stems and the

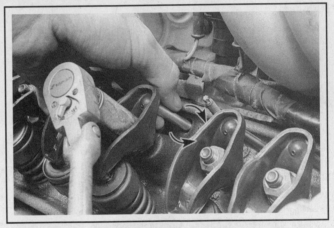

7.13 Rotate each pushrod as the rocker arm nut is tightened to determine the point at which all play is removed, then tighten each nut an additional one full turn

upper ends of the pushrods before positioning the rocker arms over the studs (see illustration).

11 Set the rocker arms in place, then install the pivot balls and nuts. Apply moly-base grease to the pivot balls to prevent damage to the mating surfaces before engine oil pressure builds up (see illustration). Be sure to install each nut with the flat side against the pivot ball.

Valve adjustment (all except VIN W engines)

➡**Note: On 1992 and later VIN W engines, there are no provisions for valve adjustment. The rocker arm studs have a positive stop shoulder for the rocker arm. After valve service, tighten the rocker arm nuts to the torque listed in this Chapter's Specifications. Unless there have been machining operations that significantly altered the valve lash, the adjustment should be correct. Some VIN Z engines came equipped with threaded-in rocker arm studs that had positive stop shoulders. Others came with pressed-in rocker arm studs, without positive stop shoulders and require the following valve adjustment procedure whenever the rocker arms have been loosened or removed.**

12 Refer to Section 9 and bring the number one piston to top dead center on the compression stroke.

13 Tighten the rocker arm nut (number one cylinder only) until all play is removed at the pushrod. This can be determined by rotating the pushrod between your thumb and index finger as the nut is tightened (see illustration). At the point where a slight drag is just felt as you spin the pushrod, all lash has been removed.

14 Tighten the nut an additional one full turn (360°) to center the lifter. Valve adjustment for cylinder number one is now complete.

15 At TDC for the number one cylinder on compression, the following valves can be adjusted using the method described above; exhaust valves on cylinders 1, 5, 6, intake valves on cylinders 1, 2 and 3.

16 Rotate the crankshaft clockwise one full rotation (360 degrees) until the timing marks align again. This is TDC for the number 4 cylinder. With the number four cylinder on TDC for compression, the following valves can be adjusted using the method described in steps 13 and 14; exhaust valves on cylinders 2, 3, 4, intake valves on cylinders 4, 5 and 6.

17 Refer to Section 3 and install the rocker arm covers. Start the engine, listen for unusual valvetrain noises and check for oil leaks at the rocker arm cover joints.

2000 AND LATER MODELS

Removal

18 Refer to Section 3 and detach the rocker arm cover(s) from the cylinder head(s).

19 Beginning with the center cylinder, loosen the rocker arm mounting bolts and remove both rocker arm assemblies from that cylinder. Next loosen and remove both rocker arm assemblies from one of the other cylinders, then repeat for the remaining cylinder. Store them separately in marked containers to ensure that they will be reinstalled in their original locations.

➡ Note: If the pushrods are the only items being removed, loosen each rocker arm bolt just enough to allow the rocker arms to be rotated to the side so the pushrods can be lifted out.

20 Pull straight up and remove the rocker arm support from the cylinder head.

21 Remove the pushrods and store them separately to make sure they don't get mixed up during installation (see illustration 7.4).

Inspection

22 Check each rocker arm for wear, cracks and other damage, especially where the pushrods and valve stems contact the rocker arm faces.

23 Make sure the hole at the pushrod end of each rocker arm is open.

24 Check the roller bearing pivot for binding or damage. If damaged, replace the rocker arm assembly.

25 Check the rocker arm bolt for thread damage. Repair the threads or replace the rocker arm assembly.

26 Inspect the pushrods for cracks and excessive wear at the ends. Roll each pushrod across a piece of plate glass to see if it's bent (if it wobbles, it's bent).

27 Inspect the rocker arm support for excessive wear or distortion.

Installation

28 Lubricate the lower end of each pushrod with clean engine oil or moly-based grease and install them in their original locations. Make sure each pushrod seats completely in the lifter.

29 Apply assembly oil to the rocker arm roller bearing pivots.

30 Position the rocker arm support with the UP arrow pointing toward the center of the engine and install both rocker arm supports. Align the holes with the threaded holes in the cylinder head.

31 Apply moly-based grease to the ends of the valve stems and the upper ends of the pushrods before installing the rocker arm assemblies.

32 Beginning with the center cylinder, install both rocker arm assemblies through the rocker arm support and into the cylinder head. Tighten the bolt finger-tight at this time. Next install both rocker arm assemblies into one of the other cylinders, then repeat for the remaining cylinder.

33 Refer to Section 9 and bring the number one piston to top dead center on the compression stroke.

34 Tighten the rocker arm bolt (number one cylinder only) until all play is removed at the pushrod. This can be determined by rotating the pushrod between your thumb and index finger as the bolt is tightened (see illustration 7.13). At the point where a slight drag is just felt as you spin the pushrod, all lash has been removed.

35 Tighten the bolt an additional one full turn (360 degrees) to center the lifter, then tighten to the torque listed in this Chapter's Specifications. Valve adjustment is now complete.

36 At TDC on the number one cylinder on the compression stoke, the following valves can be adjusted using the method described in Steps 34 and 35: exhaust valves on cylinders 1, 5 and 6, intake valves on cylinders 1, 2 and 3.

37 Rotate the crankshaft clockwise one full turn (360 degrees) until the timing marks align again. This is TDC for the number four cylinder. With the number four cylinder on TDC for compression, the following valves can be adjusted using the method described in Steps 34 and 35: exhaust valves on cylinders 2, 3 and 4, intake valves on cylinders 4, 5 and 6.

38 Refer to Section 3 and install the rocker arm covers. Start the engine, listen for unusual valve train noises and check for oil leaks at the rocker arm cover joints.

8   Cylinder heads - removal and installation

➡ Note: The engine must be completely cool when the heads are removed. Failure to allow the engine to cool off could result in head warpage.

REMOVAL

▶ Refer to illustration 8.15

1   Disconnect the negative battery cable. Remove the accessory drivebelt(s) (see Chapter 1).

2   On carbureted or TBI engines, remove the intake manifold (see Chapter 2A or 2B).

3   On CMFI or CSFI engines, remove the upper intake plenum (see Chapter 4). Also, remove the intake manifold (see Section 5).

4   Remove the engine cover. Raise the vehicle and support it securely on jackstands.

5   Remove the spark plugs (see Chapter 1).

6   Remove the exhaust manifold (see Section 6).

7   Remove the knock sensor and coolant temperature sensor (as applicable).

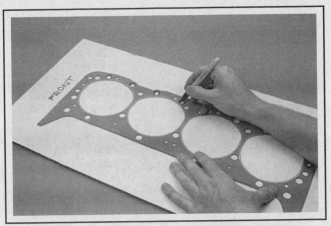

8.15 To avoid mixing up the head bolts, use a new gasket to transfer the bolt hole pattern to a piece of cardboard, then punch holes to accept the bolts

Left (driver's side) cylinder head

8 If equipped, remove the A/C compressor from its bracket (see Chapter 3) and, without disconnecting the refrigerant lines, place it out of the way.

✳✳ WARNING:

The A/C system is under high pressure. DO NOT disassemble any part of the system unless it has been properly discharged.

9 Remove the power steering pump bracket from the cylinder head (see Chapter 10 if necessary).
10 Check to see that all brackets or electrical connections have been removed from the cylinder head.

Right cylinder head

11 Remove the alternator (see Chapter 5).
12 On 1985 through 1993 carbureted and TBI models, remove the AIR pump and related components (see Chapter 6 if necessary).
13 Check to see that all brackets or electrical connections have been removed from the cylinder head.

Both cylinder heads

14 Remove the valve covers (see Section 3). Loosen the rocker arm fulcrum bolts enough to allow the rocker arms to be lifted off the pushrods and rotate them to one side. Remove the pushrods (see Section 7). Store them so they can be reinstalled in the same locations.
15 Using a new head gasket, outline the cylinders and bolt pattern on a piece of cardboard (see illustration). Be sure to indicate the front of the engine for reference. Punch holes at the bolt locations.
16 Loosen the head bolts in 1/4-turn increments until they can be removed by hand. Work from bolt-to-bolt in a pattern that's the reverse of the tightening sequence shown in illustration 8.26.
➡**Note: Don't overlook the row of bolts on the lower edge of each head, near the spark plug holes. Store the bolts in the cardboard holder as they're removed; this will ensure that the bolts are reinstalled in their original holes.**
17 Lift the head(s) off the engine. If resistance is felt, DO NOT pry between the head and block as damage to the mating surfaces will result. To dislodge the head, place a block of wood against the end of it and strike the wood block with a hammer. Store the heads on blocks of wood to prevent damage to the gasket sealing surfaces.

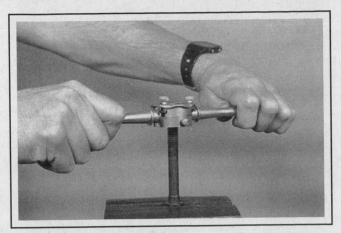

8.22 A die should be used to remove sealant and corrosion from the head bolt threads prior to installation

8.23a Locating dowels (arrows) are used to position the head gaskets on the block

18 Cylinder head disassembly and inspection procedures are covered in detail in Chapter 2, Part C.

INSTALLATION

◆ **Refer to illustrations 8.22, 8.23a, 8.23b, 8.25 and 8.26**

19 The mating surfaces of the cylinder heads and block must be perfectly clean when the heads are installed.
20 Use a gasket scraper to remove all traces of carbon and old gasket material, then clean the mating surfaces with lacquer thinner or acetone. If there's oil on the mating surfaces when the heads are installed, the gaskets may not seal correctly and leaks may develop. When working on the block, cover the lifter valley with shop rags to keep debris out of the engine. Use a vacuum cleaner to remove any debris that falls into the cylinders.
21 Check the block and head mating surfaces for nicks, deep scratches and other damage. If damage is slight, it can be removed with a file - if it's excessive, machining may be the only alternative.
22 Use a tap of the correct size to chase the threads in the head bolt holes. Mount each bolt in a vise and run a die down the threads to remove corrosion and restore the threads (see illustration). Dirt, corrosion, sealant and damaged threads will affect torque readings.
23 Position the new gaskets over the dowel pins in the block (see illustration).

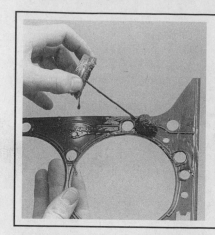

8.23b Steel gaskets should be coated with a sealant such as K&W Copper Coat before installation

8.25 The head bolts MUST be coated with a non-hardening sealant (such as Permatex no. 2) before they're installed - coolant will leak past the bolts if this isn't done

➥Note: If a steel gasket is used, apply a thin, even coat of sealant such as K&W Copper Coat to both sides prior to installation (see illustration).

Steel gaskets must be installed with the raised bead UP. The composition gasket must be installed dry - don't use sealant.

24 Carefully position the heads on the block without disturbing the gaskets.

25 Before installing the head bolts, coat the threads with a nonhardening sealant such as Permatex no. 2 (see illustration).

26 Install the bolts in their original locations and tighten them finger tight. Follow the recommended sequence and tighten the bolts in several steps to the torque listed in this Chapter's Specifications (see illustration).

27 The remaining installation steps are the reverse of removal.

28 Change the engine oil and filter (see Chapter 1), then start the engine and check carefully for oil and coolant leaks.

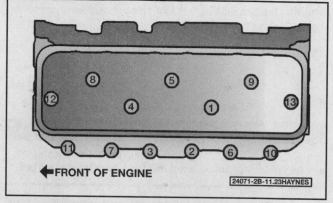

8.26 Cylinder head bolt tightening sequence

9 Top Dead Center (TDC) for number 1 piston - locating

▶ Refer to illustrations 9.4a, 9.4b, 9.6a, 9.6b and 9.7

1   Top Dead Center (TDC) is the highest point in the cylinder that each piston reaches as it travels up-and-down when the crankshaft turns. Each piston reaches TDC on the compression stroke and again on the exhaust stroke, but TDC generally refers to piston position on the compression stroke. The timing marks on the vibration damper installed on the front of the crankshaft are referenced to the number one piston at TDC on the compression stroke.

2   Positioning the piston(s) at TDC is an essential part of many procedures such as rocker arm removal, valve adjustment, timing chain and sprocket replacement and distributor removal.

3   In order to bring any piston to TDC, the crankshaft must be turned using one of the methods outlined below. When looking at the front of the engine, normal crankshaft rotation is clockwise.

✸✸ WARNING:

Before beginning this procedure, be sure to place the transmission in Neutral and unplug the wire connector at the distributor to disable the ignition system.

a) The preferred method is to turn the crankshaft with a large socket and breaker bar attached to the vibration damper bolt threaded into the front of the crankshaft.

b) A remote starter switch, which may save some time, can also be used. Attach the switch leads to the S (switch) and B (battery) terminals on the starter motor. Once the piston is close to TDC, use a socket and breaker bar as described in the previous paragraph.

c) If an assistant is available to turn the ignition switch to the Start position in short bursts, you can get the piston close to TDC without a remote starter switch. Use a socket and breaker bar as described in Paragraph a) to complete the procedure.

4   Make a mark on the distributor housing directly below the number one spark plug wire terminal on the distributor cap (see illustration).

➥Note: The terminal numbers are marked on the spark plug wires near the distributor (see illustration).

5   Remove the distributor cap as described in Chapter 1.

6   Turn the crankshaft (see Paragraph 3 above) until the line on the vibration damper is aligned with the zero mark on the timing plate (see illustrations). The timing plate and vibration damper are located low on the front of the engine, near the pulley that turns the drivebelt.

➥Note: Some 1992 and all later models have a TDC indicator located on the front cover which incorporates a notch that represents TDC.

7   The rotor should now be pointing directly at the mark on the distributor housing (see illustration). If it isn't, the piston is at TDC on the exhaust stroke.

9.4a Make a mark on the distributor housing directly below the number 1 spark plug wire (arrow)

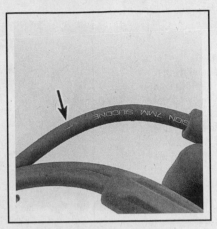

9.4b The spark plug wires are numbered to correspond to their respective cylinders (arrows)

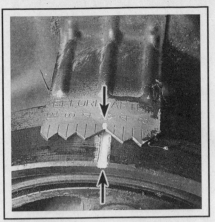

9.6a Turn the crankshaft until the line on the vibration damper is directly opposite the zero mark on the timing plate as shown here

8   To get the piston to TDC on the compression stroke, turn the crankshaft one complete turn (360°) clockwise. The rotor should now be pointing at the mark. When the rotor is pointing at the number one spark plug wire terminal in the distributor cap (which is indicated by the mark on the housing) and the ignition timing marks are aligned, the number one piston is at TDC on the compression stroke.

9   After the number one piston has been positioned at TDC on the compression stroke, TDC for any of the remaining cylinders can be located by turning the crankshaft 120° at a time and following the firing order (refer to the Specifications).

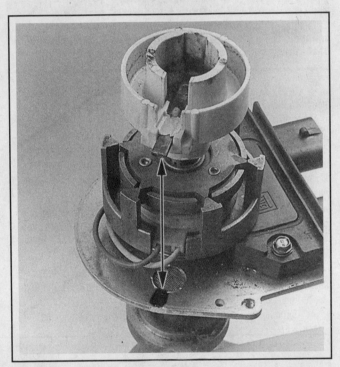

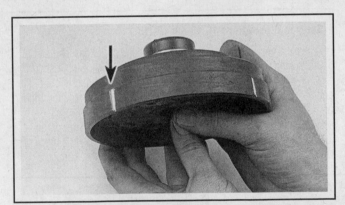

9.6b The vibration damper has two marks - the first one that approaches the timing plate can be disregarded, while the second mark (arrow) is the one that's used to locate TDC

9.7 If the rotor is pointing directly at the mark on the distributor housing, as shown here, the number one piston is at TDC on the compression stroke

10 Timing cover, chain and sprockets - removal and installation

REMOVAL

⬥ Refer to illustrations 10.2, 10.4, 10.5, 10.7, 10.8, 10.9a and 10.9b

1   Refer to Chapter 3 and remove the water pump.
2   Remove the bolts and separate the crankshaft drivebelt pulley from the vibration damper (see illustration).

3   Refer to Section 9 and position the number four piston at TDC on the compression stroke.

✳ CAUTION:

Once this has been done, DO NOT turn the crankshaft until the timing chain and sprockets have been reinstalled!

10.2 The vibration damper bolt (arrow) is usually very tight, so use a six-point socket and a breaker bar to loosen it (the three other bolts hold the pulley to the vibration damper)

4 Remove the bolt from the front of the crankshaft, then use a puller to detach the vibration damper (see illustration).

✸✸ CAUTION:

Don't use a puller with jaws that grip the outer edge of the damper. The puller must be the type shown in the illustration that utilizes bolts to apply force to the damper hub only.

5 On 1995 and later models, remove the Crankshaft Position Sensor from the timing chain cover and detach the wiring harness (see illustration).

6 Remove the oil pan (see Section 13).

7 Remove the bolts and separate the timing chain cover from the engine block. It will probably be stuck - if so, use a putty knife or screwdriver to break the seal (see illustration). Metal type covers can be easily distorted, so be careful when prying it off.

➡Note: On 1995 and later models, the timing chain cover is a composite material and is not reusable. Replace it if removed. Measure the timing chain free play. If it is more that 5/8 inch, the chain and both sprockets should be replaced.

8 On 1995 and later models, remove the Crankshaft Position Sensor reluctor ring (see illustration). On 2004 and later models, remove the timing chain tensioner.

9 Remove the 3 bolts (or 2 bolts and 1 nut on some engines)

10.4 Use the recommended puller to remove the vibration damper - if a puller that applies force to the outer edge is used, the damper will be damaged!

securing the timing chain sprocket to the camshaft, then detach the camshaft sprocket and chain as an assembly (see illustrations).

➡Note: On engines equipped with a balancer shaft, the balancer shaft drive gear will remain attached to the camshaft.

10 To remove the crankshaft sprocket, use a two or three jaw type gear puller. Be careful not to damage the threads in the end of the crankshaft. If the crankshaft front seal has been leaking, refer to Section 15 and replace it.

INSTALLATION

◆ **Refer to illustrations 10.13 and 10.15**

11 Use a gasket scraper to remove all traces of old gasket material and sealant from the cover (if metal) and engine block. Stuff a shop rag into the opening at the front of the oil pan to keep debris out of the engine. Clean the cover and block sealing surfaces with lacquer thinner or acetone.

12 On metal covers, check the cover flange for distortion, particularly around the bolt holes. If necessary, place the cover on a block of wood and use a hammer to flatten and restore the gasket surface.

10.5 On 1995 and later vehicles, remove the crankshaft position sensor (large arrow) and detach the wiring harness from the timing chain cover (small arrow)

10.7 A putty knife or screwdriver can be used to break the timing chain cover-to-block seal, but be careful when prying it off as damage to the cover may result

10.8 Removing the crankshaft position sensor reluctor ring (1995 and later models)

13 On engines equipped with a balancer shaft, make sure the balancer shaft gears are properly aligned (see illustration). The balancer shaft driven gear (small gear) timing mark should be located at the 6 o'clock position and the drive gear (large gear) should be directly under it at the 12 o'clock position.

14 Install the crankshaft timing chain sprocket onto the crankshaft, making sure to align the keyway in the sprocket with the Woodruff key on the crankshaft. Press the crankshaft sprocket onto the crankshaft using the damper pulley bolt, a large socket and some washers or gently tap it into place until it's completely seated.

✳✳ CAUTION:

If resistance is encountered, DO NOT hammer the sprocket onto the crankshaft. It may crack in the process and fail later causing extensive engine damage.

15 Position the crankshaft so that the timing mark on the sprocket is in the 12 o'clock position. Loop the timing chain over the camshaft sprocket, mesh the chain with the crankshaft sprocket so the timing marks on both sprockets are together and install the camshaft sprocket onto the camshaft. When correctly installed, the timing marks on the sprockets should be aligned as shown (see illustration).

➠Note: With engine on TDC for the number 4 cylinder.

On 2004 and later models, install the timing chain tensioner.

16 On 1995 and later models, install the Crankshaft Position Sensor reluctor onto the crankshaft. Install the reluctor with the dished side facing OUT.

17 Apply a thread locking compound to the camshaft sprocket retaining fasteners, then install and tighten them to the torque listed in this Chapter's Specifications.

18 Lubricate the timing chain with clean engine oil.

19 On engines with metal timing chain covers, apply a thin layer of RTV sealant to both sides of the timing chain cover gasket, then position it on the engine block (the dowel pins and sealant will hold it in place).

20 On engines with a composite type timing chain cover, make sure you obtain a NEW timing chain cover with a new front seal. If you attempt to reuse the old cover, be prepared to perform this job again

10.9a Carefully note the positions of the timing marks on the camshaft sprocket and the crankshaft gear

because it will leak. Also install a new crankshaft front oil seal into the timing chain cover (see Section 15). Apply a thin bead of RTV sealant approximately 3/16 inch wide on the engine block-to-cover mating surface.

21 Install the timing chain cover onto the engine block and tighten the bolts to the torque listed in this Chapter's Specifications.

22 Install the oil pan (see Section 13).

23 On 1995 and later models, install the Crankshaft Position Sensor (with a new seal) into the timing chain cover. Secure the wiring harness to the timing chain cover.

24 Lubricate the crankshaft front oil seal and crankshaft vibration damper sealing surfaces with moly-base grease or clean engine oil. Install the damper onto the crankshaft aligning the keyway in the damper with the Woodruff key in the crankshaft. If the vibration damper cannot be seated by hand, use a bolt of appropriate size installed in the crankshaft to press the damper into place. Once the damper is bottomed, replace the bolt with the proper bolt and washer. Tighten the vibration damper bolt to the torque listed in this Chapter's Specifications.

25 The remaining installation steps are the reverse of removal.

10.9b Remove the camshaft sprocket and timing chain together

10.13 Before installing the camshaft sprocket, make sure the timing marks on the balancer shaft gears are properly aligned (arrows)

➠Note: Engine set on TDC at number 4 piston

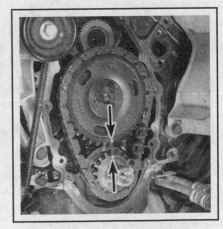

10.15 When correctly installed, the camshaft and crankshaft sprocket timing marks should be together (arrows)

➠Note: Engine set on TDC at number 4 piston

11 Camshaft, bearings and lifters - removal, inspection and installation

CAMSHAFT LOBE LIFT CHECK

▶ **Refer to illustration 11.3**

1   In order to determine the extent of cam lobe wear, the lobe lift should be checked prior to camshaft removal. Refer to Section 3 and remove the rocker arm covers.

2   Position the number one piston at TDC on the compression stroke (see Section 9).

3   Beginning with the number one cylinder valves, loosen the rocker arm nuts and pivot the rocker arms sideways. Mount a dial indicator on the engine and position the plunger against the top of the first pushrod. The plunger should be directly above and in line with the pushrod (see illustration).

4   Zero the dial indicator, then very slowly turn the crankshaft in the normal direction of rotation until the indicator needle stops and begins to move in the opposite direction. The point at which it stops indicates maximum cam lobe lift.

5   Record this figure for future reference, then reposition the piston at TDC on the compression stroke.

6   Move the dial indicator to the remaining number one cylinder rocker arm and repeat the check. Be sure to record the results for each valve.

7   Repeat the check for the remaining valves. Since each piston must be at TDC on the compression stroke for this procedure, work from cylinder-to-cylinder following the firing order sequence.

8   After the check is complete, compare the results to the Specifications listed in this Chapter. If camshaft lobe lift is less than specified, cam lobe wear has occurred and a new camshaft should be installed.

REMOVAL

▶ **Refer to illustrations 11.11a, 11.11b, 11.12, 11.14 and 11.15**

9   Refer to the appropriate Sections and remove the intake manifold, the rocker arms, the pushrods and the timing chain and camshaft

11.3 The dial indicator plunger must be positioned directly above and in-line with the pushrod (use a short length of vacuum hose to hold the plunger over the pushrod end, if you encounter difficulty keeping the plunger on the pushrod)

sprocket. The fan shrouds, water pump, radiator and A/C condenser must be removed as well (see Chapter 3).

10   There are several ways to extract the lifters from the bores. A special tool designed to grip and remove lifters is manufactured by many tool companies and is widely available, but it may not be required in every case. On newer engines without a lot of varnish buildup, the lifters can often be removed with a small magnet or even with your fingers. A machinist's scribe with a bent end can be used to pull the lifters out by positioning the point under the retainer ring in the top of each lifter.

❊❊ CAUTION:

Don't use pliers to remove the lifters unless you intend to replace them with new ones (along with the camshaft). The pliers will damage the precision machined and hardened lifters, rendering them useless.

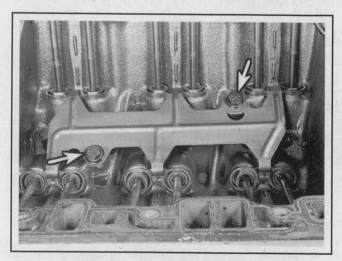

11.11a On 1993 and earlier models with roller lifters, the guide plate is retained by two bolts (arrows). The guide plate presses against the six restrictors that pair the lifters together - be sure all components are reassembled in the same orientation

11.11b On 1994 and later models, the roller lifters are held in place by a bar type retainer - note its position before removing it so you can install it in the same position

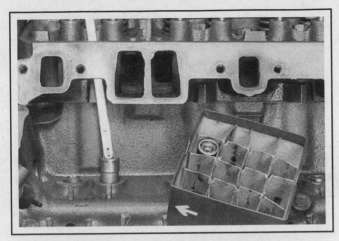

11.12 The lifters on an engine that has accumulated many miles may have to be removed with a special tool - be sure to store the lifters in an organized manner to make sure they are installed in their original locations

11 On engines equipped with roller lifters, the retainer and restrictors (restrictors are used on 1993 and earlier engines only) must be removed first before the lifters can be withdrawn (see illustrations).

12 Before removing the lifters, make sure you have a place to store them so that you can maintain their proper positions for reinstallation. Use a segregated box or egg carton for this purpose. Withdraw the lifters and place them into the box where they won't get mixed up or contaminated with dirt (see illustration). DO NOT attempt to remove the camshaft with the lifters in place.

13 On engines equipped with a balancer shaft, remove the balancer shaft drive gear from the camshaft (see Section 12 if necessary).

14 On engines equipped with a camshaft retainer, note which side of the retainer is facing out, then remove the 2 bolts securing the retainer to the block and remove it from the vehicle (see illustration).

15 To get a handle on the camshaft, thread 2 or 3 six inch long 5/16 - 18 bolts into the camshaft sprocket bolt holes. Carefully pull the camshaft out of the block (see illustration). Support the camshaft near the block so the lobes do not nick or gouge the bearings as it is removed.

INSPECTION

16 Refer to Chapter 2 Part A, Section 15 for the camshaft and bearing inspection procedures. Lifter inspection procedures (for conventional and roller types) are covered in Chapter 2 Part A, Section 7. The illustrations shown apply to both four cylinder and V6 engines.

BEARING REPLACEMENT

17 Camshaft bearing replacement requires special tools and expertise that place it outside the scope of the home mechanic. If bearing replacement is required, remove the engine from the vehicle and take it to an automotive machine shop to ensure the job is performed correctly.

INSTALLATION

▸ **Refer to illustration 11.20**

18 Lubricate the camshaft bearing journals and lobes with moly-base grease or engine assembly lube (see illustration 15.34 in Chapter 2A).

19 Slide the camshaft into the engine. Support the camshaft near the

11.14 Removing the camshaft retainer - note which side is facing out so you can install it in the same position

11.15 Long bolts can be threaded into the camshaft bolt holes to provide a handle for removal and installation of the camshaft - support the cam near the block as it's withdrawn

11.20 After the camshaft is in place, turn it until the dowel pin (arrow) is in the 3 o'clock position as shown here

engine block, being careful not to nick or scrape the bearings.

20 Once it is seated in the block, rotate the camshaft until the dowel pin is located at the 3 o'clock position (see illustration).

21 On engines so equipped, install the camshaft retainer maintaining

the original position noted in Step 14. Tighten the camshaft retainer bolts to the torque listed in this Chapter's Specifications.

22 On engines equipped with a balancer shaft, install the balancer shaft drive gear onto the camshaft, aligning the dowel pin. Make sure the timing marks on the balancer shaft gears are properly aligned after installation (see illustration 10.13).

23 Refer to Section 10 and install the timing chain and sprockets.

24 Lubricate the lifters with clean engine oil and install them in the block. If the original lifters are being reinstalled, be sure to return them to their original locations. If a new camshaft was installed, be sure to install new lifters as well (except for engines with roller lifters).

25 The remaining installation steps are the reverse of removal.

26 Before starting and running the engine, change the oil and install a new oil filter (see Chapter 1).

12 Balancer shaft - removal and installation (1992 and later VIN W and X and all 1994 and later VIN Z engines)

➡Note: The balancer shaft front bearing is not serviceable. Do not attempt to remove it from the shaft. If the bearing has excessive wear, the entire balancer shaft assembly must be replaced as a unit.

REMOVAL

1 Raise the hood and remove the engine cover (see Chapter 11).
2 Perform the fuel pressure relief procedure (see Chapter 4).
3 Remove the air cleaner assembly and ducting (see Chapter 4).
4 Remove the hood latch (see Chapter 11).
5 Remove the radiator grille (see Chapter 11).
6 On 1992 to 1994 models, remove the headlight bezels (see Chapter 11).
7 Remove the radiator and its support braces (see Chapter 3).
8 If equipped, have the air conditioning properly discharged and remove the A/C condenser, compressor and on 1996 and later models remove the compressor bracket (see Chapter 3).
9 Remove the intake manifold, timing chain cover, timing chain and camshaft sprocket (see Section 10).

➡Note: Before removing the camshaft timing chain sprocket, loosen the balancer shaft bolt. To keep the camshaft from rotating, counteract the torque using one of the camshaft sprocket bolts.

10 Remove the stud and the balancer shaft drive gear from the camshaft.
11 Remove the retaining bolt and balancer shaft driven gear from the balancer shaft.
12 Remove the 2 bolts and balancer shaft retainer.
13 If applicable, remove the hydraulic lifter retainer (see illustration 11.11a) or, on later engines, the valve lifter pushrod guides.
14 Using a soft-faced mallet, carefully tap the balancer shaft out of the block.

12.24 Position the balancer shaft drive and driven gears with the timing marks aligned (arrows)

INSPECTION

15 After the balancer shaft has been removed from the engine, cleaned with solvent and dried, inspect the rear bearing journal for uneven wear, pitting and evidence of seizure. If the journal is damaged, the rear bearing in the block is probably damaged as well. The bearing will have to be replaced.

16 Check the balancer drive and driven gears for cracks, missing teeth and excessive wear. If the teeth are highly polished, pitted or galled, or if the outer hardened surface of the teeth is flaking off, new parts will be required. If one gear is worn or damaged, replace both gears as a set. Never install one new gear and one used gear.

REAR BEARING REPLACEMENT

17 Balancer shaft rear bearing replacement requires special tools and expertise that place it outside the scope of the home mechanic. Take the engine block to a dealer service department or an automotive machine shop to ensure that the job is done correctly.

INSTALLATION

▶ **Refer to illustration 12.24**

18 Lubricate the balancer shaft bearing journals with moly-base grease or engine assembly lube.

19 Slide the balancer shaft into the engine. Support the balancer near the block and be careful not to scrape or nick the bearing. It may be necessary to gently tap on the shaft with a soft-face mallet.

20 Install the balance shaft retainer and two bolts and tighten them to the torque listed in this Chapter's Specifications.

21 Install the lifter retainer and bolts, then tighten the bolts securely.

22 Rotate the balancer shaft by hand to make sure there is sufficient clearance between the balancer shaft and the lifter retainer. Replace the lifter retainer if necessary.

23 Install the balancer shaft driven gear and tighten the bolt to the torque listed in this Chapter's Specifications.

24 Rotate the camshaft so that, with the drive gear temporarily installed, the timing mark is straight up at the 12 o'clock position (see illustration). Remove the drive gear.

25 Rotate the balancer shaft until the timing mark is facing straight down at the 6 o'clock position (see illustration 12.24).

26 Install the balancer drive gear onto the camshaft and install the retaining stud, then tighten it to the torque listed in this Chapter's Specifications.

27 Install the timing chain, sprockets and cover (see Section 10).

28 The remaining installation steps are the reverse of removal.

13 Oil pan - removal and installation

➡Note: Lubricants and fluids used on your vehicle are hazardous waste and must be disposed of properly. When disposal is required, place the waste into a sealable container and deliver it to a service station, auto parts store or other facility which accepts hazardous waste for recycling.

REMOVAL

1 Disconnect the cable from the negative terminal of the battery.

❋❋ CAUTION:

On models equipped with an anti-theft audio system, disable the anti-theft feature before performing any operation that requires disconnecting the battery or disrupting power to the stereo (see the Anti-theft audio system procedures at the front of this manual).

2 Raise the vehicle and support it securely on jackstands.
3 Drain the engine oil and remove the oil filter (see Chapter 1).
4 If necessary for clearance, detach the exhaust pipes from the exhaust manifolds.
➡Note: Use penetrating oil spray to ease nut removal.
5 If the engine is equipped with strut rods, remove them.
6 Remove the bellhousing lower cover from the transmission.
7 Remove the starter motor and shield (if equipped) (see Chapter 5).
8 Using the vibration damper bolt, rotate the crankshaft clockwise until the timing mark on the damper is in the 6 o'clock position. This places the crankshaft counter-balance weights at their nominal position.
9 Disconnect the transmission oil cooler lines and/or engine oil cooler lines from the oil pan and adapter as necessary. If removal is

required, cap the lines to prevent contamination and leakage.
10 On 1996 and later models, remove the oil filter adapter.
11 On 1996 and later models, remove the rubber bellhousing plugs.
12 On models so equipped, remove and discard the oil level sensor and remove the crankshaft position sensor (CKP) wiring harness from the retainer.
13 Remove the oil pan bolts, nuts and studs as applicable, noting their locations. On 1985 to 1995 models, steel reinforcement strips are used on each side of the oil pan.
14 Carefully separate the oil pan from the block. DO NOT pry between the oil pan and block or damage to the sealing surfaces may result. If a hammer and block of wood are used to break the gasket seal, use a rawhide or rubber mallet and only strike the pan where it is reinforced. Remove the oil pan by tilting the rear portion down and working it away from the crankshaft throws, oil pump pick-up and front crossmember.

INSTALLATION

▶ **Refer to illustrations 13.22a, 13.22b, 13.22c and 13.22d**

15 Clean the gasket sealing surfaces with lacquer thinner or acetone. Make sure the bolt holes are clean.

❋❋ CAUTION:

1996 and later models use an aluminum oil pan. DO NOT use any type of metal implement to clean the gasket surfaces. Use a plastic scraper only.

16 On steel-type oil pans, check the flanges for distortion, particularly around the bolt holes. If necessary, place the pan on a block of wood and use a mallet to restore the gasket surface.
17 On 1985 through 1995 models, apply a small amount of RTV sealant to the corners of the semi-circular cutouts at both ends of the pan, then install the new gasket.
18 On 1996 and later models, apply a 1 inch long x 3/16 inch wide bead of RTV sealant to the corners of the block-to-timing chain cover joint and the rear main seal-to-block joint. Install the new gasket onto the oil pan.
19 Carefully position the oil pan against the engine block and install the fasteners in their appropriate locations finger tight (don't forget the reinforcement strips, if used).
20 On 1985 through 1995 models, tighten oil pan bolts/nuts in three equal steps to the torque listed in this Chapter's Specifications. Start at the center of the pan and work out towards the ends in a spiral pattern.
21 On models so equipped, install a new oil level sensor and tighten securely.
22 On 1996 and later models, the oil pan is used as a structural reinforcement between the engine and transmission. The alignment of the rear face of the pan to the engine block is important. With the oil pan fasteners finger tight, push the oil pan towards the transmission as far as possible. Tighten the bolts/nuts in three steps to the torque listed in this Chapters Specifications in the sequence shown (see illustration). Using a feeler gage, verify the gap at three positions (right/bottom/left) between the oil pan and the transmission bellhousing is 0.010 inch or less (see illustrations). If the gap is 0.011 inch or greater at any of the contact points, loosen the oil pan and reposition it until an acceptable gap is achieved. If the transmission has been removed from the engine,

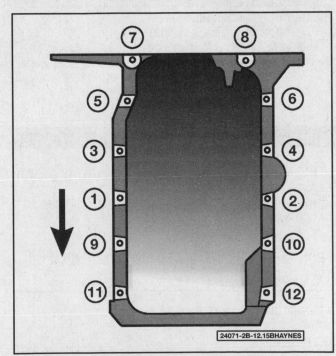

24071-2B-12.15BHAYNES

13.22a Oil pan bolt tightening sequence (1996 and later models)

13.22b Before installing the oil pan-to-bellhousing bolts, measure the gap between the oil pan and bellhousing at 3 locations - right side . . .

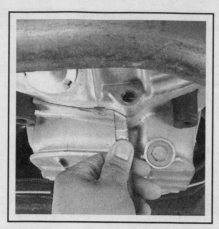

13.22c . . . bottom . . .

13.22d . . . and left side - the gap should be 0.010 inch or less at all 3 locations

use a heavy precision straightedge held against the rear of the engine block (bellhousing mating surface) and then measure between the pan and the straightedge for proper clearance. If both the transmission and oil pan are removed from the vehicle, always reinstall the transmission first, to insure proper oil pan alignment.

23 The remaining installation steps are the reverse of removal.

✳✳ CAUTION:

Don't forget to refill the engine with oil!

24 Start the engine and check carefully for leaks around the oil pan.

14 Oil pump - removal and installation

1   Remove the oil pan as described in Section 13.

2   While supporting the oil pump, remove the pump-to-rear main bearing cap bolt. On some models, the oil pan baffle must be removed first, since it's also held in place by the pump mounting bolt.

3   Lower the pump and remove it along with the pump driveshaft. Inspect the two oil pump locator pins, which project approximately 1/4 inch from the bearing cap, and replace if necessary.

4   Before reinstalling the pump, the factory recommends replacing the oil pump driveshaft retainer (the sleeve that connects the driveshaft to the pump shaft). Make sure the driveshaft is mated with the pump shaft.

5   Position the pump on the engine and make sure the slot in the upper end of the driveshaft is aligned with the tang on the lower end of the distributor shaft. The distributor drives the oil pump, so it is absolutely essential that the components mate properly.

6   Install the oil pump mounting bolt and tighten it to the torque listed in this Chapter's Specifications.

7   Install the oil pan (see Section 13).

8   Start the engine and make sure the oil pressure light goes out or the oil pressure gauge indicates the proper oil pressure, as applicable.

15 Crankshaft oil seals - replacement

FRONT SEAL - TIMING COVER IN PLACE

▶ **Refer to illustrations 15.2 and 15.4**

1   Remove the vibration damper as described in Section 10.

2   Carefully pry the seal out of the cover with a seal removal tool or a large screwdriver (see illustration). Be careful not to distort the cover or scratch the wall of the seal bore. If the engine has accumulated a lot of miles, apply penetrating oil to the seal-to-cover joint and allow it to soak in before attempting to pull the seal out.

3   Clean the bore to remove any old seal material and corrosion. Position the new seal in the bore with the open end of the seal facing IN. A small amount of oil applied to the outer edge of the new seal will make installation easier - don't overdo it!

4   Drive the seal into the bore with a large socket and hammer until it's completely seated (see illustration). Select a socket that's the same outside diameter as the seal (a section of pipe can be used if a socket isn't available).

5   Reinstall the vibration damper.

FRONT SEAL - TIMING COVER REMOVED (METAL TYPE ONLY)

▶ **Refer to illustrations 15.6 and 15.8**

6   Use a punch or screwdriver and hammer to drive the seal out of the cover from the back side. Support the cover as close to the seal bore as possible (see illustration). Be careful not to distort the cover or

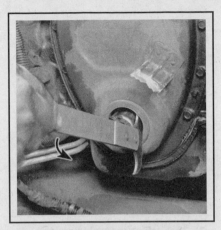

15.2 The front crankshaft oil seal can be removed in the vehicle with a seal removal tool (shown here) or a large screwdriver

15.4 Installing the front crankshaft oil seal using a deep socket

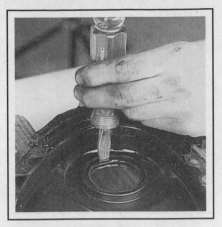

15.6 While supporting the cover near the seal bore, drive the old seal out from the inside with a hammer and punch or screwdriver

scratch the wall of the seal bore. If the engine has accumulated a lot of miles, apply penetrating oil to the seal-to-cover joint on each side and allow it to soak in before attempting to drive the seal out.

7   Clean the bore to remove any old seal material and corrosion. Support the cover on blocks of wood and position the new seal in the bore with the open end of the seal facing IN. A small amount of oil applied to the outer edge of the new seal will make installation easier - don't overdo it!

8   Drive the seal into the bore with a large socket and hammer until it's completely seated (see illustration). Select a socket that's the same outside diameter as the seal (a section of pipe can be used if a socket isn't available).

REAR SEAL

1985 models

▶ **Refer to illustrations 15.14, 15.15a, 15.15b, 15.16, 15.17 and 15.19**

9   The rear main seal on these models can be replaced with the engine in the vehicle. Refer to the appropriate Sections and remove the oil pan and oil pump.

10 Remove the bolts and detach the rear main bearing cap from the engine.

11 The seal section in the bearing cap can be pried out with a screwdriver.

12 To remove the seal section in the block, tap on one end with a hammer and brass punch or wood dowel until the other end protrudes far enough to grip it with a pair of pliers and pull it out. Be very careful not to nick or scratch the crankshaft journal or seal surface as this is done.

13 Inspect the bearing cap and engine block mating surfaces, as well as the cap seal grooves, for nicks, burrs and scratches. Remove any defects with a fine file or deburring tool.

14 A small seal installation tool is usually included when a new seal is purchased. If you didn't receive one, they can also be purchased separately at most auto parts stores or you can make one from an old feeler gauge or a piece of brass shim stock (see illustration).

15 Using the tool, install one seal section in the cap with the lip facing the front of the engine (if the seal has two lips, the one with the helix must face the front) (see illustrations). The ends should be flush with the mating surface of the cap. Make sure it's completely seated.

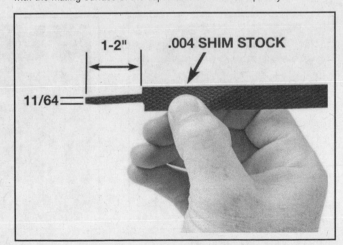

15.8 Clean the bore, then apply a small amount of oil to the outer edge of the seal and drive it squarely into the opening with a large socket and hammer - DO NOT damage the seal in the process!

15.14 If the new seal did not include an installation tool, make one from a piece of brass shim stock 0.004-inch thick

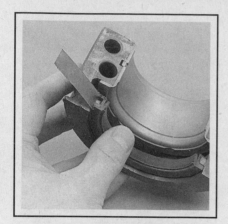

15.15a Using the tool like a shoehorn, attach the seal section to the bearing cup . . .

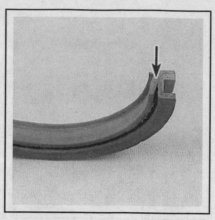

15.15b . . . with the oil seal lip pointing toward the front of the engine

15.16 Position the tool to protect the back side of the seal as it passes over the sharp edge of the ridge - note that the seal straddles the ridge

16 Position the narrow end of the tool so that it will protect the back side of the seal as it passes over the sharp edge of the ridge in the block (see illustration).

17 Lubricate the seal lips and the groove in the back side with moly-base grease or clean engine oil - don't get any lubricant on the seal ends. Insert the seal into the block, over the tool (see illustration).

✳✳ CAUTION:

Make sure that the lip points toward the front of the engine when the seal is installed.

18 Push the seal into place, using the tool like a shoehorn. Turning the crankshaft may help to draw the seal into place. When both ends of the seal are flush with the block surface, remove the tool.

19 Apply a thin, even coat of anaerobic-type gasket sealant to the areas of the cap or block indicated in the accompanying illustration. Do not get any sealant on the bearing face, crankshaft journal, seal ends or seal lips. Also, lubricate the cap seal lips with moly-base grease or clean engine oil.

20 Carefully position the bearing cap on the block, install the bolts and tighten them to 10-to-12 ft-lbs only. Tap the crankshaft forward and

backward with a lead or brass hammer to line up the main bearing and crankshaft thrust surfaces, then tighten the rear bearing cap bolts to the torque listed in Chapter 2 Part C Specifications.

21 Install the oil pump and oil pan.

1986 and later models

▶ Refer to illustrations 15.23 and 15.26

22 Later models are equipped with a one-piece seal that requires an entirely different installation procedure. The transmission (see Chapter 7) and flywheel (see Section 16) must be removed to gain access to the seal housing.

23 The old seal can be removed from the housing by inserting a large screwdriver into the notches provided and prying it out (see illustration). Be sure to note how far it's recessed into the housing bore before removing it; the new seal will have to be recessed an equal amount. Be very careful not to scratch or otherwise damage the bore in the housing or oil leaks could develop.

24 Check the seal contact surface very carefully for scratches and nicks that could damage the new seal lip and cause oil leaks. If the crankshaft is damaged, the only alternative is a new or different crankshaft.

15.17 Make sure the seal lip faces the front of the engine and hold the tool in place to protect the seal as it's installed

15.19 Apply a thin film of RTV sealant to the ends of the seal and the adjacent areas of the block (arrows) before installing the rear main bearing cap

15.23 To remove the seal from the housing, insert the tip of the screwdriver into each notch (arrows) and lever the seal out

15.26 Lubricate the seal lip, then work it over the crankshaft with a blunt tool, such as the drive end of a socket extension

25 Make sure the housing is clean, then apply a thin coat of engine oil to the outer edge of the new seal. Apply moly-based grease to the seal lips.

➡**Note: Do not allow oil or other lubricants to contact the seal lip surface. The seal must be pressed squarely into the housing bore, so hammering it into place is not recommended. If you don't have access to a special tool, remove the oil pan (see Section 13) and unbolt the seal housing from the block. Sandwich the housing and seal between two smooth pieces of wood and press the seal into place with the jaws of a large vise. The pieces of wood must be thick enough to distribute the force evenly around the entire circumference of the seal. Work slowly and make sure the seal enters the bore squarely.**

26 The seal lips must be lubricated with clean engine oil or moly-based grease before the seal/housing is slipped over the crankshaft and bolted to the block. Use a new gasket - no sealant is required - and make sure the dowel pins are in place before installing the housing (see illustration).

27 Tighten the nuts/screws a little at a time until the seal housing is fully seated. Then tighten them to the torque listed in this Chapter's Specifications.

16 Flywheel/driveplate - removal and installation

1 Refer to Chapter 7 and remove the transmission. If your vehicle has a manual transmission, the pressure plate and clutch will also have to be removed (see Chapter 8).

2 Jam a large screwdriver in the starter ring gear to keep the crankshaft from turning, then remove the mounting bolts. Since it's fairly heavy, support the flywheel as the last bolt is removed.

3 Pull straight back on the flywheel/driveplate to detach it from the crankshaft.

4 Installation is the reverse of removal. The driveplate must be mounted with the torque converter pads facing the transmission. Be sure to align the hole in the flywheel/driveplate with the dowel pin in the crankshaft. Use Locktite on the bolt threads and tighten them in a criss-cross pattern to the torque listed in this Chapter's Specifications.

17 Engine mounts - check and replacement

◗ **Refer to illustration 17.7**

1 Engine mounts seldom require attention, but broken or deteriorated mounts should be replaced immediately or the added strain placed on the driveline components may cause damage.

CHECK

2 During the check, the engine must be raised slightly to remove the weight from the mounts. Refer to Chapter 11 and remove the engine cover before raising the engine.

3 Raise the vehicle and support it securely on jackstands, then position the jack under the engine oil pan. Place a large block of wood between the jack head and the oil pan, then carefully raise the engine just enough to take the weight off the mounts.

4 Check the mounts to see if the rubber is cracked, hardened or separated from the metal plates. Sometimes the rubber will split right down the center. Rubber preservative may be applied to the mounts to slow deterioration.

5 Check for relative movement between the mount plates and the engine or frame (use a large screwdriver or pry bar to attempt to move the mounts). If movement is noted, lower the engine and tighten the mount fasteners.

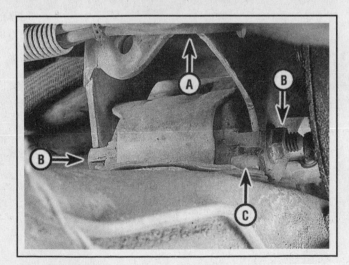

17.7 Typical engine mount components

A Bracket	C Engine mount
B Through-bolt and nut	

REPLACEMENT

6 Disconnect the negative battery cable from the battery, then raise the vehicle and support it securely on jackstands.

❄❄ CAUTION:

On models equipped with an anti-theft audio system, disable the anti-theft feature before performing any operation that requires disconnecting the battery or disrupting power to the stereo (see the Anti-theft audio system procedures at the front of this manual).

7 Remove the nut and withdraw the mount through-bolt from the frame bracket (see illustration).

8 Raise the engine slightly, then remove the mount-to-block bolts and detach the mount.

9 Installation is the reverse of removal. Use Loctite on the mount bolts and be sure to tighten them securely.

Specifications

General

Cylinder numbers (front-to-rear)	
Left (driver's) side	1-3-5
Right side	2-4-6
Firing order	1-6-5-4-3-2
Oil pressure	
1985 to 1990	30 to 35 psi @ 2000 rpm
1991 and later	18 psi @ 2000 rpm

Camshaft

Bearing journal	
Diameter	
1985 through 1995 and 1997 and later	1.8682 to 1.8696 inch
1996	1.877 to 1.8697 inch
Out-of-round limit	0.001 inch
Lobe lift	
1990 and earlier	
Intake	0.357 inch
Exhaust	0.390 inch
1991	
LB4 engine	
Intake	0.234 inch
Exhaust	0.257 inch
LU2 engine (high output)	
Intake	0.269 inch
Exhaust	0.276 inch
1992 to 1995 VIN Z	
Intake	0.234 inch
Exhaust	0.257 inch

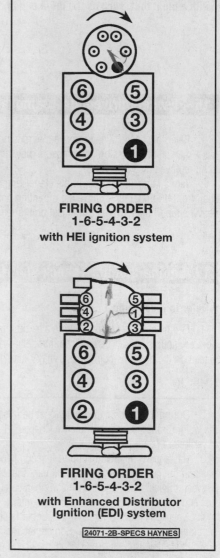

**FIRING ORDER
1-6-5-4-3-2
with HEI ignition system**

**FIRING ORDER
1-6-5-4-3-2
with Enhanced Distributor
Ignition (EDI) system**

24071-2B-SPECS HAYNES

**Cylinder numbers and distributor
spark plug wire terminal locations**

*The blackened terminal shown on the
distributor cap indicates the Number One
spark plug wire position*

Camshaft (continued)

1992 to 1995 VIN W
Intake — 0.288 inch
Exhaust — 0.294 inch
1996
Intake — 0.2763 inch
Exhaust — 0.2855 inch
1997
Intake — 0.288 inch
Exhaust — 0.294 inch
1998
Intake — 0.286 to 0.290 inch
Exhaust — 0.292 to 0.296 inch
1999
VIN W
Intake — 0.283 to 0.287 inch
Exhaust — 0.274 to 0.278 inch
VIN X
Intake — 0.266 to 0.270 inch
Exhaust — 0.257 to 0.261 inch
2000 and later
Intake — 0.274 to 0.278 inch
Exhaust — 0.283 to 0.287 inch
End play
VIN Z — 0.004 to 0.012 inch
VIN W and X — 0.001 to 0.009 inch

Balancer shaft (VIN W and X engines)

Rear bearing journal outside diameter
1985 to 1993 — 1.4209 to 1.4215 inch
1994 and later — 1.4994 to 1.5000 inch

Torque specifications — Ft-lbs (unless otherwise indicated)

Rocker arm cover bolts
1985 — 48 in-lbs
1987 — 100 in-lbs
1986 and 1988 on — 90 to 110 in-lbs
Intake manifold bolts
VIN Z engine
1985 to 1990
Step 1 — 15
Step 2 — 35
1991 to 1995
Step 1 — 15
Step 2 — 35
Step 3 - bolt A only
(see illustration 5.35b) — 41

Torque specifications Ft-lbs (unless otherwise indicated)

Intake manifold bolts (continued)
    VIN W and X engines
      Lower manifold
        1992 to 1995
          Step 1   15
          Step 2   35
        1996 and later
          Step 1   26 in-lbs
          Step 2   106 in-lbs
          Step 3   132 in-lbs
      Upper manifold (plenum)
        1992 to 1995
          Step 1   55 in-lbs
          Step 2   124 in-lbs
        1996 and later
          Step 1   44 in-lbs
          Step 2   88 in-lbs
Exhaust manifold bolts/stud   20
  1985   20
  1986 through 1995
    Center bolts   26
    All others   22
  1996 and later
    Step 1   132 in-lbs
    Step 2   22
Cylinder head bolts

➡**Note: Use a non-hardening sealer, such as Permatex Number 2, on bolt threads.**

  1985 to 1995
    Step 1   25
    Step 2   45
    Step 3   65
  1996 and later
    Step 1   22
    Step 2 (Using a torque angle meter)
      Short bolts (sequence
        numbers 11, 7, 3, 2, 6, 10)   Tighten an additional 55-degrees
      Medium bolts (sequence
        numbers 12 and 13)   Tighten an additional 65-degrees
      Long bolts (sequence
        numbers 1, 4, 8, 5, 9)   Tighten an additional 75-degrees
Timing chain cover bolts
  Metal cover   124 in-lbs
  Composite cover   106 in-lbs
Camshaft
  Sprocket bolts/nut (if equipped)   20
  Retainer bolts   106 in-lbs
Timing chain tensioner bracket bolt
  (2004 and later)   106 in-lbs

Torque specifications   **Ft-lbs (unless otherwise indicated)**

Balancer shaft (1992 and later VIN W and X and all 1994 and later VIN Z engines)
  Retainer bolts   120 in-lbs
  Driven gear bolt
    Step 1   15
    Step 2   Tighten an additional 35-degrees
  Drive gear retaining stud   144 in-lbs
Rocker arm nut (1992 and later VIN W engine)   20
Rocker arm bolt (2000 and later)   22
Crankshaft vibration damper bolt
  1985   60
  1986 and later   70
Valve lifter retainer bolts   144 in-lbs
Oil pan
  1985
    1/4-20 bolts   84 in-lbs
    5/16-18 bolts   168 in-lbs
  1986
    Bolts/studs   100 in-lbs
    Nuts   200 in-lbs
  1987 through 1997
    Bolts/studs   97 to 124 in-lbs
    Nuts   17
  1998 and later (bolt/nuts)   18
Oil pump bolt to rear crankshaft bearing cap   65
Crankshaft rear main oil seal housing bolts/nut   135 in-lbs
Flywheel bolts
  1985   65
  1986 through 1995   75
  1996 and later   74

Notes

Section

2C

GENERAL
ENGINE
OVERHAUL
PROCEDURES

1 General information

Included in this portion of Chapter 2 are the general overhaul procedures for the cylinder head(s) and internal engine components.

The information ranges from advice concerning preparation for an overhaul and the purchase of replacement parts to detailed, step-by-step procedures covering removal and installation of internal engine components and the inspection of parts.

The following Sections have been written based on the assumption that the engine has been removed from the vehicle. For information concerning in-vehicle engine repair, as well as removal and installation of the external components necessary for the overhaul, see Part A or B of this Chapter and Section 7 of this Part.

The Specifications included here in Part C are only those necessary for the inspection and overhaul procedures which follow. Refer to Parts A and B for additional Specifications.

2 Engine overhaul - general information

▶ **Refer to illustration 2.4**

It's not always easy to determine when, or if, an engine should be completely overhauled, as a number of factors must be considered.

High mileage is not necessarily an indication that an overhaul is needed, while low mileage doesn't preclude the need for an overhaul. Frequency of servicing is probably the most important consideration. An engine that's had regular and frequent oil and filter changes, as well as other required maintenance, will most likely give many thousands of miles of reliable service. Conversely, a neglected engine may require an overhaul very early in its life.

Excessive oil consumption is an indication that piston rings, valve seals and/or valve guides are in need of attention. Make sure that oil leaks aren't responsible before deciding that the rings and/or guides are bad. Have a cylinder compression or leakdown test performed by an experienced tune-up mechanic to determine the extent of the work required.

If the engine is making obvious knocking or rumbling noises, the connecting rod and/or main bearings may be at fault. Check the oil pressure with a gauge installed in place of the oil pressure sending unit (see illustration) and compare it to the pressure listed in this Chapter's Specifications. If it's extremely low, the bearings and/or oil pump are probably worn out.

Loss of power, rough running, excessive valve train noise and high fuel consumption rates may also point to the need for an overhaul, especially if they're all present at the same time. If a complete tune-up doesn't remedy the situation, major mechanical work is the only solution.

An engine overhaul involves restoring the internal parts to the specifications of a new engine. During an overhaul, the piston rings are replaced and the cylinder walls are reconditioned (rebored and/or honed). If a rebore is done, new pistons are required. The main bearings, connecting rod bearings and camshaft bearings are generally replaced with new ones and, if necessary, the crankshaft may be reground to restore the journals. Generally, the valves are serviced as well, since they're usually in less-than-perfect condition at this point. While the engine is being overhauled, other components, such as the distributor, starter and alternator, can be rebuilt as well. The end result should be a like new engine that will give many trouble free miles.

➡**Note: Critical cooling system components such as the hoses, drivebelts, thermostat and water pump MUST be replaced with new parts when an engine is overhauled. The radiator should be checked carefully to ensure that it isn't clogged or leaking; if in doubt, replace it with a new one. Also, we don't recommend overhauling the oil pump - always install a new one when an**

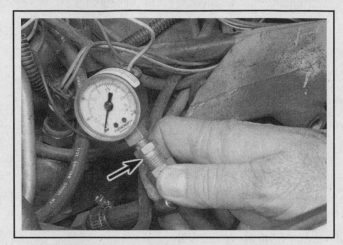

2.4 The oil pressure can be checked by attaching the gauge to the fitting block near the distributor (V6 engine shown)

engine is rebuilt.

Before beginning the engine overhaul, read through the entire procedure to familiarize yourself with the scope and requirements of the job. Overhauling an engine isn't difficult, but it is time consuming. Plan on the vehicle being tied up for a minimum of two weeks, especially if parts must be taken to an automotive machine shop for repair or reconditioning. Check on availability of parts and make sure that any necessary special tools and equipment are obtained in advance. Most work can be done with typical hand tools, although a number of precision measuring tools are required for inspecting parts to determine if they must be replaced. Often an automotive machine shop will handle the inspection of parts and offer advice concerning reconditioning and replacement.

➡**Note: Always wait until the engine has been completely disassembled and all components, especially the engine block, have been inspected before deciding what service and repair operations must be performed by an automotive machine shop. Since the block's condition will be the major factor to consider when determining whether to overhaul the original engine or buy a rebuilt one, never purchase parts or have machine work done on other components until the block has been thoroughly inspected. As a general rule, time is the primary cost of an overhaul, so it doesn't pay to install worn or substandard parts.**

As a final note, to ensure maximum life and minimum trouble from a rebuilt engine, everything must be assembled with care in a spotlessly clean environment.

3 Cylinder compression check

▶ **Refer to illustrations 3.3 and 3.4**

1   A compression check will tell you what mechanical condition the upper end (pistons, rings, valves, head gaskets) of your engine is in. Specifically, it can tell you if the compression is down due to leakage caused by worn piston rings, defective valves and seats or a blown head gasket.

➡**Note: The engine must be at normal operating temperature and the battery must be fully charged for this check. Also, if the engine is equipped with a carburetor, the choke valve must be all the way open to get an accurate compression reading (if the engine's warm, the choke should be open).**

2   Begin by cleaning the area around the spark plugs before you remove them (compressed air should be used, if available, otherwise a small brush or even a bicycle tire pump will work). The idea is to prevent dirt from getting into the cylinders as the compression check is being done. Remove all of the spark plugs from the engine (see Chapter 1).

3   Block the throttle wide open. If the ignition coil is an integral part of the distributor cap, disconnect the wire from the BAT terminal on the cap (see illustration). If the ignition coil is mounted separately, unplug the coil-to-distributor wire harness at the distributor.

4   With the compression gauge in the number one spark plug hole (see illustration), depress the accelerator pedal all the way to the floor to open the throttle valve. Crank the engine over at least four compression strokes and watch the gauge. The compression should build up quickly in a healthy engine. Low compression on the first stroke, fol-

lowed by gradually increasing pressure on successive strokes, indicates worn piston rings. A low compression reading on the first stroke, which doesn't build up during successive strokes, indicates leaking valves or a blown head gasket (a cracked head could also be the cause). Record the highest gauge reading obtained.

5   Repeat the procedure for the remaining cylinders and compare the results to the Specifications.

6   Add some engine oil (about three squirts from a plunger-type oil can) to each cylinder, through the spark plug hole, and repeat the test.

7   If the compression increases after the oil is added, the piston rings are definitely worn. If the compression doesn't increase significantly, the leakage is occurring at the valves or head gasket. Leakage past the valves may be caused by burned valve seats and/or faces or warped, cracked or bent valves.

8   If two adjacent cylinders have equally low compression, there's a strong possibility that the head gasket between them is blown. The appearance of coolant in the combustion chambers or the crankcase would verify this condition.

9   If the compression is unusually high, the combustion chambers are probably coated with carbon deposits. If that's the case, the cylinder head(s) should be removed and decarbonized.

10 If compression is way down or varies greatly between cylinders, it would be a good idea to have a leak-down test performed by an automotive repair shop. This test will pinpoint exactly where the leakage is occurring and how severe it is.

3.3 If your engine has a coil-in-cap distributor, disconnect the wire from the BAT terminal on the distributor cap when checking the compression

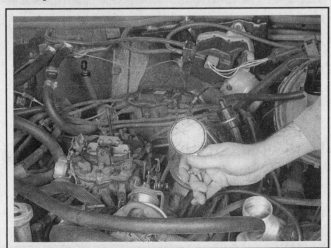

3.4 A compression gauge with a threaded fitting for the plug hole is preferred over the type that requires hand pressure to maintain the seal

4 Engine removal - methods and precautions

If you've decided that an engine must be removed for overhaul or major repair work, several preliminary steps should be taken.

Locating a suitable place to work is extremely important. Adequate work space, along with storage space for the vehicle, will be needed. If a shop or garage isn't available, at the very least a flat, level, clean work surface made of concrete or asphalt is required.

Cleaning the engine compartment and engine before beginning the removal procedure will help keep tools clean and organized.

An engine hoist or A-frame will also be necessary. Make sure the equipment is rated in excess of the combined weight of the engine and accessories. Safety is of primary importance, considering the potential hazards involved in lifting the engine out of the vehicle.

If the engine is being removed by a novice, a helper should be available. Advice and aid from someone more experienced would also be helpful. There are many instances when one person cannot simultaneously perform all of the operations required when lifting the engine out of the vehicle.

Plan the operation ahead of time. Arrange for or obtain all of the tools and equipment you'll need prior to beginning the job. Some of the equipment necessary to perform engine removal and installation safely and with relative ease are (in addition to an engine hoist) a heavy duty floor jack, complete sets of wrenches and sockets as described in the front of this manual, wooden blocks and plenty of rags and cleaning solvent for mopping up spilled oil, coolant and gasoline. If the hoist must be rented, make sure that you arrange for it in advance and perform beforehand all of the operations possible without it. This will save you money and time.

Plan for the vehicle to be out of use for quite a while. A machine shop will be required to perform some of the work which the do-it-yourselfer can't accomplish without special equipment. These shops often have a busy schedule, so it would be a good idea to consult them before removing the engine in order to accurately estimate the amount of time required to rebuild or repair components that may need work.

Always be extremely careful when removing and installing the engine. Serious injury can result from careless actions. Plan ahead, take your time and a job of this nature, although major, can be accomplished successfully.

5 Engine - removal and installation

❊❊ WARNING 1:

The engine is very heavy. Use proper lifting equipment. Never place any part of your body under the engine or transmission when it's supported only by a hoist - it could shift or fall, causing serious injury or even death! Also, the air conditioning system is under high pressure and opening the system will cause a sudden discharge of refrigerant. If the refrigerant gets in your eyes it could cause blindness, so have the system discharged by a service station before disconnecting any hoses or lines.

❊❊ WARNING 2:

Most 1993 and later models are equipped with airbags. Always disable the airbag system before working in the vicinity of the impact sensors, steering column or instrument panel to avoid the possibility of accidental deployment of the airbag(s), which could cause personal injury (see Chapter 12 for the airbag disabling procedure). The yellow wires and connectors routed through the instrument panel are for this system. Do not use electrical test equipment on these yellow wires or tamper with them in any way while working under the instrument panel.

❊❊ CAUTION:

On models equipped with an anti-theft audio system, disable the anti-theft feature before performing any operation that requires disconnecting the battery or disrupting power to the stereo (see the Anti-theft audio system procedures at the front of this manual).

REMOVAL

▶ Refer to illustrations 5.11 and 5.38

➡Note 1: On 1995 models, the radiator support upper tie-bar must be cut out and reinstalled using a factory Bracket Service Kit. Before beginning this procedure, a Bracket Service Kit must be purchased from your local GM dealer. This kit includes the brackets and fasteners necessary to properly reattach the upper tie-bar to the radiator core support.

➡Note 2: On 1996 and later models, the factory recommends removing the body from the frame to remove the engine. To perform this task a two-post side-lift hoist is required which is beyond the scope of the home mechanic; take the vehicle to your local dealer or other qualified repair shop for this operation.

5.11 Tie the horizontal wiring harness (arrow) up out of the way after removing the front body and trim components

1   Remove the engine cover (see Chapter 11).
2   On 1986 and later models, perform the fuel pressure relief procedure (see Chapter 4).
3   Disconnect the cable from the negative terminal of the battery.
4   Remove the air cleaner assembly and ducting on later models (see Chapter 4).
5   Disconnect the fuel lines from the carburetor, TBI or CPI unit (see Chapter 4) and cap them to prevent leakage and contamination.
6   Disconnect the accelerator cable from the carburetor or throttle body (see Chapter 4).
7   On models equipped with 700-R4 or 4L60 model automatic transmissions, disconnect the TV cable from the carburetor or throttle body (see Chapter 7B).
8   If equipped, remove the cruise control servo unit and bracket.
9   Raise the vehicle and support it securely on jackstands.
10 Drain the engine oil, remove the oil filter and drain the cooling system (see Chapter 1).
11 Refer to Chapter 11 and remove the following; hood, headlight bezels, radiator grille, lower grille panel, radiator support brace(s), lower tie-bar, hood latch and horns (see Chapter 12). Position the horizontal wiring harness out of the way (see illustration).
12 If equipped, have the A/C system discharged and remove the condenser, compressor and bracket (see Chapter 3). Cap the refrigerant lines to prevent contamination.

13 Remove the radiator, air deflector panels, fan shrouds, fan and fan clutch assembly (see Chapter 3).

14 Without disconnecting the hydraulic lines, remove the brake master cylinder from the power brake booster and secure it out of the way (see Chapter 9 if necessary).

15 On 1995 models, the upper radiator support tie-bar must be cut out to facilitate engine removal. Before beginning this procedure, a Bracket Service Kit must be purchased from your local GM dealer. This kit includes the brackets and fasteners necessary to properly reattach the upper tie-bar to the radiator core support. Perform the following operations:

a) *Place the front and rear portion of the brackets on each side of the upper tie-bar. Secure them in place with tape. Using the brackets as a template, center punch each bolt location and remove the brackets.*

b) *Using a 5/16 drill bit, drill the tie-bar as required for bracket installation.*

⁂ CAUTION:

Make sure all components and wiring have been removed or positioned out of the way prior to drilling.

c) *Scribe a cutting line perpendicular to the tie-bar and between each set of bracket bolt holes.*

d) *Using a hacksaw or Sawzall, carefully cut the tie-bar along the scribe line (2 places) and remove it from the vehicle.*

⁂ CAUTION:

Make sure all components and wiring have been removed or positioned out of the way prior to cutting.

e) *To avoid hurting yourself on the sharp edges, deburr the end cuts and drilled holes with a file.*

16 Disconnect the heater hoses from the engine.

17 Remove the distributor (see Chapter 5).

18 On models equipped with automatic transmissions, remove the transmission oil cooler lines and dipstick tube.

➡**Note 1: On 1995 models, a special tool must be used to disconnect the transmission cooler lines from the radiator.**

➡**Note 2: On 1996 and later models, the cooler lines are quick-connect type and secured with a spring clip, which cannot be reused. Cap the fittings to prevent leakage and contamination.**

19 On 1995 and earlier models equipped with an automatic transmission, disconnect the shift linkage from the transmission (see Chapter 7B).

20 If equipped, disconnect the engine oil cooler lines from the engine.

➡**Note: On 1996 and later models, the cooler lines are quick-connect type and secured with a spring clip, which cannot be reused. Pull back the plastic cover protecting the fitting, remove the clip and then separate the line from the fitting. Cap the fittings to prevent leakage and contamination.**

21 On four cylinder engines, remove the thermostat housing (see Chapter 3).

22 On V6 engines equipped with the AIR system, remove the diverter valve and check valves (see Chapter 6).

23 Remove the oil filler neck from the valve cover.

5.38 The automatic transmission torque converter is supported following engine removal by a pipe laid across the frame rails - stick bolts through the holes in each side of the bellhousing (arrows)

24 On models equipped with a vacuum powered brake booster, disconnect the vacuum line from the intake manifold.

25 On models equipped with the hydraulic powered brake booster, remove the power steering fluid reservoir from the firewall and drain the fluid. Disconnect the lines from the hydraulic booster unit. Remove the hydraulic booster unit from the vehicle.

26 Remove the power steering pump, bracket and related lines (see Chapter 10).

27 Label and disconnect all vacuum hoses and wiring harnesses from the engine. Don't forget to detach the wiring harness from the clips along the oil pan rail and any ground straps that may be used.

28 Disconnect the exhaust pipe(s) from the manifold(s).

29 Remove the starter motor and shield, if equipped (see Chapter 5).

30 If equipped, remove the engine-to-transmission struts.

31 On models equipped with automatic transmissions, match-mark the driveplate to torque converter and then remove the (3) torque converter-to-driveplate bolts. On 1995 and earlier models, access to the bolts is gained by removing the lower bellhousing cover. On 1996 and later models, access to the bolts is gained through the starter motor opening.

32 Remove the bolts securing the transmission to the engine.

33 Verify all brackets, wiring harnesses and ground straps have been removed from the underside of the engine. Lower the vehicle. Support the transmission with a floor jack.

34 Attach an engine hoist to the engine and take the weight off the engine mounts. Remove the engine mount-to-frame bolts.

35 Raise the engine slightly and separate the engine from the transmission. On models equipped with automatic transmissions, make sure the torque converter remains inside the transmission.

36 Raise the engine and, with a combination of tilting, twisting and raising, move the engine around any obstructions and remove it from the vehicle. As you are removing the engine, check to make sure everything has been disconnected.

37 After the engine has been removed, remove the flywheel/driveplate from the crankshaft and mount the engine on an engine stand.

38 To support the transmission while the engine is out, lay a section of pipe across the frame rails, install two bolts in each side of the bellhousing and rest the transmission on the pipe (see illustration).

INSTALLATION

39 While the engine is out, check the engine and transmission mounts. If they're worn, replace them. Also, before replacing the engine, the factory recommends flushing out the vehicle's engine cooling system and (if equipped) the oil cooling system. This is particularly important if the engine has been damaged internally.

40 Install the driveplate or flywheel and clutch pressure plate. Tighten the flywheel/driveplate bolts to the torque listed in Chapter 2A or 2B Specifications as applicable. Refer to Chapter 8 for clutch and pressure plate installation.

41 Carefully lower the engine into the vehicle and engage it with the transmission torque converter (auto transmission) or input shaft (manual transmission) as applicable. Install the transmission-to-engine bolts.

❋❋ CAUTION:

DO NOT use the transmission-to-engine bolts to force the engine and transmission together - they should mate together easily. If they don't, reposition the engine and try again. Tighten the bolts to the torque listed in Chapter 7.

42 Position the engine in the mounts and install the mounting bolts.

Tighten the bolts securely. The engine hoist can now be removed.

43 On models equipped with automatic transmissions, align the torque converter-to-driveplate match-marks and install the bolts. Tighten the torque converter-to-driveplate bolts to the torque listed in Chapter 7B.

44 On 1995 models, clean the tie-bar and radiator support as required. Treat the bare metal portions with corrosion resistant primer and paint. Install the tie-bar using the brackets and hardware provided. Securely tighten all fasteners.

45 Install the remaining components in the reverse order of removal. Pay special attention to connect any ground straps that may have been disconnected.

46 On 1996 and later models, the cooler lines at the radiator are quick-connect type and secured with a spring clip, which should not be reused. Begin assembly by installing the new spring clips so the spring ears engage in the retaining slots and are visible at all three places. Make sure the clips move freely once they're installed. Next, push the line into the fitting until you hear it "click" into place. When the line and fitting are correctly mated, the yellow band around the line should NOT be visible. Confirm a secure connection by trying to pull it apart. Finally, slide the plastic cover over the fitting.

47 Add engine oil, coolant, power steering fluid and transmission fluid as needed (see Chapter 1).

48 Start the engine and check for leaks and proper operation.

6 Engine rebuilding alternatives

The do-it-yourselfer is faced with a number of options when performing an engine overhaul. The decision to replace the engine block, piston/connecting rod assemblies and crankshaft depends on a number of factors, with the number one consideration being the condition of the block. Other considerations are cost, access to machine shop facilities, parts availability, time required to complete the project and the extent of prior mechanical experience on the part of the do-it-yourselfer.

Some of the rebuilding alternatives include:

Individual parts - If the inspection procedures reveal that the engine block and most engine components are in reusable condition, purchasing individual parts may be the most economical alternative. The block, crankshaft and piston/connecting rod assemblies should all be inspected carefully. Even if the block shows little wear, the cylinder bores should be surface honed.

Crankshaft kit - This rebuild package consists of a reground crankshaft and a matched set of pistons and connecting rods. The pistons will already be installed on the connecting rods. Piston rings and the necessary bearings will be included in the kit. These kits are

commonly available for standard cylinder bores, as well as for engine blocks which have been bored to a regular oversize.

Short block - A short block consists of an engine block with a crankshaft and piston/connecting rod assemblies already installed. All new bearings are incorporated and all clearances will be correct. The existing camshaft, valve train components, cylinder head(s) and external parts can be bolted to the short block with little or no machine shop work necessary.

Long block - A long block consists of a short block plus an oil pump, oil pan, cylinder head(s), rocker arm cover(s), camshaft and valve train components, timing sprockets and chain or gears and timing cover. All components are installed with new bearings, seals and gaskets incorporated throughout. The installation of manifolds and external parts is all that's necessary.

Give careful thought to which alternative is best for you and discuss the situation with local automotive machine shops, auto parts dealers and experienced rebuilders before ordering or purchasing replacement parts.

7 Engine overhaul - disassembly sequence

1 It's much easier to disassemble and work on the engine if it's mounted on a portable engine stand. A stand can often be rented quite cheaply from an equipment rental yard. Before the engine is mounted on a stand, the flywheel/driveplate should be removed from the engine.

2 If a stand isn't available, it's possible to disassemble the engine with it blocked up on a sturdy workbench or on the floor. Be extra careful not to tip or drop the engine when working without a stand.

3 If you're going to obtain a rebuilt engine, all external components must come off first, to be transferred to the replacement engine, just as they will if you're doing a complete engine overhaul yourself. These include:

Alternator and brackets
Emissions control components
Distributor, spark plug wires and spark plugs
Thermostat and housing cover
Water pump
EFI components or carburetor
Intake/exhaust manifolds
Oil filter
Engine mounts
Clutch and flywheel/driveplate

➡ Note: When removing the external components from the engine, pay close attention to details that may be helpful or important during installation. Note the installed position of gaskets, seals, spacers, pins, brackets, washers, bolts and other small items.

4   If you're obtaining a short block, which consists of the engine block, crankshaft, pistons and connecting rods all assembled, then the cylinder head(s), oil pan and oil pump will have to be removed as well. See *Engine rebuilding alternatives* for additional information regarding the different possibilities to be considered.

5   If you're planning a complete overhaul, the engine must be disassembled and the internal components removed in the following order:

Rocker arm cover(s)
Intake and exhaust manifolds
Rocker arms and pushrods
Valve lifters
Cylinder head(s)
Timing cover
Timing chain and sprockets (V6 engine only)
Camshaft
Oil pan
Oil pump
Piston/connecting rod assemblies
Crankshaft and main bearings

6   Before beginning the disassembly and overhaul procedures, make sure the following items are available:

Common hand tools
Small cardboard boxes or plastic bags for storing parts
Gasket scraper
Ridge reamer
Vibration damper puller
Micrometers
Telescoping gauges
Dial indicator set
Valve spring compressor
Cylinder surfacing hone
Piston ring groove cleaning tool
Electric drill motor
Tap and die set
Wire brushes
Oil gallery brushes
Cleaning solvent

8   Cylinder head - disassembly

▶ **Refer to illustrations 8.2, 8.3a and 8.3b**

➡ Note: New and rebuilt cylinder heads are commonly available for most engines at dealerships and auto parts stores. Due to the fact that some specialized tools are necessary for the disassembly and inspection procedures, and replacement parts may not be readily available, it may be more practical and economical for the home mechanic to purchase replacement head(s) rather than taking the time to disassemble, inspect and recondition the original(s).

1   Cylinder head disassembly involves removal of the intake and exhaust valves and related components. If they're still in place, remove the rocker arm nuts, pivot balls and rocker arms from the cylinder head studs. Label the parts or store them separately so they can be reinstalled in their original locations.

2   Before the valves are removed, arrange to label and store them, along with their related components, so they can be kept separate and reinstalled in the same valve guides they are removed from (see illustration).

3   Compress the springs on the first valve with a spring compressor and remove the keepers (see illustration). Carefully release the valve spring compressor and remove the retainer (V6 exhaust valves have rotators), the shield, the springs and the spring seat. Next, remove the O-ring seal from the upper end of the valve stem (just under the keeper groove) and the umbrella-type seal from the guide (not used on V6 exhaust valves), then pull the valve out of the head. If the valve binds in the guide (won't pull through), push it back into the head and deburr the area around the keeper groove with a fine file or whetstone (see illustration).

8.2 A small plastic bag, with an appropriate label, can be used to store the valve train components so they can be kept together and reinstalled in the correct guide

8.3a Use a valve spring compressor to compress the spring, then remove the keepers from the valve stem

8.3b If the valve won't pull through the guide, deburr the edge of the stem end and the area around the top of the keeper groove with a file

4   Repeat the procedure for the remaining valves. Remember to keep all the parts for each valve together so they can be reinstalled in the same locations.

5   Once the valves and related components have been removed and stored in an organized manner, the head should be thoroughly cleaned and inspected. If a complete engine overhaul is being done, finish the engine disassembly procedures before beginning the cylinder head cleaning and inspection process.

9   Cylinder head - cleaning and inspection

▶ **Refer to illustrations 9.12, 9.14, 9.15, 9.16, 9.17, 9.18 and 9.19**

1   Thorough cleaning of the cylinder head(s) and related valve train components, followed by a detailed inspection, will enable you to decide how much valve service work must be done during the engine overhaul.

CLEANING

2   Scrape all traces of old gasket material and sealing compound off the head gasket, intake manifold and exhaust manifold sealing surfaces. Be very careful not to gouge the cylinder head. Special gasket removal solvents that soften gaskets and make removal much easier are available at auto parts stores.

3   Remove all built up scale from the coolant passages.

4   Run a stiff wire brush through the various holes to remove deposits that may have formed in them.

5   Run an appropriate size tap into each of the threaded holes to remove corrosion and thread sealant that may be present. If compressed air is available, use it to clear the holes of debris produced by this operation.

6   Clean the rocker arm pivot stud threads with a wire brush.

7   Clean the cylinder head with solvent and dry it thoroughly. Compressed air will speed the drying process and ensure that all holes and recessed areas are clean.

➡**Note: Decarbonizing chemicals are available and may prove very useful when cleaning cylinder heads and valve train components. They are very caustic and should be used with caution. Be sure to follow the instructions on the container.**

8   Clean the rocker arms, pivot balls, nuts and pushrods with solvent and dry them thoroughly (don't mix them up during the cleaning process). Compressed air will speed the drying process and can be used to clean out the oil passages.

9   Clean all the valve springs, shields, keepers and retainers (or rotators) with solvent and dry them thoroughly. Do the components from one valve at a time to avoid mixing up the parts.

10 Scrape off any heavy deposits that may have formed on the valves, then use a motorized wire brush to remove deposits from the valve heads and stems. Again, make sure the valves don't get mixed up.

INSPECTION

Cylinder head

11 Inspect the head very carefully for cracks, evidence of coolant leakage and other damage. If cracks are found, a new cylinder head should be obtained.

12 Using a straightedge and feeler gauge, check the head gasket mating surface for warpage (see illustration). If the warpage exceeds the specified limit, it can be resurfaced at an automotive machine shop.

➡**Note: If the V6 engine heads are resurfaced, the intake manifold flanges will also require machining.**

13 Examine the valve seats in each of the combustion chambers. If they're pitted, cracked or burned, the head will require valve service that is beyond the scope of the home mechanic.

14 Check the valve stem-to-guide clearance by measuring the lateral movement of the valve stem with a dial indicator attached securely to the head (see illustration). The valve must be in the guide and approxi-

9.12 Check the cylinder head gasket surface for warpage by trying to slip a feeler gauge under the straightedge (see the Specifications for the maximum warpage allowed and use a feeler gauge of that thickness)

9.14 A dial indicator can be used to determine the valve stem-to-guide clearance (move the valve stem as indicated by the arrows)

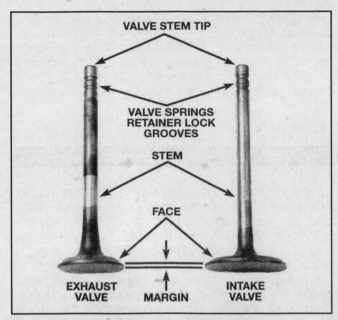

9.15 Check for valve wear at the points shown here

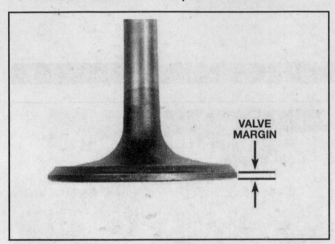

9.16 The margin width on each valve must be as specified (if no margin exists, the valve cannot be reused)

mately 1/16-inch off the seat. The total valve stem movement indicated by the gauge needle must be divided by two to obtain the actual clearance. After this is done, if there's still some doubt regarding the condition of the valve guides they should be checked by an automotive machine shop (the cost should be minimal).

Valves

15 Carefully inspect each valve face for uneven wear, deformation, cracks, pits and burned areas (see illustration). Check the valve stem for scuffing and galling and the neck for cracks. Rotate the valve and check for any obvious indication that it's bent. Look for pits and excessive wear on the end of the stem. The presence of any of these conditions indicates the need for valve service by an automotive machine shop.

16 Measure the margin width on each valve (see illustration). Any valve with a margin narrower than 1/32-inch will have to be replaced with a new one.

Valve components

17 Check each valve spring for wear (on the ends) and pits. Measure the free length and compare it to the Specifications (see illustration). Any springs that are shorter than specified have sagged and should not be reused. The tension of all springs should be checked with a special fixture before deciding that they're suitable for use in a rebuilt engine (take the springs to an automotive machine shop for this check).

18 Stand each spring on a flat surface and check it for squareness (see illustration). If any of the springs are distorted or sagged, replace all of them with new parts.

19 Check the spring retainers (or rotators) and keepers for obvious wear and cracks. Any questionable parts should be replaced with new ones, as extensive damage will occur if they fail during engine operation. Make sure the rotators operate smoothly with no binding or excessive play (see illustration).

Rocker arm components

20 Check the rocker arm faces (the areas that contact the pushrod ends and valve stems) for pits, wear, galling, score marks and rough spots. Check the rocker arm pivot contact areas and pivot balls as well. Look for cracks in each rocker arm and nut.

21 Inspect the pushrod ends for scuffing and excessive wear. Roll each pushrod on a flat surface, like a piece of plate glass, to determine if it's bent.

9.17 Measure the free length of each valve spring with a dial or vernier caliper

9.18 Check each valve spring for squareness

9.19 The exhaust valve rotators can be checked by turning the inner and outer sections in opposite directions - feel for smooth movement and excessive play

22 Check the rocker arm studs in the cylinder heads for damaged threads and secure installation.

23 Any damaged or excessively worn parts must be replaced with new ones.

24 If the inspection process indicates that the valve components are in generally poor condition and worn beyond the limits specified, which is usually the case in an engine that's being overhauled, reassemble the valves in the cylinder head and refer to Section 10 for valve servicing recommendations.

25 If the inspection turns up no excessively worn parts, and if the valve faces and seats are in good condition, the valve train components can be reinstalled in the cylinder head without major servicing. Refer to the appropriate Section for the cylinder head reassembly procedure.

10 Valves - servicing

1 Because of the complex nature of the job and the special tools and equipment needed, servicing of the valves, the valve seats and the valve guides, commonly known as a valve job, is best left to a professional.

2 The home mechanic can remove and disassemble each head, do the initial cleaning and inspection, then reassemble and deliver it to a dealer service department or an automotive machine shop for the actual valve servicing.

3 The dealer service department, or automotive machine shop, will remove the valves and springs, recondition or replace the valves and valve seats, recondition the valve guides, check and replace the valve springs, spring retainers or rotators and keepers (as necessary), replace the valve seals with new ones, reassemble the valve components and make sure the installed spring height is correct. The cylinder head gasket surface will also be resurfaced if it's warped.

4 After the valve job has been performed by a professional, the head will be in like new condition. When the head is returned, be sure to clean it again before installation on the engine to remove any metal particles and abrasive grit that may still be present from the valve service or head resurfacing operations. Use compressed air, if available, to blow out all the oil holes and passages. Also, check the installed spring height to make sure it's correct.

11 Cylinder head - reassembly

▶ **Refer to illustrations 11.5, 11.6a, 11.6b, 11.7, 11.9 and 11.10**

1 Regardless of whether or not a head was sent to an automotive repair shop for valve servicing, make sure it's clean before beginning reassembly.

2 If a head was sent out for valve servicing, the valves and related components will already be in place. Begin the reassembly procedure with Step 9.

3 Apply moly-based grease or clean engine oil to the valve stem. Beginning at one end of the head, install the first valve. A tennis ball placed under the head will help hold the valve in place until the seal is installed.

4 Three different types of valve stem oil seals are used on these engines, depending on year, engine type and size. Some models use a small O-ring which simply fits around the valve stem just above the guide boss. A second type is a flat O-ring which fits into a groove in the valve stem just below the valve stem keeper groove. The most common seal used is an umbrella type seal which extends down over the valve guide and is usually used in conjunction with the flat O-ring type seal.

➡ **Note: On 1998 and later models, a new exhaust valve oil seal was introduced. It has the letters "EX" molded into the top of the seal and is either brown with a white stripe around its circumference or solid red. They are not interchangeable with the intake valve seals, which are black.**

5 Install the new valve seal onto the guide and, using an appropriate seal installation tool or deep socket and hammer, drive the seal onto the valve guide (see illustration). On all engines except 1998 models, drive the seal until it's bottomed on the guide. On 1998 engines, drive the seal onto the guide until the gap between the bottom of the seal and the start of the chamfer on the guide is 0.040 to 0.078 inch.

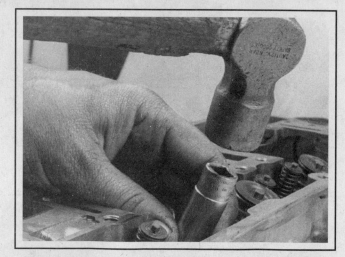

11.5 Make sure the new valve stem seals are seated against the tops of the valve guides

6 Install the damper spring (if used), valve spring, oil shield (if used), spring retainer or rotator (as applicable) onto the valve (see illustrations).

7 Compress the valve spring(s) with a suitable valve spring compressor and if a flat type O-ring is used, lubricate it with clean engine oil and carefully install it in the lower groove on the valve stem (see illustration). Make sure the seal is not twisted - it must lie perfectly flat in the groove. Next, place a small dab of grease onto each keeper, place them in the upper groove on the valve stem and slowly release the valve spring compressor, making sure the keepers remain in position.

8 Repeat the procedure for the remaining valves. Be sure to return

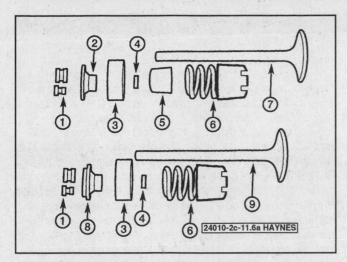

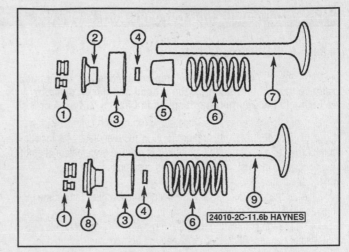

11.6a Four-cylinder valve spring components - exploded view

1	Keepers	6	Valve spring and damper
2	Retainer	7	Intake valve
3	Oil shield	8	Retainer/rotator
4	O-ring oil seal	9	Exhaust valve
5	Umbrella seal		

11.6b V6 engine valve spring components - exploded view

1	Keepers	6	Spring
2	Retainer	7	Intake valve
3	Oil shield	8	Retainer/rotator
4	O-ring oil seal	9	Exhaust valve
5	Umbrella seal		

the components to their original locations - don't mix them up!

9   Once all the valves are in place in both heads, the valve stem O-ring seals must be checked to make sure they don't leak. This procedure requires a vacuum pump and special adapter, so it may be a good idea to have it done by a dealer service department, repair shop or automotive machine shop. The adapter is positioned on each valve retainer or rotator and vacuum is applied with the hand pump (see illustration). If the vacuum can't be maintained, the seal is leaking and must be checked/replaced before the head is installed on the engine.

10 Check the installed valve spring height with a ruler graduated in 1/32-inch increments or a dial caliper. If the head was sent out for service work, the installed height should be correct (but don't automatically assume that it is). The measurement is taken from the top of each spring seat or shim(s) to the top of the oil shield (or the bottom of the

retainer/rotator, the two points are the same) (see illustration). If the height is greater than specified, shims can be added under the springs to correct it.

> ※※ **CAUTION:**
>
> **Don't, under any circumstances, shim the springs to the point where the installed height is less than specified.**

11 Apply moly-base grease to the rocker arm faces and the pivot balls, then install the rocker arms and pivots on the cylinder head studs. Thread the nuts on three or four turns only (the heads must be installed and the pushrods in position before the nuts can be tightened).

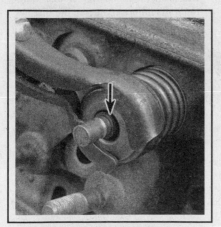

11.7 Make sure the O-ring seal under the retainer is seated in the groove and not twisted before installing the keepers

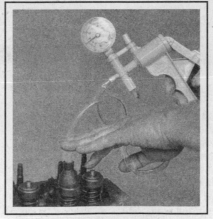

11.9 A special adapter and vacuum pump are required to check the O-ring valve stem seals for leaks

11.10 Be sure to check the valve spring installed height (the distance from the top of the seat/shims to the top of the shield)

12 Piston/connecting rod assembly - removal

▶ Refer to illustrations 12.1, 12.3 and 12.5

➡ Note: Prior to removing the piston/connecting rod assemblies, remove the cylinder head(s), the oil pan and the oil pump by referring to the appropriate Sections in Chapter 2, Part A or B.

1   Completely remove the ridge at the top of each cylinder with a ridge reaming tool (see illustration). Follow the manufacturer's instructions provided with the tool. Failure to remove the ridge before attempting to remove the piston/connecting rod assemblies may result in piston breakage.

2   After the cylinder ridges have been removed, turn the engine upside-down so the crankshaft is facing up.

3   Before the connecting rods are removed, check the end play with feeler gauges. Slide them between the first connecting rod and the crankshaft throw until the play is removed (see illustration). The end play is equal to the thickness of the feeler gauge(s). If the end play exceeds the service limit, new connecting rods will be required. If new rods (or a new crankshaft) are installed, the end play may fall under the specified minimum (if it does, the rods will have to be machined to restore it - consult an automotive machine shop for advice if nec-

essary). Repeat the procedure for the remaining connecting rods.

4   Check the connecting rods and caps for identification marks. If they aren't plainly marked, use a small center punch to make the appropriate number of indentations on each rod and cap (1 - 4 or 6, depending on the engine type and cylinder they're associated with).

5   Loosen each of the connecting rod cap nuts 1/2-turn at a time until they can be removed by hand. Remove the number one connecting rod cap and bearing insert. Don't drop the bearing insert out of the cap. Slip a short length of plastic or rubber hose over each connecting rod cap bolt to protect the crankshaft journal and cylinder wall as the piston is removed (see illustration). Push the connecting rod/piston assembly out through the top of the engine. Use a wooden hammer handle to push on the upper bearing insert in the connecting rod. If resistance is felt, double-check to make sure that all of the ridge was removed from the cylinder.

6   Repeat the procedure for the remaining cylinders. After removal, reassemble the connecting rod caps and bearing inserts in their respective connecting rods and install the cap nuts finger tight. Leaving the old bearing inserts in place until reassembly will help prevent the connecting rod bearing surfaces from being accidentally nicked or gouged.

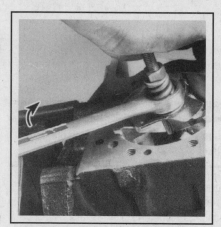

12.1 A ridge reamer is required to remove the ridge from the top of the cylinder - do this before removing the pistons!

12.3 Check the connecting rod side clearance with a feeler gauge as shown

12.5 To prevent damage to the crankshaft journals and cylinder walls, slip sections of hose over the rod bolts before removing the pistons

13 Crankshaft - removal

▶ Refer to illustrations 13.1, 13.3, 13.4a, 13.4b and 13.4c

➡ Note: The crankshaft can be removed only after the engine has been removed from the vehicle. It's assumed that the flywheel or driveplate, vibration damper/crankshaft pulley hub, timing chain (V6 engine only), oil pan, oil pump and piston/connecting rod assemblies have already been removed. If your engine is equipped with a one-piece rear main oil seal (later model V6 engines only), the seal housing must be unbolted and separated from the block before proceeding with crankshaft removal.

1   Before the crankshaft is removed, check the end play. Mount a dial indicator with the stem in line with the crankshaft and just touching

one of the crank throws (see illustration).

2   Push the crankshaft all the way to the rear and zero the dial indicator. Next, pry the crankshaft to the front as far as possible and check the reading on the dial indicator. The distance that it moves is the end play. If it's greater than specified, check the crankshaft thrust surfaces for wear. If no wear is evident, new main bearings should correct the end play.

3   If a dial indicator isn't available, feeler gauges can be used. Gently pry or push the crankshaft all the way to the front of the engine. Slip feeler gauges between the crankshaft and the front face of the thrust main bearing to determine the clearance (see illustration).

4   Check the main bearing caps to see if they're marked to indicate

13.1 Checking crankshaft end play with a dial indicator

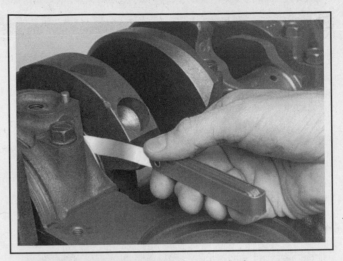

13.3 Checking crankshaft end play with a feeler gauge

13.4a Use a center punch or number stamping dies to mark the main bearing caps to ensure that they're reinstalled in their original locations on the block (make the punch marks near one of the bolt heads)

13.4b Mark the caps in order from the front of the engine to the rear (one mark for the front cap, two for the second one and so on) - the rear cap doesn't have to be marked since it can't be installed in any other location

13.4c The arrow on the main bearing cap indicates the front of the engine

their locations. They should be numbered consecutively from the front of the engine to the rear. If they aren't, mark them with number stamping dies or a center punch (see illustrations). Main bearing caps generally have a cast-in arrow, which points to the front of the engine (see illustration). Loosen each of the main bearing cap bolts 1/4-turn at a time each, until they can be removed by hand.

5   Gently tap the caps with a soft-face hammer, then separate them from the engine block. If necessary, use the bolts as levers to remove the caps. Try not to drop the bearing inserts if they come out with the caps.

6   Carefully lift the crankshaft out of the engine. It's a good idea to have an assistant available, since the crankshaft is quite heavy. With the bearing inserts in place in the engine block and main bearing caps, return the caps to their respective locations on the engine block and tighten the bolts finger tight.

14 Engine block - cleaning

▶ Refer to illustrations 14.1, 14.8 and 14.10

➡Note: The core plugs (also known as freeze or soft plugs) may be difficult or impossible to retrieve if they're driven into the block coolant passages.

1 Remove the core plugs from the engine block. To do this, knock one side of the plug into the block with a hammer and punch, then grasp them with large pliers and pull them back through the holes (see illustration).

2 Using a gasket scraper, remove all traces of gasket material from the engine block. Be very careful not to nick or gouge the gasket sealing surfaces.

3 Remove the main bearing caps and separate the bearing inserts from the caps and the engine block. Tag the bearings, indicating which cylinder they were removed from and whether they were in the cap or the block, then set them aside.

4 Using a 1/4-inch drive breaker bar or ratchet, remove all of the threaded oil gallery plugs from the rear of the block. The plugs are usually very tight - they may have to be drilled out and the holes retapped. Discard the plugs and use new ones when the engine is reassembled.

5 If the engine is extremely dirty it should be taken to an automotive machine shop to be steam cleaned or hot tanked.

6 After the block is returned, clean all oil holes and oil galleries one more time. Brushes specifically designed for this purpose are available at most auto parts stores. Flush the passages with warm water until the water runs clear, dry the block thoroughly and wipe all machined surfaces with a light, rust preventative oil. If you have access to compressed air, use it to speed the drying process and to blow out all the oil holes and galleries.

7 If the block isn't extremely dirty or sludged up, you can do an adequate cleaning job with hot soapy water and a stiff brush. Take plenty of time and do a thorough job. Regardless of the cleaning method used, be sure to clean all oil holes and galleries very thoroughly, dry the block completely and coat all machined surfaces with light oil.

8 The threaded holes in the block must be clean to ensure accurate torque readings during reassembly. Run the proper size tap into each

14.1 Use a hammer and a large punch to knock the core plugs sideways in their bores, then pull them out with pliers

of the holes to remove any rust, corrosion, thread sealant or sludge and to restore any damaged threads (see illustration). If possible, use compressed air to clear the holes of debris produced by this operation. Now is a good time to clean the threads on the head bolts and the main bearing cap bolts as well.

9 Reinstall the main bearing caps and tighten the bolts finger tight.

10 After coating the sealing surfaces of the new core plugs with RTV sealant, install them in the engine block (see illustration). Make sure they're driven in straight and seated properly or leakage could result. Special tools are available for this purpose, but equally good results can be obtained using a large socket, with an outside diameter that will just slip into the core plug, a 1/2-inch drive extension and a hammer.

11 Apply non-hardening sealant (such as Permatex number 2 or Teflon tape) to the new oil gallery plugs and thread them into the holes at the rear of the block. Make sure they're tightened securely.

12 If the engine isn't going to be reassembled right away, spray it with WD-40 and cover it with a large plastic trash bag to keep it clean.

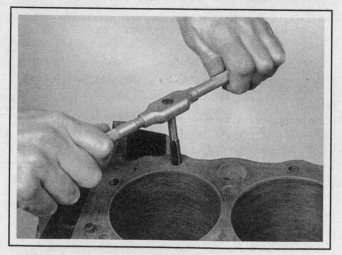

14.8 All bolt holes in the block - particularly the main bearing cap and head bolt holes - should be cleaned and restored with a tap (be sure to remove debris from the holes after this is done)

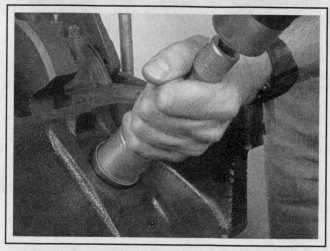

14.10 A large socket on an extension can be used to drive the new core plugs into the bores

15 Engine block - inspection

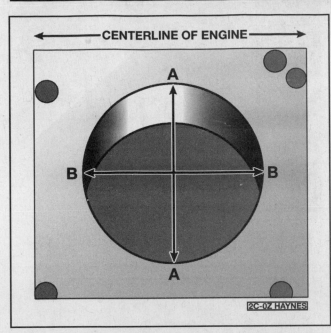

15.4a Measure the diameter of each cylinder at a right angle to the engine centerline (A), and parallel to engine centerline (B) - out-of-round is the difference between A and B; taper is the difference between A and B at the top of the cylinder and A and B at the bottom of the cylinder

▶ Refer to illustrations 15.4a, 15.4b and 15.4c

1 Before the block is inspected, it should be cleaned as described in Section 14. Double-check to make sure the ridge at the top of each cylinder has been completely removed.

2 Visually check the block for cracks, rust and corrosion. Look for stripped threads in the threaded holes. It's also a good idea to have the block checked for hidden cracks by an automotive machine shop that has the special equipment to do this type of work. If defects are found, have the block repaired, if possible, or replaced.

3 Check the cylinder bores for scuffing and scoring.

4 Measure the diameter of each cylinder at the top (just under the ridge area), center and bottom of the cylinder bore, parallel to the crankshaft axis (see illustrations).

➡ Note: These measurements should not be taken with the engine block mounted on an engine stand - the cylinders will be distorted and the measurements inaccurate.

5 Next, measure each cylinder's diameter at the same three locations except measure across the crankshaft axis. The taper of the cylinder is the difference between the bore diameter at the top of the cylinder and the diameter of the bottom. The out-of-round dimension is the difference between the parallel and perpendicular measurements. Compare your measurements to those listed in this Chapter's Specifications. If the cylinder walls are badly scuffed or scored, or if they're out-of-round or tapered beyond the limits given in this Chapter's Specifications, have the engine block rebored and honed at an automotive machine shop. If a rebore is performed, oversized pistons and rings will be required.

6 If the cylinders are in reasonably good condition and not worn to the outside of the limits, and if the piston-to-cylinder clearances can be maintained properly, then they don't have to be rebored. Honing is all that's necessary (see Section 16).

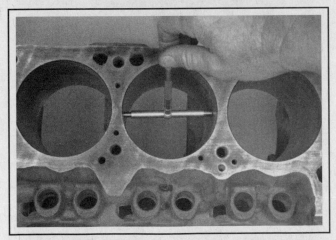

15.4b The ability to "feel" when the telescoping gauge is at the correct point will be developed over time, so work slowly and repeat the check until you're satisfied that the bore measurement is accurate

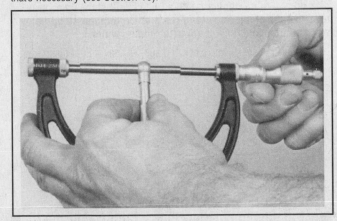

15.4c The gauge is then measured with a micrometer to determine the bore size

16 Cylinder honing

▶ Refer to illustrations 16.3a and 16.3b

1 Prior to engine reassembly, the cylinder bores must be honed so the new piston rings will seat correctly and provide the best possible combustion chamber seal.

➡ Note: If you don't have the tools or don't want to tackle the honing operation, most automotive machine shops will do it for a reasonable fee.

2 Before honing the cylinders, install the main bearing caps and tighten the bolts to the specified torque.

3 Two types of cylinder hones are commonly available - the flex hone or "bottle brush" type and the more traditional surfacing hone with spring-loaded stones. Both will do the job, but for the less experienced mechanic the "bottle brush" hone will probably be easier to use. You'll also need plenty of light oil or honing oil, some rags and an electric

drill motor. Proceed as follows:

a) *Mount the hone in the drill motor, compress the stones and slip it into the first cylinder (see illustration). Be sure to wear safety goggles or a face shield!*

b) *Lubricate the cylinder with plenty of oil, turn on the drill and move the hone up-and-down in the cylinder at a pace that will produce a fine crosshatch pattern on the cylinder walls. Ideally, the crosshatch lines should intersect at approximately a 60-degree angle (see illustration). Be sure to use plenty of lubricant and don't take off any more material than is absolutely necessary to produce the desired finish.*

➡**Note: Piston ring manufacturers may specify a smaller cross-hatch angle than the traditional 60-degree - read and follow any instructions included with the new rings.**

c) *Don't withdraw the hone from the cylinder while it's running. Instead, shut off the drill and continue moving the hone up-and-down in the cylinder until it comes to a complete stop, then compress the stones and withdraw the hone. If you're using a "bottle brush" type hone, stop the drill motor, then turn the chuck in the normal direction of rotation while withdrawing the hone from the cylinder.*

d) *Wipe the oil out of the cylinder and repeat the procedure for the remaining cylinders.*

4　After the honing job is complete, chamfer the top edges of the cylinder bores with a small file so the rings won't catch when the pistons are installed. Be very careful not to nick the cylinder walls with the end of the file.

5　The entire engine block must be washed again very thoroughly with warm, soapy water to remove all traces of the abrasive grit produced during the honing operation.

➡**Note: The bores can be considered clean when a white cloth - dampened with clean engine oil - used to wipe them out doesn't pick up any more honing residue, which will show up as gray areas on the cloth. Be sure to run a brush through all oil holes and galleries and flush them with running water.**

6　After rinsing, dry the block and apply a coat of light rust preventive oil to all machined surfaces. Wrap the block in a plastic trash bag to keep it clean and set it aside until reassembly.

16.3a A "bottle brush" hone will produce better results if you've never honed a cylinder before

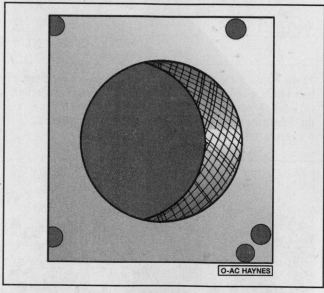

16.3b The cylinder hone should leave a smooth, crosshatch pattern with the lines intersecting at approximately a 60-degree angle

17 Piston/connecting rod assembly - inspection

▶ **Refer to illustrations 17.4a, 17.4b, 17.10 and 17.11**

1　Before the inspection process can be carried out, the piston/connecting rod assemblies must be cleaned and the original piston rings removed from the pistons.

➡**Note: Always use new piston rings when the engine is reassembled.**

2　Using a piston ring installation tool, carefully remove the rings from the pistons. Be careful not to nick or gouge the pistons in the process.

3　Scrape all traces of carbon from the top of the piston. A handheld wire brush or a piece of fine emery cloth can be used once the majority of the deposits have been scraped away. Do not, under any circumstances, use a wire brush mounted in a drill motor to remove deposits from the pistons. The piston material is soft and will be eroded away by the wire brush.

4　Use a piston ring groove cleaning tool to remove carbon deposits from the ring grooves. If a tool isn't available, a piece broken off the old ring will do the job. Be very careful to remove only the carbon deposits - don't remove any metal and do not nick or scratch the sides of the ring grooves (see illustrations).

5　Once the deposits have been removed, clean the piston/rod assemblies with solvent and dry them with compressed air (if available). Make sure the oil return holes in the back sides of the ring

17.4a The piston ring grooves can be cleaned with a special tool, as shown here . . .

17.4b . . . or a section of a broken ring

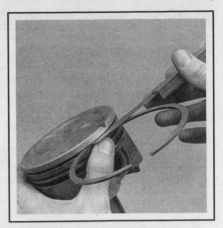

17.10 Check the ring side clearance with a feeler gauge at several points around the groove

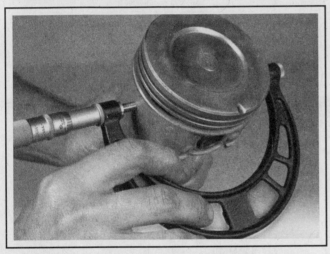

17.11 Measure the piston diameter at a 90-degree angle to the piston pin and in line with it

grooves are clear.

6   If the pistons and cylinder walls aren't damaged or worn excessively, and if the engine block is not rebored, new pistons won't be necessary. Normal piston wear appears as even vertical wear on the piston thrust surfaces and slight looseness of the top ring in its groove. New piston rings, on the other hand, should always be used when an engine is rebuilt.

7   Carefully inspect each piston for cracks around the skirt, at the pin bosses and at the ring lands.

8   Look for scoring and scuffing on the thrust faces of the skirt, holes in the piston crown and burned areas at the edge of the crown. If the skirt is scored or scuffed, the engine may have been suffering from overheating and/or abnormal combustion, which caused excessively high operating temperatures. The cooling and lubrication systems should be checked thoroughly. A hole in the piston crown is an indication that abnormal combustion (preignition) was occurring. Burned areas at the edge of the piston crown are usually evidence of spark

knock (detonation). If any of the above problems exist, the causes must be corrected or the damage will occur again.

9   Corrosion of the piston, in the form of small pits, indicates that coolant is leaking into the combustion chamber and/or the crankcase. Again, the cause must be corrected or the problem may persist in the rebuilt engine.

10   Measure the piston ring side clearance by laying a new piston ring in each ring groove and slipping a feeler gauge in beside it (see illustration). Check the clearance at three or four locations around each groove. Be sure to use the correct ring for each groove; they are different. If the side clearance is greater than those listed in this Chapter's Specifications, new pistons will have to be used.

11   Check the piston-to-bore clearance by measuring the bore (see Section 15) and the piston diameter. Make sure the pistons and bores are correctly matched. Measure the piston across the skirt, at a 90-degree angle to and in line with the piston pin (see illustration). Subtract the piston diameter from the bore diameter to obtain the clearance. If it's greater than specified, the block will have to be rebored and new pistons and rings installed.

12   Check the piston-to-rod clearance by twisting the piston and rod in opposite directions. Any noticeable play indicates excessive wear, which must be corrected. The piston/connecting rod assemblies should be taken to an automotive machine shop to have the pistons and rods rebored and new pins installed.

13   If the pistons must be removed from the connecting rods for any reason, they should be taken to an automotive machine shop. While they are there have the connecting rods checked for bend and twist, since automotive machine shops have special equipment for this purpose.

➡**Note: Unless new pistons and/or connecting rods must be installed, do not disassemble the pistons and connecting rods.**

14   Check the connecting rods for cracks and other damage. Temporarily remove the rod caps, lift out the old bearing inserts, wipe the rod and cap bearing surfaces clean and inspect them for nicks, gouges and scratches. After checking the rods, replace the old bearings, slip the caps into place and tighten the nuts finger tight.

18 Crankshaft - inspection

▶ **Refer to illustration 18.2**

1   Clean the crankshaft with solvent and dry it with compressed air (if available). Be sure to clean the oil holes with a stiff brush and flush them with solvent. Check the main and connecting rod bearing journals for uneven wear, scoring, pits and cracks. Check the rest of the crankshaft for cracks and other damage.

2   Using a micrometer, measure the diameter of the main and connecting rod journals and compare the results to the dimensions listed in this Chapter's Specifications (see illustration). By measuring the diameter at a number of points around each journal's circumference, you'll be able to determine whether or not the journal is out-of-round. Take the measurement at each end of the journal, near the crank throws, to determine if the journal is tapered.

3   If the crankshaft journals are damaged, tapered, out-of-round or worn beyond the limits given in this Chapter's Specifications, have the crankshaft reground by an automotive machine shop. Be sure to use the correct size bearing inserts if the crankshaft is reconditioned.

4   Refer to Section 19 and examine the main and rod bearing inserts.

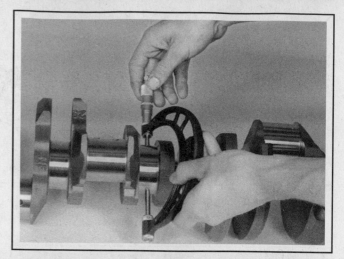

18.2 Measure the diameter of each crankshaft journal at several points to detect taper and out-of-round conditions

19 Main and connecting rod bearings - inspection

▶ **Refer to illustration 19.1**

1   Even though the main and connecting rod bearings should be replaced with new ones during the engine overhaul, the old bearings should be retained for close examination, as they may reveal valuable information about the condition of the engine (see illustration).

2   Bearing failure occurs because of lack of lubrication, the presence of dirt or other foreign particles, overloading the engine and corrosion. Regardless of the cause of bearing failure, it must be corrected before the engine is reassembled to prevent it from happening again.

3   When examining the bearings, remove them from the engine block, the main bearing caps, the connecting rods and the rod caps and lay them out on a clean surface in the same general position as their location in the engine. This will enable you to match any bearing problems with the corresponding crankshaft journal.

4   Dirt and other foreign particles get into the engine in a variety of ways. It may be left in the engine during assembly, or it may pass through filters or the PCV system. It may get into the oil, and from there into the bearings. Metal chips from machining operations and normal engine wear are often present. Abrasives are sometimes left in engine components after reconditioning, especially when parts are not thoroughly cleaned using the proper cleaning methods. Whatever the source, these foreign objects often end up embedded in the soft bearing material and are easily recognized. Large particles will not embed in the bearing and will score or gouge the bearing and journal. The best prevention for this cause of bearing failure is to clean all parts thoroughly and keep everything spotlessly clean during engine assembly. Frequent and regular engine oil and filter changes are also recommended.

5   Lack of lubrication (or lubrication breakdown) has a number of interrelated causes. Excessive heat (which thins the oil), overloading (which squeezes the oil from the bearing face) and oil leakage or throw off (from excessive bearing clearances, worn oil pump or high engine speeds) all contribute to lubrication breakdown. Blocked oil passages, which usually are the result of misaligned oil holes in a bearing shell,

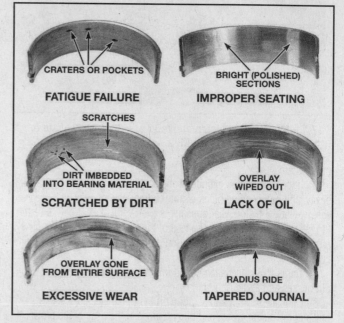

19.1 Typical bearing failures

will also oil starve a bearing and destroy it. When lack of lubrication is the cause of bearing failure, the bearing material is wiped or extruded from the steel backing of the bearing. Temperatures may increase to the point where the steel backing turns blue from overheating.

6   Driving habits can have a definite effect on bearing life. Full throttle, low speed operation (lugging the engine) puts very high loads on bearings, which tends to squeeze out the oil film. These loads cause the bearings to flex, which produces fine cracks in the bearing face (fatigue failure). Eventually the bearing material will loosen in pieces and tear away from the steel backing. Short trip driving leads to corro-

sion of bearings because insufficient engine heat is produced to drive off the condensed water and corrosive gases. These products collect in the engine oil, forming acid and sludge. As the oil is carried to the engine bearings, the acid attacks and corrodes the bearing material.

7   Incorrect bearing installation during engine assembly will lead to

bearing failure as well. Tight fitting bearings leave insufficient bearing oil clearance and will result in oil starvation. Dirt or foreign particles trapped behind a bearing insert result in high spots on the bearing which lead to failure.

20 Engine overhaul - reassembly sequence

1   Before beginning engine reassembly, make sure you have all the necessary new parts, gaskets and seals as well as the following items on hand:

Common hand tools
A 1/2-inch drive torque wrench
Piston ring installation tool
Piston ring compressor
Short lengths of rubber or plastic hose to fit over connecting rod bolts
Plastigage
Feeler gauges
A fine-tooth file
New engine oil
Engine assembly lube or moly-base grease
RTV-type gasket sealant
Anaerobic-type gasket sealant
Thread locking compound

2   In order to save time and avoid problems, engine reassembly must be done in the following general order:

New camshaft bearings (must be done by automotive machine shop)
Piston rings
Crankshaft and main bearings
Piston/connecting rod assemblies
Oil pump
Camshaft and lifters
Oil pan
Timing chain and sprockets (V6 engine only)
Cylinder head(s), pushrods and rocker arms
Timing cover
Intake and exhaust manifolds
Rocker arm cover(s)
Flywheel/driveplate

21 Piston rings - installation

▶ **Refer to illustrations 21.3, 21.4, 21.5, 21.9a, 21.9b and 21.12**

1   Before installing the new piston rings, the ring end gaps must be checked. It's assumed that the piston ring side clearance has been checked and verified correct (see Section 17).

2   Lay out the piston/connecting rod assemblies and the new ring sets so the ring sets will be matched with the same piston and cylinder during the end gap measurement and engine assembly.

3   Insert the top (number one) ring into the first cylinder and square it up with the cylinder walls by pushing it in with the top of the piston

(see illustration). The ring should be near the bottom of the cylinder, at the lower limit of ring travel.

4   To measure the end gap, slip feeler gauges between the ends of the ring until a gauge equal to the gap width is found (see illustration). The feeler gauge should slide between the ring ends with a slight amount of drag. Compare the measurement to the Specifications. If the gap is larger or smaller than specified, double-check to make sure you have the correct rings before proceeding.

5   If the gap is too small, it must be enlarged or the ring ends may

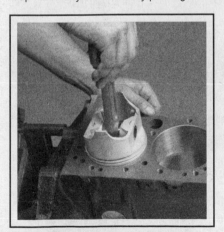

21.3 When checking piston ring end gap, the ring must be square in the cylinder bore (this is done by pushing the ring down with the top of a piston as shown)

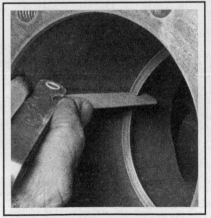

21.4 With the ring square in the cylinder, measure the end gap with a feeler gauge

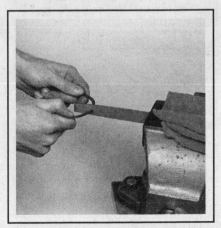

21.5 If the end gap is too small, clamp a file in a vise and file the ring ends (from the outside in only) to enlarge the gap slightly

ENGINE BEARING ANALYSIS

Debris

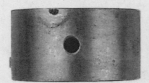

Babbitt bearing embedded with debris from machinings

Microscopic detail of debris

Microscopic detail of gouges

Overplated copper alloy bearing gouged by cast iron debris

Aluminum bearing embedded with glass beads

Microscopic detail of glass beads

Damaged lining caused by dirt left on the bearing back

Misassembly

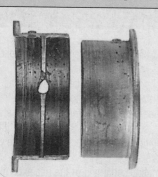

Result of a lower half assembled as an upper - blocking the oil flow

Excessive oil clearance is indicated by a short contact arc

Polished and oil-stained backs are a result of a poor fit in the housing bore

Result of a wrong, reversed, or shifted cap

Overloading

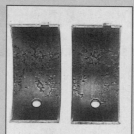

Damage from excessive idling which resulted in an oil film unable to support the load imposed

Damaged upper connecting rod bearings caused by engine lugging; the lower main bearings (not shown) were similarly affected

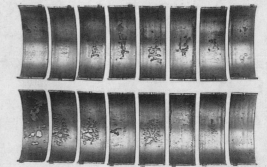

The damage shown in these upper and lower connecting rod bearings was caused by engine operation at a higher-than-rated speed under load

Misalignment

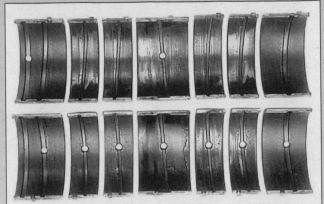

A warped crankshaft caused this pattern of severe wear in the center, diminishing toward the ends

A poorly finished crankshaft caused the equally spaced scoring shown

A tapered housing bore caused the damage along one edge of this pair

A bent connecting rod led to the damage in the "V" pattern

Lubrication

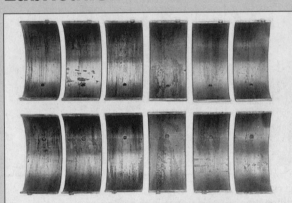

Result of dry start: The bearings on the left, farthest from the oil pump, show more damage

Result of a low oil supply or oil starvation

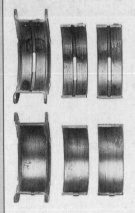

Severe wear as a result of inadequate oil clearance

Corrosion

Microscopic detail of corrosion

Corrosion is an acid attack on the bearing lining generally caused by inadequate maintenance, extremely hot or cold operation, or inferior oils or fuels

Microscopic detail of cavitation

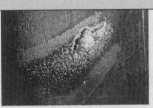

Example of cavitation - a surface erosion caused by pressure changes in the oil film

Damage from excessive thrust or insufficient axial clearance

Bearing affected by oil dilution caused by excessive blow-by or a rich mixture

2C-22 GENERAL ENGINE OVERHAUL PROCEDURES

come in contact with each other during engine operation, which can cause serious damage to the engine. The end gap can be increased by filing the ring ends very carefully with a fine file. Mount the file in a vise equipped with soft jaws, slip the ring over the file with the ends contacting the file face and slowly move the ring to remove material from the ends. When performing this operation, file only from the outside in (see illustration).

6 Excess end gap isn't critical unless it's greater than 0.040-inch. Again, double-check to make sure you have the correct rings for your engine.

7 Repeat the procedure for each ring that will be installed in the first cylinder and for each ring in the remaining cylinders. Remember to keep rings, pistons and cylinders matched up.

8 Once the ring end gaps have been checked/corrected, the rings can be installed on the pistons.

9 The oil control ring (lowest one on the piston) is installed first. It's composed of three separate components. Slip the spacer/expander into the groove (see illustration). If an anti-rotation tang is used, make sure it's inserted into the drilled hole in the ring groove. Next, install the lower side rail. Don't use a piston ring installation tool on the oil ring side rails, as they may be damaged. Instead, place one end of

the side rail into the groove between the spacer/expander and the ring land, hold it firmly in place and slide a finger around the piston while pushing the rail into the groove (see illustration). Next, install the upper side rail in the same manner.

10 After the three oil ring components have been installed, check to make sure that both the upper and lower side rails can be turned smoothly in the ring groove.

11 The number two (middle) ring is installed next. It's stamped with a mark which must face up, toward the top of the piston.

→Note: Always follow the instructions printed on the ring package or box - different manufacturers may require different approaches. Do not mix up the top and middle rings, as they have different cross sections.

12 Use a piston ring installation tool and make sure the identification mark is facing the top of the piston, then slip the ring into the middle groove on the piston (see illustration). Don't expand the ring any more than is necessary to slide it over the piston.

13 Install the number one (top) ring in the same manner. Make sure the mark is facing up. Be careful not to confuse the number one and number two rings.

14 Repeat the procedure for the remaining pistons and rings.

21.9a Installing the spacer/expander in the oil control ring groove

21.9b DO NOT use a piston ring installation tool when installing the oil ring side rails

21.12 Installing the compression rings with a ring expander - the mark (arrow) must face up

22 Crankshaft - installation and main bearing oil clearance check

♦ Refer to illustrations 22.10 and 22.14

1 Crankshaft installation is the first step in engine reassembly. It's assumed at this point that the engine block and crankshaft have been cleaned, inspected and repaired or reconditioned.

→Note: If your engine has a two-piece rear main oil seal (1985 V6 models), refer to Section 23 and install the oil seal sections in the cap and block before proceeding with crankshaft installation.

2 Position the engine with the bottom facing up.

3 Remove the main bearing cap bolts and lift out the caps. Lay them out in the proper order to ensure correct installation.

4 If they're still in place, remove the old bearing inserts from the block and the main bearing caps. Wipe the main bearing surfaces of the block and caps with a clean, lint free cloth. They must be kept spotlessly clean.

5 Clean the back sides of the new main bearing inserts and lay one

bearing half in each main bearing saddle in the block. Lay the other bearing half from each bearing set in the corresponding main bearing cap. Make sure the tab on the bearing insert fits into the recess in the block or cap. Also, the oil holes in the block must line up with the oil holes in the bearing insert. Do not hammer the bearing into place and do not nick or gouge the bearing faces. No lubrication should be used at this time.

6 The flanged thrust bearing must be installed in the rear cap and saddle.

7 Clean the faces of the bearings in the block and the crankshaft main bearing journals with a clean, lint free cloth. Check or clean the oil holes in the crankshaft, as any dirt here can go only one way - straight through the new bearings.

8 Once you're certain the crankshaft is clean, carefully lay it in position in the main bearings.

9 Before the crankshaft can be permanently installed, the main bearing oil clearance must be checked.

22.10 Lay the Plastigage strips (arrow) on the main bearing journals, parallel to the crankshaft centerline

22.14 Compare the width of the crushed Plastigage to the scale on the container to determine the main bearing oil clearance (always take the measurement at the widest point of the Plastigage); be sure to use the correct scale - standard and metric scales are included

10 Trim several pieces of the appropriate size of Plastigage (they must be slightly shorter than the width of the main bearings) and place one piece on each crankshaft main bearing journal, parallel with the journal axis (see illustration).

11 Clean the faces of the bearings in the caps and install the caps in their respective positions (don't mix them up) with the arrows pointing toward the front of the engine. Don't disturb the Plastigage.

12 Starting with the center main and working out toward the ends, tighten the main bearing cap bolts, in three steps, to the specified torque. Don't rotate the crankshaft at any time during this operation.

13 Remove the bolts and carefully lift off the main bearing caps. Keep them in order. Don't disturb the Plastigage or rotate the crankshaft. If any of the main bearing caps are difficult to remove, tap them gently from side-to-side with a soft-face hammer to loosen them.

14 Compare the width of the crushed Plastigage on each journal to the scale printed on the Plastigage container to obtain the main bearing oil clearance (see illustration). Check the Specifications to make sure it's correct.

15 If the clearance is not as specified, the bearing inserts may be the wrong size (which means different ones will be required). Before deciding that different inserts are needed, make sure that no dirt or oil was between the bearing inserts and the caps or block when the clearance was measured. If the Plastigage was wider at one end than the other, the journal may be tapered (refer to Section 18).

16 Carefully scrape all traces of the Plastigage material off the main bearing journals and/or the bearing faces. Don't nick or scratch the bearing faces.

17 Carefully lift the crankshaft out of the engine. Clean the bearing faces in the block, then apply a thin, uniform layer of clean moly-base grease or engine assembly lube to each of the bearing surfaces. Be sure to coat the thrust faces as well as the journal face of the rear bearing.

18 If not already done, refer to Section 23 and install the rear main oil seal sections in the block and bearing cap (1985 V6 engine only). Lubricate the seal faces with moly-base grease, engine assembly lube or clean engine oil.

19 Make sure the crankshaft journals are clean, then lay the crankshaft back in place in the block. Clean the faces of the bearings in the caps, then apply lubricant to them. Install the caps in their respective positions with the arrows pointing toward the front of the engine. Note that the rear cap must have a special sealant applied (see Chapter 2B). Install the bolts.

20 Tighten all except the rear cap bolts (the one with the thrust bearing) to the specified torque (work from the center out and approach the final torque in three steps). Tighten the rear cap bolts to 10-to-12 ft-lbs. Tap the ends of the crankshaft forward and backward with a lead or brass hammer to line up the main bearing and crankshaft thrust surfaces. Retighten all main bearing cap bolts to the specified torque, starting with the center main and working out toward the ends.

21 On manual transmission equipped models, install a new pilot bearing in the end of the crankshaft (see Chapter 8).

22 Rotate the crankshaft a number of times by hand to check for any obvious binding.

23 The final step is to check the crankshaft end play with a feeler gauge or a dial indicator as described in Section 13. The end play should be correct if the crankshaft thrust faces aren't worn or damaged and new bearings have been installed.

24 If you have a V6 engine with a one-piece rear main oil seal, refer to Section 23 and install the new seal, then bolt the housing to the block.

23 Rear main oil seal installation

FOUR-CYLINDER ENGINE

1 Clean the bore in the block and seal contact surface on the crankshaft. Check the crankshaft surface for scratches and nicks that could damage the new seal lip and cause oil leaks. If the crankshaft is damaged, the only alternative is a new or different crankshaft.

2 Apply a light coat of engine oil or multi-purpose grease to the outer edge of the new seal. Lubricate the seal lip with moly-base grease.

3 Press the new seal into place with a special seal installation tool (if available). The seal lip must face toward the front of the engine. If the special tool isn't available, carefully work the seal lip over the end of the crankshaft and tap the seal in with a hammer and punch until it's seated in the bore (see illustration 20.8 in Chapter 2, Part A).

V6 ENGINE

1985

4   Inspect the rear main bearing cap and engine block mating surfaces, as well as the seal grooves, for nicks, burrs and scratches. Remove any defects with a fine file or deburring tool.

5   Install one seal section in the block with the lip facing the front of the engine (if the seal has two lips, the one with the helix must face the front) (see illustration 15.15b in Chapter 2, Part B). Leave one end protruding from the block approximately 1/4 to 3/8-inch and make sure it's completely seated.

6   Repeat the procedure to install the remaining seal half in the rear main bearing cap. In this case, leave the opposite end of the seal protruding from the cap the same distance the block seal is protruding from the block.

7   During final installation of the crankshaft (after the main bearing oil clearances have been checked with Plastigage) as described in Section 22, apply a thin, even coat of anaerobic-type gasket sealant to the areas of the cap or block indicated in illustration 15.19 (Part B of Chapter 2). Don't get any sealant on the bearing face, crankshaft journal, seal ends or seal lips. Also, lubricate the seal lips with moly-base grease or clean engine oil.

1986 on

8   Later models are equipped with a one-piece seal that requires an entirely different installation procedure. The crankshaft must be installed first and the main bearing caps bolted in place, then the new seal should be installed in the housing and the housing bolted to the block (see illustration 15.26 in Chapter 2, Part B).

9   Before installing the crankshaft, check the seal contact surface very carefully for scratches and nicks that could damage the new seal lip and cause oil leaks. If the crankshaft is damaged, the only alternative is a new or different crankshaft.

10   The old seal can be removed from the housing by inserting a large screwdriver into the notches provided and prying it out (see illustration 15.23 in Part B). Be sure to note how far it's recessed into the housing bore before removing it; the new seal will have to be recessed an equal amount. Be very careful not to scratch or otherwise damage the bore in the housing or oil leaks could develop.

11   Make sure the housing is clean, then apply a thin coat of engine oil to the outer edge of the new seal. The seal must be pressed squarely into the housing bore, so hammering it into place is not recommended. If you don't have access to a press, sandwich the housing and seal between two smooth pieces of wood and press the seal into place with the jaws of a large vise. The pieces of wood must be thick enough to distribute the force evenly around the entire circumference of the seal. Work slowly and make sure the seal enters the bore squarely.

➡Note: Do not allow oil or other lubricants to contact seal lip surface. Use a new gasket - no sealant is required - and make sure the dowel pins are in place before installing the housing.

12   Tighten the nuts/screws a little at a time until the seal housing is fully seated. Then tighten them to the torque listed in the Chapter 2B Specifications.

24   Piston/connecting rod assembly - installation and rod bearing oil clearance check

▶ Refer to illustrations 24.5, 24.8, 24.9, 24.11 and 24.13

1   Before installing the piston/connecting rod assemblies, the cylinder walls must be perfectly clean, the top edge of each cylinder must be chamfered, and the crankshaft must be in place.

2   Remove the connecting rod cap from the end of the number one connecting rod. Remove the old bearing inserts and wipe the bearing surfaces of the connecting rod and cap with a clean, lint free cloth. They must be kept spotlessly clean.

3   Clean the back side of the new upper bearing half, then lay it in place in the connecting rod. Make sure the tab on the bearing fits into the recess in the rod. Don't hammer the bearing insert into place and be very careful not to nick or gouge the bearing face. Don't lubricate the bearing at this time.

4   Clean the back side of the other bearing insert and install it in the rod cap. Again, make sure the tab on the bearing fits into the recess in the cap, and don't apply any lubricant. It's critically important that the mating surfaces of the bearing and connecting rod are perfectly clean and oil free when they're assembled.

5   Position the piston ring gaps at 120-degree intervals around the piston (see illustration), then slip a section of plastic or rubber hose over each connecting rod cap bolt.

6   Lubricate the piston and rings with clean engine oil and attach a piston ring compressor to the piston. Leave the skirt protruding about 1/4-inch to guide the piston into the cylinder. The rings must be compressed until they're flush with the piston.

7   Rotate the crankshaft until the number one connecting rod journal is at BDC (bottom dead center) and apply a coat of engine oil to the cylinder walls.

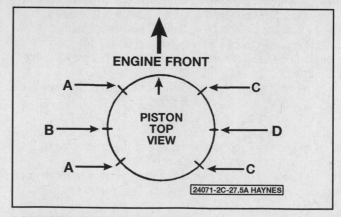

24.5 Ring end gap positions

A   Oil ring rail gaps
B   Second compression ring gap
C   Oil ring spacer gap (position in-between marks)
D   Top compression ring gap

8   With the notch on top of the piston (see illustration) facing the front of the engine, gently insert the piston/connecting rod assembly into the number one cylinder bore and rest the bottom edge of the ring compressor on the engine block. Tap the top edge of the ring compressor to make sure it's contacting the block around its entire circumference.

9   Carefully tap on the top of the piston with the end of a wooden hammer handle (see illustration) while guiding the end of the connect-

24.8 The notch in each piston must face the FRONT of the engine as the pistons are installed

24.9 The piston can be driven (gently) into the cylinder bore with the end of a wooden hammer handle

24.11 Lay the Plastigage strips on each rod bearing journal, parallel to the crankshaft centerline

ing rod into place on the crankshaft journal. The piston rings may try to pop out of the ring compressor just before entering the cylinder bore, so keep some downward pressure on the ring compressor. Work slowly, and if any resistance is felt as the piston enters the cylinder, stop immediately. Find out what's hanging up and fix it before proceeding. Do not, for any reason, force the piston into the cylinder, as you might break a ring and/or the piston.

10 Once the piston/connecting rod assembly is installed, the connecting rod bearing oil clearance must be checked before the rod cap is permanently bolted in place.

11 Cut a piece of the appropriate size Plastigage slightly shorter than the width of the connecting rod bearing and lay it in place on the number one connecting rod journal, parallel with the journal axis (see illustration).

12 Clean the connecting rod cap bearing face, remove the protective hoses from the connecting rod bolts and install the rod cap. Make sure the mating mark on the cap is on the same side as the mark on the connecting rod. Install the nuts and tighten them to the specified torque, working up to it in three steps.

➡Note: Use a thin-wall socket to avoid erroneous torque readings that can result if the socket is wedged between the rod cap and nut. Do not rotate the crankshaft at any time during this operation.

13 Remove the rod cap, being very careful not to disturb the Plastigage. Compare the width of the crushed Plastigage to the scale printed on the Plastigage container to obtain the oil clearance (see illustration). Compare it to the dimension listed in this Chapter's Specifications to make sure the clearance is correct. If the clearance is not as specified, the bearing inserts may be the wrong size (which means different ones will be required). Before deciding that different inserts are needed, make sure that no dirt or oil was between the bearing inserts and the connecting rod or cap when the clearance was measured. Also, recheck the journal diameter. If the Plastigage was wider at one end than the other, the journal may be tapered (refer to Section 18).

14 Carefully scrape all traces of the Plastigage material off the rod journal and/or bearing face. Be very careful not to scratch the bearing - use your fingernail or a piece of hardwood. Make sure the bearing faces are perfectly clean, then apply a uniform layer of clean moly-base grease or engine assembly lube to both of them. You'll have to push the piston into the cylinder to expose the face of the bearing insert in the connecting rod - be sure to slip the protective hoses over the rod bolts first.

15 Slide the connecting rod back into place on the journal, remove the protective hoses from the rod cap bolts, install the rod cap and tighten the nuts to the specified torque. Again, work up to the torque in three steps.

16 Repeat the entire procedure for the remaining piston/connecting rod assemblies. Keep the back sides of the bearing inserts and the inside of the connecting rod and cap perfectly clean when assembling them. Make sure you have the correct piston for the cylinder and that the notch on the piston faces to the front of the engine when the piston is installed. Remember, use plenty of oil to lubricate the piston before installing the ring compressor. Also, when installing the rod caps for the final time, be sure to lubricate the bearing faces adequately.

17 After all the piston/connecting rod assemblies have been properly installed, rotate the crankshaft a number of times by hand to check for any obvious binding.

18 As a final step, the connecting rod end play must be checked. Refer to Section 12 for this procedure. Compare the measured end play to the Specifications to make sure it's correct. If it was correct before disassembly and the original crankshaft and rods were reinstalled, it should still be right. If new rods or a new crankshaft were installed, the end play may be too small. If so, the rods will have to be removed and taken to an automotive machine shop for resizing.

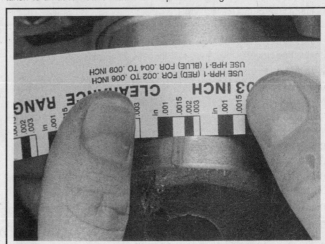

24.13 Measuring the width of the crushed Plastigage to determine the rod bearing oil clearance (be sure to use the correct scale - standard and metric scales are included)

GLOSSARY

B

Backlash - The amount of play between two parts. Usually refers to how much one gear can be moved back and forth without moving the gear with which it's meshed.

Bearing Caps - The caps held in place by nuts or bolts which, in turn, hold the bearing surface. This space is for lubricating oil to enter.

Bearing clearance - The amount of space left between shaft and bearing surface. This space is for lubricating oil to enter.

Bearing crush - The additional height which is purposely manufactured into each bearing half to ensure complete contact of the bearing back with the housing bore when the engine is assembled.

Bearing knock - The noise created by movement of a part in a loose or worn bearing.

Blueprinting - Dismantling an engine and reassembling it to EXACT specifications.

Bore - An engine cylinder, or any cylindrical hole; also used to describe the process of enlarging or accurately refinishing a hole with a cutting tool, as to bore an engine cylinder. The bore size is the diameter of the hole.

Boring - Renewing the cylinders by cutting them out to a specified size. A boring bar is used to make the cut.

Bottom end - A term which refers collectively to the engine block, crankshaft, main bearings and the big ends of the connecting rods.

Break-in - The period of operation between installation of new or rebuilt parts and time in which parts are worn to the correct fit. Driving at reduced and varying speed for a specified mileage to permit parts to wear to the correct fit.

Bushing - A one-piece sleeve placed in a bore to serve as a bearing surface for shaft, piston pin, etc. Usually replaceable.

C

Camshaft - The shaft in the engine, on which a series of lobes are located for operating the valve mechanisms. The camshaft is driven by gears or sprockets and a timing chain. Usually referred to simply as the cam.

Carbon - Hard, or soft, black deposits found in combustion chamber, on plugs, under rings, on and under valve heads.

Cast iron - An alloy of iron and more than two percent carbon, used for engine blocks and heads because it's relatively inexpensive and easy to mold into complex shapes.

Chamfer - To bevel across (or a bevel on) the sharp edge of an object.

Chase - To repair damaged threads with a tap or die.

Combustion chamber - The space between the piston and the cylinder head, with the piston at top dead center, in which air-fuel mixture is burned.

Compression ratio - The relationship between cylinder volume (clearance volume) when the piston is at top dead center and cylinder volume when the piston is at bottom dead center.

Connecting rod - The rod that connects the crank on the crankshaft with the piston. Sometimes called a con rod.

Connecting rod cap - The part of the connecting rod assembly that attaches the rod to the crankpin.

Core plug - Soft metal plug used to plug the casting holes for the coolant passages in the block.

Crankcase - The lower part of the engine in which the crankshaft rotates; includes the lower section of the cylinder block and the oil pan.

Crank kit - A reground or reconditioned crankshaft and new main and connecting rod bearings.

Crankpin - The part of a crankshaft to which a connecting rod is attached.

Crankshaft - The main rotating member, or shaft, running the length of the crankcase, with offset throws to which the connecting rods are attached; changes the reciprocating motion of the pistons into rotating motion.

Cylinder sleeve - A replaceable sleeve, or liner, pressed into the cylinder block to form the cylinder bore.

D

Deburring - Removing the burrs (rough edges or areas) from a bearing.

Deglazer - A tool, rotated by an electric motor, used to remove glaze from cylinder walls so a new set of rings will seat.

E

Endplay - The amount of lengthwise movement between two parts. As applied to a crankshaft, the distance that the crankshaft can move forward and back in the cylinder block.

F

Face - A machinist's term that refers to removing metal from the end of a shaft or the face of a larger part, such as a flywheel.

Fatigue - A breakdown of material through a large number of loading and unloading cycles. The first signs are cracks followed shortly by breaks.

Feeler gauge - A thin strip of hardened steel, ground to an exact thickness, used to check clearances between parts.

Free height - The unloaded length or height of a spring.

Freeplay - The looseness in a linkage, or an assembly of parts, between the initial application of force and actual movement. Usually perceived as slop or slight delay.

Freeze plug - See Core plug.

G

Gallery - A large passage in the block that forms a reservoir for engine oil pressure.

Glaze - The very smooth, glassy finish that develops on cylinder walls while an engine is in service.

H

Heli-Coil - A rethreading device used when threads are worn or damaged. The device is installed in a retapped hole to reduce the thread size to the original size.

I

Installed height - The spring's measured length or height, as installed on the cylinder head. Installed height is measured from the spring seat to the underside of the spring retainer.

J

Journal - The surface of a rotating shaft which turns in a bearing.

K

Keeper - The split lock that holds the valve spring retainer in position on the valve stem.

Key - A small piece of metal inserted into matching grooves machined into two parts fitted together - such as a gear pressed onto a shaft - which prevents slippage between the two parts.

Knock - The heavy metallic engine sound, produced in the combustion chamber as a result of abnormal combustion - usually detonation. Knock is usually caused by a loose or worn bearing. Also referred to as detonation, pinging and spark knock. Connecting rod or main bearing knocks are created by too much oil clearance or insufficient lubrication.

L

Lands - The portions of metal between the piston ring grooves.

Lapping the valves - Grinding a valve face and its seat together with lapping compound.

Lash - The amount of free motion in a gear train, between gears, or in a mechanical assembly, that occurs before movement can begin. Usually refers to the lash in a valve train.

Lifter - The part that rides against the cam to transfer motion to the rest of the valve train.

M

Machining - The process of using a machine to remove metal from a metal part.

Main bearings - The plain, or babbitt, bearings that support the crankshaft.

Main bearing caps - The cast iron caps, bolted to the bottom of the block, that support the main bearings.

O

O.D. - Outside diameter.

Oil gallery - A pipe or drilled passageway in the engine used to carry engine oil from one area to another.

Oil ring - The lower ring, or rings, of a piston; designed to prevent excessive amounts of oil from working up the cylinder walls and into the combustion chamber. Also called an oil-control ring.

Oil seal - A seal which keeps oil from leaking out of a compartment. Usually refers to a dynamic seal around a rotating shaft or other moving part.

O-ring - A type of sealing ring made of a special rubberlike material; in use, the O-ring is compressed into a groove to provide the sealing action.

Overhaul - To completely disassemble a unit, clean and inspect all parts, reassemble it with the original or new parts and make all adjustments necessary for proper operation.

P

Pilot bearing - A small bearing installed in the center of the flywheel (or the rear end of the crankshaft) to support the front end of the input shaft of the transmission.

Pip mark - A little dot or indentation which indicates the top side of a compression ring.

Piston - The cylindrical part, attached to the connecting rod, that moves up and down in the cylinder as the crankshaft rotates. When the fuel charge is fired, the piston transfers the force of the explosion to the connecting rod, then to the crankshaft.

Piston pin (or wrist pin) - The cylindrical and usually hollow steel pin that passes through the piston. The piston pin fastens the piston to the upper end of the connecting rod.

Piston ring - The split ring fitted to the groove in a piston. The ring contacts the sides of the ring groove and also rubs against the cylinder wall, thus sealing space between piston and wall. There are two types of rings: Compression rings seal the compression pressure in the combustion chamber; oil rings scrape excessive oil off the cylinder wall.

Piston ring groove - The slots or grooves cut in piston heads to hold piston rings in position.

Piston skirt - The portion of the piston below the rings and the piston pin hole.

Plastigage - A thin strip of plastic thread, available in different sizes, used for measuring clearances. For example, a strip of plastigage is laid across a bearing journal and mashed as parts are assembled. Then parts are disassembled and the width of the strip is measured to determine clearance between journal and bearing. Commonly used to measure crankshaft main-bearing and connecting rod bearing clearances.

Press-fit - A tight fit between two parts that requires pressure to force the parts together. Also referred to as drive, or force, fit.

Prussian blue - A blue pigment; in solution, useful in determining the area of contact between two surfaces. Prussian blue is commonly used to determine the width and location of the contact area between the valve face and the valve seat.

R

Race (bearing) - The inner or outer ring that provides a contact surface for balls or rollers in bearing.

Ream - To size, enlarge or smooth a hole by using a round cutting tool with fluted edges.

Ring job - The process of reconditioning the cylinders and installing new rings.

Runout - Wobble. The amount a shaft rotates out-of-true.

S

Saddle - The upper main bearing seat.

Scored - Scratched or grooved, as a cylinder wall may be scored by abrasive particles moved up and down by the piston rings.

Scuffing - A type of wear in which there's a transfer of material between parts moving against each other; shows up as pits or grooves in the mating surfaces.

Seat - The surface upon which another part rests or seats. For example, the valve seat is the matched surface upon which the valve face rests. Also used to refer to wearing into a good fit; for example, piston rings seat after a few miles of driving.

Short block - An engine block complete with crankshaft and piston and, usually, camshaft assemblies.

Static balance - The balance of an object while it's stationary.

Step - The wear on the lower portion of a ring land caused by excessive side and back-clearance. The height of the step indicates the ring's extra side clearance and the length of the step projecting from the back wall of the groove represents the ring's back clearance.

Stroke - The distance the piston moves when traveling from top dead center to bottom dead center, or from bottom dead center to top dead center.

Stud - A metal rod with threads on both ends.

T

Tang - A lip on the end of a plain bearing used to align the bearing during assembly.

Tap - To cut threads in a hole. Also refers to the fluted tool used to cut threads.

Taper - A gradual reduction in the width of a shaft or hole; in an engine cylinder, taper usually takes the form of uneven wear, more pronounced at the top than at the bottom.

Throws - The offset portions of the crankshaft to which the connecting rods are affixed.

Thrust bearing - The main bearing that has thrust faces to prevent excessive endplay, or forward and backward movement of the crankshaft.

Thrust washer - A bronze or hardened steel washer placed between two moving parts. The washer prevents longitudinal movement and provides a bearing surface for thrust surfaces of parts.

Tolerance - The amount of variation permitted from an exact size of measurement. Actual amount from smallest acceptable dimension to largest acceptable dimension.

U

Umbrella - An oil deflector placed near the valve tip to throw oil from the valve stem area.

Undercut - A machined groove below the normal surface.

Undersize bearings - Smaller diameter bearings used with re-ground crankshaft journals.

V

Valve grinding - Refacing a valve in a valve-refacing machine.

Valve train - The valve-operating mechanism of an engine; includes all components from the camshaft to the valve.

Vibration damper - A cylindrical weight attached to the front of the crankshaft to minimize torsional vibration (the twist-untwist actions of the crankshaft caused by the cylinder firing impulses). Also called a harmonic balancer.

W

Water jacket - The spaces around the cylinders, between the inner and outer shells of the cylinder block or head, through which coolant circulates.

Web - A supporting structure across a cavity.

Woodruff key - A key with a radiused backside (viewed from the side).

25 Pre-oiling engine after overhaul

◆ **Refer to illustrations 25.3, 25.5 and 25.6**

➡ **Note: This procedure applies to the V6 engine only.**

1   After an overhaul it's a good idea to pre-oil the engine before it is installed in the vehicle and started for the first time. Pre-oiling will reveal any problems with the lubrication system at a time when corrections can be made easily and will prevent major engine damage. It will also allow the internal engine parts to be lubricated thoroughly in the normal fashion without the heavy loads associated with combustion placed on them.

2   The engine should be completely assembled with the exception of the distributor and rocker arm covers. The oil filter and oil pressure sending unit must be in place and the specified amount of oil must be in the crankcase (see Chapter 1).

3   A modified small block Chevrolet V8 distributor will be needed for this procedure - a junkyard should be able to supply one for a reasonable price. In order to function as a pre-oil tool, the distributor must have the gear on the lower end of the shaft and, if equipped, the advance weights on the upper end of the shaft removed (see illustration).

4   Install the pre-oil distributor in place of the original distributor and make sure the lower end of the shaft mates with the upper end of the oil pump driveshaft. Turn the distributor shaft until they're aligned and the distributor body seats on the block. Install the distributor hold-down clamp and bolt.

5   Mount the upper end of the shaft in the chuck of an electric drill and use the drill to turn the pre-oil distributor shaft, which will drive the

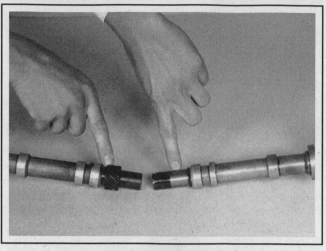

25.3 The pre-oil distributor (right) has the gear and advance weights (if equipped) removed

oil pump and circulate the oil throughout the engine (see illustration).

➡ **Note: The drill must turn in a clockwise direction.**

6   It may take two or three minutes, but oil should soon start to flow out of all of the rocker arm holes, indicating that the oil pump is working properly (see illustration). Let the oil circulate for several seconds, then shut off the drill motor.

7   Remove the pre-oil distributor, then install the rocker arm covers.

25.5 A drill motor connected to the modified distributor shaft drives the oil pump - make sure it turns clockwise as viewed from above

25.6 Oil will begin to flow from all of the rocker arm holes if the oil pump and lubrication system are functioning properly

26 Initial start-up and break-in after overhaul

1   Once the engine has been installed in the vehicle, double-check the engine oil and coolant levels.

2   With the spark plugs out of the engine and the ignition system disabled (see Section 3), crank the engine until oil pressure registers on the gauge.

3   Install the spark plugs, hook up the plug wires and restore the ignition system functions (see Section 3).

4   Start the engine. It may take a few moments for the gasoline to reach the carburetor or injectors, but the engine should start without a great deal of effort.

5   After the engine starts, it should be allowed to warm up to normal operating temperature. While the engine is warming up, make a thorough check for oil and coolant leaks.

6   Shut the engine off and recheck the engine oil and coolant levels.

7   Drive the vehicle to an area with minimum traffic, accelerate from 30 to 50 mph, then allow the vehicle to slow to 30 mph with the throttle closed. Repeat the procedure 10 or 12 times. This will load the piston rings and cause them to seat properly against the cylinder walls. Check again for oil and coolant leaks.

8   Drive the vehicle gently for the first 500 miles (no sustained high speeds) and keep a constant check on the oil level. It is not unusual for an engine to use oil during the break-in period.

9   At approximately 500 to 600 miles, change the oil and filter.

10 For the next few hundred miles, drive the vehicle normally. Do not pamper it or abuse it.

11 After 2000 miles, change the oil and filter again and consider the engine fully broken in.

Specifications

Four-cylinder engine

General

Bore and stroke	4.00 x 3.00 inches
Oil pressure	36 to 41 psi at 2000 rpm
Compression pressure	140 psi at 160 rpm

Cylinder head and valve train

Head warpage limit	0.006 inch
Valve face angle	45-degrees
Valve seat angle	
1985 and 1986	45-degrees
1987 on	46-degrees
Minimum valve margin width	1/32 inch
Valve stem-to-guide clearance	
Intake	0.001 to 0.0027 inch
Exhaust	0.001 to 0.0027 inch
Valve seat width	
Intakc	0.035 to 0.075 inch
Exhaust	
1985 and 1986	0.058 to 0.097 inch
1987 on	0.085 to 0.105 inch
Valve spring free length	1.78 inches
Valve spring installed height	
1985 and 1986	1.690 inches
1987 on	1.440 inches
Valve spring pressure and length (intake and exhaust)	
Valve closed	
1985 and 1986	78 to 86 lbs at 1.66 inches
1987 on	71 to 78 lbs at 1.44 inches
Valve open	
1985 and 1986	170 to 180 lbs at 1.26 inches
1987 on	158 to 170 lbs at 1.040 inches

Crankshaft and connecting rods

Crankshaft end play	0.0035 to 0.0085 inch
Connecting rod end play (side clearance)	0.006 to 0.022 inch
Main bearing journal diameter	2.3 inches
Main bearing oil clearance	0.0005 to 0.0022 inch
Connecting rod journal diameter	2.0 inches
Connecting rod bearing oil clearance	0.0005 to 0.0026 inch
Crankshaft journal taper/out-of-round limit	0.005 inch

Four-cylinder engine (continued)

Engine block

Cylinder bore diameter	4.0 inches
Out-of-round limit	0.001 inch
Taper limit	0.005 inch

Pistons and rings

Piston-to-bore clearance	
1985	
Top of bore	0.0025 to 0.0033 inch
Bottom of bore	0.0017 to 0.0041 inch
1986 on	0.0014 to 0.0022 inch
Ring side clearance	
Top compression ring	
1985	0.0015 to 0.003 inch
1986 on	0.002 to 0.003 inch
Second compression ring	
1985	0.0015 to 0.003 inch
1986 on	0.001 to 0.003 inch
Oil ring	0.015 to 0.055 inch
Compression ring end gap	0.010 to 0.020 inch
Oil ring end gap	
1985	0.015 to 0.055 inch
1986 on	0.020 to 0.060 inch

Torque specifications* Ft-lbs (unless otherwise indicated)

Main bearing cap bolts	70
Connecting rod cap nuts	32
Oil pump-to-block bolts	22
Oil pick-up tube bracket nut	37

*Note: Refer to Part A for additional torque specifications.

V6 engine

General

Bore and stroke	
1985 through 1998	4.000 x 3.480 inches
1999 and later	4.012 x 3.480 inches
Oil pressure (minimum)	
1985 to 1994	
all except VIN W	10 psi @ 500 rpm; 30 to 55 psi @ 2000 rpm
VIN W	25 to 50 psi @ 1200 rpm; 42 to 60 psi @ 2400 to 5000 rpm
1995 and later models	6.0 psi @ 1000 rpm; 18.0 psi @ 2000 rpm; 24.0 psi @ 4000 rpm
Compression pressure	Lowest reading cylinder must be at least 70% of highest reading cylinder (100 psi minimum)

Engine block

Cylinder bore diameter	3.9995 to 4.0025 inches
Taper limit	0.001 inch
Out-of-round limit	0.002 inch

Pistons and rings

Piston-to-cylinder bore clearance
  1985 to 1994

Standard	0.0007 to 0.0017 inch
Limit	0.0027 inch (max)

  1995 through 1998

Standard	0.0007 to 0.0020 inch
Limit	0.0024 inch (max)

  1999 and later

Standard	0.0007 to 0.0024 inch
Limit	0.0029 inch

Piston ring-to-groove side clearance
  Standard
    1985 through 1998

Compression rings	0.0012 to 0.0032 inch
Oil control ring	0.002 to 0.007 inch

    1999 through 2002

Top compression ring	0.0012 to 0.0027 inch
Second compression ring	0.0015 to 0.0031 inch
Oil ring	0.0018 to 0.0077 inch

    2003 and later

Top compression ring	0.0012 to 0.0027 inch
Second compression ring	0.0030 to 0.0110 inch
Oil ring	0.0018 to 0.0077 inch

  Service limit

Compression rings	0.0033 inch
Oil control ring	0.008 inch

Piston ring end gap
  1985 to 1995
    Top compression ring

Standard	0.010 to 0.020 inch
Service limit	0.035 inch

    Second compression ring

Standard	0.010 to 0.025 inch
Service limit	0.035 inch

    Oil control ring

Standard	0.015 to 0.055 inch
Service limit	0.065 inch

  1996 through 1998
    Top compression ring

Standard	0.010 to 0.016 inch
Service limit	0.035 inch

    Second compression ring

Standard	0.018 to 0.026 inch
Service limit	0.035 inch

Piston ring end gap (continued)
 Oil control ring
  Standard   0.011 to 0.055 inch
  Service limit   0.065 inch
 1999 and later
  Top compression ring
   Standard   0.010 to 0.016 inch
   Service limit   0.010 to 0.020 inch
  Second compression ring
   Standard   0.015 to 0.023 inch
   Service limit   0.015 to 0.031 inch
  Oil ring
   Standard   0.010 to 0.029 inch
   Service limit   0.0002 to 0.0035 inch

Crankshaft

Main journal
 1985 through 1991
  Diameter
   No. 1 journal   2.4484 to 2.4493 inches
   No. 2 and 3 journals 2.4481 to 2.4490 inches
   No. 4 journal   2.4479 to 2.4488 inches
  Taper limit   0.001 inch
  Out-of-round limit   0.001 inch
 1992 and later
  Diameter
   No. 1 journal   2.4488 to 2.4495 inches
   No. 2 and 3 journals 2.4485 to 2.4494 inches
   No. 4 journal   2.4480 to 2.4489 inches
  Taper limit   0.0003 inch
  Out-of-round limit   0.0010 inch
Main bearing oil clearance
 1985 through 1991
  Standard
   No. 1 journal   0.0008 to 0.0020 inch
   No. 2 and 3 journals 0.0011 to 0.0023 inch
   No. 4 journal   0.0017 to 0.0032 inch
  Service limit
   No. 1 journal   0.001 to 0.0015 inch
   No. 2 and 3 journals 0.001 to 0.0025 inch
   No. 4 journal   0.0025 to 0.0035 inch
Main bearing oil clearance (continued)
 1992 and later
  Standard
   No. 1 journal   0.0008 to 0.0020 inch
   No. 2, 3 and 4 journals 0.0011 to 0.0023 inch
  Service limit
   No. 1 journal   0.0010 to 0.0020 inch
   No. 2, 3 and 4 journals 0.0010 to 0.0025 inch

Connecting rod journal
    Diameter   2.2487 to 2.2497 inches
    Taper limit   0.001 inch
    Out-of-round limit   0.001 inch
Connecting rod bearing oil clearance
    Standard   0.0013 to 0.0035 inch
    Service limit   0.0035 inch
Connecting rod end play (side clearance)   0.006 to 0.014 inch
Crankshaft end play
    1985 to 1994   0.002 to 0.006 inch
    1995   0.005 to 0.018 inch
    1996 and later   0.002 to 0.008 inch

Cylinder head and valve train

Head warpage limit   0.004 inch
Valve seat angle   46 degrees
Valve seat width
    Intake   1/32 to 1/16 (0.0313 to 0.0625) inch
    Exhaust   1/16 to 3/32 (0.0625 to 0.0938) inch
Valve seat runout limit   0.002 inch
Valve face angle   45 degrees
Minimum valve margin width   1/32 inch
Valve stem-to-guide clearance
    Intake valves
        Standard   0.0010 to 0.0027 inch
        Service limit   0.0037 inch
    Exhaust valves
        Standard   0.0010 to 0.0027 inch
        Service limit   0.0037 inch
Valve spring free length
    1985 through 1998   2.030 inches
    1999 and later   2.020 inches
Valve spring damper (inner spring) free length   1.860 inches
Valve spring installed height
    1985 to 1995   1-23/32 ± 1/32 inches
    1996 through 1998   1.690 to 1.710 inches
    1999 and later   1.670 to 1.700 inches
Valve spring pressure and length
    Closed   76 to 84 lbs @ 1.70 inches
    Open
        1985 through 1995   194 to 206 lbs @ 1.25 inches
        1996 and later   187 to 203 lbs @ 1.27 inches

Torque specifications* Ft-lbs (unless otherwise indicated)

Rocker arm studs   35
Main bearing cap bolts
    1985   70
    1986 to 1994   75
    1995   81
    1996 through 1999   77

Torque specifications*	Ft-lbs (unless otherwise indicated)
Main bearing cap bolts (continued)	
2000 and later	
Step 1	15
Step 2	Tighten an additional 73-degrees
Connecting rod cap nuts	
1990 and earlier	45
1991	
Step 1	20
Step 2	Tighten an additional 60-degrees
1992 on	
Step 1	20
Step 2	Tighten an additional 70-degrees
Oil pump bolt to rear crankshaft bearing cap	65

*Note: Refer to Part B for additional torque specifications

Section

Reference to other Chapters

3

COOLING,
HEATING
AND AIR
CONDITIONING
SYSTEMS

1 General information

ENGINE COOLING SYSTEM

The Astro/Safari employs a pressurized engine cooling system with thermostatically controlled coolant circulation. An impeller type water pump mounted on the front of the block pumps coolant through the engine. The coolant flows around each cylinder and toward the rear of the engine. Cast-in coolant passages direct coolant around the intake and exhaust ports, near the spark plug areas and in close proximity to the exhaust valve guide inserts.

A wax pellet type thermostat is located at the front of the engine. During warm up, the closed thermostat prevents coolant from circulating through the radiator. When the engine reaches normal operating temperature, the thermostat opens and allows hot coolant to travel through the radiator, where it is cooled before returning to the engine. The aluminum radiator is a crossflow type, with tanks on either side of the core.

The cooling system is sealed by a pressure type radiator cap. This raises the boiling point of the coolant and the higher boiling point of the coolant increases the cooling efficiency of the radiator.

If the system pressure exceeds the cap pressure relief value, the excess pressure in the system forces the spring-loaded valve inside the cap off its seat and allows the coolant to escape through the overflow tube into a coolant reservoir. When the system cools the excess coolant is automatically drawn from the reservoir back into the radiator.

The coolant reservoir does double duty as both the point at which fresh coolant is added to the cooling system to maintain the proper fluid level and as a holding tank for overheated coolant. This type of cooling system is known as a closed design because coolant that escapes past the pressure cap is saved and reused.

HEATING SYSTEM

The heating system consists of a blower fan and heater core located under the dashboard, the inlet and outlet hoses connecting the heater core to the engine cooling system and the heater/air conditioning control head on the dashboard. Hot engine coolant is circulated through the heater core at all times. When the heater mode is activated, a flap door opens to expose the heater box to the passenger compartment. A fan switch on the control head activates the blower motor, which forces air through the core, heating the air.

AIR CONDITIONING SYSTEM

The air conditioning system consists of a condenser mounted in front of the radiator, an evaporator mounted under the dash, a compressor mounted on the engine, a filter-drier (accumulator) which contains a high pressure relief valve and the plumbing connecting all of the above.

A blower fan forces the warmer air of the passenger compartment through the evaporator core (sort of a radiator-in-reverse), transferring the heat from the air to the refrigerant. The liquid refrigerant boils off into low pressure vapor, taking the heat with it when it leaves the evaporator.

2 Antifreeze - general information

✳ WARNING:

Don't allow antifreeze to come in contact with your skin or painted surfaces of the vehicle. Flush contacted areas immediately with plenty of water. Don't store new coolant or leave old coolant lying around where it's easily accessible to children and pets - they are attracted by its sweet smell. Ingestion of even a small amount can be fatal. Wipe up garage floor and drip pan coolant spills immediately. Keep antifreeze containers covered and repair leaks in your cooling system immediately. Antifreeze is flammable - be sure to read the precautions on the container.

The cooling system should be filled with a 50/50 mixture of antifreeze and distilled water which will prevent freezing down to at least -20 degrees Fahrenheit. It also provides protection against corrosion, prevents foaming and increases the coolant boiling point.

1995 and earlier models use conventional ethylene glycol type (green colored) antifreeze which should be drained, flushed and refilled at least every other year (see Chapter 1). The use of ethylene glycol antifreeze for longer than two years is likely to cause damage to the system in the form of rust and scale.

1996 and later models use a new silicate-free type of antifreeze called DEX-COOL. This type of antifreeze is orange in color and has a service life of five years or 100,000 miles whichever comes first. Do not mix ethylene glycol and DEX-COOL type coolants since this will significantly reduce the corrosion fighting properties of DEX-COOL.

Before adding antifreeze to the system, check all hose connections. Antifreeze can leak through very minute openings.

The exact mixture of antifreeze to water which you should use depends on the relative weather conditions. The mixture should contain at least 50 percent antifreeze, but should never contain more than 70 percent antifreeze. Refer to the antifreeze ratio table on the coolant container.

3 Thermostat - check and replacement

▶ **Refer to illustration 3.9**

➡ **Note: Don't drive the vehicle without a thermostat! The computer will stay in open loop and emissions and fuel economy will suffer.**

CHECK

1 Before condemning the thermostat, check the coolant level, drivebelt tension and temperature gauge (or light) operation.

2   If the engine takes a long time to warm up, the thermostat is probably stuck open. Replace the thermostat.

3   If the engine runs too hot, check the temperature of the upper radiator hose. If the hose isn't hot, the thermostat is probably stuck closed. Replace the thermostat.

4   If the upper radiator hose is hot, it means the coolant is circulating and the thermostat is open. Refer to the *Troubleshooting* Section for the cause of overheating.

5   If an engine has been overheated, you may find damage such as leaking head gaskets, scuffed pistons and warped or cracked heads.

REPLACEMENT

☀☀ WARNING:

The engine must be completely cool before beginning this procedure!

6   Disconnect the negative cable from the battery.

☀☀ CAUTION:

On models equipped with an anti-theft audio system, disable the anti-theft feature before performing any operation that requires disconnecting the battery or disrupting power to the stereo (see the Anti-theft audio system procedures at the front of this manual).

7   Drain about two quarts of coolant from the cooling system (see Chapter 1). Remove the air cleaner assembly (see Chapter 4).

8   Loosen the hose clamp and detach the upper radiator hose from the thermostat housing cover fitting.

9   Remove the bolts, then detach the thermostat housing cover and gasket (see illustration).

➡Note: If the cover is difficult to remove, tap it gently with a soft-face hammer or a block of wood. Don't try to pry the cover loose or damage to the gasket sealing surfaces may result and leaks could develop.

3.9 Remove the thermostat housing bolts (arrows)

10 Note how it's installed (which end is facing up), then lift out the thermostat.

11 Remove all traces of old gasket material and sealant from the housing and cover with a gasket scraper, then clean the gasket mating surfaces with lacquer thinner or acetone.

12 Apply a thin layer of RTV sealant to the gasket mating surfaces of the housing and cover, then install the new thermostat in the housing. Make sure the correct end faces up - the spring is normally directed down, into the lower housing or intake manifold.

13 Position a new gasket on the housing and make sure the bolt holes line up.

14 Carefully position the cover on the housing and install the bolts. Tighten them a little at a time to the specified torque - don't overtighten them or the cover may be distorted!

15 Reattach the radiator hose to the cover fitting and tighten the hose clamp. Now may be a good time to check and replace all of the cooling system hoses and clamps (see Chapter 1).

16 Refer to Chapter 1 and refill the system, then run the engine and check carefully for leaks.

4   Engine cooling fan and clutch - check and replacement

CHECK

1   All V6 and some four-cylinder engines are equipped with thermostatically controlled fan clutches.

2   Begin the clutch check with a lukewarm engine (start it when cold and let it run for two minutes only).

3   Remove the key from the ignition switch for safety purposes.

4   Turn the fan blades and note the resistance. There should be moderate resistance, depending on temperature.

5   Drive the vehicle until the engine is warmed up. Shut it off and remove the key.

6   Turn the fan blades and again note the resistance. There should be a noticeable increase in resistance.

7   If the fan clutch fails this check or is locked up solid, replacement is indicated. If excessive fluid is leaking from the hub or lateral play over 1/4-inch is noted, replace the fan clutch.

8   If any fan blades are bent, don't straighten them! The metal will

be weakened and blades could fly off during engine operation. Replace the fan with a new one.

REPLACEMENT

1995 and earlier models

▶ **Refer to illustration 4.12**

9   Remove the hood latch mechanism from the radiator support and position it out of the way (see Chapter 11). If equipped, loosen the A/C accumulator brackets.

10 Detach any hoses or wiring attached to the upper fan shroud. Remove the upper fan shroud (see illustrations 5.6b, 5.6c and 5.6d).

11 Remove the drive belt(s) (see Chapter 1).

12 Remove the fasteners securing the fan assembly to the water pump pulley (see illustration).

13 Remove the fasteners securing the fan to the fan clutch (if used, some 2.5L engines do not use a fan clutch).

1996 and later models

▶ **Refer to illustrations 4.16 and 4.17**

14 Remove the air cleaner intake duct (see Chapter 4).
15 Detach any hoses or wiring attached to the upper fan shroud. Remove the upper fan shroud.

➡ **Note:** On some models, the fan clutch hub nut has left-hand thread (turn clockwise to loosen). Refer to the belt routing label on the fan shroud; if the hub nut has left-hand thread, it should be stated as such on the label.

16 Use a large wrench to remove the clutch retaining nut and detach the fan clutch assembly from the engine. The drivebelt should keep the pulley from turning as the fan nut is loosened. If the water pump pulley slips on the belt it will be necessary to remove the drivebelt(s) (see Chapter 1) and use a strap wrench to hold the pulley (see illustration).

17 Remove the fasteners securing the fan to the fan clutch (see illustration).

All models

18 Installation is the reverse of removal. Be sure to tighten all fasteners to the torques listed in this Chapter's Specifications.

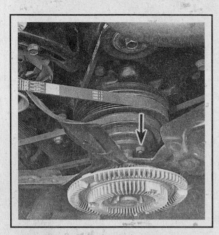
4.12 The fan is attached to the clutch assembly or water pump hub with four nuts or bolts (arrow)

4.16 Loosening the fan clutch retaining nut while holding the pulley with a strap wrench

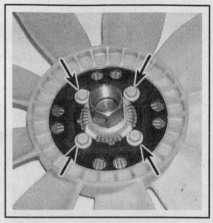
4.17 Fan retaining bolts (arrows)

5 Radiator - removal and installation

✳✳ **WARNING 1:**

The engine must be completely cool before beginning this procedure!

✳✳ **WARNING 2:**

Most 1993 and later models are equipped with airbags. Always disable the airbag system before working in the vicinity of the impact sensors, steering column or instrument panel to avoid the possibility of accidental deployment of the airbag(s), which could cause personal injury (see Chapter 12 for the airbag disabling procedure). The yellow wires and connectors routed through the instrument panel are for this system. Do not use electrical test equipment on these yellow wires or tamper with them in any way while working under the instrument panel.

➡ **Note:** Lubricants and fluids used on your vehicle are hazardous waste and must be disposed of properly. When disposal is required, place the waste into a sealable container and deliver it to a service station, auto parts store or other facility which accepts hazardous waste for recycling.

REMOVAL

▶ **Refer to illustrations 5.6a, 5.6b, 5.6c, 5.6d and 5.8**

1 Disconnect the cable from the negative terminal of the battery.

✳✳ **CAUTION:**

On models equipped with an anti-theft audio system, disable the anti-theft feature before performing any operation that requires disconnecting the battery or disrupting power to the stereo (see the Anti-theft audio system procedures at the front of this manual).

2 On 1996 and later models, remove the air cleaner assembly and intake duct (see Chapter 4).
3 Drain the coolant from the radiator (see Chapter 1).
4 Remove the hood latch mechanism (see Chapter 11).
5 Disconnect the overflow, upper and lower radiator hoses from the radiator. If a hose is really stuck to the radiator, cut it with a utility knife and peel it off the fitting. Be careful not to damage the fitting. A new hose costs a lot less than having the radiator repaired.
6 Detach any hoses or wiring that may be attached to the upper fan shroud. Remove the upper fan shroud. There are two bolts at the top hidden by the wiring harness and two bolts on each side attaching it to the lower fan shroud (see illustrations).
7 On 1998 models, remove the engine cooling fan and clutch (see Section 4) and lower fan shroud.
8 On models equipped with automatic transmissions, disconnect the transmission oil cooler lines from the radiator (see illustration).

➡ **Note 1:** On 1995 models, a special tool must be used to disconnect the transmission cooler lines from the radiator.

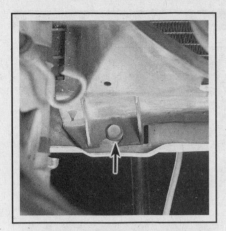

5.6a The fan shroud is secured at the bottom by a bolt in each corner

5.6b Unbolt the radiator hose bracket prior to removing the upper fan shroud

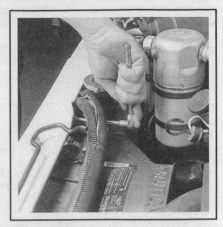

5.6c Unclip the wiring harness and remove the bolts above each end of the radiator . . .

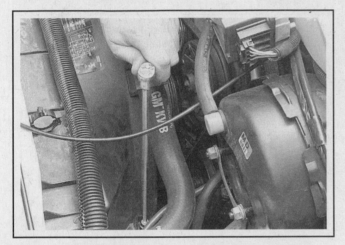

5.6d . . . then use a long extension to reach the two lower mounting bolts on each side of the shroud

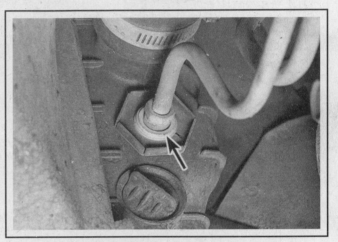

5.8 The automatic transmission cooler lines are attached to the right (passenger) side of the radiator (arrow is to lower fitting)

➡**Note 2: On 1996 and later models, the cooler lines are quick-connect type and secured with a spring clip, which cannot be reused. Pull back the plastic cover protecting the fitting, remove the clip and then separate the line from the fitting. Cap the fittings to prevent leakage and contamination.**

9   If equipped, disconnect the engine oil cooler lines from the radiator.

➡**Note: On 1996 and later models, the cooler lines are quick-connect type and secured with a spring clip, which cannot be reused. Cap the fittings to prevent leakage and contamination.**

10 On 1997 and later models equipped with A/C, remove the radiator grille (see Chapter 11) and then remove the four bolts securing the A/C condenser to the radiator.

11 On some early models the lower fan shroud was attached to the radiator. If this is the case, remove the bolts securing the lower fan shroud to the radiator.

12 Carefully lift the radiator out of the vehicle, keeping the filler neck tilted up.

13 Inspect the radiator for signs of leakage, deterioration of the tubes, crushed cooling fins or other damage. Cooling fins can be straightened easily using a tool called a fin comb which is available at most auto parts stores. For most other repairs, take the radiator to a qualified radiator shop.

INSTALLATION

➡**Note: Since the radiator was removed, now is a good time to flush out the engine block and refill the system with new coolant.**

14 Prior to installing the radiator, flush it out using a garden hose. If it was sent out to a repair shop, flush the transmission and engine oil coolers as well, using the appropriate oil for each cooler.

15 On 1996 and later models, the cooler lines at the radiator are quick-connect type and secured with a spring clip, which should not be reused. Begin assembly by installing the new spring clips so the spring ears engage in the retaining slots and are visible at all 3 places. Make sure the clips move freely once they're installed. Next, push the line into the fitting until you hear it "click" into place. When the line and fitting are correctly mated, the yellow band around the line should NOT be visible. Confirm a secure connection by trying to pull it apart. Finally, slide the plastic cover over the fitting.

16 Installation is the reverse of removal. After installation, fill the system with the proper mixture of antifreeze (refer to *Recommended lubricants and fluids* in Chapter 1) and distilled water. Also, check the transmission fluid and engine oil levels if oil coolers are incorporated into your radiator.

6 Coolant reservoir - removal and installation

✳✳ WARNING:

The engine must be completely cool before beginning this procedure!

➡ Note 1: On 1994 and earlier models, the coolant overflow and windshield washer reservoirs are separate units but attached together by a slide locking tang located between them.

➡ Note 2: On 1995 and later models, the coolant overflow and windshield washer reservoirs are a single unit and must be removed together.

REMOVAL

▶ **Refer to illustration 6.2**

1 Disconnect the overflow hose from the radiator filler neck.

2 Remove the fasteners securing the coolant overflow and windshield washer reservoirs to the radiator core support (see illustration).

➡ Note: 1994 and earlier models use bolts and 1995 and later models use push-in type fasteners.

3 On 1994 and earlier models, lift up both reservoirs and detach the coolant reservoir from the windshield washer reservoir and remove it from the vehicle.

4 On 1995 and later models, disconnect the electrical connections and washer fluid hoses from the windshield washer pumps. Remove the coolant overflow and windshield washer reservoirs from the vehicle.

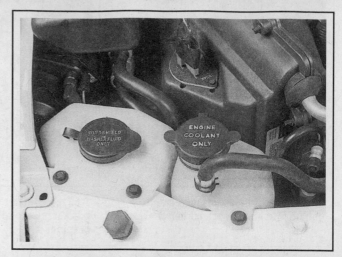

6.2 Coolant reservoir mounting details - note that the windshield washer reservoir must be raised first

INSTALLATION

5 Installation is the reverse of removal. On 1994 and earlier models, be sure to lock the reservoirs together at the side and engage the coolant reservoir in the lower bracket. On 1995 and later models, be sure to engage the three locating pins at the bottom of the reservoirs with the corresponding holes in the lower support.

7 Water pump - check

▶ **Refer to illustration 7.3**

1 Water pump failure can cause overheating of and serious damage to the engine. There are three ways to check the operation of the water pump while it's installed on the engine. If any one of the three following quick checks indicates water pump failure, it should be replaced immediately.

2 Start the engine and warm it up to normal operating temperature. Squeeze the upper radiator hose. If the water pump is working properly, a pressure surge should be felt as the hose is released.

3 A seal protects the water pump impeller shaft bearing from contamination by engine coolant. If the seal fails, weep holes in the top and bottom of the water pump snout (see illustration) will leak coolant under the vehicle. If the weep hole is leaking, shaft bearing failure will follow. Replace the water pump immediately.

4 Besides contamination by coolant after a seal failure, the water pump impeller shaft bearing can also be prematurely worn out by an improperly tensioned drivebelt. When the bearing wears out, it emits a high pitched squealing sound. If noise is coming from the water pump during engine operation, the shaft bearing has failed. Replace the water pump immediately.

5 To identify excessive bearing wear before the bearing actually

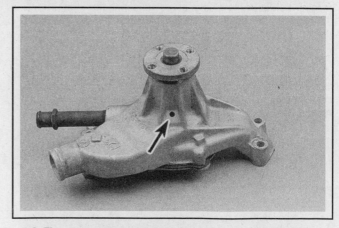

7.3 The water pump weep hole (arrow) will drip coolant when the mechanical seal for the pump shaft bearing fails

fails, grasp the water pump pulley and try to force it up-and-down or from side-to-side. If the pulley can be moved either horizontally or vertically, the bearing is nearing the end of its service life. Replace the water pump.

8 Water pump - removal and installation

▶ Refer to illustrations 8.9 and 8.10

REMOVAL

1 Disconnect the cable from the negative terminal of the battery.

2 On 1995 and earlier models, remove the hood latch mechanism (see Chapter 11).

3 Drain the coolant from the radiator (see Chapter 1).

4 On 1996 and later models, remove the air cleaner assembly and intake duct (see Chapter 4).

5 Detach any hoses or wiring that may be attached to the upper fan shroud. Remove the upper fan shroud. There are two bolts at the top hidden by the wiring harness and two bolts on each side attaching it to the lower fan shroud (see illustrations 5.6b, 5.6c and 5.6d).

6 Remove the drivebelt(s) (see Chapter 1).

7 Disconnect the hoses from the water pump. If a hose is really stuck, cut it with a utility knife and peel it off the fitting. Be careful not to damage the fitting.

8 Remove the engine cooling fan and clutch (see Section 4).

9 Remove any brackets necessary to gain access to the water pump bolts or facilitate removal (see illustration).

10 Remove the water pump-to-engine block bolts (see illustration) and detach it from the engine. If the water pump is stuck to the block, tap it with a soft-faced hammer to break it loose.

INSTALLATION

11 Using a gasket scraper, remove all traces of old gasket material and sealant from the engine block. Wipe the mating surfaces clean with a rag moistened with lacquer thinner or acetone.

12 Coat both sides of the gasket(s) (V6 engines have two gaskets) with a thin layer of RTV sealant and place it onto the water pump.

13 To prevent leakage at the bolts, coat the threads of each water pump mounting bolt with a thin layer of RTV sealant.

14 Insert the bolts into the water pump and place it against the block. Tighten the bolts hand tight. Tighten the bolts in a criss-cross pattern to the torque listed in this Chapter's Specifications.

15 The remaining installation steps are the reverse of removal.

16 Refill the cooling system, run the engine and check for leaks.

8.9 Several brackets must be removed before the water pump can be detached (early V6 engine shown)

8.10 V6 engine water pump mounting bolt locations (arrows)

9   Coolant temperature sending unit - check and replacement

▶ Refer to illustration 9.1

TEMPERATURE WARNING LIGHT SYSTEM CHECK

1   If the light doesn't come on when the ignition switch is turned on, check the bulb. If the light stays on even with the engine cold, unplug the wire at the sending unit (see illustration). If the light goes off, replace the sending unit. If the light stays on, the wire is grounded somewhere in the harness.

TEMPERATURE GAUGE SYSTEM CHECK

2   If the gauge is inoperative, check the fuse (see Chapter 12).

3   If the fuse is OK, unplug the wire connected to the sending unit and ground it with a jumper wire. Turn on the ignition switch. The gauge should now register at maximum. If it does, replace the sending unit. If it's still inoperative, the gauge or wiring may be faulty.

SENDING UNIT REPLACEMENT

4   Allow the engine to cool completely.
5   Remove the engine cover (see Chapter 11).
6   Unplug the wire connected to the sending unit.
7   Unscrew the sending unit and quickly install the new unit to prevent loss of coolant.
8   Connect the wire and install the engine cover.

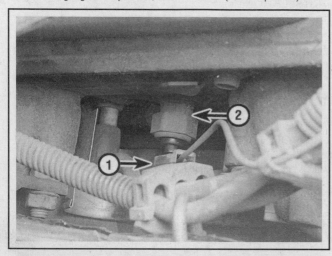
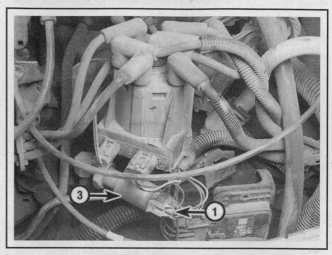

9.1 Remove the connector (1), then unscrew the sending unit. The coolant temperature sending unit (2) is located on the side of the left cylinder head on V6 models or at the rear of the cylinder head on four-cylinder models. The oil pressure sending unit (3) is located near the distributor on all models

10   Heater and air conditioner blower motors - removal and installation

FRONT BLOWER MOTOR

Removal

▶ Refer to illustrations 10.5 and 10.6

1   The front blower motor is used to force the air through the ducts for both heating and air conditioning since the front unit has both the heater core and A/C evaporator inside (if equipped). It is located on the passenger's side of the engine compartment on the firewall.
2   Raise the hood. Disconnect the cable from the negative terminal of the battery.

❋❋ **CAUTION:**

On models equipped with an anti-theft audio system, disable the anti-theft feature before performing any operation that requires disconnecting the battery or disrupting power to the stereo (see the Anti-theft audio system procedures at the front of this manual).

3   Remove the windshield washer fluid and coolant overflow reservoirs (see Section 6).
4   1996 and later models have an acoustic cover over the blower motor. To remove the cover, remove the two screws securing the cover and then cut it between the clip lands.
5   Disconnect the electrical connections and cooling tube from the blower motor (see illustration).
6   Remove the mounting bolts securing the blower motor, withdraw it from the HVAC unit and remove it from the vehicle. On some models, the fan must be removed from the motor before it can be removed from the vehicle (see illustration).

Installation

7   Installation is the reverse of removal.

REAR AUXILIARY BLOWER MOTOR (SOME MODELS)

Removal

8   Disconnect the cable from the negative terminal of the battery.

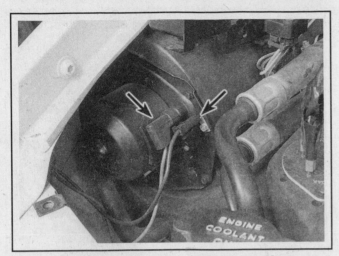

10.5 Air conditioner/heater blower motor wire harness connectors (arrows)

※※ CAUTION:

On models equipped with an anti-theft audio system, disable the anti-theft feature before performing any operation that requires disconnecting the battery or disrupting power to the stereo (see the Anti-theft audio system procedures at the front of this manual).

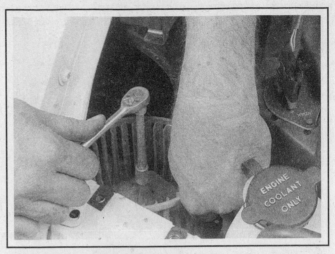

10.6 Due to limited space, the fan must be removed from the motor before the motor can be lifted out

9   Remove the auxiliary heater cover to expose the blower motor (see Section 17).

10 Disconnect the electrical connections from the blower motor.

11 Remove the mounting bolts securing the blower motor and withdraw it from the auxiliary heater/air conditioning unit.

Installation

12 Installation is the reverse of removal.

11 Heater core - removal and installation

▶ Refer to illustrations 11.5, 11.6 and 11.7

※※ WARNING:

The engine must be completely cool before beginning this procedure!

MAIN HEATER (FRONT)

1   Disconnect the negative cable from the battery.

※※ CAUTION:

On models equipped with an anti-theft audio system, disable the anti-theft feature before performing any operation that requires disconnecting the battery or disrupting power to the stereo (see the Anti-theft audio system procedures at the front of this manual).

2   Drain the coolant (see Chapter 1).

3   On 1994 and earlier models, remove the coolant reservoir (see Section 6).

4   On 1994 and earlier models, remove the windshield washer fluid reservoir mounting bolts and move it aside.

5   Loosen the clamps and detach the heater hoses from the core tubes. Plug the core tubes (see illustration).

6   Working in the passenger compartment, under the dash, remove

11.5 The heater hoses are attached to the core tubes with hose clamps - remove the top hose first, followed by the bottom one

the heater core cover screws (see illustration).

7   Remove the core mounting screws at the rear of the core (see illustration).

8   Lift the core out of the housing.

9   Installation is the reverse of removal.

10 Refill the cooling system and check for leaks.

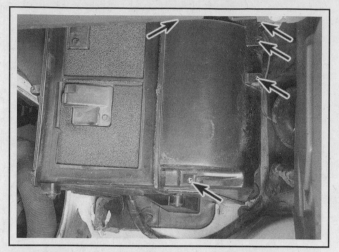

11.6 Heater core cover screw locations (arrows) - screws are removed in this photo, and the upper two are not visible

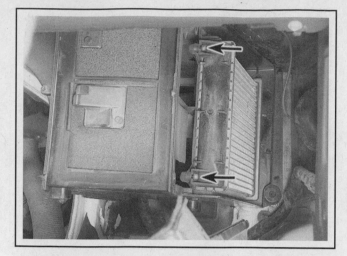

11.7 Remove the heater core mounting screws (arrows)

AUXILIARY HEATER (REAR)

11 The optional auxiliary heater is floor mounted just ahead of the left rear wheel housing.

12 Perform steps 1 and 2 above.

13 Working under the vehicle, disconnect the hoses from the heater core pipes.

14 Working inside the vehicle, remove the auxiliary heater cover.

15 Remove the band clamps on each side of the heater core and separate it from the heater module.

16 Installation is the reverse of removal. Refill the cooling system (see Chapter 1) and check for leaks.

12 Air conditioning system - check and maintenance

▶ Refer to illustrations 12.9 and 12.12

✳✳ WARNING:

The air conditioning system is under high pressure. Do not loosen any hose fittings or remove any components until after the system has been discharged by a dealer service department, automotive air conditioning shop or service station.

✳✳ CAUTION:

The air conditioning systems on 1994 and later models uses R-134a, the non-ozone depleting refrigerant. R-134a refrigerant and its lubricating oil are not compatible with earlier models that use R-12 refrigerant. Under no circumstances should the two types of refrigerants and oils be mixed - as costly compressor failure may result.

➡Note: Some models have an optional rear auxiliary A/C unit mounted at the left rear corner of the vehicle. This optional rear A/C unit has an additional evaporator and blower motor along with various hoses and pipes to transport the refrigerant from the compressor.

1 The following maintenance steps should be performed on a regular basis to ensure that the air conditioner continues to operate at peak efficiency.

 a) *Check the tension of the drivebelt and adjust it if necessary (see Chapter 1).*

 b) *Check the condition of the hoses. Look for cracks, hardening and deterioration.*

✳✳ WARNING:

Don't replace A/C hoses until the system has been discharged by a dealer service department or repair shop.

 c) *Check the fins of the condenser for leaves, bugs and any other foreign material. A soft brush and compressed air can be used to remove them.*

 d) *Maintain the correct refrigerant charge.*

2 The system should be run for about 10 minutes at least once a month. This is particularly important during the winter months because long term non-use can cause hardening and failure of the seals.

3 Because of the complexity of the air conditioning system and the special equipment necessary to service it, troubleshooting and repairs should be done by a professional mechanic. The most common cause of poor cooling is low refrigerant charge. If a noticeable drop in system cooling ability occurs, the following procedure will help pinpoint the cause.

4 Warm the engine to normal operating temperature.

5 The hood and doors should be open.

6 Operate the A/C mode control.

7 Move the temperature selector lever to the coolest position.

8 Turn the fan control to the High position.

9 With the compressor engaged, feel the evaporator inlet pipe between the orifice and the evaporator. Put your other hand on the surface of the accumulator can (see illustration).

10 If both surfaces feel about the same temperature and if both feel a little cooler than the surrounding air, the refrigerant level is probably okay. The problem is elsewhere.

11 If the inlet pipe has frost accumulation or feels cooler than the

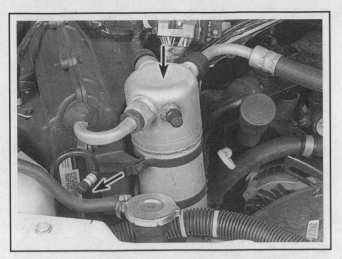

12.9 To determine if the refrigerant level is adequate, feel the evaporator inlet pipe with one hand and the surface of the accumulator (arrows) with the other and note the temperature of each

accumulator surface, the refrigerant charge is low. Add refrigerant as follows.

➡ **Note: Because of federal restrictions on the sale of R-12**

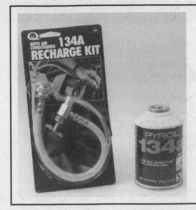

12.12 A typical aftermarket R-134a recharge kit

refrigerant, it isn't practical for refrigerant to be added by the home mechanic. When the R-12 system needs recharging, take the vehicle to a dealer service department or professional air conditioning shop for recharging.

12 Buy an automotive air conditioner recharge kit and hook it up to the evaporator inlet pipe fitting in accordance with the kit manufacturer's instructions (see illustration). Add refrigerant until both the accumulator surface and the evaporator inlet pipe feel about the same temperature. Allow stabilization time between each addition.

13 Never add more than two cans of refrigerant.

13 Air conditioner accumulator - removal and installation

▶ **Refer to illustrations 13.3 and 13.4**

❋❋ WARNING:

The air conditioning system is under high pressure. Do not loosen any hose fittings or remove any components until after the system has been discharged by a dealer service department, automotive air conditioning shop or service station. After the system has been discharged, residual pressure may still remain - be sure to wear eye protection when loosening line fittings!

1 Have the system discharged.
2 Disconnect the negative cable from the battery.

❋❋ CAUTION:

On models equipped with an anti-theft audio system, disable the anti-theft feature before performing any operation that requires disconnecting the battery or disrupting power to the stereo (see the Anti-theft audio system procedures at the front of this manual).

3 Disconnect the accumulator inlet and outlet lines (see illustration). Immediately cap the open lines.
4 Loosen the bolts and lift the accumulator out of the bracket (see illustration). Pour the oil out of the accumulator into a measuring cup, noting the amount.
5 Pour an equal amount of the appropriate refrigerant oil into the new accumulator, plus an additional two ounces.
6 Use new O-rings lubricated with refrigerant oil during installation.
7 Installation is the reverse of removal.
8 Have the system evacuated and recharged by an air conditioning shop.

13.3 Use a back-up wrench when loosening/tightening the A/C line fittings

13.4 The accumulator is held in place with two brackets - loosen the bolts to withdraw it

14 Air conditioner compressor - removal and installation

◆ Refer to illustrations 14.6, 14.7a and 14.7b

❋❋ **WARNING:**

The air conditioning system is under high pressure. Do not loosen any hose fittings or remove any components until after the system has been discharged by a dealer service department, automotive air conditioning shop or service station. After the system has been discharged, residual pressure may still remain - be sure to wear eye protection when loosening line fittings!

1 Have the system discharged.
2 Disconnect the negative cable from the battery.

❋❋ **CAUTION:**

On models equipped with an anti-theft audio system, disable the anti-theft feature before performing any operation that requires disconnecting the battery or disrupting power to the stereo (see the Anti-theft audio system procedures at the front of this manual).

3 Remove the engine cover (see Chapter 11).
4 Remove the drivebelt (see Chapter 1).
5 On 1996 and later models, remove the oil filler tube.
6 Detach the wire harness from the compressor (see illustration).
7 Unbolt the refrigerant hose manifold from the compressor and cap all the fittings (see illustrations).

14.7a Rear view of the compressor (V6 engine) showing the manifold bolt, brace bolt and mounting bolts (arrows) (2000 and earlier models)

14.6 Detach the wire harness and ground wire (arrows) before removing the compressor drivebelt and mounting bolts (arrows)

8 Unbolt the compressor and remove it from the vehicle.
9 Installation is the reverse of removal. Refer to Chapter 1 and adjust the drivebelt.
10 Have the system evacuated and recharged by an air conditioning shop.

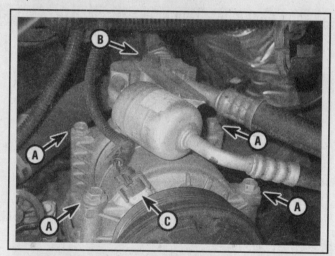

14.7b Air conditioning compressor mounting details (2001 and later models)

A Mounting bolts
B Refrigerant line manifold bolt
C Clutch connector

15 Air conditioner condenser - removal and installation

◆ Refer to illustrations 15.5 and 15.6

❋❋ **WARNING 1:**

The air conditioning system is under high pressure. Do not loosen any hose fittings or remove any components until after the system has been discharged by a dealer service department, automotive air conditioning shop or service station. After the system has been discharged, residual pressure may still remain - be sure to wear eye protection when loosening line fittings!

Most 1993 and later models are equipped with airbags. Always disable the airbag system before working in the vicinity of the impact sensors, steering column or instrument panel to avoid the possibility of accidental deployment of the airbag(s), which could cause personal injury (see Chapter 12 for the airbag disabling procedure). The yellow wires and connectors routed through the instrument panel are for this system. Do not use electrical test equipment on these yellow wires or tamper with them in any way while working under the instrument panel.

1 Have the system discharged.
2 Disconnect the cable from the negative terminal of the battery.

On models equipped with an anti-theft audio system, disable the anti-theft feature before performing any operation that requires disconnecting the battery or disrupting power to the stereo (see the Anti-theft audio system procedures at the front of this manual).

3 Remove the radiator grille (see Chapter 11).
4 On 1996 and later models, remove the air cleaner assembly and duct.
5 Remove the radiator support(s) (late models use a single vertical support) and, on models with the horns mounted to the upper radiator tie-bar, remove the horns (see illustration).
6 Disconnect the refrigerant lines from the condenser (see illustration). Use a back-up wrench to avoid twisting the condenser fitting.
7 Remove the condenser mounting bolts and carefully withdraw the condenser from the vehicle. Don't lose the rubber mounting pads. If the condenser is going to be reinstalled, store it so the oil won't run out.
8 Cap the refrigerant line fittings to prevent leakage, and keep moisture from entering the system.
9 Installation is the reverse of removal. If you are replacing the condenser, pour the oil out of it into a measuring cup and record the amount that comes out. The amount that comes out plus an additional

15.5 Remove the radiator support bolts (arrows) to gain access to the condenser

15.6 Disconnect the refrigerant lines (this is the left one) before lifting out the condenser

ounce must be introduced into the system during recharging. Be sure to use the appropriate oil for your system.
10 After installation, have the system evacuated, recharged and leak tested by your local dealer or other qualified repair shop.

16 Air conditioner orifice tube screen - removal and installation

▸ Refer to illustration 16.2

The air conditioning system is under high pressure. Do not loosen any hose fittings or remove any components until after the system has been discharged by a dealer service department, automotive air conditioning shop or service station. After the system has been discharged, residual pressure may still remain - be sure to wear eye protection when loosening line fittings!

➥Note: The orifice tube screen is a filtering device that may clog and cause insufficient cooling. It can be removed and cleaned or replaced if necessary.

1 Have the system discharged.
2 Disconnect the refrigerant line at the evaporator inlet and remove the orifice tube screen from the inlet tube (see illustration).
3 Installation is the reverse of removal. Be sure to install the end marked "shorter screen end" first.

16.2 The orifice tube screen is located inside this tube (which is taped over to seal out dirt) (the radiator has been removed for clarity)

4 Have the system evacuated, recharged and leak tested by an air conditioning shop.

17 Air conditioner evaporator - removal and installation

�֍ WARNING:

The air conditioning system is under high pressure. Do not loosen any hose fittings or remove any components until after the system has been discharged by a dealer service department, automotive air conditioning shop or service station. After the system has been discharged, residual pressure may still remain - be sure to wear eye protection when loosening line fittings!

➡Note: Some models have an optional rear auxiliary A/C unit mounted at the left rear corner of the vehicle. This optional rear A/C unit has an additional evaporator and blower motor along with various hoses and pipes to transport the refrigerant from the compressor.

MAIN EVAPORATOR (FRONT)

Removal

➡Note: On 2002 and later models, removing the evaporator core does not require draining the coolant or removing the HVAC unit from the vehicle.

1 Have the system discharged.
2 Disconnect the cable from the negative terminal of the battery.

�֍ CAUTION:

On models equipped with an anti-theft audio system, disable the anti-theft feature before performing any operation that requires disconnecting the battery or disrupting power to the stereo (see the Anti-theft audio system procedures at the front of this manual).

1985 through 2001 models

3 Drain the coolant (see Chapter 1).
4 Remove the windshield washer and coolant overflow reservoirs (see Section 6).
5 Disconnect the electrical connectors from the HVAC unit.
6 Disconnect the hoses from the heater core (see illustration 11.5). If the hoses are stuck to the heater core, cut them with a utility knife and peel them off the heater core fittings. Twisting the hoses may break the heater core pipes. Don't damage the fittings with the knife.
7 Remove the accumulator (see Section 13).
8 Remove the tube connecting the evaporator to the condenser.
9 Remove the blower motor relay bracket and place it out of the way.
10 Remove the fasteners securing the HVAC unit to the firewall.
11 Remove the HVAC unit from the vehicle.
12 Remove the clips and screws and separate the cases and then withdraw the evaporator.

2002 and later models

13 Remove the coolant overflow reservoir (See Section 6).
14 Disconnect the electrical connectors from the HVAC unit.
15 Remove the accumulator (see Section 13).
16 Remove the tube connecting the evaporator to the condenser.
17 Remove the blower motor relay bracket and place it out of the way. (On later models, the relay is an integral part of the blower motor

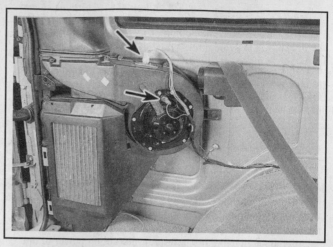

17.25 Rear auxiliary A/C unit - blower motor resistor (upper arrow) and blower motor (lower arrow) electrical connections

resistor and cannot be removed separately.) Remove the coolant bypass valve bracket (on the right side of the HVAC unit as you face it).
18 Remove the clips and/or screws from the outer evaporator case, separate the cases, and withdraw the evaporator core from the inner case.

Installation

19 Installation is the reverse of removal.
20 On 2001 and earlier models, refill the cooling system (see Chapter 1).
21 If a new evaporator was installed, add 1 ounce of the proper refrigerant oil when recharging the system. Have the system evacuated, recharged and leak tested by your local dealer or other qualified repair shop.

AUXILIARY EVAPORATOR (REAR)

Removal

▶ **Refer to illustration 17.25**

22 Perform steps 1 and 2 above.
23 Working under the vehicle, disconnect the refrigerant lines from the evaporator.
24 Working inside the vehicle, remove the left rear body side panel (see Chapter 11, Section 22).

➡Note: If equipped, detach the seat belt shoulder harness from the roof so the panel can be moved aside.

25 Disconnect the electrical connections from the blower motor and resistor (see illustration).
26 Remove the bolts connecting the auxiliary A/C unit to the air outlet duct and body and then detach it from the vehicle.
27 Remove the screws and separate the cases. Withdraw the evaporator from the case.

Installation

28 Installation is the reverse of removal.
29 If a new evaporator was installed, add 1 ounce of the proper refrigerant oil when recharging the system. Have the system evacuated, recharged and leak tested by your local dealer or other qualified repair shop.

18 Air conditioner and heater control assembly - removal and installation

♦ Refer to illustrations 18.3a, 18.3b, 18.4a, 18.4b and 18.4c

❊❊ **WARNING:**

Most 1993 and later models are equipped with airbags. Always disable the airbag system before working in the vicinity of the impact sensors, steering column or instrument panel to avoid the possibility of accidental deployment of the airbag(s), which could cause personal injury (see Chapter 12 for the airbag disabling procedure). The yellow wires and connectors routed through the instrument panel are for this system. Do not use electrical test equipment on these yellow wires or tamper with them in any way while working under the instrument panel.

1 Disconnect the cable from the negative terminal of the battery.

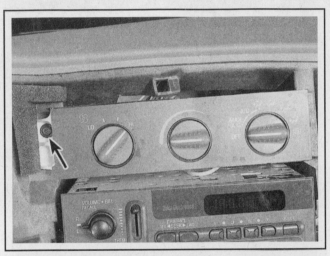

18.3b Air conditioner/heater control mounting screw on 1996 and later models (arrow) - remove the screw and pivot the control away from the instrument panel

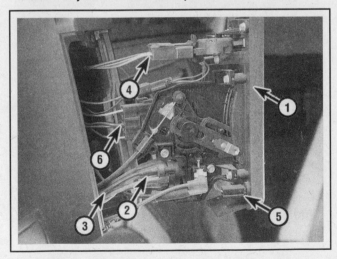

18.4b Air conditioning control assembly details (1995 and earlier models)

1	Control assembly	5	Light bulb connector
2	Vacuum harness	6	Mode selector wiring harness
3	Vacuum hoses		
4	Blower switch connector		

❊❊ **CAUTION:**

On models equipped with an anti-theft audio system, disable the anti-theft feature before performing any operation that requires disconnecting the battery or disrupting power to the stereo (see the Anti-theft audio system procedures at the front of this manual).

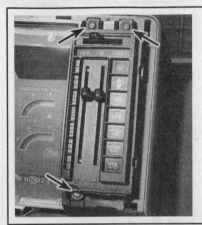

18.3a Air conditioner/ heater control mounting screws (arrows) (1985 to 1995 models)

18.4a The cable (arrow) can be detached after removing the press-on retainer from the arm (1995 and earlier models)

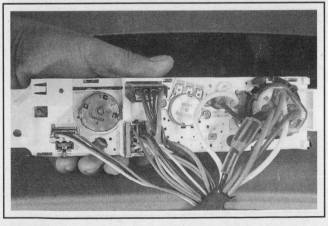

18.4c Air conditioner/heater control vacuum and electrical connections (1996 and later models)

2   On 1985 through 1992 and 1996 and later models, remove the instrument cluster trim panel (see Chapter 12, Section 16).

3   Remove the attaching screws and pull the control assembly away from the instrument panel (see illustrations).

4   Disconnect the control cable, vacuum hoses and wire harnesses (see illustrations).

5   Installation is the reverse of removal.

Specifications

General

Coolant capacity	See Chapter 1
Radiator cap pressure rating	15 psi
Thermostat rating	195° F

Torque specifications

	Ft-lbs
Compressor line manifold bolt	18
Fan assembly-to-water pump hub	
Four-cylinder engine without fan clutch	20
Four-cylinder engine with fan clutch	18
V6 engine	
1985 to 1995	22
1996 and later	41
Fan-to-fan clutch	
Four-cylinder engine	7
V6 engine	
1985 to 1995	18
1996 and later	24
Thermostat housing cover bolts	
Cast iron intake manifold	22
Aluminum intake manifold	15 to 20
Water pump-to-block bolts	
Four-cylinder engine	17
V6 engine	30

Section

Reference to other Chapters

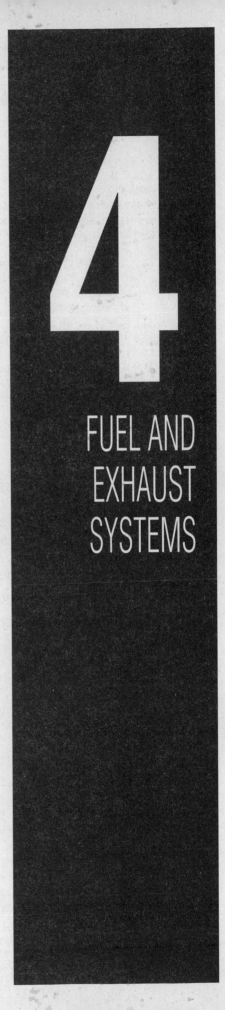

4

FUEL AND EXHAUST SYSTEMS

Several different types of fuel delivery systems are used on these models. The 2.5L four cylinder engine uses a single barrel Throttle Body Injection (TBI) system. All 2.5L engines manufactured until 1987 are equipped with Model 300 TBI. On 1988 and later engines, Model 700 TBI was installed.

The fuel delivery system on 1985 models with 4.3L V6 engines was the only engine that used a conventional carburetor. Two models of four-barrel carburetors were available, the E4ME and the E4MED Rochester Quadrajet.

From 1986 on, all 4.3L V6 VIN Z engines came equipped with the Model 220 TBI (twin injector) system. In 1992 another type of 4.3L V6 engine was introduced and is identified as VIN W. Models with this engine have a "W" in the eighth position of their Vehicle Identification Number (VIN) (refer to the VIN section at front of this manual for more information). The VIN W engine incorporated a balancer shaft and a new fuel delivery system known as Central Multiport Fuel Injection (CMFI). In 1994, the balancer shaft became standard on all 4.3L V6 engines. In 1996 an improved fuel delivery system called Central Sequential Fuel Injection (CSFI) was introduced and is installed on subsequent V6 engines.

CARBURETOR SYSTEM

The carburetor fuel system consists of the fuel tank, a mechanical fuel pump, two fuel filters (one in-tank one inside the carburetor), the carburetor and the air cleaner assembly. The mechanical fuel pump is mounted on the right side of the engine and is actuated by the camshaft.

THROTTLE BODY INJECTION (TBI) SYSTEM

The Throttle Body Injection system utilizes a single or dual fuel injectors (depending on model) which are centrally mounted in a carburetor-like housing (throttle body). The fuel injector is an electronically controlled solenoid which is actuated by a signal from the Powertrain Control Module (PCM). The amount of time the injector is held open by the PCM determines the fuel/air mixture. Pressurized fuel is delivered to the injector(s) by an electric fuel pump located inside the fuel tank. To maintain the fuel at a constant pressure, a fuel pressure regulator is incorporated into the fuel circuit which allows excess fuel to bleed off and return to the fuel tank through a separate fuel line.

CENTRAL MULTIPORT FUEL INJECTION (CMFI) SYSTEM

The CMFI system is used on all 4.3L V6 VIN W engines manufactured from 1992 to 1995. The system incorporates a two-piece intake manifold. The upper intake manifold is a variable single/split design that includes an intake tuning valve which is controlled by the Powertrain Control Module (PCM). The tuning valve switches the plenum to provide a "split" or "single" runner condition to maximize engine performance throughout all rpm ranges. Located under the upper intake manifold is a single central fuel injector which feeds each cylinder through six separate tubes and poppet nozzles. The fuel injector is an electronically controlled solenoid which is actuated by a signal from the Powertrain Control Module (PCM). The amount of time the injector is held open by the PCM determines the fuel/air mixture. Pressurized fuel is delivered to the injector by an electronic fuel pump located inside the fuel tank. To maintain the fuel at a constant pressure, a fuel pressure regulator is incorporated into the fuel circuit which allows excess fuel to bleed off and return to the fuel tank through a separate fuel line.

CENTRAL SEQUENTIAL FUEL INJECTION (CSFI) SYSTEM

Beginning in 1996, 4.3L V6 engines were equipped with the CSFI system. This system like its predecessor, CMFI, incorporates a two-piece intake manifold. The upper intake manifold is made from a new composite material and the lower manifold is cast aluminum. Located under the upper intake manifold is a central fuel meter body which houses six separate fuel injectors. Each fuel injector is connected to an individual cylinder through six separate tubes and poppet nozzles. The fuel injectors are electronically controlled solenoids which are actuated in the firing order by a signal from the Powertrain Control Module (PCM). The amount of time each injector is held open by the PCM determines the fuel/air mixture. Pressurized fuel is delivered to the injectors by an electronic fuel pump located inside the fuel tank. To maintain the fuel at a constant pressure, a fuel pressure regulator is incorporated into the fuel circuit which allows excess fuel to bleed off and return to the fuel tank through a separate fuel line.

EXHAUST SYSTEM

The exhaust system, which is similar for both 2.5L and 4.3L engines, includes an exhaust manifold fitted with an oxygen sensor, the exhaust pipe itself, a catalytic converter an a muffler. The exhaust system is supported by several rubber mountings which permit some movement and do not allow noise or vibration to be transmitted to the body or under carriage.

The catalytic converter is an emission control device incorporated into the exhaust system to reduce pollutants from the exhaust gas stream. A single-bed converter is used in combination with a three-way (reduction) catalyst. Refer to Chapter 6 for more information regarding the catalytic converter.

✳✳ WARNING:

Gasoline is extremely flammable, so take extra precautions when you work on any part of the fuel system. Don't smoke or allow open flames or bare light bulbs near the work area, and don't work in a garage where a gas-type appliance (such as a water heater or clothes dryer) is present. Since gasoline is carcinogenic, wear latex gloves when there is a possibility of being exposed to fuel, and, if you spill any on your skin, rinse it off immediately with soap and water. Mop up any spills immediately and do not store fuel soaked rags where they could ignite. When you perform any kind of work on the fuel system, wear safety glasses and have a Class B type fire extinguisher on hand.

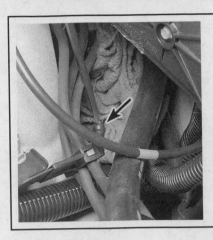

2.12a Schrader valve location on CMFI engines (arrow)

2.12b Schrader valve location on CSFI engines (as viewed from inside the vehicle - engine cover removed)

➡Note: After the fuel pressure has been relieved, it's a good idea to lay a shop towel over any fuel line connection to be disassembled, to absorb the residual fuel that will leak out.

1 Before servicing any component on a fuel injected vehicle it is necessary to relieve the fuel pressure to minimize the risk of fire or personal injury.

MODEL 300 TBI (EARLIER 4-CYLINDER ENGINE)

2 Remove the fuse marked Fuel Pump from the fuse block in the passenger compartment.

3 Crank the engine over. It will start and run until the fuel supply remaining in the fuel lines is depleted.

4 After the engine stops, engage the starter again for another three seconds to assure that any remaining pressure is dissipated.

5 Turn the ignition to Off.

6 Replace the fuel pump fuse.

MODEL 700 TBI (LATER 4-CYLINDER ENGINES)

7 Place the transmission in Park (automatic) or Neutral (manual), set the parking brake and block the drive wheels.

8 Detach the three terminal electrical connector at the fuel tank.

9 Start the engine and allow it to run until it stops for lack of fuel.

10 Engage the starter for three seconds to dissipate fuel pressure in lines.

MODEL 220 TBI (V6 ENGINES)

11 The Model 220 TBI contains a constant bleed feature in the pressure regulator that relieves pressure. No special procedure is required to relieve the fuel pressure.

CMFI AND CSFI SYSTEMS

▶ Refer to illustrations 2.12a and 2.12b

※※ WARNING:

These systems are under constant high-pressure - wear safety goggles and take extra precautions.

12 Use one of the following methods to relieve the fuel pressure:

a) Attach a fuel pressure gauge equipped with a bleeder hose to the Schrader valve on the fuel supply line (see illustrations). Place the bleeder hose in an approved gasoline container and open the bleed valve to relieve the pressure.

b) If a fuel pressure gauge is not available, use the tip of a screwdriver to depress the Schrader valve while covering the valve using shop towels to absorb the expelled fuel.

13 Unless this procedure is performed before servicing fuel lines or connections, fuel spray (and possible injury) may occur.

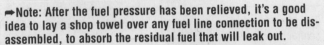

3 Fuel pump/fuel system pressure - check

※※ WARNING:

Gasoline is extremely flammable, so take extra precautions when you work on any part of the fuel system. See the Warning in Section 2.

➡Note: After the fuel pressure has been relieved, it's a good idea to lay a shop towel over any fuel line connection to be disassembled, to absorb the residual fuel that will leak out.

PRELIMINARY INSPECTION

1 First verify that there is fuel in the fuel tank.

2 With the engine running, inspect the fuel lines, hoses and connections for leaks. Examine the lines and hoses for kinks or restrictions and replace as necessary (see Section 4).

PRESSURE CHECK

▶ Refer to illustration 3.5

➡Note: In order to perform the fuel pressure check on carbureted and TBI engines, an adapter set is required to install the pressure gauge into the fuel supply line.

3 Perform the fuel pressure relief procedure (see Section 2).

4 On carbureted and TBI engines, disconnect the fuel supply line from the throttle body and install an in-line pressure gauge using a T-fitting.

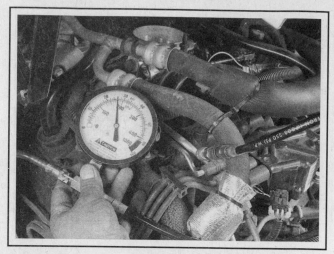

3.5 Checking the fuel pressure on a CSFI engine

5   On CMFI and CSFI engines, install a pressure gauge equipped with a bleeder hose onto the Schrader valve on the fuel supply line (see illustration).

6   On carbureted engines, start the engine and compare the fuel pressure with that listed in this Chapter's Specifications. If the pressure is too low, perform the fuel pump volume check (see Step 11). If the pressure is too high, check for a pinched or kinked fuel return line. If the line is OK, replace the fuel pump (see Section 7).

7   On TBI, CMFI and CSFI engines, turn the ignition key ON (engine OFF) several times to pressurize the system (this activates the fuel pump for 2 seconds each time the key is turned to the ON position). Compare the fuel pressure with that listed in this Chapter's Specifications. If no pressure is produced, check the fuel pump (see below). On CMFI and CSFI engines the pressure should remain constant. If it bleeds down, there is a leak in the system that must be repaired. TBI units are designed to bleed down after a period of time.

8   On TBI, CMFI and CSFI engines, if the fuel pressure is too low, check for a pinched or kinked supply line or clogged fuel filter. If the line and filter are OK, the problem may be a faulty fuel pressure regulator. To check the regulator, install a fuel line shut-off valve between the pressure regulator and the return line (this can be fabricated from high-pressure fuel line, a shut-off valve and the necessary fittings to mate with the pressure regulator and the return line, or between the return line and the fitting on the line that leads to the pressure regulator, depending on design. Or, instead of a shut-off valve, use a length of high-pressure fuel hose and the necessary fittings and clamps; the hose can be pinched with a pair of pliers instead of turning off the valve). With the pressure gauge still in place, start the engine. While observing the pressure gauge, slowly close the valve or pinch the hose. If the pressure rises above the specified pressure, replace the fuel pressure regulator (see the appropriate fuel system Section in this Chapter).

✳ WARNING:

Don't allow the fuel pressure to rise more than five psi above the maximum specified pressure. If the pressure is still low, replace the fuel pump (see Section 7).

9   On TBI, CMFI and CSFI engines, if the pressure is too high, check for a pinched or kinked fuel return line. If the line is OK, replace the fuel pressure regulator (see the appropriate fuel system Section in this Chapter).

3.19 On 1995 and earlier vehicles, disconnect the electrical connector from the fuel pump relay (arrow)

10 After testing is complete, perform the fuel pressure relief procedure (see Section 2) and remove the fuel pressure gauge (and shut-off valve, if installed).

FUEL PUMP

Volume check - carbureted and TBI engines only

11 Disable the ignition system by disconnecting the ignition coil primary wires (see Chapter 5).

12 Disconnect the fuel supply line from the throttle body and install a suitable length of fuel hose onto the supply line. Place the end of the hose into a container capable of holding at least 1 quart and secure the container as required.

13 Crank the engine for 15 seconds. Use a measuring cup or equivalent to determine the amount of fuel discharged and compare with the flow rate listed in this Chapter's Specifications. If the volume is low, check for a restricted fuel line, including the in-line fuel filter (TBI engines) and fuel pump inlet filter (strainer). If no restrictions are found, replace the fuel pump (see Section 7).

Electrical circuit check - TBI, CMFI and CSFI engines

▶ **Refer to illustration 3.19**

14 The PCM energizes the fuel pump for 2 seconds through the fuel pump relay when the ignition key is turned to the ON position. During engine cranking, the PCM receives a signal from the crankshaft position sensor which keeps the fuel pump relay energized. As a backup system, the oil pressure sensor can also energize the fuel pump. The oil pressure switch has two circuits; one operates the oil pressure light or gauge and the other connects power to the fuel pump once the oil pressure reaches approximately 4 psi. So, if the fuel pump relay fails, the oil pressure switch will operate the fuel pump. Fuel pump relay failure will result in unusually long cranking until the engine builds up enough oil pressure, particularly when the engine is cold.

15 If you suspect a problem with the fuel pump, first verify that it actuates. Remove the fuel filler cap. Have an assistant turn the ignition switch ON (engine OFF) while you listen at the fuel tank filler neck.

➡**Note: The key must remain in the OFF position for 10 seconds before the fuel pump will be energized again. You should hear a brief whirring noise as the pump comes on and pressurizes the system.**

16 If you don't hear anything, the pump itself may be faulty or the electrical circuit to the pump may have a short in it. Check the fuel pump fuse "ECM-B" in the fuse block (1995 and earlier models) or the underhood electrical center (1996 and later models). If the fuse was blown replace it and see if it blows again. If it does, check the circuit for a short (see Chapter 12).

17 If the fuse is OK, check for power to the fuel pump at the fuel pump connector located on top of the fuel tank. On some models the fuel tank must be removed or simply lowered to gain access to the connector (see Section 5 if necessary). With a test light connected to the appropriate terminal of the fuel pump connector, have an assistant turn the ignition key ON (engine OFF). Battery voltage should be supplied to the connector for approximately 2 seconds each time the ignition key is cycled to the ON position.

➡Note: The key must remain in the OFF position for 10 seconds before the fuel pump will be energized again.

18 If power is not present at the fuel pump connector, there is either a short in the wiring from the fuel pump relay to the fuel pump, the PCM to the fuel pump relay, or the fuel pump relay may be faulty.

19 To check the fuel pump relay, disconnect the electrical connector from the relay or remove the relay as applicable. On 1995 and earlier models, the fuel pump relay is located on the firewall (see illustration). On 1996 and later models, the fuel pump relay is located inside the electrical convenience center toward the rear of the engine compartment on the driver's side (see Chapter 12 if necessary). With the ignition key ON (engine OFF), probe the relay terminals to see if battery voltage is present. If voltage is present, replace the relay. If no voltage is detected, repair the circuit as required (see Chapter 12).

20 To check the oil pressure switch, start the engine and let it idle for about 3 minutes. With the engine running, remove the fuel pump relay. If the engine stops, the oil pressure switch and/or wiring is faulty.

4 Fuel lines and fittings - repair and replacement

✳✳ WARNING:

Gasoline is extremely flammable, so take extra precautions when you work on any part of the fuel system. See the Warning in Section 2.

1 Before attempting to disconnect or repair a fuel line, perform the fuel pressure relief procedure (see Section 2), then disconnect the cable from the negative terminal of the battery.

✳✳ CAUTION:

On models equipped with an anti-theft audio system, disable the anti-theft feature before performing any operation that requires disconnecting the battery or disrupting power to the stereo (see the Anti-theft audio system procedures at the front of this manual).

2 The fuel feed and return lines extend from the fuel tank to the engine compartment. The lines are secured to the underbody with clips and brackets. Both fuel and vapor lines must be occasionally inspected for leaks and damage.

3 If evidence of dirt is found in the system or fuel filter during disassembly, the line should be disconnected and flushed out. Check the fuel strainer on the fuel pick up in the fuel tank for damage and deterioration.

STEEL TUBING

4 If replacement of a fuel line or emission line is called for, use only welded steel tubing that meets or exceeds the manufacturer's specifications.

5 DO NOT use copper or aluminum tubing to replace steel tubing. These materials cannot withstand normal engine vibration.

6 Because the fuel lines used on fuel-injected models are under high pressure, they require special consideration.

7 Most metal fuel lines have threaded fittings with O-rings. Each time the fittings are loosened to service or replace components, make sure:

a) A back-up wrench is used while loosening and tightening the fittings.
b) Check all O-rings for cuts, cracks and deterioration. Replace any that appear worn or damaged.
c) If the lines are replaced, always use original equipment parts, or parts that meet the GM standards noted in this Section.

RUBBER HOSE

8 When rubber hose is used to replace/repair a metal line, use reinforced, fuel resistant hose. Hose other than this could fail prematurely and fail to meet Federal emission standards. Hose inside diameter must the same size as the line outside diameter.

✳✳ WARNING:

Don't substitute rubber hose for metal line on high-pressure systems. Use only genuine factory replacement lines or lines meeting factory specifications.

9 DO NOT use rubber hose within 4 inches of any part of the exhaust system or within 10 inches of the catalytic converter. Metal lines and rubber hoses must never be allowed to contact the frame. A minimum of 1/4 inch clearance must be maintained around the line or hose to prevent contact with the frame.

REPAIRING STEEL TUBING

10 In repairable areas, cut a piece of fuel hose 4 inches longer than the portion of the line to be removed. If more than 6 inches of line is to be replaced, use a combination of steel line and hose so that the hose lengths will not exceed 10 inches. Always follow the same routing as the original line.

11 Cut the ends of metal lines using a tube cutter. Using the first step of a double flaring tool, form a bead on the end of both line sections. If the line is too corroded to withstand the bead operation without becoming damaged, replace the line assembly.

NYLON FUEL PIPES

12 Nylon fuel pipes are used on later models. They are serviced as an assembly and are not repairable. Take care when working around these pipes. They are fragile and can be easily damaged by scraping against sharp objects. Do not kink them. Avoid getting any source of heat near them such as a torch or drop light. If they are nicked or scarred, they should be replaced.

QUICK-CONNECT FITTINGS

➡Note: Quick-connect fittings are found on later model fuel systems.

Disassembly and assembly

♦ Refer to illustration 4.16

13 Two types of quick-connect fittings are used on these models. The first is a metal collar type which requires a special removal tool to disconnect the fitting. The other type has a plastic collar and 2 locking tabs which can be disconnected by hand.

14 Before disconnecting any quick connect fitting, perform the fuel pressure relief procedure (see Section 2).

15 If equipped, slide the dust cover back, then clean the fitting and the area around the fitting. If possible use compressed air to blow any dirt or debris out of the connection.

16 On metal collar type fittings, grab each end of the fitting and twist it back and forth (approximately 1/4 turn either direction) to loosen it.

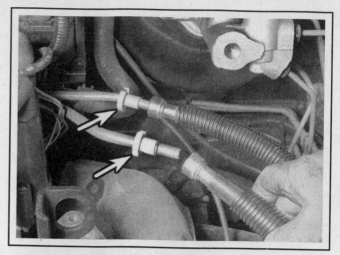

4.16 On metal collar fittings, place the special tool onto the tube and slide it into the fitting to release the locking tabs, then pull the fitting apart

Next, select the proper size removal tool (available at most auto parts stores) and install it onto the male portion of the line. Slide the tool into the female portion of the fitting to release the locking tabs and then pull the lines apart (see illustration).

17 On plastic collar type fittings, compress the locking tabs and separate the lines.

18 After disassembly inspect the fittings for damage and replace if necessary. On male fittings, minor defects can be removed using fine emery cloth or Scotchbrite.

19 Before assembling these fittings, apply a light coat of clean engine oil to the male part of the fitting.

20 After assembly, try to pull the lines apart to confirm proper connection.

5   Fuel tank - removal and installation

♦ Refer to illustrations 5.5 and 5.6

❄❄ WARNING:

Gasoline is extremely flammable, so take extra precautions when you work on any part of the fuel system. See the Warning in Section 2.

1   Relieve the system fuel pressure (see Section 2).
2   Detach the cable from the negative terminal of the battery.

❄❄ CAUTION:

On models equipped with an anti-theft audio system, disable the anti-theft feature before performing any operation that requires disconnecting the battery or disrupting power to the stereo (see the Anti-theft audio system procedures at the front of this manual).

3   Raise the vehicle and support it securely on jackstands.

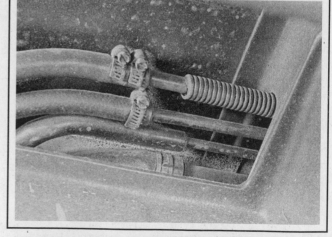

5.5 Unclip the lower shield from the fuel tank and pull it down far enough to detach the fuel feed, return, vapor canister and filler neck breather hoses

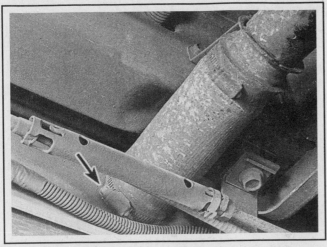

5.6 Loosen the hose clamp (arrow) and detach the filler neck hose from the fuel tank

4 Support the fuel tank with a transmission jack, a floor jack or a pair of jackstands.

5 Unclip the lower shield from the right side of the tank and pull it down far enough to remove the fuel feed, return and vapor hoses (see illustration).

6 Loosen the filler neck hose clamp (see illustration) and detach the filler neck hose from the pipe on the tank.

7 Remove the fuel tank support strap bolts and remove the support straps.

8 Carefully lower the tank enough to disconnect the fuel pump lead, sending unit lead and ground wire.

9 Remove the fuel tank from the vehicle.

10 Installation is the reverse of removal.

6 Fuel tank - cleaning and repair

1 Any repairs to the fuel tank or filler neck should be carried out by a professional who has experience in this critical and potentially dangerous work. Even after cleaning and flushing of the fuel tank, explosive fumes can remain and ignite during repair.

2 If the fuel tank is removed from the vehicle, it should not be placed in an area where sparks or open flames could ignite the fumes coming out of the tank. Be especially careful inside garages where a gas-type appliance is located, because it could cause an explosion.

7 Fuel pump - removal and installation

✳✳ WARNING:

Gasoline is extremely flammable, so take extra precautions when you work on any part of the fuel system. See the Warning in Section 2.

MECHANICAL PUMP (CARBURETOR EQUIPPED MODELS)

1 The fuel pump is a sealed unit and cannot be rebuilt. It's mounted on the right (passenger) side of the engine at the very front.

2 To remove the pump, first isolate the battery by disconnecting the negative battery cable.

3 Detach the fuel inlet hose, the outlet line and the vapor return hose (if equipped). Hold the fitting on the pump with a back-up wrench as the outlet line is disconnected. Also, if possible, use a flare-nut wrench on the fuel line fitting.

4 Remove the two mounting bolts and detach the fuel pump. As the pump is removed, the pushrod may fall out - be sure to retrieve it.

5 Remove the gasket and mounting plate.

6 Remove all traces of old gasket and sealant with a scraper, then clean the block mounting surface with lacquer thinner or acetone.

7 Apply a dab of heavy grease to the pushrod to hold it in place as the pump is installed.

8 Position the new gasket, the mounting plate and the pump on the block, then install the bolts and tighten them 1/4-turn at a time until they're secure.

9 Reattach the inlet hose, the outlet line and the vapor return hose to the pump. Be sure to tighten the fitting on the outlet line and the clamps on the hoses securely.

10 Start the engine and check for fuel leaks at the hose and line connections.

ELECTRIC PUMP (FUEL INJECTED MODELS)

▶ **Refer to illustrations 7.15 and 7.16a, 7.16b and 7.18**

11 Perform the fuel pressure relief procedure (see Section 2).

12 Disconnect the cable from the negative terminal of the battery.

✳✳ CAUTION:

On models equipped with an anti-theft audio system, disable the anti-theft feature before performing any operation that requires disconnecting the battery or disrupting power to the stereo (see the Anti-theft audio system procedures at the front of this manual).

13 Remove the fuel tank (see Section 5).

14 Match-mark the fuel pump/sending unit assembly to the fuel tank so it can be reinstalled in the same position.

15 The fuel pump/sending unit assembly on most models is secured to the fuel tank with a cam lock ring. Turn the cam lock ring counter-clockwise until the tangs and cut-outs align and then carefully withdraw the fuel pump/sending unit assembly out of the fuel tank taking care not to damage the float arm (see illustration).

❄❄ WARNING:

If the cam lock ring is very tight, use a hammer and a brass drift to drive the ring. DO NOT use a steel drift, which may cause a spark and lead to an explosion!

16 On some 1996 and later models, the fuel pump/sending assembly is secured by a snap-ring. Using snap-ring pliers, remove the snap ring and carefully withdraw the fuel pump/sending unit assembly from the fuel tank, taking care not to damage the float arm (see illustrations).

17 Inspect the condition of the rubber gasket or O-ring as applicable. If it's dried, cracked, hard or deteriorated, replace it.

18 Inspect the fuel strainer on the bottom of the unit (see illustration). If it's damaged or clogged, replace it. To remove the strainer, simply twist it off the fitting.

19 If it's necessary to replace the fuel pump itself, match-mark the components as required so they can be reinstalled in the same positions. Disassemble the fuel pump/sending unit assembly and install the

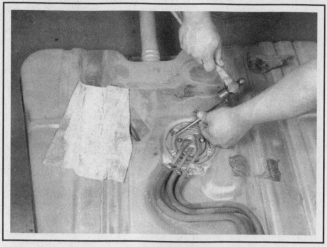

7.15 Use a brass punch to tap the lock ring counterclockwise until the tabs align with the cut-out areas of the fuel tank

new fuel pump per the manufacturer's instructions.

20 Install the fuel pump/sending unit assembly into the fuel tank and align the previously applied match-marks.

21 Install the cam lock ring or snap-ring as applicable.

22 Install the fuel tank (see Section 5).

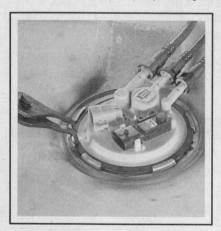

7.16a Use snap-ring pliers to remove the snap-ring securing the fuel pump/sending unit assembly (1996 and later models)

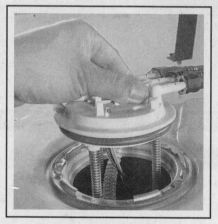

7.16b Withdraw the fuel pump/sending unit assembly from the fuel tank, being careful not to damage the float arm

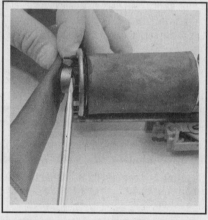

7.18 Inspect the fuel strainer for dirt; if too dirty to be cleaned carefully pry the fuel strainer from the inlet pipe

8 Air cleaner assembly - removal and installation

CARBURETOR AND TBI ENGINES

1 Detach the cable from the negative terminal of the battery.

❄❄ CAUTION:

On models equipped with an anti-theft audio system, disable the anti-theft feature before performing any operation that requires disconnecting the battery or disrupting power to the stereo (see the Anti-theft audio system procedures at the front of this manual).

2 Detach the fresh air intake duct from the mouth of the air cleaner housing assembly.

3 Remove the air cleaner cover nuts or wing nut.

4 Remove the air cleaner assembly and the gasket between the air cleaner housing and the adapter.

5 Discard the old gasket. Be sure to remove the paper from the new gasket before installing it.

6 If you are replacing the air cleaner housing assembly, installation is the reverse of removal.

7 If you are planning to remove the TBI unit, the intake manifold, a cylinder head, etc. remove the adapter nuts, the adapter assembly and the gasket between the adapter and the TBI assembly.

8.16a Air cleaner housing assembly (V6 CSFI engine, CMFI engine similar)

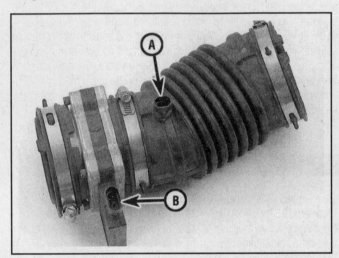

8.16b V6 CSFI engine air intake duct - IAT sensor (A) and MAF sensor (B)

8 Discard the old gasket. Be sure to remove the paper from the new gasket before installing it.
9 Installation is the reverse of removal.

CMFI AND CSFI ENGINES

◆ **Refer to illustrations 8.16a and 8.16b**

10 Disconnect the cable from the negative terminal of the battery.

❋❋ CAUTION:

On models equipped with an anti-theft audio system, disable the anti-theft feature before performing any operation that requires disconnecting the battery or disrupting power to the stereo (see the Anti-theft audio system procedures at the front of this manual).

11 Disconnect the electrical connector from the Intake Air Temperature (IAT) sensor, which is located in the air duct, just before the throttle body.
12 Detach the crankcase ventilation hose (if equipped) from the air intake duct.
13 On CSFI engines, disconnect the electrical connector from the Mass Air Flow (MAF) sensor.
14 Remove the air intake duct from the air cleaner assembly and throttle body.

❋❋ CAUTION:

On CSFI engines, the MAF sensor is part of the ducting. Treat it gently, as rough handling may damage this component.

15 Unclip the latches securing the air cleaner top cover. Remove the top cover and air filter element.
16 Remove the bolts and screws attaching the air cleaner lower housing to the radiator support and upper fan shroud (see illustrations). Remove it from the vehicle.
17 Installation is the reverse of removal.

9 Carburetor - removal and installation

❋❋ WARNING:

Gasoline is extremely flammable, so take extra precautions when you work on any part of the fuel system. See the Warning in Section 2.

1 Remove the battery (see Chapter 5).
2 Remove the air cleaner assembly (see Section 8).
3 Detach the accelerator linkage.
4 Detach the transmission detent cable (see Chapter 7 Part B).
5 Detach the cruise control cable, if equipped.
6 Clearly label, then detach, all vacuum lines.
7 Disconnect the fuel line at the carburetor inlet.
8 Clearly label, then unplug all electrical connectors.
9 Remove the four carburetor mounting bolts.
10 Remove the carburetor from the intake manifold.
11 Installation is the reverse of removal. Be sure to replace the carburetor base gasket and tighten the carburetor mounting bolts to the specified torque.

10 Carburetor - overhaul and adjustments

✷✷ WARNING:

Gasoline is extremely flammable, so extra precautions must be taken when working on any part of the fuel system. Do not smoke or allow open flames or bare light bulbs in or near the work area. Also, don't work in a garage if a gas appliance such as a water heater or clothes dryer is present.

OVERHAUL

1   If you are going to overhaul the carburetor yourself, first obtain a good quality carburetor rebuild kit (which will include all necessary gaskets, internal parts, instructions and a parts list). You will also need some solvent and a means of blowing out the internal passages of the carburetor with air.

2   Because carburetor designs are constantly modified by the manufacturer in order to meet emissions regulations, it isn't feasible for us to do a step-by-step overhaul of each type. You'll receive a detailed set of instructions with any quality carburetor overhaul kit. They will apply in a more specific manner to the carburetor on your vehicle.

3   An alternative is to obtain a new or rebuilt carburetor. They are readily available from dealers and auto parts stores. Make sure the exchange carburetor is identical to the original. A tag is usually attached to the top of the carburetor. It will aid in determining the exact type of carburetor you have. When obtaining a rebuilt carburetor or a rebuild kit, take time to make sure that the kit or carburetor matches your application exactly. Seemingly insignificant differences can make a large difference in the performance of your engine.

4   If you choose to overhaul your own carburetor, allow enough time to disassemble the carburetor carefully, soak the necessary parts in the cleaning solvent (usually for at least one-half day or according to the instructions listed on the carburetor cleaner) and reassemble it, which will usually take much longer than disassembly. When disassembling the carburetor, match each part with the illustration in the carburetor kit and lay the parts out in order on a clean work surface.

ADJUSTMENTS

5   Because there are a number of different configurations for Federal and California carburetors and because a considerable number of special tools and tuning equipment is necessary to adjust these carburetors, it is impossible to include a detailed step-by-step procedure outlining every adjustment. Aside from idle speed adjustment and other adjustments shown in the carburetor overhaul kit, do not attempt to adjust the carburetor on your vehicle. If adjustments are needed other than those listed above, take the vehicle to a professional mechanic. The procedure for adjusting the idle speed on your vehicle is located on the VECI label (see Chapter 6).

11 Fuel injection systems - general information

THROTTLE BODY INJECTION (TBI) SYSTEM

▶ **Refer to illustrations 11.2a and 11.2b**

1   Throttle Body Injection is used on all 2.5L four-cylinder engines and is a single injector type. All 2.5L engines manufactured until 1987 are equipped with Model 300 TBI, on 1988 and later 2.5L engines, Model 700 TBI was installed. From 1986 on, all 4.3L V6 VIN Z engines came equipped with the Model 220 TBI (twin injector) system. Although they somewhat differ in design, there is very little difference in function between these three TBI models.

2   The Throttle Body Injection system utilizes a single or dual fuel

11.2a Typical Model 300 TBI unit on earlier vehicles with a 4-cylinder engine

A Fuel injector	D Throttle Position Sensor
B Fuel pressure regulator	(TPS)
C Idle Air Control (IAC) valve	E Fuel meter cover

11.2b Typical Model 220 TBI unit on V6 powered vehicles

A Fuel injectors	C Idle Air Control (IAC) valve
B Fuel pressure regulator	D Throttle Position
(under fuel meter cover)	Sensor(TPS)

injectors (depending on model) which are centrally mounted in a carburetor-like housing (throttle body) (see illustrations). The fuel injector is an electronically controlled solenoid which is actuated by a signal from the Powertrain Control Module (PCM). The amount of time the injector is held open by the PCM determines the fuel/air mixture. Pressurized fuel is delivered to the injector(s) by an electronic fuel pump located inside the fuel tank. To maintain the fuel at a constant pressure, a fuel pressure regulator is incorporated into the fuel circuit which allows excess fuel to bleed off and return to the fuel tank through a separate fuel line. To control the idle speed, a device called the Idle Air Control (IAC) valve is used. The IAC valve receives a signal from the PCM which tells it how much air (if any) should be allowed to by-pass the throttle valves to increase the engine idle speed to compensate for engine temperature, air density and as additional loads are placed on the engine such as the power steering pump and A/C compressor.

CENTRAL MULTIPORT FUEL INJECTION (CMFI) SYSTEM

3   The CMFI system is used on all 4.3L V6 VIN W engines manufactured from 1992 to 1995. The system incorporates a two-piece intake manifold. The upper intake manifold is a variable single/split design that includes an intake tuning valve which is controlled by the Powertrain Control Module (PCM). The tuning valve switches the plenum to provide a "split" or "single" runner condition to maximize engine performance throughout all rpm ranges. Located under the upper intake manifold is a single central fuel injector which feeds each cylinder through six separate tubes and poppet nozzles. The fuel injector is an electronically controlled solenoid which is actuated by a signal from the Powertrain Control Module (PCM). The amount of time the injector is held open by the PCM determines the fuel/air mixture. Pressurized fuel is delivered to the injector by an electronic fuel pump located inside the fuel tank. To maintain the fuel at a constant pressure,

a fuel pressure regulator is incorporated into the fuel circuit which allows excess fuel to bleed off and return to the fuel tank through a separate fuel line. To control the idle speed, a device called the Idle Air Control (IAC) valve is used. The IAC valve receives a signal from the PCM which tells it how much air (if any) should be allowed to by-pass the throttle valves to increase the engine idle speed to compensate for engine temperature, air density and as additional loads are placed on the engine such as the power steering pump and A/C compressor.

CENTRAL SEQUENTIAL FUEL INJECTION (CSFI) SYSTEM

4   Beginning in 1996, 4.3L V6 engines were equipped with the CSFI system. This system like its predecessor CMFI, incorporates a two-piece intake manifold. The upper intake manifold is made from a new composite material and the lower manifold is cast aluminum. Located under the upper intake manifold is a central fuel meter body which houses six separate fuel injectors. Each fuel injector is connected to an individual cylinder through six separate tubes and poppet nozzles. The fuel injectors are electronically controlled solenoids which are actuated in the firing order by a signal from the Powertrain Control Module (PCM). The amount of time each injector is held open by the PCM determines the fuel/air mixture. Pressurized fuel is delivered to the injectors by an electronic fuel pump located inside the fuel tank. To maintain the fuel at a constant pressure, a fuel pressure regulator is incorporated into the fuel circuit which allows excess fuel to bleed off and return to the fuel tank through a separate fuel line. To control the idle speed, a device called the Idle Air Control (IAC) valve is used. The IAC valve receives a signal from the PCM which tells it how much air (if any) should be allowed to by-pass the throttle valves to increase the engine idle speed to compensate for engine temperature, air density and additional loads placed on the engine such as the power steering pump and A/C compressor.

12 Model 300 Throttle Body Injection (TBI) - removal, overhaul and installation

▸ **Refer to illustrations 12.5, 12.7, 12.8a, 12.8b, 12.10, 12.11, 12.17, 12.18, 12.19, 12.25, 12.26 and 12.35**

✸✸ WARNING:

Gasoline is extremely flammable, so take extra precautions when you work on any part of the fuel system. See the Warning in Section 2.

➡ **Note: Because of its relative simplicity, a throttle body assembly need be neither removed from the intake manifold nor completely disassembled during component replacement. However, for the sake of clarity, the following procedures are shown with the TBI assembly removed from the vehicle.**

1   Relieve the fuel pressure (see Section 2).
2   Detach the cable from the negative terminal of the battery

✸✸ CAUTION:

On models equipped with an anti-theft audio system, disable the anti-theft feature before performing any operation that requires disconnecting the battery or disrupting power to the stereo (see the Anti-theft audio system procedures at the front of this manual).

3   Remove the air cleaner housing assembly, adapter and gaskets (see Section 8).

FUEL METER COVER AND FUEL INJECTOR

Disassembly

4   Remove the injector electrical connector (on top of the TBI) by squeezing the two tabs together and pulling straight up.
5   Unscrew the fuel meter cover retaining screws and lockwashers securing the fuel meter cover to the fuel meter body. Note the location of the two short screws (see illustration).
6   Remove the fuel meter cover.

✸✸ CAUTION:

Do not immerse the fuel meter cover in solvent. It might damage the pressure regulator diaphragm and gasket.

7   The fuel meter cover contains the fuel pressure regulator, which is pre-set and plugged at the factory. If a malfunction occurs, it cannot be serviced, and must be replaced as a complete assembly.

✳✳ WARNING:

Do not remove the screws securing the pressure regulator to the fuel meter cover (see illustration). It has a large spring inside under heavy compression.

8   With the old fuel meter cover gasket in place to prevent damage to the casting, carefully pry the injector from the fuel meter body with a screwdriver until it can be lifted free (see illustrations).

✳✳ CAUTION:

Use care in removing the injector to prevent damage to the electrical connector terminals, the injector fuel filter, the O-ring and the nozzle.

9   The fuel meter body should be removed from the throttle body if it needs to be cleaned. To remove it, remove the fuel feed and return line fittings and the Torx screws that attach the fuel meter body to the throttle body.

10 Remove the old gasket from the fuel meter cover and discard it.

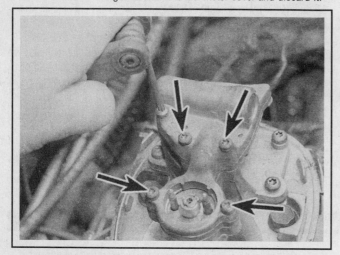

12.5 Unscrew the fuel meter cover retaining screws (arrows)

Remove the large O-ring and steel back-up washer from the upper counterbore of the fuel meter body injector cavity (see illustration). Clean the fuel meter body thoroughly in solvent and blow dry.

11 Remove the small O-ring from the nozzle end of the injector. Carefully rotate the injector fuel filter back and forth and remove the filter from the base of the injector (see illustration). Gently clean the filter in solvent and allow it to drip dry. It is too small and delicate to dry with compressed air.

✳✳ CAUTION:

The fuel injector itself is an electrical component. Do not immerse it in any type of cleaning solvent.

12 The fuel injector is not serviceable. If it is malfunctioning, replace it as an assembly.

Reassembly

13 Install the clean fuel injector nozzle filter on the end of the fuel injector with the larger end of the filter facing the injector so that the filter covers the raised rib at the base of the injector. Use a twisting motion to position the filter against the base of the injector.

14 Lubricate a new small O-ring with automatic transmission fluid. Push the O-ring onto the nozzle end of the injector until it presses against the injector fuel filter.

15 Insert the steel backup washer in the top counterbore of the fuel meter body injector cavity.

16 Lubricate a new large O-ring with automatic transmission fluid and install it directly over the backup washer. Be sure that the O-ring is seated properly in the cavity and is flush with the top of the fuel meter body casting surface.

✳✳ CAUTION:

The back-up washer and large O-ring must be installed before the injector or improper seating of the large O-ring could cause fuel to leak.

17 Install the injector in the cavity in the fuel meter body, aligning the raised lug on the injector base with the cast-in notch in the fuel

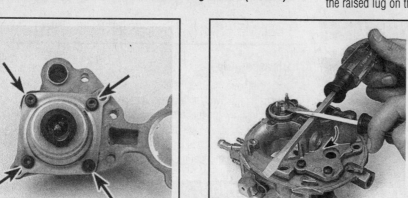

12.7 The fuel pressure regulator is installed in the fuel meter cover and pre-adjusted by the factory - do not remove the four retaining screws (arrows) or you may damage the regulator

12.8a The best way to remove the fuel injector is to pry on it with a screwdriver, using a second screwdriver as a fulcrum

12.8b Note the position of the terminals on top and the dowel pin on bottom of the injector in relation to the fuel meter cover when you lift the injector out of the cover

12.10 Carefully peel away the old outlet passage and cover gaskets with a razor blade

12.11 Gently rotate the fuel injector filter back and forth and carefully pull it off the nozzle

12.17 Make sure the lug is aligned with the groove in the bottom of the fuel injector cavity

meter body cavity. Push straight down on the injector with both thumbs (see illustration) until it is fully seated in the cavity.

➡️**Note: The electrical terminals of the injector should be approximately parallel to the throttle shaft.**

18 Install a new fuel outlet passage gasket on the fuel meter cover and a new fuel meter cover gasket on the fuel meter body (see illustration).

19 Install a new dust seal into the recess on the fuel meter body (see illustration).

20 Install the fuel meter cover onto the fuel meter body, making sure that the pressure regulator dust seal and cover gaskets are in place.

21 Apply a thread locking compound to the threads of the fuel meter cover attaching screws. Install the screws (the two short screws go next to the injector) and tighten them securely.

➡️**Note: Service repair kits include a small vial of thread locking compound with directions for use. If this material is not available, use Loctite 262, or some other non-hardening thread locking compound. Do not use a higher strength locking compound than recommended, as this may prevent subsequent removal of the attaching screws or cause breakage of the screwhead if removal becomes necessary.**

22 Plug in the electrical connector to the injector.
23 Install the air cleaner.

IDLE AIR CONTROL (IAC) VALVE

Removal

24 Unplug the electrical connector at the IAC valve.
25 Remove the IAC valve with a wrench on the hex surface only (see illustration).

Adjustment

26 Before installing a new IAC valve, measure the distance the valve is extended (see illustration). The measurement should be made from the motor housing to the end of the cone. The distance should be no greater than 1-1/8 inch. If the cone is extended too far, damage may occur to the valve when it is installed.

27 Identify the replacement IAC valve as either a Type I (with a collar at the electric terminal end) or a Type II (without a collar) (see illustration 12.26). If the measured dimension "A" is greater than 1-1/8 inch, the distance must be reduced as follows:

Type I - Exert firm pressure on the valve to retract it (a slight side-to-side movement may be helpful).

Type II - Compress the retaining spring of the valve while turning the valve in a clockwise direction. Return the spring to its original position with the straight portion of the spring aligned with the flat surface of the valve.

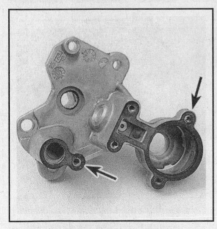

12.18 Position the fuel outlet passage gasket (A) and the fuel meter cover gasket (B) properly

12.19 Install a new dust seal into the recess of the fuel meter body

12.25 Remove the IAC valve with a large wrench, but be careful - it's a delicate device. Note that the IAC valve shown is a Type 2 (no collar)

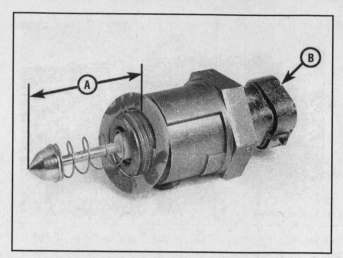

12.26 Distance (A) should be less than 1-1/8-inch - if it isn't, determine which type of valve you have and adjust it accordingly. This is a Type 1 valve since it has a collar at the electrical connector end (B). Type 2 valves do not have a collar at this location

INSTALLATION

28 Install the new IAC valve to the throttle body. Use the new gasket supplied with the assembly.

29 Plug in the electrical connector.

30 Install the air cleaner.

31 Start the engine and allow it to reach normal operating temperature. The Powertrain Control Module (PCM) will reset the idle speed when the vehicle is driven above 35 mph.

THROTTLE POSITION SENSOR (TPS)

32 The Throttle Position Sensor (TPS) is connected to the throttle shaft on the TBI unit. As the throttle valve angle is changed (as the accelerator pedal is moved), the output of the TPS also changes. At

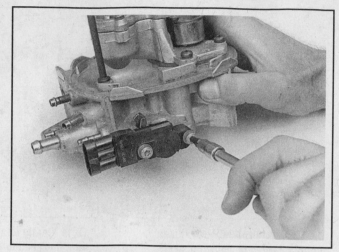

12.35 The Throttle Position Sensor (TPS) is mounted to the side of the TBI with two Torx screws

a closed throttle position, the output of the TPS is below 1.25 volts. As the throttle valve opens, the output increases so that, at wide-open throttle, the output voltage is approximately 5 volts.

33 A broken or loose TPS can cause intermittent bursts of fuel from the injector and an unstable idle, because the PCM thinks the throttle is moving. A problem in any of the TPS circuits will set either a Code 21 or 22 (see "Trouble Codes," Chapter 6).

34 The TPS is not adjustable. The PCM uses the reading at idle for the zero reading. If the TPS malfunctions, it is replaced as a unit.

35 Unscrew the two Torx screws (see illustration) and remove the TPS.

36 Install the new TPS.

➡Note: Make sure that the tang on the lever is properly engaged with the stop on the TBI.

37 Install the air cleaner assembly.

38 Attach the cable to the negative terminal of the battery.

39 With the ignition switch on and the engine off, check for fuel leaks.

13 Model 700 Throttle Body Injection (TBI) - removal, overhaul and installation

▶ Refer to illustrations 13.5, 13.6, 13.32, 13.33, 13.51 and 13.62

✳✳ WARNING:

Gasoline is extremely flammable, so take extra precautions when you work on any part of the fuel system. See the Warning in Section 2.

➡Note: Because of its relative simplicity, the throttle body does not need to be removed from the intake manifold or disassembled for component replacement. However, for the sake of clarity, the following procedures are shown with the TBI assembly removed from the vehicle.

1 Relieve the fuel pressure (see Section 2).

2 Detach the cable from the negative terminal of the battery.

13.5 Remove the injector retaining screw and retainer

※※ **CAUTION:**

On models equipped with an anti-theft audio system, disable the anti-theft feature before performing any operation that requires disconnecting the battery or disrupting power to the stereo (see the Anti-theft audio system procedures at the front of this manual).

3 Remove the air cleaner housing assembly, adapter and gaskets (see Section 8).

FUEL INJECTOR

4 Unplug the electrical connector from the fuel injector.

5 Remove the injector retainer screw and retainer (see illustration).

6 To remove the fuel injector assembly, place a screwdriver blade under the ridge (see illustration) and carefully pry it out.

7 Remove the upper and lower O-rings from the injector and from the fuel injector cavity and discard them.

8 Inspect the fuel injector filter for evidence of dirt and contamination. If present, look for the presence of dirt in the fuel lines and the fuel tank.

9 Lubricate new upper and lower O-rings with transmission fluid and install them on the injector. Make sure that the upper O-ring is in the groove and the lower one is flush up against the filter.

10 To install the injector assembly, push it straight into the fuel injector cavity. Be sure that the electrical connector end on the injector is facing in the general direction of the cut-out in the fuel meter body for the wire grommet.

➡Note: If you are installing a new injector, be sure to replace the old unit with an identical part. Injectors from other models will fit in the Model 700 TBI assembly but are calibrated for different flow rates.

11 Using thread locking compound on the retainer attaching screw, install the injector retainer and tighten the retaining screw.

12 Install the air cleaner housing assembly, adapter and gaskets.

13 With the engine off and the ignition on, check for fuel leaks.

FUEL PRESSURE REGULATOR

14 Remove the four pressure regulator attaching screws while keeping the pressure regulator compressed.

※※ **CAUTION:**

The pressure regulator contains a large spring under heavy compression. Use care when removing the screws to prevent personal injury.

15 Remove the pressure regulator cover assembly.

16 Remove the pressure regulator spring, seat, and the pressure regulator diaphragm.

17 Using a magnifying glass, if necessary, inspect the pressure regulator seat in the fuel meter body cavity for pitting, nicks or irregularities. If any damage is present, the entire fuel body casting must be replaced.

18 Install the new pressure regulator diaphragm assembly. Make sure it is seated in the groove in the fuel meter body.

19 Install the regulator spring seat and spring into the cover assembly.

20 Install the cover assembly over the diaphragm while aligning the mounting holes.

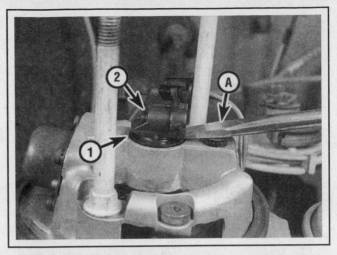

13.6 To remove the fuel injector unit (2) from the fuel meter body (1), insert a screwdriver (A) under the fuel injector flange and pry the injector loose

※※ **CAUTION:**

Use care while installing the pressure regulator to prevent misalignment and possible leaks.

21 While maintaining pressure on the regulator spring, install the four screw assemblies that have been coated with thread locking compound.

22 Reconnect the negative battery cable. With the engine off and the ignition on, check for fuel leaks.

THROTTLE POSITION SENSOR (TPS)

23 Unplug the electrical connector from the throttle position sensor.

24 Remove the two TPS attaching screws and remove the TPS from the throttle body.

25 With the throttle valve closed, install the TPS on the throttle shaft. Rotate it counterclockwise to align the mounting holes.

26 Install the two TPS attaching screws.

27 Install the air cleaner housing assembly, adapter and gaskets.

28 Attach the cable to the negative terminal of the battery.

➡Note: See non-adjustable TPS output check at end of this section.

IDLE AIR CONTROL (IAC) VALVE

29 Unplug the electrical connector from the IAC valve and remove the IAC valve mounting screws and the IAC valve.

30 Remove the O-ring from the IAC valve and discard it.

31 Clean the IAC valve seating surfaces on the throttle body to assure proper sealing of the new O-ring and proper contact of the IAC valve flange.

32 Before installing a new IAC valve, measure the distance between the tip of the valve pintle and the flange mounting surface when the pintle is fully extended (see illustration). If dimension "A" is greater than given in the Specifications, it must be reduced to prevent damage to the valve.

33 To retract the IAC valve, grasp the IAC valve as shown and exert firm pressure with your thumb, using a slight side-to-side movement on the valve pintle (see illustration).

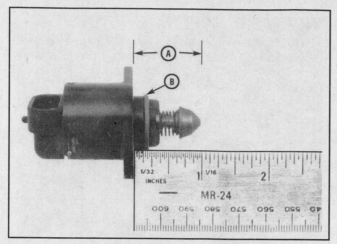

13.32 The Idle Air Control valve pintle must not extend more than 1-1/8-inch - also, replace the O-ring if it is brittle

A   *Distance of pintle extension*
B   *O-ring*

34 Lubricate a new O-ring with transmission fluid and install it on the IAC valve.

35 Install the IAC valve to the throttle body. Coat the IAC valve attaching screws with thread locking compound. Install and tighten them securely. Plug in the IAC valve electrical connector.

36 Install the air cleaner housing assembly, adapter and gaskets.

37 Attach the cable to the negative terminal of the battery.

38 Start the engine, allow it to reach operating temperature, then take the vehicle for a drive. When the engine reaches normal operating temperature the PCM will set the proper idle speed.

TUBE MODULE ASSEMBLY

39 Remove the tube module assembly attaching screws and remove the tube module.

40 Remove the tube module gasket and discard it. Clean any old gasket material from the surface of the throttle body to insure proper sealing of the new gasket.

41 Install the new tube module gasket.

42 Install the tube module.

43 Install the air cleaner housing assembly, adapter and gaskets.

13.51 Remove the throttle body mounting studs (arrows)

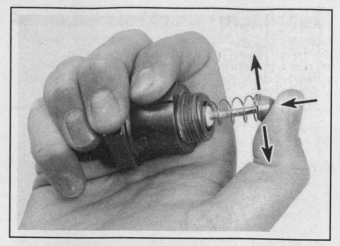

13.33 To adjust the IAC valve pintle, grasp it as shown and, using a side-to-side motion with your thumb, press it firmly down into the valve

FUEL METER ASSEMBLY

44 Remove the throttle body injection unit.

45 Remove the two fuel meter body attaching screws and washers and remove the fuel meter assembly from the throttle body.

46 If you are installing a new fuel meter, remove the fuel pressure regulator and the fuel outlet nut and transfer them to the new fuel meter.

47 Remove the fuel meter body-to-throttle body gasket and discard it.

48 Install a new fuel meter-to-throttle body gasket. Match the cutout portions of the gasket with the openings in the throttle body.

49 Install the fuel meter onto the throttle body and tighten the attaching screws.

THROTTLE BODY ASSEMBLY

50 Unplug the electrical connectors from the fuel injector, IAC and TPS.

51 Remove the throttle body mounting studs (see illustration).

52 Remove and discard the throttle body-to-intake manifold gasket. Place the TBI assembly on a clean work surface.

53 Remove the fuel injector, fuel meter assembly, TPS, IAC valve and the tube module.

54 Install the new tube module assembly, IAC valve and TPS on the new throttle body assembly.

55 Install a new fuel meter body-to-throttle body gasket.

56 Install the fuel meter assembly and fuel injector.

57 Install a new throttle body-to-intake manifold gasket and install the throttle body assembly on the intake manifold. Tighten the mounting studs to the specified torque.

MINIMUM IDLE SPEED ADJUSTMENT

➡Note: This adjustment should be performed only when the throttle body assembly has been replaced. The engine should be at normal operating temperature before making the adjustment.

58 Plug any vacuum line ports as required (see the VECI label refer to Chapter 6).

59 With the IAC valve connected, ground the diagnostic terminal of the ALDL connector (see Chapter 6).

13.62 To gain access to the factory sealed idle stop screw, remove the plug (arrow) by piercing it with an awl, then levering it loose

60 Turn the ignition on but do not start the engine. Wait at least 30 seconds to allow the IAC valve pintle to extend and seat in the throttle body. Unplug the IAC valve electrical connector.

61 Remove the ground jumper from the diagnostic terminal and start the engine.

62 The throttle stop screw used for regulating minimum idle speed is adjusted at the factory. The screw is covered with a plug to discourage unauthorized adjustments. To remove the plug, pierce it with an awl (see illustration), then apply leverage.

63 With the transmission in Neutral (manual) or Park (automatic), adjust the idle stop screw to obtain the specified rpm.

64 Turn the ignition off and reconnect the IAC valve electrical connector, unplug any plugged vacuum line ports and install the air cleaner housing assembly, adapter and gaskets.

NON-ADJUSTABLE TPS OUTPUT CHECK

➡ **Note: This check should be performed only when the throttle body or the throttle position sensor has been replaced or after the minimum idle speed has been adjusted.**

65 Connect a digital voltmeter from center terminal "B" to outside terminal "A" of the TPS connector.

66 With the ignition on and the engine stopped, the TPS voltage should be less than 1.25 volts. If the voltage is greater than specified, replace the TPS.

14 Model 220 Throttle Body Injection (TBI) - removal, overhaul and installation

▶ **Refer to illustrations 14.6, 14.7, 14.14, 14.16, 14.21, 14.22, 14.23, 14.24, 14.25, 14.37, 14.40, 14.41a, 14.41b, 14.41c, 14.50, 14.52, 14.65 and 14.66**

✳ **WARNING:**

Gasoline is extremely flammable, so take extra precautions when you work on any part of the fuel system. See the Warning in Section 2.

➡ **Note: Because of its relative simplicity, the throttle body assembly does not need to be removed from the intake manifold or disassembled for component replacement. However, for the sake of clarity, the following procedures are shown with the TBI assembly removed from the vehicle.**

1 Relieve system fuel pressure (see Section 2).
2 Detach the cable from the negative terminal of the battery.

✳ **CAUTION:**

On models equipped with an anti-theft audio system, disable the anti-theft feature before performing any operation that requires disconnecting the battery or disrupting power to the stereo (see the Anti-theft audio system procedures at the front of this manual).

3 Remove the air cleaner housing assembly, adapter and gaskets.

FUEL METER COVER/FUEL PRESSURE REGULATOR ASSEMBLY

➡ **Note: The fuel pressure regulator is housed in the fuel meter cover. Whether you are replacing the meter cover or the regulator itself, the entire assembly must be replaced. The regulator must not be removed from the cover.**

14.6 Carefully peel away the old fuel meter outlet passage gasket and fuel meter cover gasket with a razor blade

4 Unplug the electrical connectors to the fuel injectors.
5 Remove the long and short fuel meter cover screws and remove the fuel meter cover.
6 Remove the fuel meter outlet passage gasket, cover gasket and pressure regulator seal. Carefully remove any old gasket material that is stuck with a razor blade (see illustration).

✳ **CAUTION:**

Do not attempt to re-use either of these gaskets.

7 Inspect the cover for dirt, foreign material and casting warpage. If it is dirty, clean it with a clean shop rag soaked in solvent. Do not immerse the fuel meter cover in cleaning solvent - it could damage the pressure regulator diaphragm and gasket.

14.7 Never remove the four pressure regulator screws (arrows) from the fuel meter cover

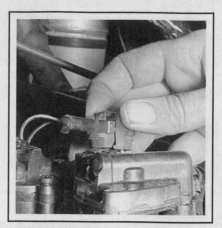

14.14 To remove either injector electrical connector, depress the two tabs on the front and rear of each connector and lift straight up

14.16 To remove an injector, slip the tip of a flat-bladed screwdriver under the lip of the lug on top of the injector and, using another screwdriver as a fulcrum, carefully pry the injector up and out

⁑ WARNING:

Do not remove the four screws (see illustration) securing the pressure regulator to the fuel meter cover. The regulator contains a large spring under compression which, if accidentally released, could cause injury. Disassembly might also result in a fuel leak between the diaphragm and the regulator housing. The new fuel meter cover assembly will include a new pressure regulator.

8   Install the new pressure regulator seal, fuel meter outlet passage gasket and cover gasket.

9   Install the fuel meter cover assembly using Loctite 262 or equivalent on the screws.

➡Note: The short screws go next to the injectors.

10   Attach the electrical connectors to both injectors.

11   Attach the cable to the negative terminal of the battery.

12   With the engine off and the ignition on, check for leaks around the gasket and fuel line couplings.

13   Install the air cleaner, adapter and gaskets.

FUEL INJECTOR ASSEMBLY

14   To unplug the electrical connectors from the fuel injectors, squeeze the plastic tabs and pull straight up (see illustration).

15   Remove the fuel meter cover/pressure regulator assembly.

➡Note: Do not remove the fuel meter cover assembly gasket - leave it in place to protect the casting from damage during injector removal.

16   Use a screwdriver and fulcrum (see illustration) to pry out the injector.

17   Remove the upper (larger) and lower (smaller) O-rings and filter from the injector.

18   Remove the steel backup washer from the top of the injector cavity.

19   Inspect the fuel injection filter for evidence of dirt and contamination. If present, check for the presence of dirt in the fuel lines and fuel tank.

20   Be sure to replace the fuel injector with an identical part. Injectors from other models can fit in the Model 220 TBI assembly but are calibrated for different flow rates.

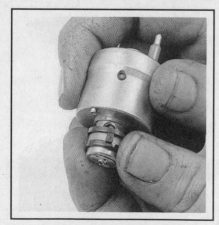

14.21 Slide the new filter onto the nozzle of the fuel injector

14.22 Lubricate the lower O-ring with transmission fluid then place it on the shoulder in the bottom of the injector cavity

14.23 Place the steel back-up washer on the shoulder near the top of the injector cavity

14.24 Lubricate the upper O-ring with transmission fluid then install it on top of the steel washer

14.25 Make sure that the lug is aligned with the groove in the bottom of the fuel injector cavity

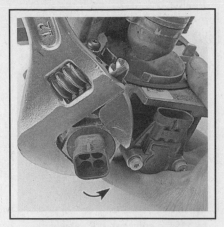

14.37 The IAC valve can be removed with an adjustable wrench (shown) or a 1-1/4 inch wrench

21 Slide the new filter into place on the nozzle of the injector (see illustration).

22 Lubricate the new lower (smaller) O-ring with automatic transmission fluid and place it on the small shoulder at the bottom of the fuel injector cavity in the fuel meter body (see illustration).

23 Install the steel back-up washer in the injector cavity (see illustration).

24 Lubricate the new upper (larger) O-ring with automatic transmission fluid and install it on top of the steel back-up washer (see illustration).

➡**Note: The back-up washer and the large O-ring must be installed before the injector. If they aren't, improper seating of the large O-ring could cause fuel leakage.**

25 To install the injector, align the raised lug on the injector base with the notch in the fuel meter body cavity (see illustration). Push down on the injector until it is fully seated in the fuel meter body.

➡**Note: The electrical terminals should be parallel with the throttle shaft.**

26 Install the fuel meter cover assembly and gasket.

27 Attach the cable to the negative terminal of the battery.

28 With the engine off and the ignition on, check for fuel leaks.

29 Attach the electrical connectors to the fuel injector(s).

30 Install the air cleaner housing assembly, adapter and gaskets.

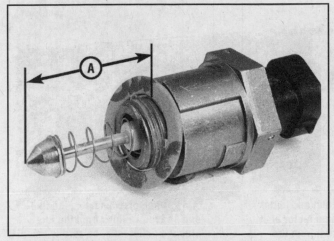

14.40 Measuring the pintle extension on a flange-type IAC valve

THROTTLE POSITION SENSOR (TPS)

31 Remove the two TPS attaching screws and retainers and remove the TPS from the throttle body.

32 If you intend to re-use the same TPS, do not attempt to clean it by soaking it in any liquid cleaner or solvent. The TPS is a delicate electrical component and can be damaged by solvents.

33 Install the TPS on the throttle body while lining up the TPS lever with the TPS drive lever.

34 Install the two TPS attaching screws and retainers.

35 Install the air cleaner housing assembly, adapter and gaskets.

36 Attach the cable to the negative terminal of the battery.

➡**Note: See non-adjustable TPS output check at end of this section.**

IDLE AIR CONTROL (IAC) VALVE

37 Unplug the electrical connector from the IAC valve and remove the IAC valve (see illustration).

38 Remove and discard the old IAC valve gasket. Clean any old gasket material from the surface of the throttle body assembly to insure proper sealing of the new gasket.

39 All pintles in IAC valves on Model 220 TBI units have the same dual taper. However, the pintles on some units have a 12mm diameter and the pintles on others have a 10mm diameter. A replacement IAC valve must have the appropriate pintle taper and diameter for proper seating of the valve in the throttle body.

40 Measure the distance between the tip of the pintle and the housing mounting surface with the pintle fully extended (see illustration). If dimension "A" is greater than the specified dimension, it must be reduced to prevent damage to the valve.

41 If the pintle must be adjusted, determine whether your valve is a Type I (collar around the electrical terminal) or a Type II (no collar around the electrical terminal) (see illustration 12.26).

   a) To adjust the pintle of an IAC valve with a collar, grasp the valve and exert firm pressure on the pintle with the thumb. Use a slight side-to-side movement on the pintle as you press it in with your thumb (see illustration).

   b) To adjust the pintle of an IAC valve without a collar, compress the retaining spring while turning the pintle clockwise (see illustration). Return the spring end to its original position with the

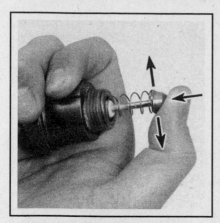

14.41a To adjust an IAC valve with a collar, retract the valve pintle by exerting firm pressure while using a slight side-to-side movement on the pintle

14.41b To adjust an IAC valve without a collar, compress the valve retaining spring while turning the valve clockwise . . .

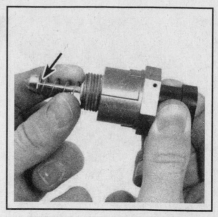

14.41c . . . then return the spring to its original position with the straight portion aligned in the slot under the flat surface of the valve

straight portion aligned in the slot under the flat surface of the valve (see illustration).

42 Install the IAC valve and tighten it to the specified torque. Attach the electrical connector.

43 Install the air cleaner housing assembly, adapter and gaskets.

44 Attach the cable to the negative terminal of the battery.

45 Start the engine and allow it to reach operating temperature, then turn it off. No adjustment of the IAC valve is required after installation. The IAC valve is reset by the PCM when the engine is turned off.

FUEL METER BODY ASSEMBLY

46 Unplug the electrical connectors from the fuel injectors.

47 Remove the fuel meter cover/pressure regulator assembly, fuel meter cover gasket, fuel meter outlet gasket and pressure regulator seal.

48 Remove the fuel injectors.

49 Unscrew the fuel inlet and return line threaded fittings, detach the lines and remove the O-rings.

50 Remove the fuel inlet and outlet nuts and gaskets from the fuel meter body assembly (see illustration). Note the locations of the nuts to ensure proper reassembly. The inlet nut has a larger passage than the

outlet nut.

51 Remove the gasket from the inner end of each fuel nut.

52 Remove the fuel meter body-to-throttle body attaching screws and remove the fuel meter body from the throttle body (see illustration).

53 Install the new throttle body-to-fuel meter body gasket. Match the cut-out portions in the gasket with the openings in the throttle body.

54 Install the fuel meter body on the throttle body. Coat the fuel meter body-to-throttle body attaching screws with thread locking compound before installing them.

55 Install the fuel inlet and outlet nuts, with new gaskets, in the fuel meter body and tighten the nuts to the specified torque. Install the fuel inlet and return line threaded fittings with new O-rings. Use a backup wrench to prevent the nuts from turning.

56 Install the fuel injectors.

57 Install the fuel meter cover/pressure regulator assembly.

58 Attach the cable to the negative terminal of the battery.

59 Attach the electrical connectors to the fuel injectors.

60 With the engine off and the ignition on, check for leaks around the fuel meter body, the gasket and around the fuel line nuts and threaded fittings.

61 Install the air cleaner housing assembly, adapters and gaskets.

14.50 Remove the fuel inlet and outlet nuts from the fuel meter body

14.52 Once the fuel inlet and outlet nuts are off, pull the fuel meter body straight up to separate it from the throttle body

14.65 When disconnecting the fuel feed and return lines from the fuel inlet and outlet nuts, be sure to use a backup wrench to prevent damage to the lines

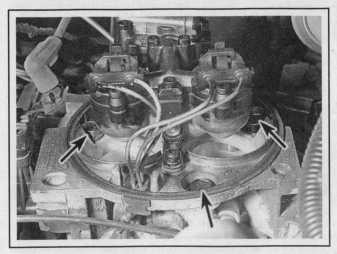

14.66 To remove the Model 220 throttle body from the intake manifold, remove the three bolts (arrows)

THROTTLE BODY ASSEMBLY

62 Unplug all electrical connectors - the IAC valve, TPS and fuel injectors. Detach the grommet with the wires from the throttle body.

63 Detach the throttle linkage, return spring(s), transmission control cable (automatics) and, if equipped, cruise control.

64 Clearly label, then detach, all vacuum hoses.

65 Using a backup wrench, detach the inlet and outlet fuel line nuts (see illustration). Remove the fuel line O-rings from the nuts and discard them.

66 Remove the TBI mounting bolts (see illustration) and lift the TBI unit from the intake manifold. Remove and discard the TBI manifold gasket.

67 Place the TBI unit on a holding fixture, if available.

➡Note: If you don't have a holding fixture, and decide to place the TBI directly on a work bench surface, be extremely careful when servicing it. The throttle valve can be easily damaged.

68 Remove the fuel meter body-to-throttle body attaching screws and separate the fuel meter body from the throttle body.

69 Remove the throttle body-to-fuel meter body gasket and discard it.

70 Remove the TPS.

71 Invert the throttle body on a flat surface for greater stability and remove the IAC valve.

72 Clean the throttle body assembly in a cold immersion cleaner. Clean the metal parts thoroughly and blow dry with compressed air. Be sure that all fuel and air passages are free of dirt or burrs.

✳✳ CAUTION:

Do not place the TPS, IAC valve, pressure regulator diaphragm, fuel injectors or other components containing rubber in the solvent or cleaning bath. If the throttle body requires cleaning, soaking time in the cleaner should be kept to a minimum. Some models have throttle shaft dust seals that could lose their effectiveness by extended soaking.

73 Inspect the mating surfaces for damage that could affect gasket sealing. Inspect the throttle lever and valve for dirt, binds, nicks and other damage.

74 Invert the throttle body on a flat surface for stability and install the IAC valve and the TPS.

75 Install a new throttle body-to-fuel meter body gasket and place the fuel meter body assembly on the throttle body assembly. Coat the fuel meter body-to-throttle body attaching screws with thread locking compound and tighten them securely.

76 Install the TBI unit and tighten the mounting bolts to the specified torque. Use a new TBI-to-manifold gasket.

77 Install new O-rings on the fuel line nuts. Install the fuel line and outlet nuts by hand to prevent stripping the threads. Using a backup wrench, tighten the nuts to the specified torque once they have been correctly threaded into the TBI unit.

78 Attach the vacuum hoses, throttle linkage, return spring(s), transmission control cable (automatics) and, if equipped, cruise control cable. Attach the grommet, with wire harness, to the throttle body.

79 Plug in all electrical connectors, making sure that the connectors are fully seated and latched.

80 Check to see if the accelerator pedal is free by depressing the pedal to the floor and releasing it with the engine off.

81 Connect the negative battery cable, and, with the engine off and the ignition on, check for leaks around the fuel line nuts.

82 Adjust the minimum idle speed and check the TPS output (see Steps 84 through 93).

83 Install the air cleaner housing assembly, adapter and gaskets.

MINIMUM IDLE SPEED ADJUSTMENT

➡Note: This adjustment should be performed only when the throttle body has been replaced. The engine should be at normal operating temperature before making the adjustment.

84 Remove the air cleaner housing assembly, adapter and gaskets.

85 Plug any vacuum ports as required by the VECI label.

86 With the IAC valve connected, ground the diagnostic terminal of the ALDL connector (see Chapter 6). Turn on the ignition but do not start the engine. Wait at least 30 seconds to allow the IAC valve pintle to extend and seat in the throttle body. Disconnect the IAC valve electrical connector. Remove the ground from the diagnostic terminal and start the engine.

87 Remove the plug by first piercing it with an awl (see illustration 13.62), then applying leverage.

88 Adjust the idle stop screw to obtain the specified rpm in neutral (manual) or in Drive (automatic).

89 Turn the ignition off and reconnect the IAC valve electrical connector.

90 Unplug any plugged vacuum line ports.

91 Install the air cleaner housing assembly, adapter and new gaskets (see Section 8).

NON-ADJUSTABLE TPS OUTPUT CHECK

➡Note: This check should be performed only when the throttle body or the TPS has been replaced or after the minimum idle speed has been adjusted.

92 Connect a digital voltmeter from the TPS connector center terminal "B" to outside terminal "A" (you'll have to fabricate jumpers for terminal access).

93 With the ignition on and the engine off, TPS voltage should be less than 1.25 volts. If it's more than the specified voltage, check the minimum idle speed before replacing the TPS.

15 Central Multiport Fuel Injection (CMFI) and Central Sequential fuel Injection (CSFI) systems - component check and replacement

GENERAL INFORMATION

1 The function of the CMFI or CSFI unit is to precisely control fuel delivery to the engine. These systems are called "central" because the fuel injection unit (CMFI) or fuel meter body (CSFI) is located in the center of the lower intake manifold. The major differences between these systems are as follows; the CMFI system has an adjustable aluminum air intake plenum and a single (non-serviceable) fuel injector unit which feeds all six cylinders. The CSFI system has a non-adjustable air intake plenum which is made from a composite material, uses six individually controlled fuel injectors (one for each cylinder) and incorporates a Mass Air Flow sensor in the air intake duct. Each system is controlled by the Powertrain Control Module (PCM) (see Chapter 6 for more information).

2 The PCM monitors various engine parameters via several sensors to determine how much fuel the engine requires. When the ignition key is first turned ON, the PCM energizes the fuel pump for 2 seconds to pressurize the system.

3 Pressurized fuel is supplied by a high pressure electric fuel pump located inside the fuel tank. The pump is designed to provide pressure above that required by the system. To maintain a constant regulated fuel supply, a fuel pressure regulator is incorporated in the fuel circuit and mounted directly on the fuel injection unit (CMFI) or fuel meter body (CSFI). The fuel pressure regulator maintains the correct pressure to the fuel injector(s) by allowing fuel to return to the fuel tank when the pressure exceeds system requirements (refer to this Chapter's Specifications).

4 The throttle valve located in the throttle body is used to control the air flow into the engine and consequently engine power output. When the throttle valve is closed, additional air requirements are determined by the PCM based on engine load(s), temperature and then compensated for via the Idle Air Control (IAC) valve. The IAC valve receives a signal from the PCM which tells it how much air (if any) should be allowed to by-pass the throttle valves to increase the engine idle speed to compensate for engine temperature, air density and as additional loads are placed on the engine such as the power steering pump and A/C compressor.

5 Fuel is supplied to each cylinder through a tube connecting the fuel injection unit (CMFI) or fuel meter body (CSFI) to each cylinder intake runner in the lower intake manifold. Located at the end of each fuel tube is a poppet nozzle which snaps into the intake manifold and holds the fuel tube in place. Inside the poppet nozzle is a spring loaded ball seal which allows fuel to pass through the nozzle and into the engine when the fuel pressure exceeds approximately 52 psi.

UPPER INTAKE MANIFOLD (PLENUM)

> ※※ **WARNING:**
>
> **Gasoline is extremely flammable, so take extra precautions when you work on any part of the fuel system. See the Warning in Section 2.**

Removal

▶ **Refer to illustrations 15.11, 15.12, 15.17a, 15.17b, 15.20, 15.22a, 15.22b and 15.23**

6 Disconnect the cable from the negative terminal of the battery.

> ※※ **CAUTION:**
>
> **On models equipped with an anti-theft audio system, disable the anti-theft feature before performing any operation that requires disconnecting the battery or disrupting power to the stereo (see the Anti-theft audio system procedures at the front of this manual).**

7 Perform the fuel pressure relief procedure (see Section 2).

8 Remove the air cleaner assembly and intake duct (see Section 8).

9 Remove the engine cover (see Chapter 11).

10 On some models equipped with air conditioning the A/C compressor and/or refrigerant lines will have to be removed from their mounting and placed out of the way or completely removed from the vehicle (see Chapter 3 for more information).

> ※※ **WARNING:**
>
> **The air conditioning system is under high pressure. Do not loosen any hose fittings or remove any components until after the system has been properly discharged by a dealer service department or other qualified repair shop. Even after the system has been discharged, residual pressure may still remain - be sure to wear eye protection when loosening A/C fittings!**

11 On CSFI engines, working from inside the vehicle, disconnect the fuel supply and return lines and remove the bolt attaching the fuel line assembly to the lower intake manifold (see illustration).

12 On CSFI engines, remove the retaining clip and disconnect the fuel injector electrical connector from the fuel meter body (see illustration).

13 On CSFI engines, remove the EVAP system canister purge solenoid and bracket (see Chapter 6).

14 On 1998 and later CSFI engines, detach the transmission and engine oil dipstick tubes from the accelerator cable bracket.

15 On CMFI engines, if equipped, remove the plastic "VORTEC" cover from the upper intake manifold.

16 On 1992 and earlier CMFI engines equipped with an automatic

15.11 CSFI engine fuel return and supply lines (arrows)

15.12 Removing the fuel injector electrical connector retaining clip (CSFI engine)

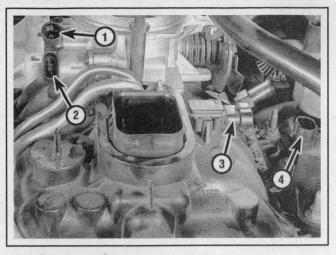

15.17a CSFI components

1	TPS sensor	4	EVAP canister purge
2	IAC valve		solenoid
3	MAP sensor		

transmission, disconnect the transmission TV cable from the throttle body (see Chapter 7B if necessary).

17 Clearly label and disconnect all electrical connectors from the upper intake manifold components, such as the TPS sensor, IAC valve, MAP sensor and the intake tuning valve (CMFI engines only) (see illustrations). Remove the nuts holding the engine wiring harness and move the harness out of the way.

18 Disconnect the PCV hose from the upper intake manifold (see Chapter 1).

19 Label and disconnect any vacuum hoses attached to the upper intake manifold.

20 Disconnect the accelerator and (if equipped) cruise control cables from the throttle body (see Section 16). Remove the bracket securing the cables to the upper intake manifold (see illustration).

21 Remove the ignition coil (see Chapter 5).

➡Note: This does not apply to coil-in-cap distributor models.

22 Remove the 10 upper intake manifold nuts, bolts or studs as applicable (see illustrations).

➡Note: Mark the location of each bolt or stud for proper placement at installation.

15.17b Disconnect the MAP sensor electrical connector (CMFI engine shown)

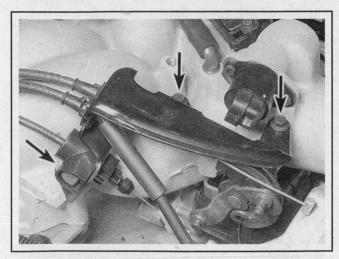

15.20 Remove the three bolts (arrows) from the plenum and lift the cable bracket off the plenum (CMFI engine shown)

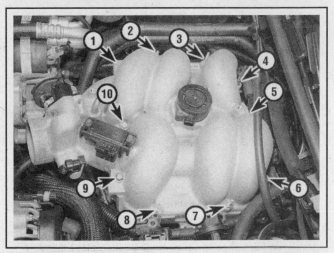

15.22a CMFI engine upper intake manifold bolts/nuts (arrows) (the numbers indicate tightening sequence for assembly)

15.22b CSFI engine upper intake manifold bolt/stud locations (arrows) (upper intake manifold removed for clarity)

23 Remove the upper intake manifold from the engine (see illustration).

24 Remove the gasket. If necessary, scrape all traces of old gasket material from the upper and lower intake manifolds. Be careful not to gouge the sealing surfaces.

Installation

▶ Refer to illustrations 15.25 and 15.26

25 On CSFI engines, inspect the upper intake manifold, gasket and fuel meter body O-ring for cracks or other damage (see illustration). Replace if necessary.

26 On CMFI engines, place the new gasket onto the lower intake manifold (see illustration).

27 On CSFI engines, place the new (or used if it's OK) gasket into the groove on the upper intake manifold. The factory recommends installing a new gasket.

28 Install the upper intake manifold and secure with the nuts, bolts and studs as required, placing them in their proper locations. Tighten the upper intake manifold fasteners evenly - proceeding in a clockwise direction to the torque listed in this Chapter's Specifications (see illustration 15.22a).

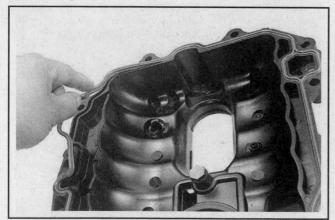

15.25 On CSFI engines, examine the upper intake manifold rubber gasket and O-rings for damage. If they look OK, you can reuse them. Since it's made of plastic, also inspect the upper intake manifold for cracks

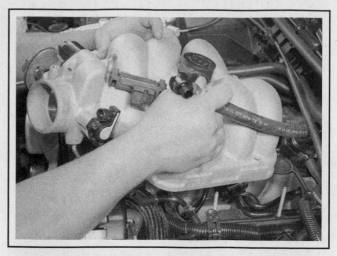

15.23 Removing the upper intake manifold (CMFI engine shown)

✳✳ CAUTION:

Do not overtighten the fasteners on CSFI engines - the composite material may crack.

29 The remaining installation steps are the reverse of removal. On CSFI engines, pressurize the fuel system by turning the ignition key ON (engine OFF) for 2 seconds and then OFF for 10 seconds - repeat this several times and check for fuel leaks.

CMFI FUEL INJECTION UNIT AND CSFI FUEL METER BODY

✳✳ WARNING:

Gasoline is extremely flammable, so take extra precautions when you work on any part of the fuel system. See the Warning in Section 2.

➡Note: The CMFI fuel injection unit has no serviceable parts. If any part of the unit is faulty, the entire unit must be replaced.

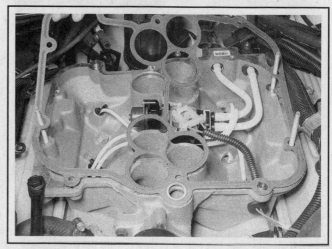

15.26 Installing the new gasket onto the lower intake manifold (CMFI engine)

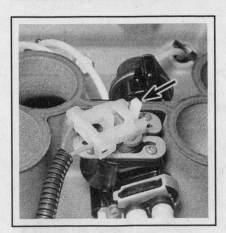

15.34 Squeeze the tab (arrow) to disconnect the injector electrical connector

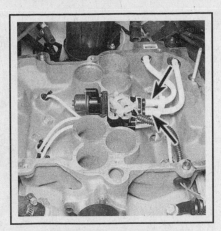

15.35 Carefully remove the inlet and return lines from the injector assembly

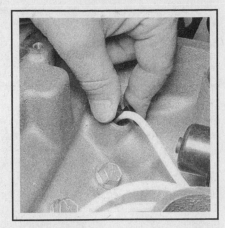

15.38a Squeeze the two tabs and lift to remove the poppet nozzle from the intake manifold (CMFI engine shown)

Removal

▶ Refer to illustrations 15.34, 15.35, 15.38a, 15.38b, 15.40a and 15.40b

30 Disconnect the cable from the negative terminal of the battery.

⁂ CAUTION:

On models equipped with an anti-theft audio system, disable the anti-theft feature before performing any operation that requires disconnecting the battery or disrupting power to the stereo (see the Anti-theft audio system procedures at the front of this manual).

---

31 Perform the fuel pressure relief procedure (see Section 2).

32 Remove the air cleaner assembly and intake duct (see Section 8).

33 Remove the engine cover (see Chapter 11).

34 On CMFI engines, disconnect the electrical connector from the fuel injection unit (see illustration).

35 On CMFI engines, remove the fuel line retaining clip and detach the fuel lines from the injector assembly (see illustration). Discard the retaining clip and fuel line O-ring seals as new ones should be installed at reassembly.

36 Remove the upper intake manifold (see above, this Section).

37 Before removing the poppet nozzles, clearly label each poppet nozzle and its respective location in the intake manifold. This will help avoid mixing them up at assembly. This is especially important on CSFI engines.

38 Remove the 6 poppet nozzles by squeezing the tabs together to release the locking feature and then withdraw them from the intake manifold socket (see illustrations). If they're stuck, wiggle the nozzle to help break the seal.

39 On CMFI engines, slide the fuel injection unit assembly out of it's retaining slot and remove it from the engine.

40 On CSFI engines, detach the vacuum hose from the fuel pressure regulator. Next, gently pry the locking tabs on the bracket away from the tangs on the fuel meter body and then remove the fuel meter body from the engine (see illustrations).

Installation

41 Install the fuel injection unit (CMFI) or fuel meter body (CSFI) onto the lower intake manifold. On CSFI engines, make sure the fuel meter bracket retainers are completely engaged.

42 On CMFI engines, connect the fuel injection unit electrical connector.

43 Install the poppet nozzles into their proper locations.

15.38b Removing a poppet nozzle (CSFI engine shown)

15.40a Use 2 screwdrivers (arrows) to spread the fuel meter body bracket . . .

15.40b . . . and then remove the fuel meter body from the engine - be careful not to snag the nozzle lines during removal

> ✳✳ **CAUTION:**
> Be sure the poppet nozzles are completely seated and secured in the manifold sockets. Check by pulling on each poppet nozzle to make sure it's locked in place.

44 On CMFI engines, obtain a new fuel pipe retaining clip and new fuel pipe O-ring seals. Lightly coat the seals with clean engine oil and install them onto the fuel pipe fittings. Insert the fuel pipes into the fuel injection unit and secure them with the retaining clip.

45 On CMFI engines only, temporarily connect the negative cable to the battery and pressurize the fuel system by turning the ignition key ON (engine OFF) for 2 seconds and then OFF for 10 seconds - repeat this several times. Disconnect the cable from the negative terminal of the battery and inspect the fuel injection unit and poppet nozzles for leaks. Repair if necessary.

46 Install the upper intake manifold (see above, this Section).

47 The remaining installation steps are the reverse of removal. Start the engine and check for leaks.

IDLE AIR CONTROL (IAC) VALVE

Check

▶ **Refer to illustrations 15.54a and 15.54b**

48 The IAC valve controls the engine idle speed. This output actuator is mounted on the throttle body and is controlled by the PCM. To increase engine speed, the PCM retracts the IAC pintle away from the seat and allows more air to bypass the throttle valve. To decrease the idle speed, the PCM extends the IAC pintle towards the seat, the bypass air flow.

49 Run the engine until it reaches normal operating temperature, then shut off the engine.

50 Locate a vacuum fitting on the intake manifold. Remove the hose from the fitting and attach a vacuum gauge in its place. Position the vacuum gauge so you can see it from inside the vehicle.

51 If the vehicle is not equipped with a tachometer, connect one to the engine per the manufacturer's instructions and place it so you can see the meter from inside the vehicle.

52 Inside the vehicle, start the engine and note the reading on the vacuum gauge and the engine speed on the tachometer.

53 If the vehicle is equipped with an automatic transaxle, apply the

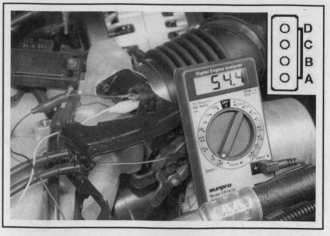

15.54a Measuring the IAC valve resistance (CMFI engine)

A   Coil B low
B   Coil B high
C   Coil A low
D   Coil A high

brakes and place the transaxle in DRIVE. If the vehicle is equipped with air conditioning, turn it on. If the vehicle does not have either of the previously mentioned options, turn the steering wheel to the left and hold it in there. Observe the reading on the vacuum gauge. It should drop slightly as the IAC motor opens the throttle by-pass airway. The engine speed should drop momentarily and then rise to the previously noted rpm. If the vacuum reading remains the same and the engine speed drops, turn off the engine and proceed to the next step.

54 To check the IAC valve, disconnect the electrical connector from the IAC valve.

➡**Note: On CSFI engines, you'll need to remove the air cleaner assembly (see Section 8) to access the IAC valve.**

Using an ohmmeter, measure the resistance across the IAC valve terminals A and B (see illustrations). Note the resistance. Next, measure the resistance between terminals C and D. The resistance should be 40 to 80 ohms for each check. If not, replace the IAC valve.

55 There is an alternate method for testing the IAC valve. Various SCAN tools are available from auto parts stores and specialty tool companies that can be connected to the Assembly Line Data Link (ALDL) connector to read trouble codes and actuate the IAC valve (see Chapter 6 for more information).

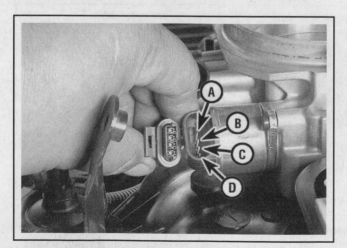

15.54b IAC valve (CSFI engine)

15.58 Removing the IAC valve (CMFI engine shown, CSFI engine similar)

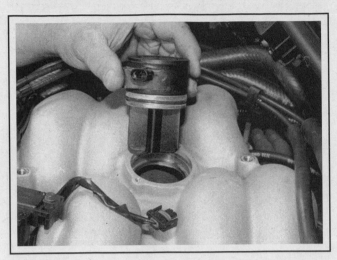

15.69 Intake manifold tuning valve (CMFI engines only)

Removal

▶ **Refer to illustration 15.58**

56 On CSFI engines, remove the air cleaner assembly and intake duct (see Section 8).

57 Disconnect the electrical connector from the IAC valve.

58 Remove the 2 screws securing the IAC valve (see illustration) and withdraw the valve.

59 Check the condition of the O-ring seal. If it's damaged or hardened, replace it.

60 Clean carbon deposits from the throttle body and IAC valve using a shop rag moistened with a spray type carburetor and choke cleaner.

✳✳ CAUTION:

DO NOT soak the IAC valve in any cleaning solvent!

Installation

61 When replacing the IAC valve, it is very important to purchase a part which is identical to the original. IAC valve pintle shapes and diameters are designed for specific engine applications. Although they may look the same, they may be different.

62 Before installing a new IAC valve, the position of the pintle must be checked. If it sticks out too far, it may be damaged during installation.

63 On new IAC valves only - measure the distance from the mounting surface of the IAC valve to the tip of the pintle. If the distance is greater than 1-1/8 inch, use thumb pressure to press the pintle into the valve until the dimension is acceptable.

✳✳ CAUTION:

Do not use thumb pressure to move the pintle on a used IAC valve - you'll damage the valve.

64 If you are replacing the O-ring seal, coat the new seal lightly with clean engine oil and install it onto the IAC valve.

65 If installing a used valve and O-ring seal, coat the O-ring seal lightly with clean engine oil and insert the valve into the bore.

66 Tighten the 2 screws to the torque listed in this Chapter's Specifications.

67 The remaining installation steps are the reverse of removal.

68 After installation, reset the IAC valve as follows;

a) *Disconnect the cable from the negative terminal of the battery for 10 seconds.*

✳✳ CAUTION:

On models equipped with an anti-theft audio system, disable the anti-theft feature before performing any operation that requires disconnecting the battery or disrupting power to the stereo (see the Anti-theft audio system procedures at the front of this manual).

b) *Next, turn the ignition key ON (engine OFF) for 5 seconds.*

c) *Turn the ignition key OFF for 10 seconds and then start the engine and check for proper idle operation.*

INTAKE MANIFOLD TUNING VALVE (CMFI ENGINES ONLY)

Check and replacement

▶ **Refer to illustration 15.69**

69 The intake manifold tuning valve is controlled by the PCM to change the intake air passageway (shortening or lengthening the distance the air travels before entering the combustion chamber) for optimum engine torque and peak horsepower based on engine speed and throttle position (see illustration). During low and high rpms the PCM de-energizes the valve to create a "split" plenum condition. When the engine is at mid-range rpm, the PCM closes the valve to create a "single" plenum condition.

70 Remove the engine cover (see Chapter 11).

71 If equipped, remove the "VORTEC" plastic cover from the upper intake manifold.

72 With the engine running at idle, use a test light to check for battery voltage at the tuning valve electrical connector PURPLE wire. No voltage should be present at idle. Next, while observing the test light, quickly open the throttle to the midway position. Battery voltage should become present at the connector when the engine speed and throttle are in mid-range. If battery voltage is not present as described, check the tuning valve relay and circuit for a bad connection or short (see Chapter 12).

73 Next, shut off the engine. Remove the 2 screws securing the valve and withdraw it from the intake manifold.

74 To check valve operation, use fused jumper wires to apply battery voltage to the valve.

➡**Note: The terminal which corresponds to the PURPLE wire on the electrical connector should get the positive (+) feed.**

When battery voltage is applied the valve should actuate. If it doesn't, replace it.

75 Before installing the valve, check the O-ring seal to make sure it's free from damage. Replace it if it's cracked or hardened.

76 To avoid breaking the mounting ears, tighten the screws evenly.

FUEL INJECTOR(S)

Check

77 On 1995 and earlier models, check for stored trouble codes in the PCM using the On Board Diagnostic (OBD) system (see Chapter 6).

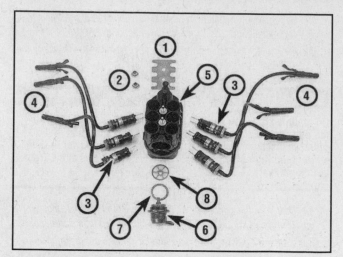

15.85 CSFI engine fuel meter body components

1	Fuel injector hold- down plate	6	Fuel pressure regulator
2	Nuts		➤Note: Retaining clip not shown
3	Fuel injectors	7	O-ring seal
4	Poppet nozzles	8	Filter screen
5	Fuel meter body		

78 Due to the complexity of this system, if a fuel injector problem is suspected, take the vehicle to a dealer service department or other qualified repair shop for further diagnosis. However, before taking the vehicle in for service, check the electrical connections related to the system to make sure they are clean and tight. Also, make sure the fuel system pressure is correct (see Section 3).

Replacement (CSFI engines only)

♦ **Refer to illustrations 15.85 and 15.86**

❈❈ WARNING:

Gasoline is extremely flammable, so take extra precautions when you work on any part of the fuel system. See the Warning in Section 2.

➤Note: On CMFI engines, the fuel injection unit is not serviceable and must be replaced as a unit if faulty.

79 Disconnect the cable from the negative terminal of the battery.

❈❈ CAUTION:

On models equipped with an anti-theft audio system, disable the anti-theft feature before performing any operation that requires disconnecting the battery or disrupting power to the stereo (see the Anti-theft audio system procedures at the front of this manual).

80 Perform the fuel pressure relief procedure (see Section 2).
81 Remove the air cleaner assembly and intake duct (see Section 8).
82 Remove the engine cover (see Chapter 11).
83 Remove the upper intake manifold (see above, this Section).
84 Remove the fuel meter body (see above, Steps 30 through 40).
85 Unscrew the nuts and remove the fuel injector hold down plate (see illustration).

15.86 Use needle nose pliers to grip the injector at the poppet nozzle line fitting and then pivot the pliers on the stud shoulder to extract the injector from the fuel meter body (CSFI engine)

86 Using needle nose pliers, applied on the poppet nozzle tube at the injector, withdraw the injector from the fuel meter body. Pivot the pliers on the stud shoulder as shown (see illustration).

87 If replacing the fuel injector, it is very important to purchase a part which is identical to the original as each injector is calibrated for a specific flow rate.

88 If you are installing a used fuel injector, obtain new O-ring seals. Lightly coat them with clean engine oil and install them onto the injector.

89 Lightly coat the O-ring seals with clean engine oil and install the injector into the fuel meter body making sure the electrical terminals are properly aligned.

90 Install the hold down plate and nuts. Tighten the fuel injector hold down plate nuts to the torque listed in this Chapter's Specifications.

91 Install the fuel meter body into the manifold (see above, this Section).

92 The remaining installation steps are the reverse of removal.

FUEL PRESSURE REGULATOR (CSFI ENGINES ONLY)

❈❈ WARNING:

Gasoline is extremely flammable, so take extra precautions when you work on any part of the fuel system. See the Warning in Section 2.

➤Note 1: On CMFI engines, the fuel pressure regulator is part of the fuel injection unit and is not serviceable. The entire unit must be replaced if faulty.

➤Note 2: Component check is part of fuel pressure check (see Section 3).

Replacement

♦ **Refer to illustration 15.99**

93 Disconnect the cable from the negative terminal of the battery.

94 Perform the fuel pressure relief procedure (see Section 2).

95 Remove the air cleaner assembly and intake duct (see Section 8).

96 Remove the engine cover (see Chapter 11).

97 Remove the upper intake manifold (see above, this Section).

98 Detach the vacuum hose from the fuel pressure regulator.

99 Remove the fuel pressure regulator retaining clip and withdraw the pressure regulator from the fuel meter body (see illustration).

100 If equipped, inspect the filter screen. If it's dirty, remove it and install a new one (see illustration 15.85).

101 Remove the O-rings and discard them, new ones should be installed at installation.

102 Coat the new O-ring seals lightly with clean engine oil.

103 Install the backup O-ring, the large O-ring, the filter screen (if equipped) and the small O-ring.

104 Install the pressure regulator onto the fuel meter body and secure with the retaining clip. Make sure the vacuum fitting is pointing down through the retaining clip.

105 Connect the vacuum hose to the pressure regulator.

106 Install the upper intake manifold (see above, this Section).

107 The remaining installation steps are the reverse of removal.

THROTTLE BODY (CSFI ENGINES ONLY)

➡Note: The throttle body on CMFI engines is an integral part of the upper intake manifold casting and cannot be separated.

Removal

▸ Refer to illustration 15.113

108 Disconnect the cable from the negative terminal of the battery.

109 Remove the air cleaner assembly and intake duct (see Section 8).

110 Disconnect the electrical connectors from the TPS and IAC valve.

111 Detach the accelerator and (if equipped) cruise control cable from the throttle body (see Section 16).

112 Remove the accelerator bracket (with cable(s) attached) and place it out of the way.

113 Remove the 3 bolts and lift the throttle body off the intake manifold (see illustration).

114 Remove the throttle body gasket and discard. A new gasket should be used at installation. Scrape all traces of old gasket material from the throttle body and intake manifold. Be careful not to gouge the sealing surfaces.

115 While it is removed, clean the throttle body using a clean

15.99 Removing the fuel pressure regulator retaining clip

15.113 CSFI engine throttle body mounting bolt locations (arrows) (upper intake manifold removed for clarity)

shop towel moistened with spray type carburetor and choke cleaner to remove all carbon deposits.

116 Clean the throttle body mounting bolt threads to remove any thread locking compound.

Installation

117 Install the new gasket and throttle body onto the intake manifold.

118 Apply a thread locking compound (Loctite 262 or equivalent) to the bolt threads.

119 Install the bolts and tighten them evenly to the torque listed in this Chapter's Specifications.

120 Install the accelerator cable bracket and attach the cable(s) (including cruise control if equipped) to the throttle body (see Section 16). Check accelerator pedal operation. Make sure the cable(s) do not bind and that the throttle valve lever is seated against its stop when the accelerator pedal is released.

121 The remaining installation steps are the reverse of removal.

THROTTLE POSITION SENSOR (TPS)

General description

122 The TPS is located on the end of the throttle shaft on the throttle body. By monitoring the output voltage from the TPS, the PCM can determine fuel delivery based on the throttle valve angle (driver

15.124a Measuring the TPS reference voltage between the GRAY (+) and BLACK (-) wires (CMFI engine shown). Note the straight pins used as probes

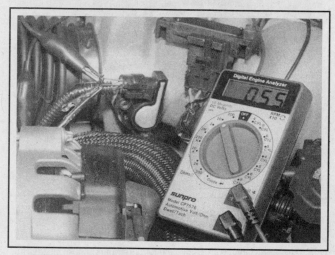

15.125a Use straight pins to backprobe the TPS electrical connector DARK BLUE (+) and BLACK (-) wires. With the throttle closed, the voltage should be approximately 0.5 volt (CMFI engine shown)

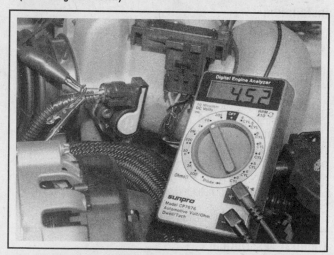

15.125b With the throttle wide open, the voltage should increase smoothly to approximately 4.5 volts (CMFI engine shown)

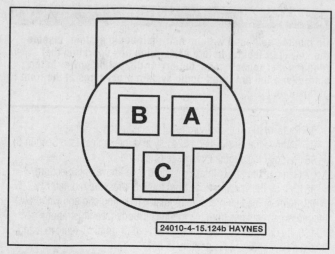

24010-4-15.124b HAYNES

15.124b On CSFI models, measure the TPS reference voltage between connectors A and B. Measure the signal between connectors B and C

A 5-volt reference   C TPS signal
B Ground

demand). A broken or loose TPS can cause intermittent bursts of fuel from the fuel injector(s) and an unstable idle because the PCM thinks the throttle is moving.

Check

▶ Refer to illustrations 15.124a, 15.124b, 15.125a and 15.125b

➡ Note: As with other sensors, an alternative to the following procedure is to use a scan tool.

123 On CSFI engines, remove the air cleaner assembly and intake duct (see Section 8) to gain access to the TPS electrical connector.

124 Locate the TPS sensor on the throttle body. Using a digital voltmeter, backprobe the TPS electrical connector gray wire (positive probe) and black wire (negative probe). (On later models, both wires are black.) Turn the ignition key ON (engine OFF). The voltmeter should indicate a reference voltage of approximately 5 volts (see illustration). Turn off the ignition.

125 Next, backprobe the dark blue wire (TPS signal) with the voltmeter positive (+) probe and the negative probe on the black wire (-).

15.128 TPS mounting screws (arrows) - CSFI engine shown, CMFI engine similar

Turn the ignition key ON (engine OFF). The voltmeter should indicate approximately 0.5 volts (see illustration). While observing the voltmeter, manually operate the throttle to the fully open position. The voltmeter should graduate smoothly to approximately 4.5 volts (see illustration). If the voltage readings are out of tolerance, replace the TPS sensor.

Replacement

▶ **Refer to illustration 15.128**

126 On CSFI engines, remove the air cleaner assembly (see Section 8) to gain access to the TPS connector.

127 Disconnect the electrical connector from the TPS sensor.
128 Remove the 2 mounting screws and withdraw the TPS from the throttle body (see illustration).
129 When installing the TPS, be sure to align the socket locating tangs on the TPS with the throttle valve shaft. The TPS may have to be rotated slightly to properly engage the shaft.
130 Before installing the mounting screws, apply a coat of thread locking compound (Loctite 262 or equivalent) to the threads. Tighten the screws securely.
131 On CSFI engines, install the air cleaner assembly (see Section 8).

16 Accelerator cable - removal and installation

▶ **Refer to illustrations 16.8, 16.9 and 16.11**

1 The accelerator control system is the cable type. There are no linkage adjustments. Because there are no adjustments, only the specific replacement part will work.
2 When work has been performed on the accelerator controls, always check to ensure that all components are installed the same way they were removed and that all linkage and cables are neither rubbing nor binding.

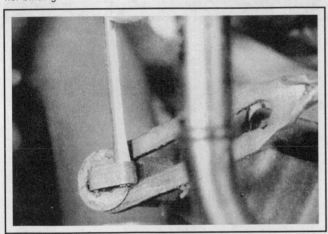

16.8 To remove the clip, pop it loose with a small screwdriver and slide the cable end off the linkage stud

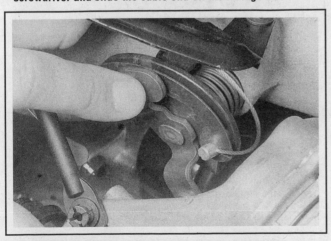

16.9 While pulling the throttle back, lift the accelerator cable from the notch on the throttle valve (CMFI engine shown)

ACCELERATOR CABLE

3 When performing service on the accelerator cable, observe the following:

a) The retainer must be installed with the tangs secured over head of the stud.
b) The conduit fitting at both ends of the cable must have its locking tangs expanded and locked in the attaching holes.
c) The braided portion of the accelerator cable must not come in contact with the front of the dash sealer during assembly, repair or replacement.
d) Flexible components (hoses, wires, conduits, etc.) must not be routed within two inches of the moving parts of the accelerator linkage outboard of the support bracket unless the routing is positively controlled.

ACCELERATOR PEDAL

4 When performing service on the accelerator pedal, observe the following:

a) The mounting surface between the support and the dash panel must be free of insulation. The carpet and padding in the pedal and tunnel area must be positioned to lay flat and be free of the wrinkles and bunches.
b) Slip the accelerator control cable through the slot in the rod and then install the retainer in the rod, being sure that it is seated. Care must be utilized in pressing the retainer into the hole in the rod to insure that the cable is not kinked or damaged in any way.
c) After all components of the accelerator linkage are secured, the linkage must operate freely without binding between closed throttle and wide open throttle.

REPLACEMENT (TYPICAL)

5 Detach the cable from the negative terminal of the battery.

❊❊ CAUTION:

On models equipped with an anti-theft audio system, disable the anti-theft feature before performing any operation that requires disconnecting the battery or disrupting power to the stereo (see the Anti-theft audio system procedures at the front of this manual).

6 On carbureted and TBI models, remove the air filter housing and adapter (see Section 8).

7   Locate the forward end of the throttle cable at the engine.

8   On carbureted and TBI models, pop the retaining clip on the end of the throttle linkage stud loose with a small screwdriver (see illustration).

9   On CMFI and CSFI models, disconnect the retainer from the throttle lever stud, then disconnect the throttle cable from the notch on the throttle valve (see illustration).

10   The cable is held in the cable bracket with a small plastic ferrule. Pinch the two locking tabs of the ferrule together and pull the cable through the hole in the bracket.

11   Inside the vehicle locate the upper end of the accelerator pedal lever. Pull the small plastic brush out of the top of the lever, push the lever forward and lift the cable free of the slot in the top of the lever (see illustration).

12   Locate the spot immediately above and forward of the upper end of the accelerator pedal level where the cable comes through the firewall. Note that the cable comes through a large rubber washer (for insulation against water). Pull this washer loose. The cable is retained at its hole in the firewall with a ferrule similar to the one used with the cable bracket. Pinch the locking tabs together and pull the cable through the firewall.

13   Installation is the reverse of removal.

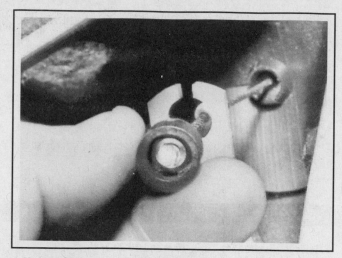

16.11 To detach the throttle cable from the accelerator pedal, pull the small plastic bushing loose from the upper end of the pedal lever then lift the cable up and out of the groove in the lever

17 Exhaust system - removal and installation

◆ **Refer to illustrations 17.3 and 17.4**

1   Detach the cable from the negative terminal of the battery.

✳✳ CAUTION:

On models equipped with an anti-theft audio system, disable the anti-theft feature before performing any operation that requires disconnecting the battery or disrupting power to the stereo (see the Anti-theft audio system procedures at the front of this manual).

2   Raise the vehicle and place it securely on jackstands.

3   Detach the exhaust manifold-to-exhaust pipe nuts (see illustration).

4   Remove the exhaust pipe hanger located in front of the muffler (see illustration) and the hanger behind the muffler.

5   Unbolt and remove the rear exhaust hanger.

6   Remove the exhaust system.

7   Installation is the reverse of removal.

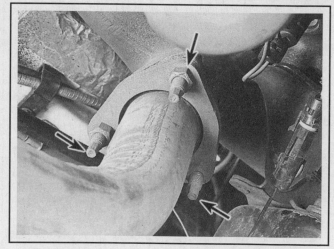

17.3 To detach the forward end of the exhaust pipe from the exhaust manifold, remove the three nuts (V6 engine shown)

17.4 To detach the forward exhaust hanger from its insulator, remove the hanger mounting bolt (arrow)

Specifications

General

Fuel pump pressure	
Carbureted system (engine running at idle)	Approximately 5 to 9 psi
Throttle Body Injection (TBI) system	9 to 13 psi
Central Multiport Fuel Injection (CMFI) system	
Key ON, engine OFF	Approximately 58 to 64 psi
Engine running (at idle)	Pressure should decrease by 3 to 10 psi
Central Sequential Fuel Injection (CSFI) system	
Key ON, engine OFF	Approximately 55 to 61 psi
Engine running (at idle)	Pressure should decrease by 3 to 10 psi
Fuel pump flow rate	
Carbureted and TBI systems (only)	1/2 pint or more in 15 seconds
Non-adjustable TPS output check	
Model 300 and 700	less than 1.25 volt
Model 220	less than 1.00 volt
IAC valve pintle protrusion	less than 1-1/8 inch
Minimum idle speed adjustment (fuel injected models)	
Model 700	650 rpm ± 25 rpm
Model 220	
Manual transmission	600 to 650 rpm
Automatic transmission	500 to 550 rpm
Injector resistance	
Model 220 TBI system	1.16 to 1.36 ohms
CMFI system	1.37 to 1.77 ohms

Torque specifications   Ft-lbs (unless otherwise indicated)

Carburetor mounting bolts	
Long bolts	84 in-lbs
Short bolts	132 in-lbs
Model 300 TBI assembly	
TBI mounting bolts	120 to 180 in-lbs
TBI mounting stud	36 to 72 in-lbs
TBI mounting nut	120 to 180 in-lbs
Model 700 TBI mounting studs	144 in-lbs
Model 220 TBI assembly	
IAC valve	156 in-lbs
Fuel inlet nut	30
Fuel outlet nut	21
Mounting bolts	144 in-lbs
CMFI and CSFI systems	
Fuel injector hold-down plate nuts (CSFI only)	27 in-lbs
Fuel pipe nuts (supply/return pipe-to-pipe)	22
Fuel pipe-to-fuel meter body nuts (CSFI only)	27 in-lbs
IAC valve bolts	27 in-lbs

CMFI and CSFI systems (continued)	
Throttle body studs (CSFI only)	
1996 through 1998	18 in-lbs
1999 and 2000	80 in-lbs
2001 and later	89 in-lbs
Upper intake manifold (air plenum) bolts/studs	
CMFI engines (aluminum manifold)	
First pass	55 in-lbs
Second pass	124 in-lbs
CSFI engine (composite manifold)	
First pass	44 in-lbs
Second pass	88 in-lbs

Section

Reference to other Chapters

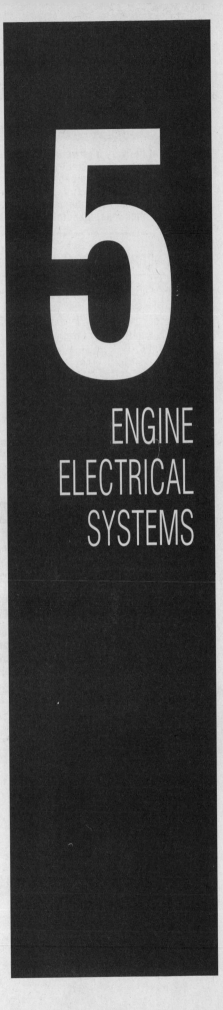

5

ENGINE ELECTRICAL SYSTEMS

1 Ignition system - general information

✳✳ WARNING:

Because of the very high voltage generated by the ignition system, extreme care should be taken whenever an operation involving ignition components is performed. This not only includes the distributor, coil, module and spark plug wires, but related items that are connected to the systems as well, such as the plug connections, tachometer and testing equipment.

HIGH ENERGY IGNITION (HEI) SYSTEM

All 1994 and earlier models are equipped with the High Energy Ignition (HEI) system. The HEI system uses a conventional distributor which houses an electronic ignition module that is triggered by distributor shaft rotation and a pick-up coil assembly (rotor/stator type). On early HEI systems the ignition coil was located on top of the distributor cap. On late models, a remote coil was incorporated which was mounted near the distributor. All spark timing changes are controlled by the Powertrain Control Module (PCM), no mechanical or vacuum advance systems are employed. Some late model HEI distributors are equipped with a Hall effect switch which is located above the pick-up coil assembly. The Hall effect switch is used by the PCM to determine engine speed.

ENHANCED DISTRIBUTOR IGNITION (EDI) SYSTEM

In 1995 the EDI system was introduced. Although the EDI system uses a rotor and distributor cap, it does not operate like a conventional system. Besides delivering the high voltage to the spark plugs, the distributor also houses the camshaft position sensor, a Hall effect device that provides the PCM with cylinder identification. This camshaft position sensor is not directly involved in the ignition process. Its sole purpose is to provide engine misfire information to the PCM. The EDI system depends on the crankshaft position sensor to provide the correct spark timing. The ignition timing is not adjustable on the EDI system.

2 Battery - removal and installation

▸ **Refer to illustration 2.3**

✳✳ WARNING:

Hydrogen gas is produced by the battery, so keep open flames and lighted cigarettes away from it at all times. Always wear eye protection when working around the battery. Rinse off spilled electrolyte immediately with large amounts of water.

✳✳ CAUTION:

On models equipped with an anti-theft audio system, disable the anti-theft feature before performing any operation that requires disconnecting the battery or disrupting power to the stereo (see the Anti-theft audio system procedures at the front of this manual).

1 Disconnect the cable from the negative terminal of the battery. Always disconnect the negative cable first and hook it up last or the battery may be shorted by the tool being used to loosen the cable fasteners.
2 Disconnect the cable from the positive terminal of the battery.
3 Remove the battery hold-down clamp (see illustration).
4 Lift out the battery. Be careful - it's heavy.

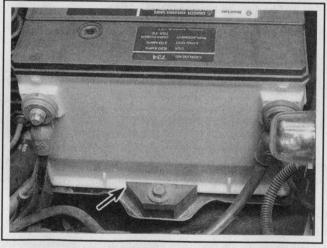

2.3 To remove the battery, disconnect the cables (negative first), then remove the hold down bolt and clamp (arrow)

5 While the battery is out, inspect the battery tray for corrosion and damage. Replace if required.
6 Installation is the reverse of removal.

3 Battery - emergency jump starting

Refer to the *Booster battery (jump) starting* procedure at the front of this manual.

4 Battery cables - check and replacement

▶ **Refer to illustration 4.1**

1   Periodically inspect the entire length of each battery cable for damage, cracked or burned insulation and corrosion (see illustration). Poor battery cable connections can cause starting problems and decreased engine performance.

2   Check the cable-to-terminal connections at the ends of the cables for cracks, loose wire strands and corrosion. The presence of white, fluffy deposits under the insulation at the cable terminal connection is a sign the cable is corroded and should be replaced. Check the terminals for distortion, missing mounting bolts or nuts and corrosion.

3   If only the positive cable is to be replaced, be sure to disconnect the negative cable from the battery first.

✳✳ CAUTION:

On models equipped with an anti-theft audio system, disable the anti-theft feature before performing any operation that requires disconnecting the battery or disrupting power to the stereo (see the Anti-theft audio system procedures at the front of this manual).

4   Disconnect and remove the cable. Make sure the replacement cable is the same length and diameter.

5   Clean the threads of the starter or ground connection with a wire brush to remove rust and corrosion. Apply a light coat of petroleum jelly to the threads to ease installation and prevent future corrosion.

6   Attach the cable to the starter or ground connection and tighten the mounting nut securely.

7   Before connecting the new cable to the battery, make sure it reaches the terminals without having to be stretched.

8   Connect the positive cable first, followed by the negative cable. Tighten the fasteners and apply a thin coat of petroleum jelly to the terminal and cable connection.

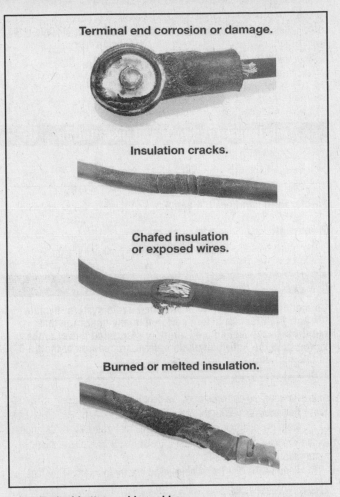

Terminal end corrosion or damage.

Insulation cracks.

Chafed insulation or exposed wires.

Burned or melted insulation.

4.1 Typical battery cable problems

5 Ignition system - check

▶ **Refer to illustration 5.3**

✳✳ WARNING:

Because of the very high voltage generated by the ignition system, extreme care should be taken whenever an operation involving ignition components is performed. This not only includes the distributor, coil, module and spark plug wires, but related items that are connected to the systems as well, such as the plug connections, tachometer and testing equipment.

1   With the ignition switch turned to the ON position, a "Service Engine Soon" light is a basic check for ignition and battery supply to the ECM.

2   Check all ignition wires and connections for tightness, cuts, corrosion or any signs of a bad connection.

3   Use a spark tester to verify adequate available secondary voltage at the spark plug (see illustration).

4   Check for carbon tracking on the coil. If carbon tracking is evident, replace the coil and be sure the high tension wire to the coil is clean and tight. Excessive wire resistance or faulty connections could

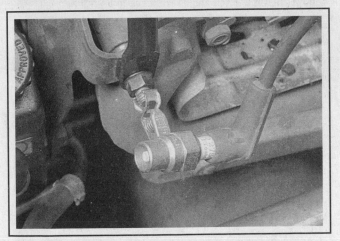

5.3 To use a calibrated ignition tester, simply disconnect a spark plug wire, attach the wire to the tester and clamp it to a good ground and crank the engine - if the secondary voltage is reaching the tester, blue sparks will be seen jumping the gap between the electrode tip and tester body

cause coil damage.

5   Using an ohmmeter, check the resistance of the coil (see Section 9). If a problem is found, replace the coil.

6   Using an ohmmeter, check the resistance of the spark plug wires, which should not exceed approximately 3,000 ohms per foot and 10,000 ohms total for any one wire.

7   If there is still no spark or a weak spark, check the ignition coil, ignition module (1994 and earlier models) and the pick-up coil or Hall-effect switch (see Sections 7 through 10).

8   Additional checks should be performed by a dealer service department or an automotive repair shop.

6   Distributor - removal and installation

FOUR-CYLINDER ENGINE

♦ Refer to illustrations 6.4 and 6.5

Removal

1   Disconnect the cable from the negative battery terminal.

❊❊ CAUTION:

On models equipped with an anti-theft audio system, disable the anti-theft feature before performing any operation that requires disconnecting the battery or disrupting power to the stereo (see the Anti-theft audio system procedures at the front of this manual).

2   Remove the coil wire from the distributor cap.
3   Remove the distributor cap.
4   Note the position of the rotor and the distributor-to-block alignment. Make an alignment mark on the distributor to indicate the position of the rotor (see illustration).
5   Remove the distributor hold-down clamp bolt and clamp (see illustration). Remove the distributor from the engine.

❊❊ CAUTION:

Do not turn the crankshaft while the distributor is removed from the engine. If the crankshaft is turned, the position of the rotor will be altered and the engine will have to be retimed.

Installation (crankshaft not turned after distributor removal)

6   Insert the distributor into the engine in exactly the same relation to the block in which it was removed. To mesh the gears, it may be necessary to turn the rotor slightly. At this point the distributor may not seat down against the block completely. This is due to the lower end of the distributor shaft not mating properly with the oil pump shaft. If this is the case, check again to make sure the distributor is aligned with the block in the same position it was in before removal and that the rotor is correctly aligned with the distributor body. The gear on the distributor shaft is engaged with the gear on the camshaft, and this relationship cannot change as long as the distributor is not lifted from the engine. Use a socket and breaker bar on the crankshaft bolt to turn the engine over in the normal direction of rotation. The rotor will turn, but the oil pump shaft will not because the two shafts are not engaged. When the proper alignment is reached the distributor will drop down over the oil pump shaft, and the distributor body will seat properly against the block.

7   Install the hold-down clamp and tighten the bolt securely.
8   Install the distributor cap and coil wire.
9   Connect the cable to the negative terminal of the battery.

Installation (crankshaft turned after distributor removal)

10 Remove the number one spark plug.
11 Place your finger over the spark plug hole while turning the crankshaft in the normal direction of rotation with a wrench on the pulley bolt at the front of the engine.

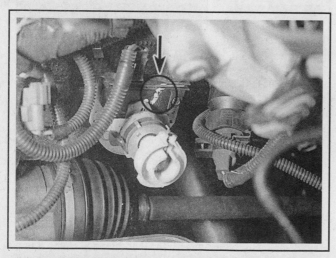

6.4   Mark the position of the rotor with respect to the distributor before removing the rotor

6.5   The distributor hold-down clamp and bolt must be removed before the distributor can be removed from the engine

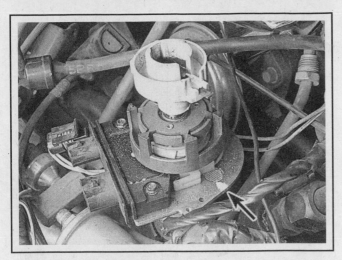

6.23 Make a mark on the distributor housing base (arrow) to show the direction the rotor is pointing before removing the distributor

6.24 Mark the position of the distributor in relation to the engine (arrow) before loosening the distributor hold-down clamp

12 When you feel compression, continue turning the crankshaft slowly until the timing mark on the crankshaft pulley is aligned with the "0" on the engine timing indicator.

13 Position the rotor to point at the number one distributor terminal.

14 Insert the distributor into the engine in exactly the same relation to the block in which it was removed. To mesh the gears, it may be necessary to turn the rotor slightly. If the distributor does not seat fully against the block it is because the oil pump shaft has not seated in the distributor shaft. Make sure the distributor drive gear is fully engaged with the camshaft gear, then use a socket on the crankshaft bolt to turn the engine over in the normal direction of rotation until the two shafts engage and the distributor seats against the block.

15 Install the hold-down clamp and tighten the bolt securely.

16 Install the distributor cap and coil wire.

17 Connect the cable to the negative terminal of the battery.

V6 ENGINE

▶ **Refer to illustrations 6.23 and 6.24**

Note: The ignition timing is not adjustable on models equipped with the EDI system.

Removal

18 Disconnect the cable from the negative terminal of the battery.

❋❋ **CAUTION:**

On models equipped with an anti-theft audio system, disable the anti-theft feature before performing any operation that requires disconnecting the battery or disrupting power to the stereo (see the Anti-theft audio system procedures at the front of this manual).

19 Remove the engine cover (see Chapter 11).

20 On carbureted and TBI engines, remove the air cleaner assembly (see Section 8).

21 Unplug the wiring harness connectors from the side of the distributor base.

22 Remove the distributor cap (see Chapter 1) and move it out of the way.

23 Scribe a mark on the distributor housing base to show the direction the rotor is pointing (see illustration).

24 Mark the position of the distributor housing in relation to the engine (see illustration).

25 Remove the distributor hold-down bolt and clamp.

26 Remove the distributor.

❋❋ **CAUTION:**

Avoid turning the crankshaft while the distributor is removed. Turning the crankshaft while the distributor is removed will change the timing position of the rotor and require retiming the engine.

Installation (crankshaft not turned after distributor removal)

27 Position the rotor in the exact location it was in when the distributor was removed.

28 Lower the distributor into the engine. To mesh the gears at the bottom of the distributor it may be necessary to turn the rotor slightly. It is possible that the distributor may not seat down fully against the block because the lower part of the distributor shaft has not properly engaged the oil pump shaft. Make sure the distributor and rotor are properly aligned with the marks made earlier, then use a large socket and breaker bar on the crankshaft bolt to turn the engine in the normal direction of rotation until the two shafts engage and the distributor drops down against the block.

29 With the base of the distributor seated against the engine block turn the distributor housing to align the marks made on the distributor base and the engine block.

30 Place the hold-down clamp in position and loosely install the holddown bolt.

31 Reconnect the ignition wiring harness.

32 Install the distributor cap.

33 Reconnect the coil connector.

34 With the distributor in its original position, tighten the hold-down bolt.

35 Check the ignition timing (see Chapter 1).

Installation (crankshaft turned after distributor removal)

36 Remove the number one spark plug.

37 Place your finger over the spark plug hole while turning the crankshaft with a wrench on the pulley bolt at the front of the engine.

38 When you feel compression, continue turning the crankshaft slowly until the timing mark on the vibration damper is aligned with the "0" on the engine timing indicator.

39 Position the rotor between the number one and six spark plug terminals on the cap.

40 Lower the distributor into the engine. To mesh the gears at the bottom of the distributor, it may be necessary to turn the rotor slightly. If the distributor does not drop down flush against the block it is because the distributor shaft has not mated to the oil pump shaft.

Place a large socket and breaker bar on the crankshaft bolt and turn the engine over in the normal direction of rotation until the two shafts engage properly, allowing the distributor to seat flush against the block.

41 With the base of the distributor properly seated against the engine block, turn the distributor housing to align the marks made on the distributor base and the engine block.

42 Place the hold-down clamp in position and loosely install the holddown bolt.

43 Reconnect the ignition wiring harness.

44 Install the distributor cap. If the secondary wiring harness was removed from the cap, reinstall it.

45 Reconnect the coil connector.

46 With the distributor in its original position, tighten the hold-down bolt and check the ignition timing.

7 Ignition module (1994 and earlier models) - check and replacement

▶ Refer to illustrations 7.4, 7.6, 7.11 and 7.15

➡ Note: It is not necessary to remove the distributor to check or replace the module.

CHECK

1 Disconnect the tachometer lead (if so equipped) at the distributor.

2 Check for a spark at the coil and spark plug wires (see Section 5).

3 If there is no spark, remove the distributor cap. Remove the ignition module from the distributor but leave the connector plugged in.

4 With the ignition switch turned On, check for voltage at the module positive terminal (see illustration).

5 If the reading is less than ten volts, there is a fault in the wire between the module positive (+) terminal and the ignition coil positive connector or the ignition coil and primary circuit-to-ignition switch.

6 If the reading is ten volts or more, check the "C" terminal on the module (see illustration).

7 If the reading is less than one volt, there is an open or grounded lead in the distributor-to-coil "C" terminal connection or ignition coil or

an open primary circuit in the coil itself.

8 If the reading is one to ten volts, replace the module with a new one and check for a spark (see Section 5). If there is a spark the module was faulty and the system is now operating properly. If there is no spark, there is a fault in the ignition coil.

9 If the reading in Step 4 is 10 volts or more, unplug the pick-up coil connector from the module. Check the "C" terminal voltage with the ignition switch On and watch the voltage reading as a test light is momentarily (five seconds or less) connected between the battery positive (+) terminal and the module "P" terminal (see illustration 7.6).

10 If there is no drop in voltage, check the module ground and, if it is good, replace the module with a new one.

11 If the voltage drops, check for spark at the coil wire as the test light is removed from the module terminal. If there is no spark, the module is faulty and should be replaced with a new one. If there is a spark, the pick-up coil or connections are faulty or not grounded (see illustration).

REPLACEMENT

12 Detach the cable from the negative terminal of the battery.

7.4 To check the module for voltage, touch the voltmeter probe to the module positive terminal

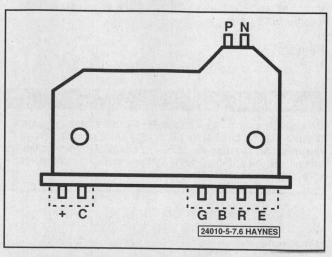

7.6 Ignition module terminal identification

7.11 As the test light is removed, check for a spark at the coil wire (arrow)

✳✳ CAUTION:

On models equipped with an anti-theft audio system, disable the anti-theft feature before performing any operation that requires disconnecting the battery or disrupting power to the stereo (see the Anti-theft audio system procedures at the front of this manual).

13 Remove the distributor cap and rotor (see Chapter 1).

14 Remove both module attaching screws and lift the module up and away from the distributor.

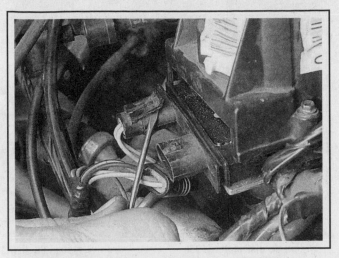

7.15 Unplug both electrical connectors from the module

15 Disconnect both electrical leads from the module (see illustration). Note that the leads cannot be interchanged.

16 Do not wipe the grease from the module or the distributor base if the same module is to be reinstalled. If a new module is to be installed, a package of silicone grease will be included with it. Wipe the distributor base and the new module clean, then apply the silicone grease on the face of the module and on the distributor base where the module seats. This grease is necessary for heat dissipation.

17 Install the module and attach both electrical leads.

18 Install the distributor rotor and cap (see Chapter 1).

19 Attach the cable to the negative terminal of the battery.

8   Ignition pick-up coil (1994 and earlier vehicles) - check and replacement

▶ **Refer to illustrations 8.4, 8.5a, 8.5b, 8.6, 8.10a, 8.10b, 8.10c and 8.11**

1   Detach the cable from the negative terminal of the battery.

8.4 Before removing the pick-up coil, unplug the lead from the ignition module

✳✳ CAUTION:

On models equipped with an anti-theft audio system, disable the anti-theft feature before performing any operation that requires disconnecting the battery or disrupting power to the stereo (see the Anti-theft audio system procedures at the front of this manual).

2   Remove the distributor cap and rotor (see Chapter 1).

3   Remove the distributor from the engine (see Section 6).

4   Detach the pick-up coil leads from the module (see illustration).

CHECK

5   Connect one lead of an ohmmeter to the terminal of the pick-up coil lead and the other to ground as shown (see illustrations). Flex the leads by hand to check for intermittent opens. The ohmmeter should indicate infinite resistance at all times. If it doesn't, the pick-up coil is defective and must be replaced.

6   Connect the ohmmeter leads to both terminals of the pick-up coil lead. Flex the leads by hand to check for intermittent opens. The ohmmeter should read one steady value between 500 and 1500 ohms as the leads are flexed by hand. If it doesn't, the pick-up coil is defective and must be replaced (see illustration).

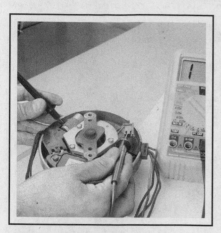

8.5a Testing the pick-up coil on a typical V6 distributor (1990 and earlier)

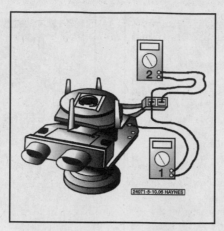

8.5b Testing the pick-up coil on a typical V6 distributor (1991 and later)

8.6 Measure the resistance of the pick-up coil. It should be between 500 and 1,500 ohms

REPLACEMENT

7   Remove the spring from the distributor shaft.

8   Mark the distributor tang drive and shaft so that they can be reassembled in the same position.

9   If the distributor is equipped with a Hall Effect switch attached to the top of the pick-up coil, remove the two retaining screws.

10 Carefully mount the distributor in a soft-jawed vise and, using a hammer and punch, remove the roll pin from the distributor shaft and gear (see illustrations). Remove the distributor shaft (see illustration).

11 To remove the pick-up coil, remove the thin "C" washer (on four-cylinder models) or the retaining clip (see illustration).

12 Lift the pick-up coil assembly straight up and remove it from the distributor. Note the order in which you remove the pieces.

13 Reassembly is the reverse of disassembly.

14 Installation is the reverse of removal.

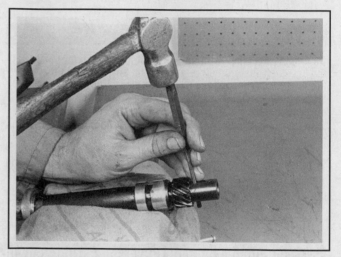

8.10a To remove the pick-up coil, mount the distributor shaft in a soft-jawed vise and, using a drift punch and hammer, knock out the roll pin

8.10b Remove the driven gear and spacer washers from the end of the shaft, making sure to note the order in which you remove any spacers

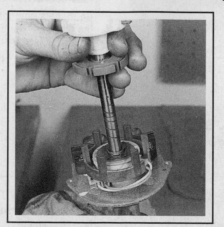

8.10c Remove the shaft from the distributor

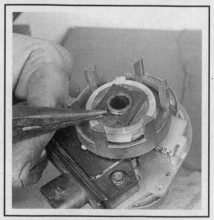

8.11 To remove the pick-up coil from a V6 engine distributor, remove the retaining clip

9 Ignition coil - removal, check and installation

HIGH ENERGY IGNITION (HEI) SYSTEMS (1994 AND EARLIER MODELS)

▶ Refer to illustrations 9.10, 9.11a, 9.11b and 9.12

1   Disconnect the cable from the negative terminal of the battery.

✳✳ CAUTION:

On models equipped with an anti-theft audio system, disable the anti-theft feature before performing any operation that requires disconnecting the battery or disrupting power to the stereo (see the Anti-theft audio system procedures at the front of this manual).

Removal

2   On models with a separately mounted coil, unplug the coil high tension wire and both electrical leads from the coil.

3   Remove both mounting nuts and remove the coil from the engine.

4   On models with the coil in the distributor cap, remove the coil cover screws and lift off the cover.

➡**Note: If you are just checking the coil, it is not necessary to remove the coil from the cap. Refer to the checking procedure below.**

5   Push the coil electrical leads through the top of the hood with a small screwdriver.

6   Remove the coil mounting screws and lift the coil, with the leads, from the cap.

Check

7   It is not necessary to remove the coil from the distributor cap on coil-in-cap models to test the coil.

8   Disconnect the negative cable from the battery.

9   Remove the distributor cap (see Chapter 1).

10   On coil in-the-cap models, position the cap upside down and, using an ohmmeter on the low scale, measure the coil primary resistance between the TACH and the BAT terminals (see illustration). The resistance should almost zero. If it's higher, replace the coil.

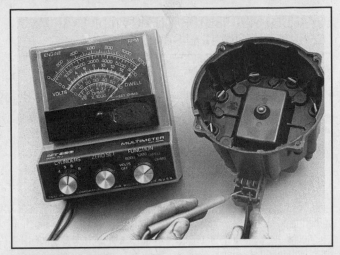

9.10 Measuring the HEI ignition coil (in cap) primary resistance, between the TACH and BAT terminals - the resistance should be nearly zero

11   On coil in-the-cap models, with the ohmmeter on the high scale, measure the coil secondary resistance between the distributor cap rotor button and the TACH terminal and note the value. Next, measure the resistance between the distributor cap rotor button and the BAT terminal (see illustrations). If both ohmmeter readings indicate infinite resistance, replace the coil.

12   On models with a separately mounted coil, check the coil for opens and grounds by performing the following three tests with an ohmmeter (see illustration).

13   Using the ohmmeter's high scale, hook up the ohmmeter leads as illustrated (see test #1 in illustration 9.12). The ohmmeter should indicate a very high, or infinite, resistance value. If it doesn't, replace the coil.

14   Using the low scale, hook up the leads as illustrated (see test #2 in illustration 9.12). The ohmmeter should indicate a very low, or zero, resistance value. If it doesn't, replace the coil.

15   Using the high scale, hook up the leads as illustrated (see test #3 in illustration 9.12). The ohmmeter should not indicate an infinite resistance. If it does, replace the coil.

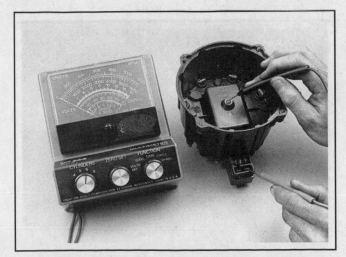

9.11a Measuring the HEI ignition coil (in cap) secondary resistance between the rotor button and the TACH terminal

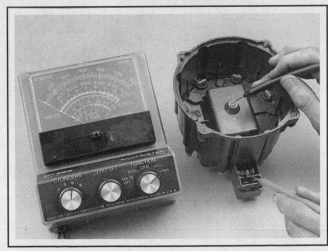

9.11b Measuring the HEI ignition coil (in cap) secondary resistance between the rotor button and the BAT terminal

Installation

16 Installation of the coil is the reverse of the removal procedure.

ENHANCED DISTRIBUTOR IGNITION (EDI) SYSTEMS (1995 AND LATER MODELS)

Check

▶ **Refer to illustration 9.21**

➡**Note: It is not necessary to remove the coil from the engine for this test.**

17 Remove the engine cover (see Chapter 11).
18 Remove the air cleaner assembly (see Chapter 4).
19 Detach the primary electrical connector from the ignition coil.
20 Using a voltmeter or test light, check to make sure the coil is receiving battery voltage. Turn the ignition key ON (engine OFF) and probe the coil primary connector. Battery voltage should be present. If it isn't check the fuses and the related wiring for a short (see Chapter 12). Turn the ignition key OFF.
21 Using an ohmmeter, measure the coil resistances (see illustration). If it fails any of the checks, replace the coil.

Removal

▶ **Refer to illustration 9.27**

22 Disconnect the cable from the negative terminal of the battery.

✳✳ CAUTION:

On models equipped with an anti-theft audio system, disable the anti-theft feature before performing any operation that requires disconnecting the battery or disrupting power to the stereo (see the Anti-theft audio system procedures at the front of this manual).

---

23 Remove the engine cover (see Chapter 11).
24 Remove the air cleaner assembly (see Chapter 4).

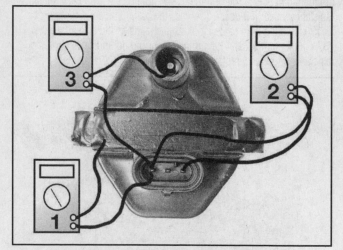

9.21 To check the EDI ignition coil, use an ohmmeter to perform the following 3 checks as shown - replace the coil if readings are different than those listed

1   The ohmmeter reading should be infinity
2   The ohmmeter reading should be approximately 0.1 ohm
3   The ohmmeter reading should be between 5,000 to 25,000 ohms

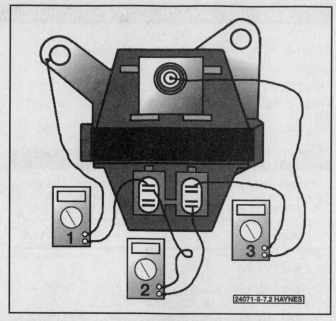

9.12 To check the HEI ignition coil, use an ohmmeter to perform the following three checks

1   On high scale, the ohmmeter should read infinity
2   On low scale, should read very low or zero
3   On high scale, should not read infinite.

25 Detach the electrical connectors from the ignition coil.
26 Disconnect the coil secondary wire leading to the distributor.
27 Unbolt the studs securing the coil to the intake manifold (see illustration).
28 If the coil is being replaced, the old coil must be removed from the bracket (the new coil does not include a bracket). Use a drill and punch to remove the rivets holding the coil to the bracket.

➡**Note: The replacement kit comes with the necessary hardware to secure the coil to the bracket.**

Installation

29 If you are installing a new coil, attach the new coil to the bracket.
30 Install the coil assembly onto the manifold and secure with the studs. Tighten the studs securely.
31 The remaining installation steps are the reverse of removal.

9.27 EDI system ignition coil (arrow) - V6 CSFI engine shown

10 Hall effect switch - check and replacement

◗ Refer to illustration 10.2

1   Some HEI distributors are equipped with a Hall effect switch which is located above the pickup coil assembly. The Hall effect switch is used in place of the R terminal of the HEI distributor to send engine RPM information to the ECM.

2   Test the switch by connecting a 12-volt power supply and voltmeter as shown (see illustration). Check the polarity markings carefully before making any connections.

3   When the feeler gauge blade is not inserted as shown, the voltmeter should read less than 0.5 volts. If the reading is more, the Hall effect switch is faulty and must be replaced by a new one.

4   With the feeler gauge blade inserted, the voltmeter should read within 0.5 volts of battery voltage. Replace the switch with a new one if the reading is more.

5   Remove the Hall effect switch by unplugging the connector and removing the retaining screws.

6   Installation is the reverse of removal.

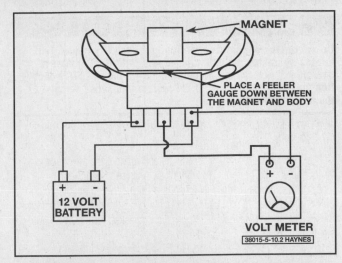

10.2 To test the Hall Effect switch, connect a 12-volt power supply and voltmeter as shown (check the polarity markings carefully before making any connections)

11 Charging system - general information and precautions

The charging system consists of a belt-driven alternator with an integral voltage regulator and the battery. These components work together to supply electrical power for the ignition system, the lights and all accessories.

There are two types of alternators used. Earlier vehicles use the SI type and later models are equipped with the CS type. There are two types of CS alternators in use, the CS-130 and the CS-144. All types use a conventional pulley and fan.

To determine which type of alternator is installed on your vehicle, look at the fasteners employed to attach the two halves of the alternator housing. All CS models use rivets instead of screws. CS alternators are rebuildable once the rivets are drilled out. However, we don't recommend this practice. For all intents and purposes, CS types should be considered non-serviceable and, if defective, should be exchanged as cores for new or rebuilt units.

The purpose of the voltage regulator is to limit the alternator's voltage to a preset value. This prevents power surges, circuit overloads, etc., during peak voltage output. On all models with which this manual is concerned, the voltage regulator is contained within the alternator housing.

The charging system does not ordinarily require periodic maintenance. The drivebelts, electrical wiring and connections should, however, be inspected at the intervals suggested in Chapter 1.

Take extreme care when making circuit connections to a vehicle equipped with an alternator and note the following. When making connections to the alternator from a battery, always match correct polarity. Before using arc welding equipment to repair any part of the vehicle, disconnect the wires from the alternator and the battery terminal. Never start the engine with a battery charger connected. Always disconnect both battery leads before using a battery charger.

The charging indicator lamp on the dash lights when the ignition switch is turned on and goes out when the engine starts. If the lamp stays on or comes on once the engine is running, a charging system problem has occurred. See Section 12 for the proper diagnosis procedure for each type of alternator.

12 Charging system - check

◗ Refer to illustration 12.5

1   If a malfunction occurs in the charging circuit, do not immediately assume that the alternator is causing the problem. First check the following items:

 a) *The battery cables where they connect to the battery. Make sure the connections are clean and tight.*

 b) *The battery electrolyte specific gravity. If it is low, charge the battery.*

 c) *Check the external alternator wiring and connections. They must be in good condition.*

 d) *Check the drivebelt condition and tension (see Chapter 1).*

 e) *Make sure the alternator mounting bolts are tight.*

 f) *Run the engine and check the alternator for abnormal noise (may be caused by a loose drive pulley, loose mounting bolts, worn or dirty bearings, defective diode or defective stator).*

2   Using a voltmeter, check the battery voltage with the engine off. It should be approximately 12 volts.

3   Start the engine and check the battery voltage again. It should now be approximately 14 to 15 volts.

4   Locate the test hole in the back of the alternator.

➡Note: If there is no test hole, your vehicle is equipped with a newer CS type alternator. Further testing of this type of alternator must be done by a dealer or automotive electrical shop.

5   Ground the tab that is located inside the hole by inserting a

screwdriver blade into the hole and touching the tab and the case at the same time (see illustration).

⁂ CAUTION:

Do not run the engine with the tab grounded any longer than necessary to obtain a voltmeter reading. If the alternator is charging, it is running unregulated during the test. This condition may overload the electrical system and cause damage to the components.

6   The reading on the voltmeter should be 15 volts or higher with the tab grounded in the test hole.

7   If the voltmeter indicates low battery voltage, the alternator is faulty and should be replaced with a new one (see Section 15).

8   If the voltage reading is 15 volts or higher and a no charge condition is present, the regulator or field circuit is the problem. Remove the alternator (see Section 13) and have it checked further by an auto electric shop.

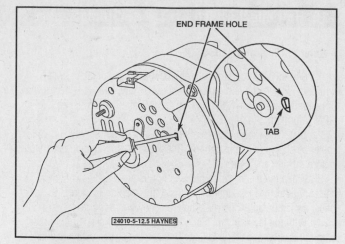

12.5   To full field an SI-type alternator, ground the tab located inside the hole by inserting a screwdriver blade into the hole and touching the tab and case at the same time

13   Alternator - removal and installation

REMOVAL

1   Disconnect the cable from the negative terminal of the battery.

⁂ CAUTION:

On models equipped with an anti-theft audio system, disable the anti-theft feature before performing any operation that requires disconnecting the battery or disrupting power to the stereo (see the Anti-theft audio system procedures at the front of this manual).

2   On CMFI and CSFI engines, remove the air cleaner assembly (see Chapter 4).

3   On carbureted and TBI engines, remove the upper fan shroud (see Chapter 3, Section 5).

4   Detach the electrical connectors from the alternator.

5   Remove the drivebelt from the alternator pulley (see Chapter 1).

6   Remove the alternator mounting bolts and braces (if used).

7   Remove the alternator from the engine.

INSTALLATION

8   Installation is the reverse of removal. On models equipped with multiple V-belts, perform the drivebelt adjustment procedure (see Chapter 1).

14   Alternator brushes - replacement

▶ **Refer to illustrations 14.2, 14.3a, 14.3b, 14.4, 14.5, 14.6, 14.7 and 14.10**

➡ **Note: The following procedure applies only to SI type alternators. CS types have riveted housings and cannot be disassembled.**

1   Remove the alternator from the vehicle (see Section 13).

2   Scribe or paint marks on the front and rear end frame housings of the alternator to facilitate reassembly (see illustration).

3   Remove the four through-bolts holding the front and rear end frames together, then separate the drive end frame from the rectifier end frame (see illustrations).

4   Remove the nuts retaining the stator leads to the rectifier bridge and separate the stator from the end frame (see illustration).

5   Remove the screw attaching the diode trio to the end frame and remove the trio (see illustration).

6   Remove the screws attaching the resistor (not used on all models), brush holder and voltage regulator to the end frame and remove the brush holder and voltage regulator (see illustration).

7   Remove the brushes from the brush holder by slipping the brush

14.2   Mark the drive end frame and rectifier end frame assemblies with a scribe or paint before separating the two halves

retainer off the brush holder (see illustration).

8   Remove the springs from the brush holder.

9   Installation is the reverse of the removal procedure, noting the following:

10   When installing the brushes in the brush holder, install the brush

14.3a Carefully separate the drive end frame and the rectifier end frame assemblies

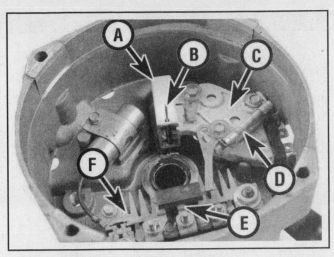

14.3b Inside the typical Si-type alternator

A	Brush holder	D	Resistor (not all models)
B	Paper clip retaining brushes	E	Diode trio
C	Regulator	F	Rectifier bridge

closest to the end frame first. Slip the paper clip through the rear of the end frame to hold the brush, then insert the second brush and push the paper clip in to hold both brushes while reassembly is completed. The paper clip should not be removed until the front and rear end frames have been bolted together (see illustration).

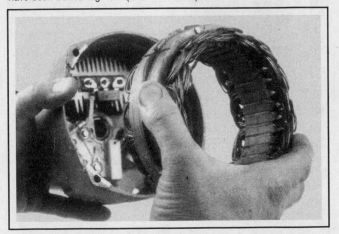

14.4 After removing the nuts retaining the stator leads to the rectifier bridge, remove the stator

14.5 Remove the screw attaching the diode trio to the end frame and remove the trio

14.6 After removing the screws that attach the brush holder, the resistor (if equipped) and voltage regulator to the end frame, remove the brush holder and voltage regulator

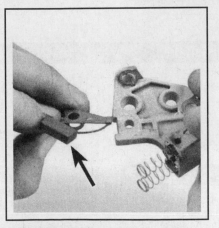

14.7 Slip the brush retainer off the brush holder and remove the brushes

14.10 To hold the brushes in place during reassembly, insert a paper clip like this through the hole in the end frame nearest the rotor shaft

15 Starting system - general information

The function of the starting system is to crank the engine. The starting system is composed of a starting motor, solenoid and battery. The battery supplies the electrical energy to the solenoid, which then completes the circuit to the starting motor, which does the actual work of cranking the engine.

The solenoid and starting motor are mounted together at the lower front side of the engine. No periodic lubrication or maintenance is required.

The electrical circuitry of the vehicle is arranged so that the starter motor can only be operated when the clutch pedal is depressed (manual transmission) or the transmission selector lever is in Park or Neutral (automatic transmission).

Never operate the starter motor for more than 15 seconds at a time without pausing to allow it to cool for at least two minutes. Excessive cranking can cause overheating, which can seriously damage the starter.

16 Starter motor - testing in vehicle

◆ **Refer to illustration 16.6**

1   If the starter motor does not turn at all when the switch is operated, make sure that the shift lever is in Neutral or Park (automatic transmission) or that the clutch pedal is depressed (manual transmission).

2   Make sure that the battery is charged and that all cables, both at the battery and starter solenoid terminals, are secure.

3   If the starter motor spins but the engine is not cranking, the overrunning clutch in the starter motor is slipping and the motor must be removed from the engine for replacement.

4   If, when the switch is actuated, the starter motor does not operate at all but the solenoid clicks, then the problem lies with either the battery, the main solenoid contacts or the starter motor itself.

➡**Note: Before diagnosing starter problems, make sure that the battery is fully charged.**

5   If the solenoid plunger cannot be heard when the switch is actuated, the solenoid itself is defective or the solenoid circuit is open.

6   To check the solenoid, connect a jumper lead between the battery (+) and the "S" terminal on the solenoid (see illustration). If the starter motor now operates, the solenoid is OK and the problem is in the ignition switch, neutral start switch or in the wiring.

7   If the starter motor still does not operate, remove the starter/solenoid assembly for disassembly, testing and repair.

8   If the starter motor cranks the engine at an abnormally slow speed, first make sure that the battery is charged and that all terminal connections are tight. If the engine is partially seized, or has the wrong viscosity oil in it, it will crank slowly.

16.6 Typical starter solenoid "S" terminal location (arrow)

9   Run the engine until normal operating temperature is reached, then disconnect the coil wire from the distributor cap and ground it on the engine.

10 Connect a voltmeter positive lead to the starter motor terminal of the solenoid and then connect the negative lead to ground.

11 Crank the engine and take the voltmeter readings as soon as a steady figure is indicated. Do not allow the starter motor to turn for more than 15 seconds at a time. A reading of 9 volts or more, with the starter motor turning at normal cranking speed, is normal. If the reading is 9 volts or more but the cranking speed is slow, the motor is faulty. If the reading is less than 9 volts and the cranking speed is slow, the solenoid contacts are probably burned.

17 Starter motor - removal and installation

REMOVAL

◆ **Refer to illustration 17.5**

1   Disconnect the cable from the negative terminal of the battery.

✳✳ CAUTION:

On models equipped with an anti-theft audio system, disable the anti-theft feature before performing any operation that requires disconnecting the battery or disrupting power to the stereo (see the Anti-theft audio system procedures at the front of this manual).

2   Raise the vehicle and support it securely on jackstands.
3   Working under the vehicle, remove any starter motor braces or shields, if used.

4   Label the electrical wires and then disconnect them from the starter motor solenoid.

5   Remove the 2 starter motor bolts (see illustration) and slowly lower the starter from the engine, noting the locations of any shims, if used.

➡**Note: Shims are used on some engines to establish the correct clearance between the starter motor pinion and the flywheel/driveplate starter ring gear (see below).**

INSTALLATION

6   Installation is the reverse of removal. Be sure to install the shims (if any) that were noted during removal. Tighten the starter mounting bolts securely.

STARTER SHIM ADJUSTMENT PROCEDURE

7   If the starter whines during or after engine starting, shimming may be necessary.

8   Disconnect the cable from the negative terminal of the battery.

✳✳ CAUTION:

On models equipped with an anti-theft audio system, disable the anti-theft feature before performing any operation that requires disconnecting the battery or disrupting power to the stereo (see the Anti-theft audio system procedures at the front of this manual).

9   Raise the vehicle and support it securely on jackstands.

10 Remove the bellhousing lower cover (see Chapter 7).

11 Inspect the flywheel or driveplate for signs of unusual wear such as chipped or missing gear teeth.

12 Using a wire type feeler gauge (round diameter), measure the clearance between the top of the flywheel/driveplate ring gear tooth and the bottom of the starter pinion tooth.

➡Note: Use a screwdriver to manually engage the starter pinion with the ring gear. The clearance should be between 0.01 to 0.06 inches.

13 If the clearance is less than 0.02 inch and the starter whined after engagement, the starter is too close to the ring gear and must be shimmed to increase the clearance.

17.5 To remove the starter from the engine, detach the electrical wires from the solenoid terminals, then remove both mounting bolts (arrows)

14 If the clearance is greater than 0.06 inches and the starter whines during engagement, the starter is too far away from the ring gear and a shim(s) must be removed. Remove a shim, tighten the starter motor and measure the clearance again.

15 To install a shim, loosen the inner starter mounting bolt (the one closest to the engine), remove the outer bolt and then slide the shim between the starter and the engine without removing the starter.

18   Starter solenoid - removal and installation

▶ **Refer to illustration 18.5**

1   Disconnect the cable from the negative terminal of the battery.

✳✳ CAUTION:

On models equipped with an anti-theft audio system, disable the anti-theft feature before performing any operation that requires disconnecting the battery or disrupting power to the stereo (see the Anti-theft audio system procedures at the front of this manual).

2   Remove the starter motor (see Section 17).

REMOVAL

3   Disconnect the strap from the solenoid to the starter motor terminal.

4   Remove the two screws which secure the solenoid to the starter motor.

5   Twist the solenoid in a clockwise direction to disengage the flange from the starter body (see illustration).

INSTALLATION

6   To install, first make sure the return spring is in position on the

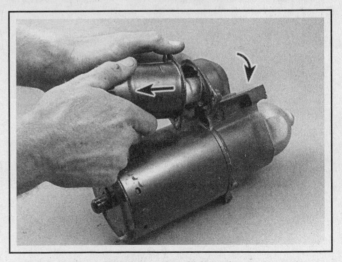

18.5 To remove the starter solenoid from the starter motor, remove the solenoid-to-starter frame mounting screws, then turn the solenoid in a clockwise direction and pull

plunger, then insert the solenoid body into the starter housing and turn the solenoid counterclockwise to engage the flange.

7   Install the two solenoid screws and connect the motor strap.

Specifications

General

Cylinder numbers	See Chapter 2
Firing order	See Chapter 2
Distributor rotation	clockwise

Torque specifications   Ft-lbs

Alternator mounting bolts	
Four-cylinder engine	
Upper bolt	20
Lower bolt.	37
V6 engine	
1985 through 1995	
Upper bolt	18
Lower bolt	35
1996 through 1998	
Front mounting bolts	37
Rear mounting bolt	18
1998 and later	37

Section

Reference to other Chapters

SERVICE ENGINE SOON light - See Section 2

6

EMISSIONS AND ENGINE CONTROL SYSTEMS

1 General information

▶ **Refer to illustration 1.6**

To prevent pollution of the atmosphere from incompletely burned and evaporating gases, and to maintain good driveability and fuel economy, a number of emission control devices are incorporated.

They include the:

Electronic Spark Timing (EST)
Electronic Spark Control (ESC) system
Exhaust Gas Recirculation (EGR) system
Evaporative Emission Control System (EECS)
Positive Crankcase Ventilation (PCV) system
Transmission Converter Clutch (TCC)
Catalytic converter
Air Injection Reaction (AIR) (carbureted V6)
Air management system (V6 with fuel injection)
Thermostatic air cleaner

All of these systems are linked, directly or indirectly, to the SERVICE ENGINE SOON light (see Section 2).

The Sections in this Chapter include general descriptions, checking procedures within the scope of the home mechanic and component replacement procedures (when possible) for each of the systems listed above.

Before assuming that an emissions control system is malfunctioning, check the fuel and ignition systems carefully. The diagnosis of some emission control devices requires specialized tools, equipment and training. If checking and servicing become too difficult or if a procedure is beyond the scope of your skills, consult your dealer service department.

This doesn't mean, however, that emission control systems are particularly difficult to maintain and repair. You can quickly and easily perform many checks and do most (if not all) of the regular maintenance at home with common tune-up and hand tools.

➡**Note: The most frequent cause of emissions problems is simply a loose or broken vacuum hose or wiring connection, so always check the hose and wiring connections first.**

Pay close attention to any special precautions outlined in this Chapter. It should be noted that the illustrations of the various systems may not exactly match the system installed on your vehicle because of

1.6 The Vehicle Emission Control Information (VECI) label is located on the fan shroud and contains information on idle speed adjustment, ignition timing, location of emission devices on your vehicle, vacuum line routing, etc.

changes made by the manufacturer during production or from year-to-year.

A Vehicle Emissions Control Information label is located in the engine compartment (see illustration). This label contains important emissions specifications and ignition timing procedures, as well as a vacuum hose schematic and emissions components identification guide. When servicing the engine or emissions systems, the VECI label in your particular vehicle should always be checked for up-to-date information.

✳✳ CAUTION:

On models equipped with an anti-theft audio system, disable the anti-theft feature before performing any operation that requires disconnecting the battery or disrupting power to the stereo (see the Anti-theft audio system procedures at the front of this manual).

2 On Board Diagnostic (OBD) system and trouble codes

DIAGNOSTIC TOOL INFORMATION

▶ **Refer to illustrations 2.1, 2.2 and 2.4**

1   A digital multimeter is necessary for checking fuel injection and emission related components (see illustration). A digital meter is preferred over the older style analog meter for several reasons. The analog multimeter cannot display the volt-ohms or amps measurement in hundredths and thousandths increments. When working with electronic circuits which are often very low voltage, this accurate reading is most important. Another good reason for the digital multi-meter is the high impedance circuit. The digital multimeter is equipped with a high resistance internal circuitry (10 million ohms). Because a voltmeter is hooked up in parallel with the circuit when testing, it is vital that none of the voltage being measured should be allowed to travel the parallel

path set up by the meter itself. This dilemma does not show itself when measuring larger amounts of voltage (9 to 12 volt circuits) but if you are measuring a low voltage circuit such as the oxygen sensor signal voltage, a fraction of a volt may be a significant amount when diagnosing a problem. Obtaining the diagnostic trouble codes is one exception where using an analog voltmeter is necessary.

2   Hand-held scanners are the most powerful and versatile tools for analyzing engine management systems used on later model models (see illustration). Each brand scan tool must be examined carefully to match the year, make and model of the vehicle you are working on. Often interchangeable cartridges are available to access the particular manufacturer (Chrysler, Ford, GM, etc.). Some manufacturers will even specify by continent (Asia, Europe, USA, etc.).

3   With the arrival of the Federally mandated emission control system (OBD-II), a specially designed scanner has been developed.

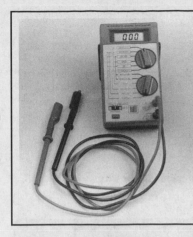

2.1 Digital multimeters can be used for testing all types of circuits; because of their high impedance, they are much more accurate than analog meters for measuring millivoltage-voltage in computer circuits

Aftermarket manufacturers have released OBD-II scan tools for the home mechanic. Ask the parts salesman at a local auto parts store for additional information.

➡ Note: Although OBD II codes cannot be accessed on 1996 models without a Scan tool, follow the simple component checks in Section 4.

4 Another type of code reader is available at auto parts stores (see illustration). These tools simplify the procedure for extracting codes from the engine management computer by simply "plugging-in" to the diagnostic connector on the vehicle wiring harness.

OBD SYSTEM GENERAL DESCRIPTION

5 The OBD system consists of an on-board computer, known as the Powertrain Control Module (PCM) and information sensors which monitor various functions of the engine and then relay the data to the PCM. Based on the data received and the information programmed into the computer's memory, the PCM then generates output signals to control various engine functions via control relays, solenoids and other output actuators.

6 The PCM is the "brain" of the OBD system. The PCM is specifically calibrated to reduce emissions, optimize fuel economy and driveability of the vehicle. There are two types of PCM's used. Models

equipped with a 2.5L four cylinder engine are equipped with a PCM which has a MEM-CAL (Memory and Calibration) unit. Models equipped with 4.3L V6 engines are equipped with a PCM which has a PROM and a CALPAK.

7 Because of a federally mandated extended warranty which covers the OBD system components and because any owner-induced damage to the PCM, the sensors and/or the control devices may void the warranty, it is not recommended to attempt diagnosis or replacement of the PCM at home while the vehicle is under warranty. Take the vehicle to your local dealer service department if the PCM or a system component malfunctions.

INFORMATION SENSORS

8 The following is a list of the OBD system information sensors. For complete information and service procedures, refer to Section 3 (unless otherwise specified).

Camshaft Position sensor (1995 and later models)
Crankshaft Position sensor (1995 and later models)
Engine Coolant Temperature (ECT) sensor
Intake Air Temperature (IAT) sensor
Knock sensor
Fuel tank pressure sensor (see Section 10)
Manifold Absolute Pressure (MAP) sensor
Mass Air Flow (MAF) sensor (1996 and later models)
Oxygen sensors
PARK/NEUTRAL switch (see Chapter 7B)
Power steering pressure switch
Throttle Position Sensor (TPS)
Vehicle Speed Sensor (VSS)

Output actuators

9 The following is a list of the OBD system output actuators. For complete information and service procedures, refer to the Chapter or Section as specified.

Air conditioning clutch relay (see Section 3)
Canister purge control solenoid (see Section 7)
EGR solenoid (see Section 9)
EGR valve (see Section 9)
Fuel injector(s) (see Chapter 4)
Fuel pump relay (see Chapter 4)
Idle Air Control (IAC) valve (see Chapter 4)
Ignition Control Module (see Chapter 5)
Transmission Converter Clutch (see Chapter 7B)

2.2 Scanners like these from Actron and AutoXray are powerful diagnostic aids - they can tell you just about anything you want to know about your engine management system

2.4 Code readers simplify the task of extracting the trouble codes but do not perform any diagnostics

SERVICE ENGINE SOON LIGHT OR MALFUNCTION INDICATOR LIGHT (MIL)

General description

10 The SERVICE ENGINE SOON light or Malfunction Indicator Light (MIL), is located in the instrument panel and should illuminate for three seconds as a bulb test each time the engine is started. When the Powertrain Control Module (PCM) detects a fault in the emissions or engine control system it sets a trouble code in the PCM's memory. If the PCM detects a fault, it illuminates the SERVICE ENGINE SOON light and logs the source of the problem in its memory as a trouble code. On OBD-II models, in the event the PCM detects an active engine misfire, the SERVICE ENGINE SOON light will flash continuously. If this occurs, turn off the engine as soon as possible and diagnose/correct the problem immediately or severe catalytic converter damage may occur.

11 On 1996 and later models (OBD-II system), the EVAP system (see Section 10) is equipped with a fuel tank pressure sensor which detect changes in the fuel tank pressure and vacuum. If abnormal vacuum or pressure is sensed by the fuel tank pressure sensor, which indicates a fuel vapor leak, the PCM will store the appropriate fault code and illuminate the SERVICE ENGINE SOON light on the instrument panel. The most common cause of SERVICE ENGINE SOON light illumination is EVAP system pressure loss due to a loose or poor sealing gas cap. Before accessing the trouble codes and trying to determine the faulty component, make sure your gas cap seal is free from defects and is tightened securely.

12 On 1995 and earlier models, in addition to notifying the driver when an emissions fault has occurred, the SERVICE ENGINE SOON light can be used to display the stored trouble codes from the PCM's memory (see below).

OBTAINING TROUBLE CODES

1995 and earlier models

▶ **Refer to illustration 2.14**

13 Before obtaining the trouble codes, start the engine and let it reach operating temperature. Turn off all electrical loads and accessories; lights, air conditioning, heater, blower fan, radio etc., and turn off the engine.

14 To retrieve the trouble codes from the PCM memory, you must use a short jumper wire to ground the diagnostic terminal. This terminal is part of a wiring connector known as the Assembly Line Data Link (ALDL) (see illustration). The ALDL is located underneath the dashboard on the left hand side and sometimes has a cover with the words DIAGNOSTIC CONNECTOR embossed on it. To remove the cover, simply pull it off.

15 Install one end of the jumper wire into the ALDL terminal "A" and the other end into terminal "B."

➡ **Note: A paper clip makes a good jumper wire.**

16 Turn the ignition key ON (engine OFF).

> ※※ **CAUTION:**
>
> **Do not start the engine! The SERVICE ENGINE SOON light should flash trouble code 12, indicating that the system is operating. Trouble code 12 is simply 1 flash, followed by a brief pause, then 2 flashes in quick succession. This code will be flashed three times before displaying any stored trouble codes. If no trouble codes have been stored, code 12 will continue to repeat until the jumper wire is removed.**

17 After flashing code 12 three times, the PCM will display any stored trouble codes. Each code will be flashed three times. After all trouble codes have been shown, code 12 will be flashed again indicating that the display of all trouble codes has been completed.

18 Once the code(s) have been noted, use the trouble code identification which follows in this Section to locate the source of the fault. Also included are simplified troubleshooting procedures. If the problem persists after these checks have been made, more detailed diagnosis will have to be performed by a dealer service department or other qualified repair shop.

19 After you have made a repair, be sure to clear the trouble codes in the PCM by disconnecting the cable from the negative terminal of the battery or removing the PCM-B fuse for at least 30 seconds.

> ※※ **CAUTION:**
>
> **On models equipped with an anti-theft audio system, disable the anti-theft feature before performing any operation that requires disconnecting the battery or disrupting power to the stereo (see the Anti-theft audio system procedures at the front of this manual).**

To confirm the repair was successful, start the engine and take the vehicle for a test drive, then check the trouble codes again.

1996 and later models

20 In 1996 the OBD-II system was introduced to comply with Federal laws that required standardization between vehicle manufacturers with respect to engine electrical system diagnostics. The heart of the system is still the Powertrain Control Module (PCM) and, like the OBD system before it (1995 and earlier models), uses information from many sensors, analyzes them and then compensates accordingly via various output actuators. If the PCM detects a system fault, it will illuminate the SERVICE ENGINE SOON light and store a trouble code in its memory (see above for more information).

21 The OBD-II system however will not flash the trouble codes on the SERVICE ENGINE SOON light as in the previous OBD system. A special computer diagnostic SCAN tool must be used to communicate

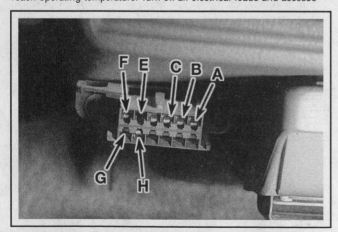

2.14 On 1995 and earlier models, the Assembly Line Data Link (ALDL) is located under the dash, on the left side

A	Ground	F	TCC (if used)
B	Diagnostic Terminal	G	Fuel pump (CK)
C	AIR (if used)	H	Brake sense speed input
E	Serial data		(CK)

with the system. If the SERVICE ENGINE SOON light comes on while you're driving the vehicle, take it to your local dealer or other qualified repair shop as soon as possible. If you have access to a SCAN tool, a trouble code chart is provided to identify the codes extracted when using the tool.

TROUBLE CODE IDENTIFICATION (1995 AND EARLIER MODELS)

22 The following is a list of typical Trouble Codes which may be encountered while diagnosing the OBD system. If the probable cause listed is followed by an asterisk (*), that means that component replacement may not cure the problem in all cases. For this reason you may want to seek professional advice before purchasing expensive replacement parts.

23 There are additional trouble codes that relate only to models equipped with the 4L60-E electronically controlled automatic transmission. They are as follows; 37, 38, 39 and 58 thru 87. If these codes are displayed, take your vehicle to a dealer service department or other qualified transmission repair shop for further diagnosis and/or repair.

TROUBLE CODE CHART FOR 1995 AND EARLIER MODELS (OBD-I)

Trouble codes	Circuit or system	Probable cause
Code 12 (1 flash, pause, 2 flashes)	Diagnostic	This code will flash whenever the diagnostic terminal is grounded with the ignition turned On and the engine not running. If additional trouble codes are stored in the PCM they will appear after this code has flashed three times. If this code appears while the engine is running, no reference pulses from the distributor are reaching the PCM.
Code 13 (1 flash, pause, 3 flashes)	Oxygen sensor circuit	Check for a sticking or misadjusted throttle position sensor. Check the wiring and connectors from the oxygen sensor. Replace the oxygen sensor.
Code 14 (1 flash, pause, 4 flashes)	Coolant sensor/high temp	If the engine is experiencing overheating problems the problem must be rectified before continuing. Check all wiring and connectors associated with the coolant temperature sensor. Replace the coolant temperature sensor.*
Code 15 (1 flash, pause, 5 flashes)	Coolant sensor/low temp	See above, then check the wiring connections at the PCM.
Code 16 (1 flash, pause, 6 flashes)	Vehicle Speed Sensor buffer fault	Check the PCM and VSS for proper operation (see Section 4).
Code 21 (2 flashes, pause, 1 flash)	Throttle position sensor/voltage high	Check for a sticking or misadjusted TPS plunger. Check all wiring and connections between the TPS and the PCM. Adjust or replace the TPS (see Chapter 4).*
Code 22 (2 flashes, pause, 2 flashes)	Throttle position sensor/voltage low	Check the TPS adjustment (Chapter 4). Check the PCM connector. Replace the TPS (Chapter 4).*
Code 23 (carbureted) (2 flashes, pause, 3 flashes)	Mixture control solenoid	The mixture control solenoid is open or grounded.
Code 23 (fuel injected) (2 flashes, pause, 3 flashes)	IAT low temp indication	Sets if the air temperature sensor, connections or wires are open for 3 seconds.
Code 24 (2 flashes, pause, 4 flashes)	Vehicle speed sensor	A fault in this circuit should be indicated only when the vehicle is in motion. Disregard Code 24 if it is set when the drive wheels are not turning. Check the connections at the PCM. Check the TPS setting.
Code 25 (2 flashes, pause, 5 flashes)	IAT high temperature indication	High temperature indication. Sets if the sensor or signal line becomes grounded for 3 seconds.
Code 32 (carbureted) (3 flashes, pause, 2 flashes)	BARO circuit low	Barometric pressure sensor circuit low.

TROUBLE CODE CHART FOR 1995 AND EARLIER MODELS (OBD-I) (CONTINUED)

Trouble codes	Circuit or system	Probable cause
Code 32 (fuel injected) (3 flashes, pause, 2 flashes)	EGR	Vacuum switch shorted to ground on start-up, switch not closed after the PCM has commanded the EGR for a specified period of time or the EGR solenoid circuit is open for specified period of time. Replace the EGR valve.*
Code 33 (3 flashes, pause, 3 flashes)	MAP sensor	Check the vacuum hoses from the MAP sensor. Check the electrical connections at the PCM. Replace the MAP sensor.*
Code 34 (3 flashes, pause, 4 flashes)	Vacuum sensor or MAP sensor	Code 34 will set when the signal voltage from the MAP sensor is too low. Instead the PCM will substitute a fixed MAP value and use the TPS to control fuel delivery. Replace the MAP sensor.*
Code 35 (carbureted) (3 flashes, pause, 5 flashes)	ISC valve	Idle Speed Control error. Replace the ISC.*
Code 35 (fuel injected) (3 flashes, pause, 5 flashes)	IAC valve	Idle Air Control error. Code will set when closed throttle speed is 50 rpm above or below the correct idle speed for 30 seconds. Replace the IAC.*
Code 41 (4 flashes, pause, 1 flash)	EST circuit or CMP sensor	No distributor reference pulses to the PCM at specified engine vacuum.
Code 42 (4 flashes, pause, 2 flashes)	Electronic Spark Timing or Ignition Control Module	Electronic Spark Timing bypass circuit or EST circuit is grounded or open. A malfunctioning ignition module can cause this code.
Code 43 (4 flashes, pause, 3 flashes)	Electronic spark control unit	The ESC retard signal has been on for too long or the system has failed a functional check.
Code 44 (4 flashes, pause, 4 flashes)	Lean exhaust	Check the PCM wiring connections. Check for vacuum leakage at the TBI base gasket, vacuum hoses or the intake manifold gasket. Replace the oxygen sensor.*
Code 45 (4 flashes, pause, 5 flashes)	Rich exhaust	Check the evaporative charcoal canister and its components for the presence of fuel. Replace the oxygen sensor.*
Code 51 (5 flashes, pause, 1 flash)	PROM or MEM-CAL	Make sure that the PROM or MEM-CAL is properly installed in the PCM. Replace the PROM or MEM-CAL.*
Code 52 (5 flashes, pause, 2 flashes)	CALPAK	Check the CALPAK to insure proper installation. Replace the CALPAK.*
Code 53 (5 flashes, pause, 3 flashes)	System voltage	Code 53 will set if the voltage at the PCM is greater than 17.1 volts or less than 10 volts. Check the charging system (see Chapter 5).
Code 54 (5 flashes, pause, 4 flashes)	Fuel pump	Low fuel pump voltage. Sets when the fuel pump voltage is less than 2 volts when reference pulses are being received.
Code 55 (5 flashes, pause, 5 flashes)	PCM	Be sure that the PCM ground connections are tight. If they are, replace the PCM.*

* Component replacement may not cure the problem in all cases. For this reason, you may want to seek professional advice before purchasing replacement parts.

TROUBLE CODE CHART FOR 1996 AND LATER MODELS (OBD-II)*

Code	Code Definition
P0016	Camshaft position (CMP) or crankshaft position (CKP) sensor signals not present or unsynchronized
P0053	HO2 sensor heater resistance (bank 1, sensor 1)
P0054	HO2 sensor heater resistance (bank 1, sensor 2)
P0059	HO2 sensor heater resistance (bank 2, sensor 1)
P0100	Mass Air Flow (MAF) insufficient activity
P0101	Mass Air Flow (MAF) sensor range error
P0102	Mass Air Flow (MAF) sensor circuit low input
P0103	Mass Air Flow (MAF) sensor circuit high input
P0106	Manifold absolute pressure (MAP) sensor range error
P0107	Manifold absolute pressure (MAP) sensor circuit low input
P0108	Manifold absolute pressure (MAP) sensor circuit high input
P0112	Intake air temperature (IAT) sensor signal circuit voltage low
P0113	Intake air temperature (IAT) sensor signal circuit voltage high
P0116	Engine coolant temperature (ECT) and intake air temperature (IAT) difference range error at restart
P0117	Engine Coolant Temperature (ECT) sensor circuit low input
P0118	Engine Coolant Temperature (ECT) sensor circuit high input
P0121	Throttle Position Sensor (TPS) range error
P0122	Throttle Position Sensor (TPS) circuit low input
P0123	Throttle Position Sensor (TPS) high input
P0125	Engine Coolant Temperature (ECT) insufficient for closed loop
P0128	Engine coolant temperature (ECT) range error
P0131	O2/HO2 sensor signal low voltage (bank 1, sensor 1)
P0132	O2/HO2 sensor signal high voltage (bank 1, sensor 1)
P0133	O2/HO2 sensor slow response (bank 1, sensor 1)
P0134	O2/HO2 sensor insufficient or no activity detected (bank 1, sensor 1)
P0135	O2/HO2 sensor heater circuit malfunction (bank 1, sensor 1)
P0137	HO2 sensor voltage low (bank 1, sensor 2)
P0138	HO2 sensor voltage high (bank 1, sensor 2)
P0140	O2/HO2 circuit no activity detected (bank 1, sensor 2)
P0141	O2/HO2 sensor heater circuit malfunction
P0143	HO2S circuit low voltage (bank 1, sensor 3)
P0144	HO2S circuit high voltage (bank 1, sensor 3)
P0146	HO2S circuit insufficient activity, bank 1
P0147	HO2S heater circuit (bank 1, sensor 3)
P0151	O2/HO2 sensor signal low voltage (bank 2, sensor 1)

TROUBLE CODE CHART FOR 1996 AND LATER MODELS (OBD-II)* (CONTINUED)

Code	Code Definition
P0152	O2/HO2 sensor high voltage (bank 2, sensor 1)
P0153	O2/HO2 sensor slow response (bank 2, sensor 1)
P0154	O2/HO2 sensor no activity detected (bank 2, sensor 1)
P0155	O2/HO2 sensor heater circuit malfunction (bank 2, sensor 1)
P0171	Fuel trim system lean
P0172	Fuel trim system rich, bank 1
P0174	Fuel trim system lean, bank 2
P0175	Fuel trim system rich, bank 2
P0200	Injector circuit malfunction
P0218	Transaxle fluid over temperature
P0230	Fuel pump primary circuit malfunction
P0300	Engine misfire detected
P0301	Cylinder number 1 misfire detected
P0302	Cylinder number 2 misfire detected
P0303	Cylinder number 3 misfire detected
P0304	Cylinder number 4 misfire detected
P0305	Cylinder number 5 misfire detected
P0306	Cylinder number 6 misfire detected
P0307	Cylinder number 7 misfire detected
P0308	Cylinder number 8 misfire detected
P0315	Crankshaft position (CKP) sensor performance problem
P0325	Knock sensor (KS) circuit malfunction
P0327	Knock sensor (KS) range error or no activity detected
P0335	Crankshaft position (CKP) sensor - no activity detected
P0336	Crankshaft position (CKP) sensor circuit range error
P0337	Crankshaft position (CKP) sensor circuit low frequency
P0338	Crankshaft position (CKP) sensor circuit high frequency
P0339	Crankshaft position (CKP) sensor circuit intermittent
P0340	Camshaft position (CMP) sensor circuit malfunction
P0341	Camshaft position (CMP) sensor system - no activity detected
P0351	Ignition control module (ICM) output signal range error
P0401	EGR circuit malfunction
P0410	AIR system fault
P0420	Catalyst system efficiency below threshold
P0430	Catalyst system efficiency below threshold
P0441	Evaporative (EVAP) emission control system incorrect purge flow

Code	Code Definition
P0442	Evaporative (EVAP) emission control system, small leak detected
P0443	Evaporative (EVAP) emission control system, purge control circuit malfunction
P0446	Evaporative (EVAP) emission control system, restricted or blocked vent path
P0449	Evaporative (EVAP) emission control system, vent control circuit malfunction
P0451	Fuel Tank Pressure (FTP) sensor performance problem
P0452	Evaporative (EVAP) emission control system, pressure sensor low input
P0453	Evaporative (EVAP) emission control system, pressure sensor high input
P0454	Fuel Tank Pressure (FTP) sensor circuit intermittent problem
P0455	Evaporative (EVAP) emission control system, large leak detected
P0461	Fuel level sensor circuit malfunction
P0462	Fuel level sensor circuit, low input
P0463	Fuel level sensor circuit, high input
P0464	Fuel level sensor circuit intermittent problem
P0496	Evaporative (EVAP) emission control system, excess manifold vacuum
P0500	Vehicle speed sensor (VSS) or circuit malfunction
P0506	Idle air control (IAC) system rpm low
P0507	Idle air control (IAC) system rpm high
P0502	Vehicle speed sensor (VSS) circuit low output
P0503	Vehicle speed sensor (VSS) circuit malfunction
P0562	System voltage low
P0563	System voltage high
P0601	Internal microprocessor control error within PCM
P0602	PCM programming error
P0603-P0607	Internal microprocessor control error within PCM
P0608	Vehicle speed sensor (VSS) circuit range error
P0609	Vehicle speed sensor (VSS) circuit low output
P0641	5-volt reference voltage range error - to throttle position (TP) or manifold absolute pressure (MAP) sensor
P0650	Malfunction indicator light (MIL) control circuit
P0651	5-volt reference voltage range error - to fuel tank pressure (FTP) sensor
P0706	Transaxle range (TR) switch circuit malfunction
P0711	Transaxle fluid temperature (TFT) sensor circuit range error
P0712	Transaxle fluid temperature (TFT) sensor circuit low input
P0713	Transaxle fluid temperature (TFT) sensor circuit high input
P0719	Torque converter clutch (TCC) brake switch circuit low input
P0724	Torque converter clutch (TCC) brake switch circuit high input
P0740	Torque converter clutch (TCC) solenoid valve circuit malfunction
P0741	Torque converter clutch (TCC) stuck off

TROUBLE CODE CHART FOR 1996 AND LATER MODELS (OBD-II)* (CONTINUED)

Code	Code Definition
P0742	Torque converter clutch (TCC) stuck on
P074	Transaxle pressure control (PC) solenoid valve circuit malfunction
P0751	1-2 shift solenoid performance (2-2-3-3 shift pattern)
P0752	1-2 shift solenoid performance (1-1-4-4 shift pattern)
P0753	1-2 shift solenoid circuit malfunction
P0756	2-3 shift solenoid performance (4-3-3-4 shift pattern)
P0757	2-3 shift solenoid performance (1-2-2-1 shift pattern)
P0758	2-3 shift solenoid circuit malfunction
P0785	3-2 shift solenoid circuit malfunction
P0894	Torque converter clutch (TCC) stuck off
P1106	Manifold absolute pressure (MAP) sensor circuit intermittent high voltage
P1107	Manifold absolute pressure (MAP) sensor circuit intermittent low voltage
P1111	Intake air temperature (IAT) sensor circuit intermittent high voltage
P1112	Intake air temperature (IAT) sensor circuit intermittent low voltage
P1114	Engine coolant temperature (ECT) sensor circuit low input
P1115	Engine coolant temperature (ECT) sensor circuit high input
P1121	Throttle position (TP) sensor circuit intermittent high voltage
P1122	Throttle position (TP) sensor circuit intermittent low voltage
P1133	Heated oxygen sensor (HO2S) circuit insufficient switching (upstream sensor)
P1134	HO2 sensor transition time ratio (bank 1, sensor 1)
P1153	Heated oxygen sensor (HO2S) circuit insufficient switching (upstream sensor)
P1154	HO2 sensor transition time ratio (bank 2, sensor 1)
P1258	Engine coolant over-temperature protection mode active
P1380	Rough road sensing error due to antilock braking system (ABS) system malfunction
P1381	Rough road sensing error due to antilock braking system (ABS) system malfunction and serial data malfunction
P1600	Internal microprocessor control error within PCM
P1621	Internal microprocessor control error within PCM
P1626	Vehicle theft deterrent (VTD) controller serial data communication failure with PCM (no password)
P1627	Internal microprocessor control error within PCM
P1631	Vehicle theft deterrent (VTD) password incorrect
P1637	Alternator turn-on signal circuit malfunction
P1638	Alternator-field duty-cycle signal circuit malfunction
P1680	Internal microprocessor control error within PCM
P1681	Internal microprocessor control error within PCM
P1683	Internal microprocessor control error within PCM
P1810	Transaxle fluid pressure (TFP) manual valve position switch circuit malfunction

Code	Code Definition
P1860	Torque converter clutch pulse width modulation (TCC PWM) solenoid valve circuit malfunction
P2610	Internal microprocessor control error within PCM
P2761	Torque Converter Clutch (TCC) Pressure Control (PC) solenoid control circuit
P2A01	HO2 sensor performance (bank 1, sensor 2)
UXXXX	Codes beginning with "U" indicate that the PCM is unable to communicate with another module

Not all codes apply to all models.

3 Powertrain Control Module (PCM)/Programmable Read Only Memory (PROM)/CALPAK/MEM-CAL

❊❊ CAUTION:

The ignition key must always be in the OFF position whenever you connect or disconnect the electrical connections on the PCM or internal damage may occur. The PCM is an Electro-Static Discharge (ESD) sensitive electronic device, meaning a static electricity discharge from your body could possibly damage internal electrical components. Make sure to properly ground yourself and the PCM before handling it. Avoid touching the electrical terminals of the PCM unless absolutely necessary.

➡ Note 1: Because of a federally mandated extended warranty which covers the OBD system components and because any owner-induced damage to the PCM, the sensors and/or the control devices may void the warranty, it is not recommended to attempt diagnosis of, or replace the PCM at home while the vehicle is under warranty.

➡ Note 2: The PCM has the ability to learn the individual driving characteristics of the vehicle. When power has been removed from the PCM, its memory is lost and it must relearn the vehicle parameters. This can adversely effect driveability during the first few miles of driving. To accomplish the relearning process, drive the vehicle for about 10 miles.

1995 AND EARLIER MODELS

▶ Refer to illustrations 3.4, 3.5, 3.6a, 3.6b and 3.7

1 The PCM on these models is located inside the passenger compartment, in the passenger side footwell behind the kick panel.
2 Disconnect the cable from the negative terminal of the battery.

❊❊ CAUTION:

On models equipped with an anti-theft audio system, disable the anti-theft feature before performing any operation that requires disconnecting the battery or disrupting power to the stereo (see the Anti-theft audio system procedures at the front of this manual).

3.4 Remove this door lock control relay (if equipped) bracket bolt and drop the relay down where it will be out of the way

3.5 Remove the right kick panel by pulling it out from the bulkhead (it has two pop fasteners) and straight back

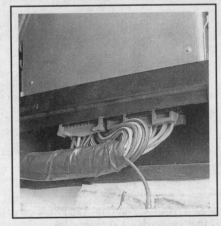

3.6a Remove the two retainer bracket bolts and the retainer bar

3.6b Pull the PCM out of the housing assembly

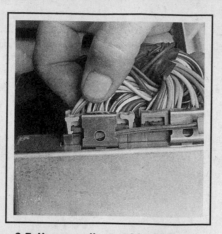

3.7 Use a small screwdriver to pry back the clips and unplug the PCM electrical connectors

3.8a If you are replacing the PCM compare the service numbers on the label on the old unit to the numbers on the new one

3 Remove the screws from the trim panel under the right end of the dash.

4 Remove the mounting bolt from the door lock relay bracket (see illustration).

5 Remove the right kick panel retaining screw and detach the panel (see illustration).

6 Remove the retaining bolts (see illustration) and carefully slide the PCM out far enough to unplug the electrical connector (see illustration).

7 Unplug both electrical connectors (see illustration) from the PCM.

PROM

▶ Refer to illustrations 3.8a and 3.8b

8 To allow one model of PCM to be used for many different vehicles (see illustration), a device called a PROM (Programmable Read Only Memory) is used. To access the PROM remove the cover. The PROM (see illustration) is located inside the PCM and contains information on the vehicle's weight, engine, transmission, axle ratio,

etc. One PCM part number can be used by many GM vehicles but the PROM is very specific and must be used only in the vehicle for which it was designed. For this reason, it is essential to check the latest parts book and Service Bulletin information for the correct part number when replacing a PROM. A PCM purchased at the dealer is purchased without a PROM. The PROM from the old PCM must be carefully removed and installed in the new PCM.

CALPAK (V6)

9 A device known as a CALPAK (see illustration 3.8b) is used to allow fuel delivery if other parts of the PCM are damaged. The CALPAK has an access door in the PCM and replacement is the same as that described for the PROM.

MEM-CAL (four-cylinder)

10 The MEM-CAL contains the functions of the PROM, CALPAK and ESC module used on other GM applications. Like the PROM, it contains the calibrations needed for a specific vehicle as well as the back-up fuel control circuitry required if the rest of the PCM becomes damaged or faulty.

3.8b The CALPAK (top) and PROM (bottom) inside a PCM for a V6 (PCMs for the 4-cylinder are similar but have a MEM-CAL instead of a PROM and CALPAK

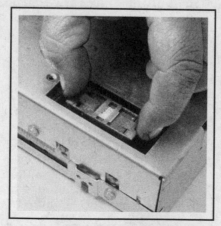

3.13 Grasp the PROM carrier at the narrow ends and gently rock the removal tool until the PROM is disconnected

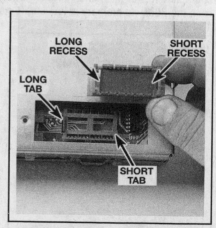

3.14 When installing the PROM, be sure to align the long and short tabs with their corresponding recesses

PCM/PROM/CALPAK replacement

▶ **Refer to illustrations 3.13, 3.14 and 3.17**

11 Turn the PCM so that the bottom cover is facing up and place it on a clean work surface.

12 Remove the PROM/CALPAK access cover.

13 Using a PROM removal tool (available at your dealer), grasp the PROM carrier at the narrow ends (see illustration). Gently rock the carrier from end to end while applying firm upward force. The PROM carrier and PROM should lift off the PROM socket easily.

> ⁂ **CAUTION:**
>
> **The PROM carrier should only be removed with the special rocker-type PROM removal tool. Removal without this tool or with any other type of tool may damage the PROM or the PROM socket.**

14 Note the reference end of the PROM carrier (see illustration) before setting it aside.

15 If you are replacing the PCM, remove the new PCM from its container and check the service number to make sure that it is the same as the number on the old PCM.

16 If you are replacing the PROM, remove the new PROM from its container and check the service number to make sure that it is the same as the number of the old PROM.

17 Position the PROM/PROM carrier assembly squarely over the PROM socket with the small notched end of the carrier aligned with the small notch in the socket at the pin 1 end. Press on the PROM carrier until it seats firmly in the socket (see illustration).

18 If the PROM is new, make sure that the notch in the PROM is matched to the small notch in the carrier.

> ⁂ **CAUTION:**
>
> **If the PROM is installed backwards and the ignition switch is turned on, the PROM will be destroyed.**

19 Using the tool, install the new PROM carrier in the PROM socket of the PCM. The small notch of the carrier should be aligned with the small notch in the socket. Press on the PROM carrier until it is firmly seated in the socket.

> ⁂ **CAUTION:**
>
> **Do not press on the PROM - press only on the carrier.**

20 Attach the access cover to the PCM and tighten the two screws.

21 Install the PCM in the support bracket, plug in the electrical connectors to the PCM and install the kick panel.

22 Start the engine.

23 Enter the diagnostic mode by grounding the diagnostic lead of the ALDL (see Section 2). If no trouble codes occur, the PROM is correctly installed.

24 If Trouble Code 51 occurs, or if the Service Engine Soon light comes on and remains constantly lit, the PROM is not fully seated, is installed backwards, has bent pins or is defective.

25 If the PROM is not fully seated, pressing firmly on both ends of the carrier should correct the problem.

26 If the pins have been bent, remove the PROM, straighten the

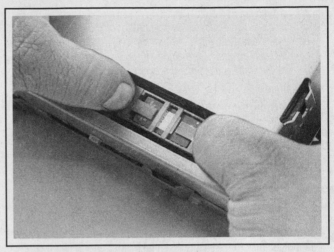

3.17 Press only on the ends of the PROM carrier - pressure on the area in between could result in bent or broken pins or damage to the PROM

pins and reinstall the PROM. If the bent pins break or crack when you attempt to straighten them, discard the PROM and replace it with a new one.

27 If careful inspection indicates that the PROM is fully seated, has not been installed backwards and has no bent pins, but the Service Engine Soon light remains lit, the PROM is probably faulty and must be replaced.

MEM-CAL replacement

➡**Note: With respect to removal of the PCM from the vehicle, MEM-CAL replacement is similar to the procedure described for the PROM or CALPAK. However, the actual removal and installation steps for the MEM-CAL and the functional check to ensure that the MEM-CAL is installed properly differ in detail from the Steps for the PROM and CALPAK.**

28 Remove the MEM-CAL access cover.

29 Using two fingers, push both retaining clips back away from the MEM-CAL. At the same time, grasp the MEM-CAL at both ends and lift it up out of its socket. Do not remove the MEM-CAL cover itself.

> ⁂ **CAUTION:**
>
> **Use of unapproved removal or installation methods may damage the MEM-CAL or socket.**

30 Verify that the numbers on the old PCM and new PCM match up (or that the numbers of the old and new MEM-CALs match up, depending on what component(s) you're replacing) as described in the procedure for removing and installing the PROM and CALPAK.

31 To install the MEM-CAL in the MEM-CAL socket, press only on the ends of the MEM-CAL.

32 The small notches in the MEM-CAL must be aligned with the small notches in the MEM-CAL socket. Press on the ends of the MEM-CAL until the retaining clips snap into the ends of the MEM-CAL. Do not press on the middle of the MEM-CAL - press only on the ends.

33 The remainder of the installation is similar to that for the PROM/CALPAK.

34 Once the new MEM-CAL is installed in the old PCM (or the old MEM-CAL is installed in the new PCM), check your installation to verify that it has been installed properly by doing the following test:

a) *Turn the ignition switch on.*
b) *Enter the diagnostics mode at the ALDL (see Section 2).*
c) *Allow Code 12 to flash four times to verify that no other codes are present. This indicates that the MEM-CAL is installed properly and the PCM is functioning properly.*

35 If trouble codes 41, 42, 43, 51 or 55 occur, or if the Service Engine Soon light is on constantly but is flashing no codes, the MEM-CAL is either not fully seated or is defective. If it's not fully seated, press firmly on the ends of the MEM-CAL. If it is necessary to remove the MEM-CAL, follow the above Steps again.

1996 AND LATER MODELS

▸ **Refer to illustration 3.36**

36 The PCM on these models is located in the engine compartment and mounted next to the battery (see illustration).

37 The PCM can be removed and replaced easily; however, the new PCM must be programmed by a dealer service department before it will operate properly.

➡**Note: The new PCM does not come with a Knock sensor PROM, so the Knock Sensor PROM from the old PCM must be installed into the new one.**

38 A special SCAN tool and dealer information are required to program the PCM, therefore this procedure should be performed by a dealer service department.

3.36 On 1996 and later models, the Powertrain Control Module is located in the engine compartment between the steering column and the battery

4   Information sensors

❋❋ CAUTION 1:

When performing the following checks, use only a high-imped-ance (10 meg-ohm) digital multimeter to prevent damage to the PCM.

❋❋ CAUTION 2:

On models equipped with an anti-theft audio system, disable the anti-theft feature before performing any operation that requires disconnecting the battery or disrupting power to the stereo (see the Anti-theft audio system procedures at the front of this manual).

➡**Note 1: After performing a check procedure a trouble code may be set. Be sure to clear the PCM of all trouble codes by removing the PCM-B fuse or disconnecting the cable from the negative terminal of the battery for at least 30 seconds.**

➡**Note 2: During some checks it may be necessary to use pins or straightened out paper clips to backprobe the sensor electri-cal connector wires. Be careful not to short the probes during this process.**

AIR CONDITIONING COMPRESSOR CLUTCH CONTROL

General description

1 When the air conditioning is turned on, an additional load is placed on the engine by the A/C compressor. The PCM controls the A/C compressor clutch relay in order to delay clutch engagement until after the PCM has increased the engine idle speed (via the IAC valve) to compensate for the additional load. The PCM will also disengage the A/C compressor clutch when the throttle is placed in the wide-open position or the PCM senses a part-throttle acceleration condition. Before attempting to diagnose an A/C electrical problem, make sure the refrigerant system is fully charged (see Chapter 3).

Check

2 First check for battery voltage to the A/C clutch with the ignition key ON (engine OFF) and the A/C turned ON. Detach the electrical con-nector from the A/C compressor and, using a test light or a voltmeter, check for battery voltage at the connector. If no voltage is indicated, check for a blown fuse and examine the wires for obvious damage. If the circuit looks OK, have the PCM checked out at your local dealer service department or other qualified repair shop.

3 If the fuse is OK, check for voltage to the A/C compressor clutch relay, which is located in the engine compartment under the cover of the electrical center (1996 and later models) or mounted on the firewall (1995 and earlier models). Remove the relay. With the ignition key ON (engine OFF) and the A/C turned ON, check for battery voltage to the relay connector. If voltage is present, replace the relay. If voltage is not present, refer to Chapter 3 for more information on troubleshooting the A/C system. In most cases, if the A/C does not operate, a problem exists with the A/C pressure switches or controls and not the PCM.

4.4 The camshaft position sensor is located under the distributor cap (CSFI engines only) - the arrow is to the sensor's electrical connector

4.13 To remove the camshaft sensor, the cut-out (arrow) in the vane wheel must be aligned with the sensor

4.15 The crankshaft position sensor (arrow) is located at the bottom of the timing chain cover (V6 engines only)

CAMSHAFT POSITION SENSOR (1995 AND LATER MODELS WITH V6 ENGINES)

General description

▶ Refer to illustration 4.4

4 The camshaft position sensor is located under the distributor cap (see illustration). This sensor is a Hall-effect device that provides the PCM with cylinder identification. The camshaft position sensor is not directly involved in the ignition process. Its sole purpose is to provide engine misfire information to the PCM. In the event the signal is lost or interrupted, the PCM will continue to pulse the fuel injectors using a default setting, but the synchronizing effect may be slightly off (stumble or misfire) due to the lack of precision.

Check

5 Remove the engine cover (see Chapter 11).
6 With the ignition key ON (engine OFF), use a voltmeter to verify the sensor is receiving battery voltage at the harness connector PINK or RED wire. Battery voltage should be present. If no voltage is indicated, check for a blown fuse and examine the wires for obvious damage. If the circuit looks OK, have the PCM checked out at your local dealer service department or other qualified repair shop.
7 Disable the ignition by grounding the ignition coil secondary wire to the engine. Backprobe the sensor connector using the positive probe of the voltmeter on the BROWN/WHITE wire and the negative probe on the PINK/BLACK wire. Crank the engine while watching the voltmeter. The voltage should pulse as the engine turns over.

➡ Note: The voltage will only pulse if battery voltage is present at the PINK or RED wire (see Step 6). If it doesn't, replace the camshaft position sensor.

Replacement

▶ Refer to illustration 4.13

8 Disconnect the cable from the negative terminal of the battery.
9 Remove the engine cover (see Chapter 11).
10 Disconnect the harness connector from the camshaft position sensor.
11 Remove the distributor cap and rotor (see Chapter 1).

12 On 1995 through 1998 models, remove the distributor (see Chapter 5). On 1999 and later models, the camshaft position sensor can be removed in the vehicle once the distributor cap has been removed.
13 Rotate the distributor shaft as required to remove the 2 screws securing the Hall effect switch to the distributor. Align the square cut-out in the vane wheel with the switch and then remove the switch from the distributor (see illustration).
14 Installation is the reverse of removal. The factory recommends using a thread-locking compound on the screws.

CRANKSHAFT POSITION SENSOR (1995 AND LATER MODELS WITH V6 ENGINES)

General description

▶ Refer to illustration 4.15

15 The crankshaft position sensor is located on the engine timing chain cover (see illustration) and mounted perpendicular to the crankshaft reluctor ring. This sensor detects three slots in the reluctor ring which are located 60 degrees apart. The air gap between the sensor and the reluctor ring is not adjustable. As the reluctor ring passes the sensor, changes in the magnetic field create an electronic pulse that is sent to the PCM. These pulses combined with the camshaft sensor input allow the PCM to precisely control the ignition timing and fuel injection.

Check

▶ Refer to illustration 4.16

16 Locate the crankshaft sensor and disconnect the harness connector. With the ignition key ON (engine OFF), use a voltmeter to verify the sensor is receiving battery voltage at the harness connector PINK or LIGHT GREEN wire (see illustration). If no voltage is indicated, check for a blown fuse and examine the wires for obvious damage.
17 If the circuit looks OK, have the PCM checked out at your local dealer service department or other qualified repair shop.

Replacement

18 Disconnect the cable from the negative terminal of the battery.
19 Disconnect the harness connector from the sensor.

4.16 Check for battery voltage at the crankshaft position sensor wiring harness

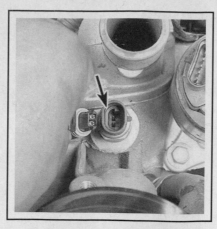

4.22 Engine Coolant Temperature sensor location (arrow) (V6 CSFI engine shown)

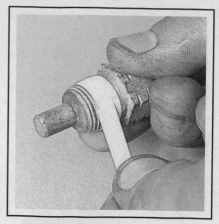

4.25 To prevent leakage, wrap the threads of the coolant temperature sensor with Teflon tape before installing it

20 Remove the bolt securing the sensor to the timing chain cover and withdraw it from the cover.

21 Installation is the reverse of removal. If you are reinstalling a used sensor, be sure to inspect the O-ring seal for damage or hardness and replace it if necessary.

ENGINE COOLANT TEMPERATURE (ECT) SENSOR

General description

▶ Refer to illustration 4.22

22 On earlier models, the ECT sensor is located in an intake manifold waterjacket, usually close to the thermostat housing (see illustration). On later models, it's on the outside of the left cylinder head, about two-thirds of the way back. The ECT is a thermistor (a resistor which varies the value of its resistance in accordance to temperature). This change in resistance will directly affect the voltage signal to the PCM. As the temperature DECREASES, the resistance INCREASES and as the temperature INCREASES, the resistance DECREASES.

Check

23 First, verify that the ECT is receiving reference voltage from the PCM. With the ignition key ON (engine OFF), probe the ECT harness connector using a voltmeter. The voltmeter should indicate approximately 5 volts. If no voltage is indicated, check for a blown fuse and examine the wires for obvious damage. If the circuit looks OK, have the PCM checked out at your local dealer service department or other qualified repair shop.

24 To check the sensor operation, use an ohmmeter to measure the resistance across the sensor terminals. With the engine cold the resistance should be high (approximately 3500 ohms at 68 degrees F). Next, start the engine and let it run until it reaches operating temperature. Now measure the ECT resistance again. This time the resistance should be low (approximately 240 ohms at 194 degrees F). If the ECT does not respond as described, replace it.

➡ Note: Due to inaccessibility on some models, it may be necessary to remove the ECT sensor and simulate the warm-up condition by suspending it in a pan of heated water. If the ECT checks out OK, check the wiring and connections.

Replacement

▶ Refer to illustration 4.25

✳✳ WARNING:

Wait until the engine is completely cool before beginning this procedure.

25 Before installing the new sensor, wrap the threads with Teflon tape to prevent leakage (see illustration).

26 To remove the sensor, release the locking tab on the harness connector and disconnect the connector. Using a deep socket, unscrew the sensor. Be prepared to install the new sensor immediately to minimize coolant loss.

✳✳ CAUTION:

Handle this sensor with care. Damage to this sensor will affect the operation of the fuel injection system.

27 Thread the new sensor into the manifold and tighten it securely. Attach the harness connector and add the appropriate coolant type (see Recommended lubricants and fluids in Chapter 1).

IDLE AIR CONTROL (IAC) VALVE

General description

28 All engine idle speeds are controlled by the PCM via the IAC valve, which is mounted on the throttle body. The PCM sends voltage pulses (counts) to the IAC valve, which then moves the pintle in or out as required, to allow intake air to bypass the throttle valve. When higher idle speeds are required due to a load placed on the engine, the IAC valve opens, allowing additional air to enter the engine.

Check and replacement

29 For IAC valve check and replacement procedures, refer to the appropriate Section in Chapter 4 that addresses your type of fuel delivery system.

INTAKE AIR TEMPERATURE (IAT) SENSOR

General description

▶ **Refer to illustration 4.30**

30 On TBI engines, the IAT is threaded into one of the intake manifold runners. On CMFI and CSFI engines, the IAT sensor is located on the air intake duct, directly downstream from the air filter housing (CMFI engines) or MAF sensor (CSFI engines) (see illustration) and is a push fit into the duct. This sensor is a thermistor (a resistor which varies the value of its resistance in accordance to temperature). This change in resistance will directly affect the voltage signal to the PCM. As the temperature DECREASES, the resistance INCREASES and as the temperature INCREASES, the resistance DECREASES.

Check

31 First, verify that the IAT sensor is receiving reference voltage from the PCM. With the ignition key ON (engine OFF), probe the sensor harness connector using a voltmeter. The voltmeter should indicate approximately 5 volts. If no voltage is indicated, check for a blown fuse and examine the wires for obvious damage. If the circuit looks OK, have the PCM checked out at your local dealer service department or other qualified repair shop.

32 To check the sensor operation, use an ohmmeter to measure the resistance across the sensor terminals. With the engine cold, the resistance should be high (approximately 3500 ohms at 68 degrees F).

➥ **Note: On TBI engines, you'll need to remove the air cleaner assembly to gain access to the sensor.**

33 On TBI engines, attach the sensor harness connector. Start the engine and let it warm up to operating temperature, then turn off the engine.

34 On CMFI and CSFI engines, remove the air cleaner cover. Use a hair dryer pointed down into the air duct to warm the IAT sensor for a few minutes.

35 Now measure the IAT sensor resistance again. This time the resistance should be low (approximately 240 ohms at 194 degrees F). If the sensor does not respond as described, replace it. If the IAT sensor checks out OK, check the wiring and connections.

Replacement

36 On TBI engines, remove the air cleaner assembly (see Chapter 4).
37 Disconnect the harness connector from the sensor.
38 On TBI engines, unscrew the sensor from the intake manifold.
39 On CMFI and CSFI engines, carefully use a pulling and twisting motion to remove it from the air duct.
40 Installation is the reverse of removal. On TBI engines, apply Teflon tape to the sensor threads prior to installation.

KNOCK SENSOR (V6 ENGINES ONLY)

General description

41 The knock sensor is used by the PCM to detect engine knock or pinging, which is caused by gasoline with a low octane rating (see illustration 7.3). Most early models used 2 sensors threaded into each side of the engine block. Later models employ 1 sensor, mounted at the top rear of the engine block, near the distributor. By sensing pinging, the PCM can advance the ignition timing to increase engine performance for any octane rated fuel used. When knock is detected, the PCM retards the ignition timing to eliminate or reduce the knock.

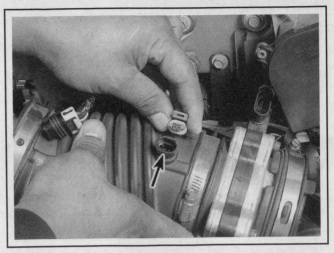

4.30 Intake Air Temperature sensor location (arrow) (V6 CSFI engine shown)

On 1995 and earlier models, a system called Electronic Spark Control (ESC) was used to control the spark advance (see Section 7). On 1996 and later models the PCM controls this feature via a removable knock sensor PROM installed in the PCM. The knock sensor produces an alternating current (AC) voltage signal which increases with the severity of the knock.

Check

42 On 1995 and earlier models, using an ohmmeter, measure the resistance between the knock sensor electrical terminal and the engine block. It should be approximately 3,900 ohms on 1993 and earlier and 2,100 ohms on 1994 and 1995 models. If the resistance is not within range, replace the knock sensor.

43 All models, disconnect the harness connector from the knock sensor. With the ignition key ON (engine OFF), use a voltmeter to check for reference voltage at the connector. It should be approximately 5.0 volts. If no voltage is indicated, check for a blown fuse and examine the wires for obvious damage. If the circuit looks OK, have the PCM checked out at your local dealer service department or other qualified repair shop.

Replacement

44 Disconnect the cable from the negative terminal of the battery.
45 If necessary, raise the vehicle and support it securely on jackstands.
46 Disconnect the knock sensor electrical harness connector.
47 Unscrew the sensor from the engine block.
48 Installation is the reverse of removal.

MANIFOLD ABSOLUTE PRESSURE (MAP) SENSOR

General description

▶ **Refer to illustrations 4.49a, 4.49b and 4.49c**

49 On TBI engines, the MAP sensor is mounted on the air cleaner assembly. On CMFI and CSFI engines, it's mounted directly on the intake manifold (see illustrations). This sensor monitors the intake manifold pressure changes resulting from changes in engine load, speed and altitude, then converts the information into a voltage output. The PCM uses the MAP sensor to control fuel delivery and ignition timing.

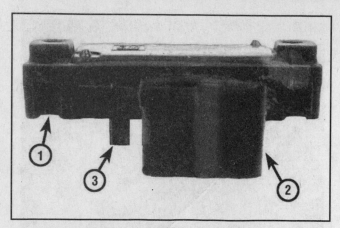

4.49a A typical Manifold Absolute Pressure (MAP) sensor

1 Sensor assembly   3 Manifold vacuum port
2 Electrical connector

Check

▶ **Refer to illustrations 4.50, 4.51a and 4.51b**

50 First, check the MAP sensor reference voltage; disconnect the electrical connector and turn the ignition key ON (engine OFF). Using a voltmeter, check for voltage between the GRAY wire (positive) and the BLACK or ORANGE/BLACK wire (ground) at the sensor connector (see illustration). There should be approximately 4 to 5 volts present. If no voltage is indicated, check for a blown fuse and examine the wires for obvious damage. If the circuit looks OK, have the PCM checked out at your local dealer service department or other qualified repair shop.

51 Next check the MAP sensor signal voltage; reconnect the electrical connector to the sensor and turn the ignition key ON (engine OFF). Using a voltmeter, backprobe the BLACK or ORANGE/BLACK wire (ground) and the LIGHT GREEN wire (positive) at the sensor connector (see illustration). There should be approximately 4 to 5 volts present. If reference voltage is detected but there is no signal voltage, replace the MAP sensor. Start the engine. The voltage should drop to approximately 1 to 2 volts with the engine idling (see illustration). If the MAP sensor voltage readings are incorrect, replace the MAP sensor.

Replacement

52 To replace the sensor, detach the vacuum hose, disconnect the wiring harness connector and remove the mounting screws.

53 Installation is the reverse of removal.

MASS AIR FLOW (MAF) SENSOR (1996 AND LATER MODELS WITH V6 ENGINES)

General description

▶ **Refer to illustration 4.54**

54 The MAF is used on all CSFI engines to measure the quantity of air entering the engine. This sensor is located in the intake air duct downstream of the air cleaner housing. This sensor is a delicate measuring instrument - handle it with care and do not touch the honeycomb grid inside (see illustration).

Check

55 With the engine running, lightly tap the MAF with the handle of a screwdriver. If the engine stumbles, the sensor is faulty. Before replacing this component, have a dealer service department or other qualified

4.49b Location of the MAP sensor on V6 engine with CMFI

repair shop confirm this diagnosis.

Replacement

56 Disconnect the wiring harness from the MAF sensor.
57 Loosen the air duct clamps and carefully remove the MAF sensor.
58 Installation is the reverse of removal.

✳✳ CAUTION:

Do not use any type of silicone spray lubricant to aid with installation. Silicone vapors will damage the oxygen sensors when ingested by the engine.

OXYGEN SENSOR

General description

59 The oxygen sensor is mounted in the exhaust system where it can monitor the oxygen content of the exhaust gas stream. By monitoring the voltage output of the sensor, the PCM will know what fuel mixture command to give the fuel injector(s).

4.49c MAP sensor location on the V6 engine with CSFI (arrow)

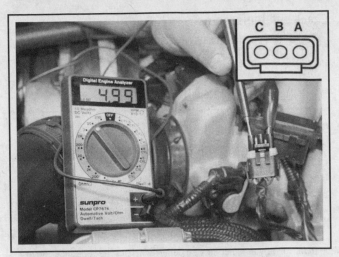

4.50 Checking the MAP sensor reference voltage - the voltage reading should be approximately 5.0 volts (V6 CMFI engine shown)

❋❋ CAUTION:

Do not use any type of silicone spray lubricant around the intake air duct, air cleaner or throttle body. Silicone vapors will damage the oxygen sensors when ingested by the engine.

60 For replacement procedures, see Section 5.

PARK/NEUTRAL SWITCH (AUTOMATIC TRANSMISSION MODELS)

General description

61 The Park/Neutral switch prevents the engine from starting in any gear other than Park or Neutral (on 1995 and later models, it also closes the circuit for the back-up lights when the shift lever is moved to Reverse). This information is also used to control the Torque Converter Clutch (TCC). If the engine starts in any position other than Park or Neutral, it's either out of adjustment or defective.

❋❋ CAUTION:

The vehicle should not be driven with the Park/Neutral switch disconnected because idle quality will adversely be affected and a false trouble code may be set.

Check, adjustment and replacement

62 For check, adjustment and replacement procedures, refer to Chapter 7B.

POWER STEERING PRESSURE SWITCH

General description

63 The power steering pressure switch is located on the steering gear box. Turning the steering wheel increases the power steering fluid pressure and the load placed upon the engine by the power steering pump. The pressure switch will close and cause the idle speed to be increased before the load causes an idle problem. A pressure switch that will not close or an open circuit may cause the engine to stall when the power steering system is used heavily. Stalling under idle load can also be caused by a malfunctioning Idle Air Control (IAC) valve (see Chapter 4).

Check

64 First, check for battery voltage at the switch wiring harness connector; disconnect the electrical connector and turn the ignition key ON (engine OFF). Using a voltmeter, check for voltage between the sensor connector terminals. Voltage should be present. If no voltage is indicated, check for a blown fuse and examine the wires for obvious damage. If the circuit looks OK, have the PCM checked out at your local dealer service department or other qualified repair shop.

65 Next, using an ohmmeter, measure the resistance across the switch terminals. It should read high (open switch).

66 Now, start the engine and turn the steering wheel all the way to one side. The ohmmeter should indicate low resistance (switch closed). If the resistances are not as noted, replace the switch.

Replacement

67 To replace the switch, disconnect the wiring harness connector and unscrew the switch from the steering gear box.

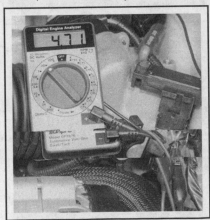

4.51a Checking the MAP sensor signal voltage - the signal voltage with the ignition ON (engine OFF) should be approximately 4 to 5 volts (V6 CMFI engine shown)

4.51b Start the engine and observe the MAP sensor signal voltage at idle. With the engine idling, the signal voltage should drop to approximately 1 to 2 volts (V6 CMFI engine shown)

4.54 The Mass Air Flow sensor (arrow) is used on V6 CSFI engines only (air cleaner housing removed for clarity)

➡ **Note: Be prepared to catch any power steering fluid that may leak out as the switch is removed. Install a plug or have the new switch ready for immediate installation to minimize leakage.**

68 Prior to installation, wrap the threads with Teflon tape to prevent fluid leakage. Installation is the reverse of removal. If the vehicle is equipped with a hydraulic power brake booster, bleed the system (see Chap-ter 10).

THROTTLE POSITION SENSOR (TPS)

General description

69 The TPS is mounted on the throttle body and engaged with the throttle valve shaft. By monitoring the throttle valve position, the PCM can determine fuel delivery based on driver demand. A loose or broken TPS can cause intermittent bursts of fuel from the injectors and an unstable idle because the PCM thinks the throttle valve is moving. Once a trouble code has been set, the PCM will use a default value for the TPS until the problem has been corrected.

Check and replacement

70 For TPS check and replacement procedures, refer to the appropriate Section in Chapter 4 that addresses your type of fuel delivery system.

VEHICLE SPEED SENSOR (VSS)

General description

71 The VSS is a permanent magnet generator which, when rotated, sends a pulsing voltage signal to the PCM. The PCM then converts these pulses to miles per hour. This sensor is used to control the TCC system (see Chapter 7B).

Check and replacement

72 Due to the complexity of this circuit, diagnosis of the VSS should be performed by a dealer service department or other qualified repair shop.

73 For replacement procedures, refer to Chapter 7 Part B.

5   Oxygen sensor(s)

♦ **Refer to illustrations 5.1a and 5.1b**

GENERAL DESCRIPTION

1   The oxygen sensor(s), which is located in the exhaust manifold (see illustrations), monitors the oxygen content of the exhaust gas stream. The oxygen content in the exhaust reacts with the oxygen sensor to produce a voltage output which varies from 0.1 volt (high oxygen, lean mixture) to 0.9 volts (low oxygen, rich mixture). The PCM constantly monitors this variable voltage output to determine the ratio of oxygen to fuel in the mixture. The PCM alters the air/fuel mixture ratio by controlling the pulse width (open time) of the fuel injectors. A ratio of 14.7 parts air to 1 part fuel is the ideal for minimizing exhaust emissions, allowing the catalytic converter to operate at maximum efficiency. It is this ratio of 14.7 to 1 which the PCM and the oxygen sensor attempt to maintain at all times. Four-cylinder engines and V6 engines equipped with TBI have a single oxygen sensor which is located in the exhaust manifold (left manifold on V6 engines). V6 engines equipped with CMFI and CSFI have 3 oxygen sensors; 1 in each exhaust manifold (left and right) and 1 installed in the exhaust pipe downstream of the catalytic converter which monitors the efficiency of the converter.

2   The oxygen sensor produces no voltage when it is below its normal operating temperature of about 600-degrees F (360-degrees C). During this initial period before warm-up, the PCM operates in open loop mode.

3   If the engine reaches normal operating temperature and/or has been running for two or more minutes, and if the oxygen sensor is producing a steady signal voltage between 0.35 and 0.55-volts, even though the TPS indicates that the engine is not at idle, the PCM will set a trouble code.

4   A delay of two minutes or more between engine start-up and normal operation of the sensor, followed by a low or high voltage signal or a short in the sensor circuit, will cause the PCM to set a trouble code.

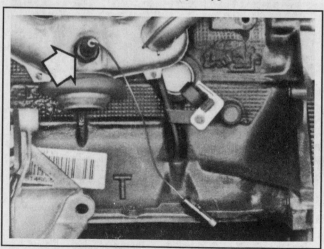

5.1a The oxygen sensor for the 4-cylinder engine is located in the exhaust manifold (engine removed from vehicle for clarity)

5.1b The oxygen sensor on the V6 engine is located in the left exhaust manifold (TBI engine shown)

5 When any of the above codes occur, the PCM operates in the open loop mode - that is it controls fuel delivery in accordance with a programmed default value instead of feedback information from the oxygen sensor.

6 The proper operation of the oxygen sensor depends on four conditions:

a) **Electrical** - *The low voltages generated by the sensor depend upon good, clean connections which should be checked whenever a malfunction of the sensor is suspected or indicated.*

b) **Outside air supply** - *The sensor is designed to allow air circulation to the internal portion of the sensor. Whenever the sensor is removed and installed or replaced, make sure the air passages are not restricted.*

c) **Proper operating temperature** - *The PCM will not react to the sensor signal until the sensor reaches approximately 600-degrees F (315-degrees C). This factor must be taken into consideration when evaluating the performance of the sensor.*

d) **Unleaded fuel** - *The use of unleaded fuel is essential for proper operation of the sensor. Make sure the fuel you are using is of this type.*

7 In addition to observing the above conditions, special care must be taken whenever the sensor is serviced.

a) *The oxygen sensor has a permanently attached pigtail and connector which should not be removed from the sensor. Damage or removal of the pigtail or connector can adversely affect operation of the sensor.*

b) *Grease, dirt and other contaminants should be kept away from the electrical connector and the louvered end of the sensor.*

c) *Do not use cleaning solvents of any kind on the oxygen sensor.*

d) *Do not drop or roughly handle the sensor.*

e) *The silicone boot must be installed in the correct position to prevent the boot from being melted and to allow the sensor to operate properly.*

REPLACEMENT

➡**Note: Because it is installed in the exhaust manifold or pipe,**

which contracts when cool, the oxygen sensor may be very difficult to loosen when the engine is cold. Rather than risk damage to the sensor (assuming you are planning to reuse it in another manifold or pipe), start and run the engine for a minute or two, then shut it off. Be careful not to burn yourself during the following procedure.

8 Disconnect the cable from the negative terminal of the battery.

✳✳ CAUTION:

On models equipped with an anti-theft audio system, disable the anti-theft feature before performing any operation that requires disconnecting the battery or disrupting power to the stereo (see the Anti-theft audio system procedures at the front of this manual).

9 Raise the vehicle and support it securely on jackstands.
10 Follow the oxygen sensor electrical wire(s) to the harness connector and disconnect the sensor wire(s) from the harness.

➡**Note: The sensor wiring is permanently attached to the sensor - do not try to disconnect it at the sensor.**

11 Carefully unscrew the sensor from the exhaust manifold.

✳✳ CAUTION:

Excessive force may damage the threads.

12 Anti-seize compound must be used on the threads of the sensor to facilitate future removal. The threads of new sensors will already be coated with this compound, but if an old sensor is removed and reinstalled, recoat the threads.
13 Install the sensor and tighten it securely.
14 Reconnect the electrical connector of the pigtail lead to the main engine wiring harness.
15 Lower the vehicle and reconnect the cable to the negative terminal of the battery.

6 Electronic Spark Timing (EST) (1994 and earlier models)

GENERAL DESCRIPTION

➡**Note: Always consult the VECI label for the exact timing procedure for your vehicle.**

1 To provide improved engine performance, fuel economy and control of exhaust emissions, the Powertrain Control Module (PCM) controls distributor spark advance (ignition timing) with the Electronic Spark Timing (EST) system.

2 The EST system consists of the distributor HEI module, the PCM and the connecting wires. The four terminals for the EST are lettered on the module. The distributor four-terminal connector is lettered left-to-right, A-B-C-D. These circuits perform the following functions:

a) **Terminal A** - *Reference Ground Low. This wire is grounded in the distributor and insures that the ground circuit has no voltage drop which could affect performance. It is open, it may cause poor performance.*

b) **Terminal B** - *Bypass. At about 400 rpm the PCM applies 5-volts to this circuit to switch spark timing control from the HEI module to the PCM. An open or grounded bypass circuit will set a Code 42 and the engine will run at base timing, plus a small amount of advance built into the HEI module.*

c) **Terminal C** - *Distributor Reference High. This provides the PCM with rpm and crankshaft position information.*

d) **Terminal D** - *EST. This circuit triggers the HEI module. The PCM doesn't know what the actual timing is, but it does know when it gets the reference signal. It advances or retards the spark from that point. If the base timing is set incorrectly, the entire spark curve will be incorrect.*

3 The PCM receives a reference pulse from the distributor, which indicates both engine rpm and crankshaft position. The PCM then determines the proper spark advance for the engine operating conditions and sends an EST pulse to the distributor.

CHECKING

4   The PCM will set spark timing at a specified value when the diagnostic "Test" terminal in the ALDL connector is grounded. To check for EST operation, the timing should be checked at 2000 rpm with the terminal ungrounded. Then ground the test terminal. If the timing changes at 2000 rpm, the EST is operating. A fault in the EST system will usually set Trouble Code 42.

SETTING BASE TIMING

5   To set the initial base timing, locate, then disconnect the timing connector (the location and wire color of the timing connector is on the VECI label).

6   Set the timing as specified on the VECI label. This will cause a Code 42 to be stored in the PCM memory. Be sure to clear the memory after setting the timing (see Section 2).

7   For further information regarding the testing and component replacement procedures for either the HEI/EST distributor, refer to Chapter 5.

---

7   Electronic Spark Control (ESC) system (V6 engine) (1995 and earlier models)

▶ **Refer to illustrations 7.3 and 7.13**

GENERAL DESCRIPTION

1   Irregular octane levels in modern gasoline can cause detonation in an engine. Detonation is sometimes referred to as "spark knock."

2   The Electronic Spark Control (ESC) system is designed to retard spark timing up to 20-degrees to reduce spark knock in the engine. This allows the engine to use maximum spark advance to improve driveability and fuel economy.

3   The ESC knock sensor (see illustration) sends a voltage signal of 8 to 10-volts to the PCM when no spark knock is occurring and the PCM provides normal advance. When the knock sensor detects abnormal vibration (spark knock), the ESC module turns off the circuit to the PCM and the voltage at PCM terminal B7 drops to zero. The PCM then retards the EST distributor until spark knock is eliminated.

4   Failure of the ESC knock sensor signal or loss of ground at the ESC module will cause the signal to the PCM to remain high. This condition will result in the PCM controlling the EST as if no spark knock is occurring. Therefore, no retard will occur and spark knock may become severe under heavy engine load conditions.

5   Loss of the ESC signal to the PCM will cause the PCM to constantly retard EST. This will result in sluggish performance.

CHECKING

6   Be sure all system electrical connectors are tight and in good condition. Also check the ignition timing (see Chapter 1), as this is responsible for most spark-knock problems. If the timing is OK, make the following checks:

a) *Hook up a timing light according to the manufacturer's instructions and run the engine while observing the timing marks with the timing light.*

b) *Using a large wrench or similar tool, tap the engine block in the area near the knock sensor. The ignition timing should retard slightly when this is done. If the timing retards, the system is working properly.*

c) *If the ignition timing does not retard, turn off the engine, remove the electrical connector from the knock sensor and measure resistance between the center terminal and outer body of the sensor. Resistance should be approximately 3,900 ohms.*

d) *If the timing did not retard and the sensor resistance is out of range, replace the sensor and retest.*

e) *If the timing did not retard and the resistance is within range, take the vehicle to a dealer service department or other qualified shop for further diagnosis, as in-depth diagnosis of this system is best done with a SCAN tool.*

7.3 The Electronic Spark Control (ESC) knock sensor on the V6 engine is located on the left side of the block, forward of the oil filter (early models)

7.13 The ESC module is located at the rear of the engine compartment

COMPONENT REPLACEMENT

ESC sensor

7 Detach the cable from the negative terminal of the battery.

※※ CAUTION:

On models equipped with an anti-theft audio system, disable the anti-theft feature before performing any operation that requires disconnecting the battery or disrupting power to the stereo (see the Anti-theft audio system procedures at the front of this manual).

8 Disconnect the wiring harness connector from the ESC sensor.
9 Remove the ESC sensor from the block.
10 Installation is the reverse of the removal procedure.

ESC module

11 Detach the cable from the negative terminal of the battery.

※※ CAUTION:

On models equipped with an anti-theft audio system, disable the anti-theft feature before performing any operation that requires disconnecting the battery or disrupting power to the stereo (see the Anti-theft audio system procedures at the front of this manual).

12 Remove the engine cover.
13 Locate the module (see illustration) at the rear of the engine compartment.
14 Detach the wiring harness electrical connector from the module.
15 Remove the module mounting bolts and remove the module.
16 Installation is the reverse of removal.

8   Air management system (V6 - TBI models California and all manual transmissions)

▶ **Refer to illustrations 8.27, 8.31, 8.43, 8.45 and 8.51**

GENERAL DESCRIPTION

1   The air management system is used to reduce carbon monoxide and hydrocarbon emissions. The system used on these vehicles, the Air Injection Reaction (AIR) system, adds air to the exhaust manifold to continue combustion after the exhaust gases leave the combustion chamber.

2   The AIR system consists of an air pump, a diverter valve (Federal/carbureted) or an Electric Air Control (EAC) valve (California/carbureted and all fuel injected vehicles), a pair of check valves and the plumbing between these components.

3   A belt-driven air pump supplies air through a centrifugal filter fan to the EAC valve.

4   A check valve on either side of the engine prevents the back flow of exhaust into the air pump if there is an exhaust backfire or pump drivebelt failure.

5   To help prevent backfiring during high vacuum conditions, Federal/carbureted engines utilize a deceleration (gulp) valve to allow air to flow into the intake manifold. This air enters the air/fuel mixture to lean the rich condition created by high vacuum when the throttle valve closes on deceleration.

CHECKING

Electric air control (EAC) valve

6   During cold starting, the PCM completes the ground circuit, the EAC solenoid is energized and air is directed to the exhaust ports. As the coolant temperature increases the PCM opens the ground circuit, the EAC solenoid is de-energized and air goes to the air cleaner.

7   Thorough diagnosis of this system is best left to a dealer service department, since special diagnostic equipment is necessary. However, the home mechanic can perform some simple diagnostics at home:

a) *The check valve(s), located near the exhaust manifold, are a common problem area. Disconnect the rubber hose from the valve(s) and run the engine. If you can hear exhaust noise and there is exhaust escaping through the valve, the valve is defective and*

must be replaced. A failed valve will also damage the rubber hose.
b) *With the hose removed from the check valve, start the engine while it's cold and feel for a gentle flow of air out of the hose while the engine is idling.*
c) *If you can't feel a flow of air at the check-valve hose, repeat this check at the hose leading from the air pump. If you can now feel an air flow, the problem lies in the remaining system valves and hoses. If you still cannot feel a flow of air, the air pump is not functioning or the belt is slipping.*

8   The AIR system is not completely noiseless. Under normal conditions, noise rises in pitch as engine speed increases. To determine if the excessive noise is the fault of the AIR system pump, operate the engine with the pump drivebelt removed.

9   If the noise is caused by the AIR system pump, check for a seized air pump, proper mounting and bolt torque of the pump and the proper routing and connections of the hoses.

Air pump

10 The air pump is a permanently-lubricated positive displacement vane type design which requires no periodic maintenance. If it is making noise, replace it. Do not attempt to lubricate it.

11 To check air flow from the hoses, accelerate the engine to about 1500 rpm and note the air flow from the hoses. If air flow increases as the engine is accelerated, the pump is operating satisfactorily. If air flow does not increase or is not present, proceed as follows:

12 Check the drivebelt for proper tension (see Chapter 1).

13 Inspect the pressure relief valve for air leaks. If the valve is leaking, you can hear it leak when the pump is running.

Check valve

14 Inspect the check valve(s) whenever the corresponding hose is disconnected or whenever check valve failure is suspected. A pump that has become inoperative and shows indications of having exhaust gases in the pump indicates check valve failure.

15 Blow through the check valve (toward the cylinder head), then attempt to suck back through the check valve. Flow should only be in one direction (toward the exhaust manifold). Replace the valve if it does not operate properly.

8.27 To remove the drivebelt loosen the adjusting bolt and mounting brace bolts (arrows)

Deceleration valve

16 Remove the air cleaner housing assembly and adapter (see Chapter 4), plug the air cleaner vacuum source and connect a tachometer.

17 With the engine running at the specified idle speed, remove the small deceleration valve signal hose from the manifold vacuum source.

18 Reconnect the signal hose and listen for air flow through the ventilation pipe and into the deceleration valve. There should also be a noticeable speed drop when the signal hose is reconnected.

19 If the air flow does not continue for at least one second, or the engine speed does not drop noticeably, check the deceleration valve hoses for restrictions or leaks.

20 If no restrictions or leaks are found, replace the deceleration valve.

Hoses and pipes

21 Inspect the hoses and pipes for deterioration and holes.

22 Inspect all hose and pipe clamps for tightness.

23 Check the routing of all hoses and pipes. Interference can cause wear.

24 If a leak is suspected on the pressure side of the system, or if a hose or pipe has been disconnected on the pressure side, the connections should be checked for leaks with a soapy water solution. With the pump running, bubbles will form if a leak exists.

Drivebelt

25 Inspect the drivebelt for wear, cracks or deterioration (see Chapter 1) and replace as necessary. When installing a new belt, make sure that it is fully seated in the V-belt grooves of the A/C compressor, air pump, alternator and crankshaft pulleys.

COMPONENT REPLACEMENT

Air pump and centrifugal filter fan

26 Detach the cable from the negative terminal of the battery.

✳✳ CAUTION:

On models equipped with an anti-theft audio system, disable the anti-theft feature before performing any operation that requires disconnecting the battery or disrupting power to the stereo (see the Anti-theft audio system procedures at the front of this manual).

8.31 The filter can be pulled out of the air pump with needle-nose pliers

27 Immobilize the pump pulley by compressing the drivebelt and loosen the pump pulley bolts. Loosen the drivebelt tension adjusting bolt and, if necessary, the mounting brace bolts (see illustration). Pivot the air pump toward the block to remove belt tension and remove the drivebelt.

28 Clearly label, then detach, the pump hoses, vacuum lines and electrical connectors.

29 Remove the belt tension adjusting bolt and the mounting brace bolt and remove the air pump.

30 Remove the pulley bolts, the pulley and the pulley spacer.

31 Use needle nose pliers to pull the filter fan from the pump hub (see illustration).

✳✳ CAUTION:

Do not allow any filter fragments to enter the air pump intake hole during removal. Do not remove the filter fan by inserting a screwdriver between the pump and the filter fan. You will damage the pump sealing lip. Do not attempt to clean the centrifugal filter fan with either compressed air or solvents. If it is dirty, it must be replaced.

32 Install the new filter fan on the pump hub.

33 Install the spacer and pump pulley against the centrifugal fan.

34 Install the pump pulley bolts and snug them finger tight.

35 Install the air pump and snug the bolts finger tight.

36 Install the drivebelt and adjust the belt tension (see Chapter 1).

37 Tighten the air pump mounting brace bolt and belt tension adjusting bolt securely.

38 Tighten the pulley bolts to the specified torque. This compresses the centrifugal filter fan onto the pump hole. Do not attempt to drive the filter fan on with a hammer.

➡Note: A slight amount of interference with the housing bore is normal.

39 Attach the cable to the negative terminal of the battery.

40 Check the air management system for proper operation.

➡Note: After a new filter fan has been installed, it may squeal upon initial operation until the sealing lip has worn in.

EAC valve

41 Detach the cable from the negative terminal of the battery.

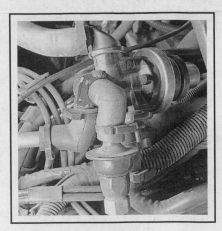

8.43 A typical EAC valve

8.45 To detach the manifold vacuum signal tube from the EAC valve, pry it off with a small screwdriver

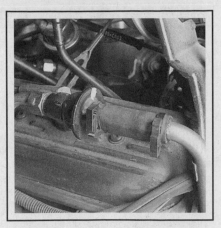

8.51 Typical air management system check valve - V6 engine shown

❊❊ CAUTION:

On models equipped with an anti-theft audio system, disable the anti-theft feature before performing any operation that requires disconnecting the battery or disrupting power to the stereo (see the Anti-theft audio system procedures at the front of this manual).

42 Remove the engine cover (see Chapter 11).
43 Locate the EAC valve (see illustration) on the right side of the engine bay.
44 Detach the electrical connector from the terminal on the lower front of the valve.
45 Detach the manifold vacuum signal hose (see illustration).
46 Detach the air inlet and outlet hoses from the valve.
47 Remove the EAC valve.
48 Installation is the reverse of removal.

Check valve

49 Detach the cable from the negative terminal of the battery.

❊❊ CAUTION:

On models equipped with an anti-theft audio system, disable the anti-theft feature before performing any operation that requires disconnecting the battery or disrupting power to the stereo (see the Anti-theft audio system procedures at the front of this manual).

50 Remove the engine cover (see Chapter 11).
51 Locate the check valve you wish to replace (see illustration).
52 Release the clamp and detach the hose from the valve.
53 Using a backup wrench, unscrew the valve from the threaded fitting on the air injection pipe.
54 Installation is the reverse of removal.

Air injection pipe assembly

55 Detach the cable from the negative terminal of the battery.

❊❊ CAUTION:

On models equipped with an anti-theft audio system, disable the anti-theft feature before performing any operation that requires disconnecting the battery or disrupting power to the stereo (see the Anti-theft audio system procedures at the front of this manual).

56 Remove the engine cover (see Chapter 11).
57 Locate the pipe assembly you wish to replace.
58 Unscrew the check valve.
59 Unscrew the pipe assembly threaded fitting from the manifold.
60 Remove the pipe assembly mounting bracket bolts.
61 Remove the pipe assembly.
62 Installation is the reverse of removal.

Deceleration valve

63 Detach the cable from the negative terminal of the battery.

❊❊ CAUTION:

On models equipped with an anti-theft audio system, disable the anti-theft feature before performing any operation that requires disconnecting the battery or disrupting power to the stereo (see the Anti-theft audio system procedures at the front of this manual).

64 Remove the engine cover (see Chapter 11).
65 Detach the vacuum hoses from the valve.
66 Remove the screws securing the valve to the engine bracket.
67 Remove the deceleration valve.
68 Installation is the reverse of removal.

9   Exhaust Gas Recirculation (EGR) system (2002 and earlier models)

GENERAL DESCRIPTION

1   The EGR system is used to lower Nox (oxides of nitrogen) emission levels caused by high combustion temperatures. The EGR valve controls the amount of exhaust gases allowed to enter into the fuel/air mixture which in turn lowers the combustion temperature.

2   Too much EGR flow tends to weaken combustion, causing the engine to run rough or stall. When EGR flow is excessive, the engine can stall after a cold start or at idle, after deceleration or the engine may surge at cruising speeds. If the EGR remains open all the time, the engine may not idle at all. Too little or no EGR flow allows combustion temperatures to get too high during acceleration and when a load is placed on the engine. This can cause spark knock (pinging), engine damage and emission test failure. Most EGR valve failure is related to carbon build-up around the valve inside the intake manifold. A fault in the EGR circuit will set a trouble code 32.

CARBURETED AND THROTTLE BODY INJECTION ENGINES

3   The EGR valve is mounted on the intake manifold, usually next to the carburetor or throttle body and is easily recognized by its disc shape. The EGR valve on these engines is a diaphragm type and operated by a controlled vacuum source. When vacuum is applied to the diaphragm, it overcomes the spring pressure and lifts the valve off the exhaust port. By varying the amount of vacuum applied, the EGR flow is regulated. On four-cylinder engines the vacuum is controlled by a ported fitting on the intake manifold that changes in relation to throttle position. On V6 engines vacuum to the EGR valve is controlled by the PCM via the EGR valve solenoid. The solenoid is turned on-and-off by the PCM in response to input signals from the Engine Coolant Temperature sender, Vehicle Speed Sensor and Intake Air Temperature sensors.

4   A procedure for performing a check of the EGR valve is included in Chapter 1.

CENTRAL MULTIPORT FUEL INJECTION (CMFI) AND CENTRAL SEQUENTIAL FUEL INJECTION (CSFI) ENGINES

5   The EGR valve on these engines (through 2002) is mounted on the front part of the intake manifold near the throttle body next to the thermostat housing. The EGR valve is an electronically controlled linear motor that moves a pintle in-and-out to control exhaust gas flow into the fuel/air mixture, much like the IAC valve controls the engine idle. No vacuum is used for EGR valve operation. The PCM monitors several engine sensor inputs and then commands the EGR valve to respond accordingly. Since this is performed with an electronic signal instead of vacuum like previous EGR systems, the EGR flow can be regulated with much more precision. Engines on 2003 and later models do not use an EGR valve

CHECK

4-cylinder engines

6   Unplug the vacuum hose from the EGR valve, then raise the engine speed to approximately 2,000 rpm while you feel the end of the hose - you should feel strong vacuum. If no vacuum is felt, check for cracked, plugged, collapsed or disconnected hoses. If vacuum is felt but the valve does not open, replace the valve. If the valve opens but has no effect on the engine at idle (see the check in Chapter 1), the passages are plugged in the valve or intake manifold; remove the valve and clean.

V6 carbureted and TBI engines

7   First, inspect the vacuum hoses between the intake manifold, EGR solenoid and the EGR valve. Look for cracking, hardening, general deterioration, loose fitting or misrouted hoses (refer to the vacuum hose routing diagram on the VECI label). Replace any suspect vacuum hoses. Make sure the hoses have the word "Fluoroelastomer" on them or they may fail prematurely.

8   When the EGR solenoid is energized by the PCM, it allows vacuum to open the EGR valve. The PCM monitors the EGR flow via the EGR temperature switch. The PCM actuates the EGR valve only when three conditions are present; engine coolant is above 113 degrees F, the TPS is at part-throttle and the MAP sensor is in mid-range. No EGR valve operation should occur with the vehicle in PARK or NEUTRAL, at engine idle or wide open throttle.

9   To check the basic system operation, see Chapter 1 and Step 6 of this Section. If no vacuum is present at the EGR valve and all hoses and connections are OK, take the vehicle to a dealer service department or other qualified shop for further diagnosis.

CMFI and CSFI engines

10 Make sure the electrical connection is clean and tight. Remove the valve, clean the pintle and intake manifold port (see below). Also, if applicable, make sure the pipe connecting the exhaust manifold to the intake manifold is not damaged and the fittings are tight.

11 To completely check out the linear EGR valve operation and its relationship with the PCM a special diagnostic interface tool is needed to communicate with the PCM. Therefore diagnosis of EGR system problems should be referred to a dealer service department or other qualified repair shop.

COMPONENT REPLACEMENT

▶ **Refer to illustration 9.27**

EGR valve (carbureted and TBI models)

12 When buying a new EGR valve, make sure that you have the right EGR valve.

13 Detach the cable from the negative terminal of the battery.

✳✳ CAUTION:

On models equipped with an anti-theft audio system, disable the anti-theft feature before performing any operation that requires disconnecting the battery or disrupting power to the stereo (see the Anti-theft audio system procedures at the front of this manual).

14 Remove the engine cover (see Chapter 11).

15 Remove the air cleaner housing assembly and adapter (see Chapter 4).

16 Detach the vacuum line from the EGR valve vacuum tube.

17 If you are replacing the EGR valve itself, trace the temperature switch pigtail lead to the main engine wire harness and unplug the

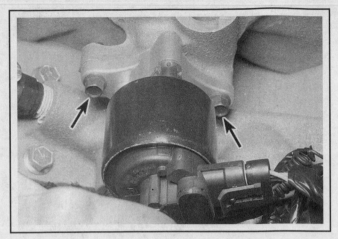

9.27 Disconnect the electrical connector from the EGR valve Then remove the mounting bolts (arrows) (CMFI engine shown)

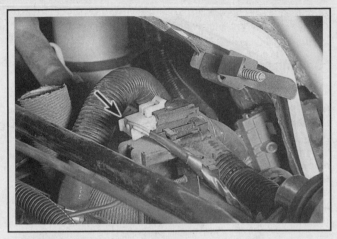

9.36 EGR valve solenoid electrical connector (arrow) (carbureted and TBI models)

electrical connector, then remove the switch from the valve. If you are replacing the EGR valve gasket, unplug the lead but do not remove the switch from the valve.

18 Remove the EGR valve mounting bolts.

19 Remove the EGR valve and gasket from the manifold. Discard the gasket.

20 With a wire wheel, buff the exhaust deposits from the EGR valve mounting surface on the manifold and, if you plan to reuse the same valve, the mounting surface of the valve itself. Look for exhaust deposits in the valve outlet. Remove deposit build-up with a screwdriver.

✸✸ CAUTION:

Never wash the valve in solvents or degreaser - both agents will permanently damage the diaphragm. Sand blasting is also not recommended because it will affect the operation of the valve.

21 If the EGR passage evinces an excessive build-up of deposits, clean it out with a wire wheel. Make sure that all loose particles are completely removed to prevent them from clogging the EGR valve or from being ingested into the engine.

22 Installation is the reverse of removal.

EGR valve (CMFI and 2002-and-earlier CSFI models)

23 When buying a new EGR valve, make sure you get the correct replacement part.

24 Detach the cable from the negative battery cable.

✸✸ CAUTION:

On models equipped with an anti-theft audio system, disable the anti-theft feature before performing any operation that requires disconnecting the battery or disrupting power to the stereo (see the Anti-theft audio system procedures at the front of this manual).

25 Remove the air cleaner assembly (see Chapter 4).

26 Disconnect the electrical connector from the EGR valve.

27 Remove the EGR mounting bolts (see illustration).

28 Remove the EGR valve and gasket from the intake manifold. Discard the gasket.

29 With the wire wheel, buff the exhaust deposits from the EGR valve mounting surface on the intake manifold and, if you plan to reuse the same valve, the mounting surface of the valve itself. Look for exhaust deposits in the valve outlet. Remove deposit build-up with a screwdriver.

✸✸ CAUTION:

Never wash the valve in solvent or degreaser and sandblasting is not recommended because it will affect the operation of the valve.

30 If the valve was cleaned, make sure all loose particles are completely removed to prevent them from clogging the EGR valve or from being ingested into the engine.

31 Installation is the reverse of removal.

Hoses (carbureted and TBI models)

32 When replacing hoses, use the VECI label (on the fan shroud) as a hose routing guide. For further information regarding hose inspection and service, see Chapter 1.

EGR vacuum solenoid (carbureted and TBI models)

▶ **Refer to illustration 9.36**

33 Detach the cable from the negative terminal of the battery.

✸✸ CAUTION:

On models equipped with an anti-theft audio system, disable the anti-theft feature before performing any operation that requires disconnecting the battery or disrupting power to the stereo (see the Anti-theft audio system procedures at the front of this manual).

34 Remove the engine cover (see Chapter 11).

35 Remove the air cleaner housing assembly and adapter (see Chapter 4).

36 Unplug the electrical connector from the solenoid (see illustration).

37 Clearly label, then detach, both vacuum hoses.
38 Remove the solenoid mounting screw and remove the solenoid.
39 Installation is the reverse of removal.

EGR temperature switch (carbureted and TBI models)

40 Detach the cable from the negative terminal of the battery.

⁂ CAUTION:

On models equipped with an anti-theft audio system, disable the anti-theft feature before performing any operation that requires disconnecting the battery or disrupting power to the stereo (see the Anti-theft audio system procedures at the front of this manual).

41 Remove the engine cover (see Chapter 11).
42 Trace the pigtail lead to the main engine wire harness and unplug the electrical connector.
43 Remove the switch.
44 Installation is the reverse of removal. Be sure to use anti-seize compound on the switch threads.

EGR tube (2002 and earlier CSFI engines only)

♦ **Refer to illustration 9.49**

45 Perform this operation when the engine is completely cool to avoid burning yourself.
46 Remove the engine cover (see Chapter 11).
47 Disconnect spark plug wires Nos. 1, 3 and 5 from the distributor

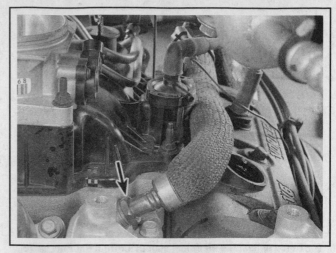

9.49 EGR tube fitting (arrow) at the intake manifold (V6 CSFI engines only)

cap and place them out of the way.
48 Remove the oil filler tube. Cover the opening to prevent contamination.
49 Loosen the EGR tube fittings at the intake and exhaust manifolds (see illustration). If difficulty is experienced, apply penetrating oil to the fittings to ease disassembly.
50 Working inside the vehicle, carefully maneuver the EGR tube from the engine.

10 Evaporative Emission Control System (EVAP)

GENERAL DESCRIPTION

♦ **Refer to illustration 10.1**

1   The function of the Evaporative Emissions Control System (EVAP), is to prevent fuel vapors from escaping the fuel system and being released into the atmosphere. Fuel vapors are trapped inside the fuel tank until the pressure overcomes the pressure relief valve and is then routed to a carbon canister via hoses for temporary storage (see illustration). When certain engine parameters are met, vacuum from the manifold opens the purge control valve and allows the fuel vapors from the canister to be drawn into the intake manifold and burned in the combustion process. To control this process, three methods are employed, depending on the engine and type of fuel delivery system used;

Ported vacuum source - Uses a Thermo Vacuum Switch (TVS) threaded into an intake manifold waterjacket which allows vacuum to pass through it only after the engine reaches approximately 120 degrees F.

Manifold vacuum with EVAP purge control valve - Uses a manifold vacuum signal to a vacuum operated purge control valve that is mounted on the engine or on the canister itself.

EVAP purge control solenoid - On later models, the EVAP purge control solenoid and valve are mounted on the right side of the upper intake manifold, in front of the distributor. The Powertrain Control Module (PCM) monitors engine operating parameters and then ener-

gizes the solenoid to allow purging of the canister.

2   The carbon canister on 1985 through 1995 models is located in the engine compartment, on the driver's side near the steering gear box. On 1996 and later models, it's located underneath the vehicle on the right frame rail or above the fuel tank. The canisters on 1996 and later models are maintenance-free and should last the life of the vehicle.

3   The most common symptom of a fault in the EVAP is a strong fuel odor or raw fuel leaking from the canister. These indications are usually more prevalent during hot temperatures.

4   On 1996 and later models, the fuel tank is equipped with a fuel tank pressure sensor which is mounted on top of the fuel tank on the fuel pump module. This sensor operates much like the MAP sensor and is used by the PCM to detect a fuel vapor leak. When the pressure inside the fuel tank is outside the parameters programmed into the PCM, a trouble code is set and the SERVICE ENGINE SOON light is illuminated on the instrument panel. The most common cause of system pressure loss is a loose or poor sealing gas cap. If the SERVICE ENGINE SOON light is illuminated, check the gas cap first! Also incorporated on these models is an EVAP canister vent valve which is mounted near the carbon canister and controlled by the PCM. On later models, the EVAP vent solenoid and valve are mounted on the crossmember, under the vehicle. This vent valve allows fresh air to enter the canister as the vapors are released to the engine (purged), and closes to allow the PCM to detect a leak in the system.

5   The fuel filler cap is equipped with a two-way pressure-vacuum relief valve as a safety device. If the pressure inside the tank exceeds

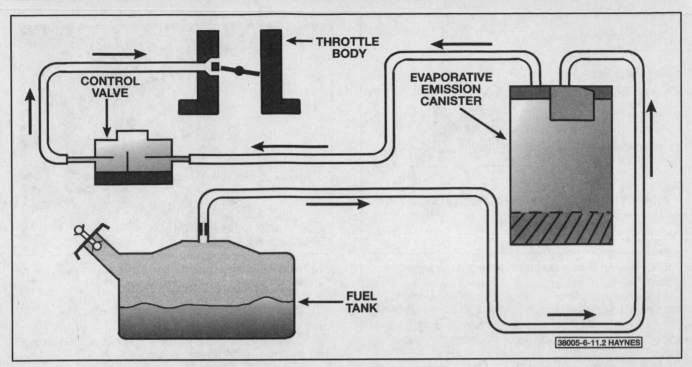

10.1 Typical EECS system flow diagram

approximately 1.58 to 1.98 psi, the relief valve vents the fuel vapors to the atmosphere. If the vacuum pressure inside the tank becomes greater than approximately 0.14 psi, the relief valve allows fresh air to be drawn into the tank.

CHECK

General

♦ **Refer to illustration 10.8**

6   First, inspect all vacuum and vapor hoses between the intake manifold, purge control valve or solenoid and the carbon canister. Look for cracking, hardening, general deterioration, loose fitting, kinked or misrouted hoses. Replace hoses as required. Make sure the hoses have the word "Fluoroelastomer" on them or they may fail prematurely.

7   On models with electronically controlled purge solenoids, check the wires for obvious damage and make sure the connections are clean and tight.

8   Examine the carbon canister for cracks or other damage. Replace if necessary. On early models, the carbon canister has a serviceable filter located at the bottom of the canister. If equipped, check the filter and replace it if it's dirty (see illustration).

9   Check the gas cap seal for defects or other damage that may affect its sealing function. Replace if necessary.

Thermo Vacuum Switch and EVAP purge control solenoid

10   With the engine cold, locate the Thermo Vacuum Switch (TVS) or the electronic purge control solenoid. Start the engine, disconnect the vacuum hose leading to the carbon canister and check for vacuum at the TVS or solenoid vacuum fitting. No vacuum should be present. Let the engine reach operating temperature and perform this check again. Vacuum should now be present.

➡**Note: The solenoid will not be energized by the PCM unless the engine temperature is over 115 degrees F and the air pres-**

10.8 To replace the canister filter, peel it out and install a new one

sure inside the fuel tank exceeds 0.7 psi. If the solenoid does not operate when the engine reaches normal operating temperature, park the vehicle in the hot sun for awhile to create pressure in the fuel tank and try this test again. If the TVS does not operate as described, replace it. If the electronic solenoid does not operate as described, further diagnosis should be performed by a dealer service department or other qualified repair shop.

Vacuum controlled purge control valve

♦ **Refer to illustration 10.12**

11   Locate the vacuum operated purge control valve. Label and disconnect the hoses.

12   Attach a section of hose to the canister tube (see illustration). Try to blow through the tube - little or no air should pass through.

13   Next, connect a hand-held vacuum pump to the control vacuum tube and apply approximately 15 in-Hg of vacuum. Try and blow

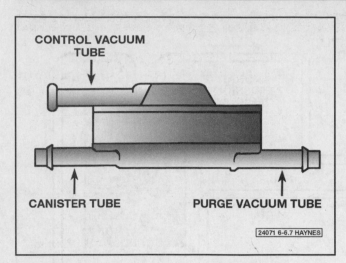

10.12 Identification of the tubes on the canister purge control valve

through the canister tube again: an increase of air flow should be noticed. If vacuum will not develop or hold steady, or a significant change in air flow is not detected, replace the purge control valve.

COMPONENT REPLACEMENT

Thermo Vacuum Switch

14 Locate the TVS, label and disconnect the vacuum hoses, then unscrew the TVS from the intake manifold. Before removing the TVS, be prepared to install a plug or the new TVS immediately to minimize coolant loss.

15 Wrap the TVS threads with Teflon tape, install the new sensor into the manifold and tighten it securely. Attach the vacuum hoses.

EVAP purge control or vent solenoid

▶ **Refer to illustration 10.16**

16 Disconnect the electrical harness connector from the solenoid (see illustration). Label and detach the vacuum and vapor hoses as applicable.

10.19 To remove the carbon canister, label and detach the vacuum lines, then remove the canister clamp bolt (arrow) and lift the canister out (grille, radiator and condenser removed for clarity)

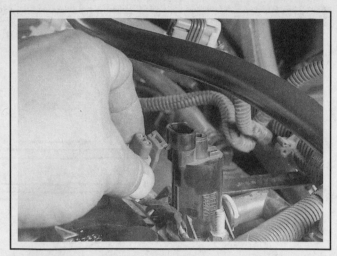

10.16 Disconnecting the EVAP purge control solenoid electrical connector (V6 CSFI engine shown)

17 Remove the mounting screws (if used) and separate the solenoid from the bracket or manifold.

18 Installation is the reverse of removal. On 1996 and later models, check the O-ring seal for damage and replace if necessary.

Carbon canister

▶ **Refer to illustrations 10.19 and 10.20**

19 On 1995 and earlier models, working in the engine compartment, clearly label and then detach the hoses from the canister. Loosen the canister clamp bolt and withdraw the canister (see illustration).

20 On 1996 and later models, raise the vehicle and support it securely on jackstands. Clearly label and then detach the hoses from the canister. Remove the mounting bolt securing the canister to the bracket and then slide it out of the pinch-clamp bracket (see illustration).

21 On early models, if the canister is to be reinstalled, check the filter (if equipped) and replace it, if necessary (see Step 8).

22 Installation is the reverse of removal.

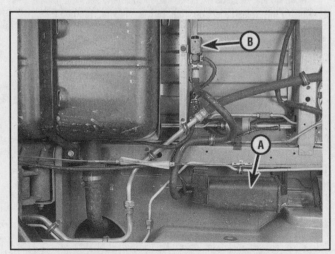

10.20 On 1996 through 2000 models, the carbon canister (arrow A) and canister vent valve (arrow B) are located under the vehicle in front of the fuel tank (on 2001 through 2005 the carbon canister is located on top of the fuel tank)

11 Positive Crankcase Ventilation (PCV) system

▶ Refer to illustration 11.1

1   The Positive Crankcase Ventilation (PCV) system reduces hydrocarbon emissions by scavenging crankcase vapors. It does this by circulating fresh air from the air cleaner through the crankcase, where it mixes with blow-by gases and is then rerouted through a PCV valve to the intake manifold (see illustration).

2   The main components of the PCV system are the PCV valve, a fresh air filtered inlet and the vacuum hoses connecting these two components with the engine and the EECS system.

3   To maintain idle quality, the PCV valve restricts the flow when the intake manifold vacuum is high. If abnormal operating conditions arise, the system is designed to allow excessive amounts of blow-by gases to flow back through the crankcase vent tube into the air cleaner to be consumed by normal combustion.

4   Checking and replacement of the PCV valve and filter is covered in Chapter 1.

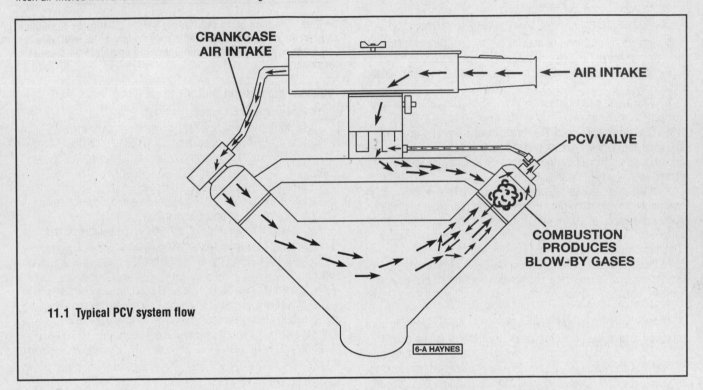

11.1 Typical PCV system flow

12 Thermostatic air cleaner (THERMAC) (carbureted and TBI models)

▶ Refer to illustrations 12.2, 12.18, 12.27 and 12.28

GENERAL DESCRIPTION

1   A heated air intake system is used to provide good driveability under varying climatic conditions. By having a uniform inlet air temperature, the fuel system can be calibrated to reduce exhaust emissions and to eliminate throttle valve icing.

2   The THERMAC air cleaner (see illustration) is operated by heated air and manifold vacuum. Air can enter the air cleaner from outside the engine compartment or from a heat stove built around the exhaust manifold. A temperature sensor located inside the air cleaner housing determines the operational mode of the THERMAC.

Checking

3   If the engine hesitates during warm-up:
   a) *The heat stove tube could be disconnected.*

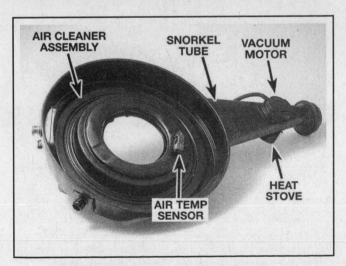

12.2 Typical THERMAC system component layout

b) *The vacuum diaphragm motor could be inoperative, leaving the snorkel (mouth) of the air cleaner housing open to outside air.*
c) *There may be no manifold vacuum.*
d) *The damper door does not move.*
e) *The air cleaner housing assembly-to-TBI adapter seal may be missing.*
f) *The air cleaner housing assembly cover seal may be missing.*
g) *The air cleaner housing assembly cover may be loose.*
h) *The air cleaner housing assembly may be loose.*

4   Lack of power and/or sluggish or spongy throttle response can be caused by:

a) *A stuck damper door that does not open to outside air.*
b) *A temperature sensor that does not bleed off vacuum.*

5   Inspect the system to be sure that all hoses and the heat stove tube are connected. Check for kinked, plugged or deteriorated hoses.

6   Check for the presence and condition of air cleaner to carburetor gasket seal.

7   With the air cleaner assembly installed, the damper door should be open to outside air.

8   Start the engine. Watch the damper door in the air cleaner snorkel. When the engine is first started, the damper door should move and close off outside air. As the air cleaner warms up, the damper door should open slowly to the outside air.

9   If the air cleaner fails to operate as described above, the door may not be moving at the right temperature. If the driveability problem is during warm-up, make the temperature sensor check below.

10 With the engine off, disconnect the vacuum hose at the vacuum diaphragm motor.

11 Apply at least 7 inches of vacuum to the vacuum diaphragm motor. The damper door should completely block off outside air when vacuum is applied. If it doesn't, check to see if the linkage is hooked up correctly.

12 With vacuum still applied, trap vacuum in the vacuum diaphragm motor by bending the hose. The damper door should remain closed. If it doesn't, replace the vacuum diaphragm motor assembly. Failure of the vacuum diaphragm motor is more likely to be caused from binding linkage or a corroded snorkel than from a failed diaphragm. Check this first before replacing the diaphragm.

13 If the vacuum motor checks out okay, check the vacuum hoses and connections. If they're okay, replace the temperature sensor.

14 Start the test with the air cleaner temperature below 86-degrees F. If the engine has been run recently, remove the air cleaner cover and place the thermometer as close as possible to the sensor. Let the air cleaner cool until the thermometer reads below 86-degrees F for about

5 to 10 minutes. Reinstall the air cleaner.

15 Start and idle the engine. The damper door should move to close off outside air immediately if the engine is cool enough. When the damper door starts to open the snorkel passage (in a few minutes), remove the air cleaner and read the thermometer. It must read about 131-degrees F.

16 If the damper door is not open to outside air at the indicated temperature, the temperature sensor is malfunctioning and must be replaced.

COMPONENT REPLACEMENT

→**Note: Check on parts availability before attempting component replacement. Some components may no longer be available for all models, requiring replacement of the complete air cleaner assembly.**

Vacuum diaphragm motor

17 Remove the air cleaner assembly (see Chapter 4).

18 Detach the vacuum tube from the motor (see illustration).

19 Drill out the two spot welds with a 1/16-inch drill, then enlarge as required to remove the retaining strap. Do not damage the snorkel tube.

20 Bend the strap up and out of the way.

21 Lift up the motor, cocking it to one side to unhook the motor linkage at the control damper assembly.

22 To install a new motor, drill a 7/64-inch hole in the snorkel tube at the center of the vacuum motor retaining strap.

23 Install the vacuum motor linkage into the control damper assembly.

24 Use the motor retaining strap and sheet metal screw provided in the motor service package to secure the motor to the snorkel tube. Make sure that the screw does not interfere with the operation of the damper assembly. Shorten the screw if necessary.

25 Install the air cleaner housing assembly (see Chapter 4). Be sure to attach the vacuum hose to the motor.

Sensor

26 Remove the air cleaner housing assembly (see Chapter 4).

27 Note the position (see illustration) of the sensor in the air cleaner housing to facilitate reinstallation.

28 Pry up the tabs on the sensor retaining clip (see illustration). Remove the clip and sensor from the air cleaner.

29 Installation is the reverse of removal. Be sure to attach the two vacuum lines to the sensor pipes.

13 Catalytic converter

GENERAL DESCRIPTION

1   The catalytic converter is an emission control device added to the exhaust system to reduce pollutants from the exhaust gas stream. A single-bed converter design is used in combination with a three-way (reduction) catalyst. The catalytic coating on the three-way catalyst contains platinum and rhodium, which lowers the levels of oxides of nitrogen (NOx) as well as hydrocarbons (HC) and carbon monoxide (CO).

CHECKING

2   The test equipment for a catalytic converter is expensive and highly sophisticated. If you suspect that the converter on your vehicle is malfunctioning, take it to a dealer or authorized emissions inspection facility for diagnosis and repair.

3   Whenever the vehicle is raised for servicing of underbody components, check the converter for leaks, corrosion and other damage. If damage is discovered, the converter should be replaced.

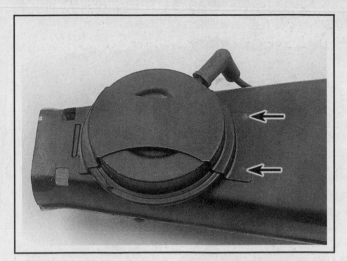

12.18 Once the air cleaner housing assembly is removed from the TBI unit, turn it upside down and detach the vacuum line from the temperature sensor, drill out the spot welds (arrows) with a 1/16-inch drill, then enlarge as necessary to remove the retaining strap

4   Because the converter is part of the exhaust system, converter replacement requires removal of the exhaust pipe assembly (see Chapter 4). Take the vehicle, or the exhaust pipe system, to a dealer or a muffler shop.

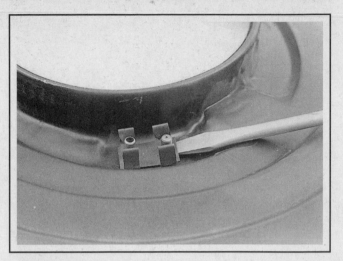

12.27 Pry off the sensor retaining clips with a small screwdriver

12.28 Note the position of the sensor in the air cleaner housing

Notes

Section

7A

MANUAL
TRANSMISSION

1 General information

The vehicles covered in this manual equipped with manual transmissions are available with either a 76mm 4-speed or a 77mm 5-speed (the figure represents the distance between the mainshaft and the countershaft centerlines). The 4-speed transmission utilizes an adjustable external gearshift linkage, while the 5-speed shift control is in a housing bolted to the rear of the transmission.

2 Shift lever (5-speed) - removal and installation

REMOVAL

1   Remove the screws from the shift lever boot retainer and slide the boot and retainer up the lever as far as possible.
2   Using a wrench on the slots provided at the bottom of the upper shift lever, loosen the lever locknut and unscrew the upper lever from the transmission shift control lever.

INSTALLATION

3   Turn the nut on the transmission shift control lever to the bottom of its thread travel.
4   Install the upper shift lever and thread it down the shaft until it seats against the nut. Turn the lever counterclockwise to align the shift pattern on the knob, then, while holding the upper shift lever with a wrench, tighten the lower lever nut securely against the shift lever.
5   Install the shift lever boot and retainer.

3 Shift linkage and control (4-speed) - removal, installation and adjustment

REMOVAL

1   Disconnect the cable from the negative terminal of the battery.

✳✳ CAUTION:

On models equipped with an anti-theft audio system, disable the anti-theft feature before performing any operation that requires disconnecting the battery or disrupting power to the stereo (see the Anti-theft audio system procedures at the front of this manual).

2   Unscrew the shift knob from the shift lever.
3   Remove the shift boot attaching bolts and carefully pull the boot off the shift lever.
4   Raise the vehicle and support it securely on jackstands.
5   Remove the clips and washers from the shift linkage rods and disconnect the rods from the shifter control levers. It is a good idea to mark the rods with tape to ensure correct installation.
6   Remove the shift control mounting bolts and guide the assembly downward until the shift lever is clear of the floorpan.
7   To remove the shift linkage rods from the transmission, remove the clips and washers that retain the rods to the transmission levers. It is not necessary to remove the nuts from the threaded portion of the rods. Again, be sure to mark the rods as to their relative positions so they can be reinstalled properly.

INSTALLATION AND ADJUSTMENT

8   To install the shift linkage rods and/or the shift control, reverse the removal procedure then adjust the linkage as described in the next two Steps.
9   Place the shift control and the transmission levers in the Neutral positions (see illustration). If the linkage rods are too short or too long, loosen the adjuster nuts on the shift rods and back them off so the levers will remain in the Neutral position.
10   Insert a 1/4-inch drill bit into the gauge pin hole. It should pass through the hole freely. If it does not, lengthen or shorten the shift rods accordingly (see illustration).
11   Tighten the shift rod adjuster nuts.

4 Shift control assembly (5-speed) - removal and installation

REMOVAL

1   Remove the transmission as outlined in Section 5.
2   Carefully pry off the clips that retain the dust cover to the control assembly.
3   Remove the shift control-to-transmission housing bolts and lift the control from the transmission.
4   Clean all traces of old RTV sealer from the dust cover and control.

INSTALLATION

5   Mount the shift control assembly to the transmission and tighten the bolts securely.
6   Apply a bead of RTV sealer to the groove in the dust cover, position the cover on the control assembly and install the retaining clips.
7   Install the transmission.

5 Transmission - removal and installation

REMOVAL

1 Disconnect the cable from the negative terminal of the battery.

✳✳ CAUTION:

On models equipped with an anti-theft audio system, disable the anti-theft feature before performing any operation that requires disconnecting the battery or disrupting power to the stereo (see the Anti-theft audio system procedures at the front of this manual).

2 Drain the transmission lubricant (Chapter 1).
3 Remove the shift lever (5-speed models). Refer to Section 2.
4 Raise the vehicle and support it securely on jackstands.
5 Disconnect the shift linkage rods from the transmission levers and the shift control assembly (4-speed models). Refer to Section 3.
6 Remove the shift control assembly (4-speed models).
7 Remove the driveshaft (refer to Chapter 8).
8 Disconnect the speedometer cable from the transmission.
9 Remove the exhaust system (refer to Chapter 4).
10 Remove the transmission-to-engine support braces.
➡**Note the position of any washers and spacers.**

11 Support the transmission with a floor jack, unbolt the transmission mount from the crossmember then remove the crossmember from the vehicle.

12 Slowly lower the floor jack slightly (maintain contact with the transmission) then remove the transmission-to-bellhousing bolts. Do not let the transmission hang unsupported, as damage to the input shaft may result.
13 Pull the transmission straight back and out of the clutch hub splines and lower it to the ground.

INSTALLATION

14 Fill the transmission with the recommended lubricant (Chapter 1).
15 Apply a light coat of high temperature grease to the input shaft splines.
16 Set the transmission on the floor jack and raise it into position with the input shaft in alignment with the clutch disc splined hub. Slide the transmission forward until it is seated against the bellhousing. If it doesn't slide in easily, don't apply excessive force. Try engaging the transmission in gear and turning the output shaft to line up the input shaft splines with those in the clutch disc hub.
17 Install the transmission-to-bellhousing bolts and tighten them to the specified torque.
18 Raise the transmission on the jack and install the rear mount and the crossmember.
19 The remainder of installation is the reverse of the removal procedure.

6 Transmission overhaul - general information

Overhauling a manual transmission is a difficult job for a do-it-yourselfer. It involves the disassembly and reassembly of many small parts. Numerous clearances must be precisely measured and, if necessary, changed with select fit spacers and snap-rings. As a result, if transmission problems arise, it can be removed and installed by a competent do-it-yourselfer, but overhaul should be left to a transmission repair shop. Rebuilt transmissions may be available - check with your dealer parts department and auto parts stores. At any rate, the time and money involved in an overhaul is almost sure to exceed the cost of a rebuilt unit.

Nevertheless, it's not impossible for an inexperienced mechanic to rebuild a transmission if the special tools are available and the job is done in a deliberate step-by-step manner so nothing is overlooked.

The tools necessary for an overhaul include internal and external snap-ring pliers, a bearing puller, a slide hammer, a set of pin punches, a dial indicator and possibly a hydraulic press. In addition, a large, sturdy workbench and a vise or transmission stand will be required.

During disassembly of the transmission, make careful notes of how each piece comes off, where it fits in relation to other pieces and what holds it in place.

Before taking the transmission apart for repair, it will help if you have some idea what area of the transmission is malfunctioning. Certain problems can be closely tied to specific areas in the transmission, which can make component examination and replacement easier. Refer to the *Troubleshooting* section at the front of this manual for information regarding possible sources of trouble.

Specifications

Torque specifications

	Ft-lbs
Drain and fill plugs	17
Transmission-to-bellhousing bolts	50
Extension housing-to-case bolts	
4-speed	45
5-speed	25
Control assembly mounting bolts (5-speed)	23
Shifter-to-shift bracket bolts (4-speed)	33
Cover-to-case bolts (4-speed)	15

7B

AUTOMATIC TRANSMISSION

Section

Reference to other Chapters

1 General information

The automatic transmission installed in this vehicle is a 4-speed overdrive unit with a clutch-type torque converter. The clutch is designed to engage at speeds above 25 mph and provides a direct connection between the engine and the drive wheels for better efficiency and fuel economy. When the shift lever is in the "Overdrive D" position, the overdrive gear range is automatically selected when the vehicle reaches a steady speed above 40 mph.

Due to the complexity of the clutches and the hydraulic control system, and because of the special tools and expertise required to perform an automatic transmission overhaul, it should not be undertaken by the home mechanic. Therefore, the procedures in this Chapter are limited to general diagnosis, routine maintenance, adjustment and transmission removal and installation.

If the transmission requires major repair work, it should be left to a dealer service department or a transmission repair shop. You can, however, remove and install the transmission yourself and save the expense, even if the repair work is done by a transmission specialist.

Adjustments that the home mechanic can perform include those involving the throttle valve (TV) cable and the shift linkage.

❊❊ CAUTION 1:

Never tow a disabled vehicle with an automatic transmission at speeds greater than 30 mph or distances over 50 miles.

❊❊ CAUTION 2:

On models equipped with an anti-theft audio system, disable the anti-theft feature before performing any operation that requires disconnecting the battery or disrupting power to the stereo (see the Anti-theft audio system procedures at the front of this manual).

2 Diagnosis - general

1 Automatic transmission malfunctions may be caused by a number of conditions, such as poor engine performance, improper adjustments, hydraulic malfunctions and mechanical problems.

2 The first check should be of the transmission fluid level and condition. Refer to Chapter 1 for more information. Unless the fluid and filter have been recently changed, drain the fluid and replace the filter (also in Chapter 1).

3 Road test the vehicle and drive in all the gear ranges, noting discrepancies in operation. Check as follows:

Overdrive range: While stopped, position the lever in the overdrive range and accelerate. Check for a 1-2 shift, 2-3 shift and 3-4 shift. Also, the converter clutch should apply in 2nd or 3rd gear, depending on calibration. Check for part-throttle downshift by depressing the accelerator 3/4 of the way - the transmission should downshift. Check for a full detent downshift by depressing the throttle all the way - the transmission should immediately downshift.

Drive range: At road speed in Fourth gear (Overdrive range), manually shift the transmission to Drive range. The transmission should shift to Third gear immediately. It should not shift back to Overdrive. Also check for part throttle and full throttle downshifts in this range.

Drive 2 range: While in the Third gear range, shift to Second gear.

The transmission should downshift immediately. While in the Second gear range, check for downshift at part and full throttle.

Low range: Position the lever in Low (1) position and check the operation.

Overrun braking: This can be checked by manually shifting to a lower range. Engine rpm should increase and a braking effect should be noticed.

Reverse: Position the shifter in Reverse and check operation.

4 Verify that the engine is not at fault. If the engine has not received a tune-up recently, refer to Chapter 1 and make sure all engine components are functioning properly.

5 Check the adjustment of the throttle valve (TV) cable (see Section 3).

6 Check the condition of all electrical wires and connectors at the transmission, or leading to it.

7 Check for proper adjustment of the shift linkage (see Section 5).

8 If at this point a problem remains, there is one final check before the transmission is removed for overhaul. The vehicle should be taken to a specialist who will connect a special oil pressure gauge and check the line pressure in the transmission under different driving conditions.

3 Throttle valve (TV) cable (1995 and earlier) (700-R4 and 4L60 models) - description, inspection and adjustment

DESCRIPTION

1 The throttle valve cable used on these transmissions should not be thought of as merely a "downshift" cable, as in earlier transmissions. The TV cable controls line pressure, shift points, shift feel, part throttle downshifts and detent downshifts.

➡Note: Later models use a 4L60-E (E for "electronic") or 4L65-E automatic transmission and do not have a throttle valve or cable. The PCM performs the same functions by means of various solenoids.

2 If the TV cable is broken, sticky, misadjusted or is the incorrect part, the vehicle will experience a number of problems.

INSPECTION

▸ **Refer to illustration 3.4**

3 Remove the engine cover (see Chapter 11). Inspection should be made with the engine running at idle speed with the selector lever in Neutral. Set the parking brake firmly and block the wheels to prevent any vehicle movement. As an added precaution, have an assistant in the driver's seat applying the brake pedal.

4 Grab the inner cable a few inches behind where it attaches to the throttle linkage and pull the cable forward. It should easily slide through the cable housing with no binding or jerky operation (see illustration).

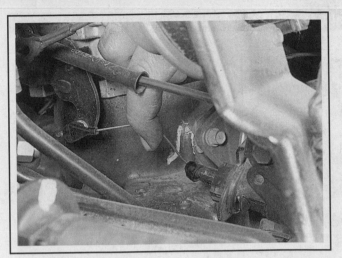

3.4 To check for free operation, pull forward on the inner cable, feeling for smooth operation through the full range of travel - the cable should retract evenly and rapidly when released

5 Release the cable and it should return to its original location with the cable stop against the cable terminal.

6 If the TV cable does not operate as above, the cause is a defective or misadjusted cable or damaged components at either end of the cable.

ADJUSTMENT

▶ Refer to illustration 3.8

7 The engine should not be running during this adjustment.

8 Depress the re-adjust tab and move the slider back through the fitting away from the throttle linkage until the slider stops against the fitting (see illustration).

3.8 Depress the throttle valve (TV) cable re-adjust tab (A) and pull the slider back (arrow) until it rests on its stop - release the tab and open the throttle completely

9 Release the re-adjust tab.

10 Turn the throttle lever to the "wide open throttle" position, which will automatically adjust the cable. Release the throttle lever.

❋❋ CAUTION:

Don't use excessive force at the throttle lever to adjust the TV cable. If great effort is required to adjust the cable, disconnect the cable at the transmission end and check for free operation. If it's still difficult, replace the cable. If it's now free, suspect a bent TV link in the transmission or a problem with the throttle lever.

11 After adjustment check for proper operation as described in Steps 3 through 6 above.

4   Throttle valve (TV) cable (700 R4 and 4L60 models) - replacement

1 Disconnect the cable from the negative terminal of the battery.

❋❋ CAUTION:

On models equipped with an anti-theft audio system, disable the anti-theft feature before performing any operation that requires disconnecting the battery or disrupting power to the stereo (see the Anti-theft audio system procedures at the front of this manual).

2 Lift off the engine cover and remove the air cleaner housing to gain access to the TV cable end (refer to Chapter 4 if necessary).

3 Disconnect the cable terminal from the throttle lever by pushing forward and then off.

4 Push in on the re-adjust tab and move the slider back through the fitting in the direction away from the throttle lever (towards the rear of the vehicle).

5 Compress the locking tabs and disconnect the cable assembly from the bracket.

6 Raise the vehicle and place it securely on jackstands.

7 Release the cable from the mounting clips, noting how it's routed.

8 Remove the screw and washer securing the cable to the transmission. Pull the cable up and disconnect it from the link at the end of the cable.

9 Place a new seal in the transmission case hole.

10 Connect the transmission end of the cable to the link and secure it to the transmission case with the screw. Tighten the screw securely.

11 Route the new cable up and secure it to the mounting clips.

12 Pass the cable through the bracket and secure it at the bracket with the locking tabs.

13 Connect the cable terminal to the throttle lever and adjust the cable as described in the previous Section.

14 Connect the negative battery cable and check for proper operation.

5   Shift linkage (1995 and earlier models) - removal, installation and adjustment

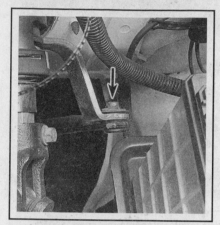

5.2 To disconnect the shift linkage rod from the shift lever on the steering column (arrow), pull out the retaining clip, remove the washer and pry the rod from the lever

5.3 Loosen the nut securing the shift rod swivel to the equalizer lever then slide the rod out of the swivel

5.4 To remove the equalizer lever from the transmission shift lever, pull out the retaining clip (arrow) and push the equalizer lever up and out

REMOVAL

▶ **Refer to illustrations 5.2, 5.3 and 5.4**

1   Apply the parking brake, raise the vehicle and support it securely on jackstands.

2   Pull the retaining clip from the shift linkage rod at the lower end of the steering column and separate the rod from the column lever (see illustration).

3   Loosen the nut or the swivel on the linkage rod and separate the rod from the equalizer lever (see illustration).

4   Remove the retaining clip from the equalizer lever and push the equalizer lever out of the shift lever on the transmission (see illustration).

INSTALLATION

5   Before reassembling, clean the rubber or nylon parts with soapy water. Check them for cracks and wear, replacing them with new ones if necessary. The metal parts can be washed in solvent.

6   To install the shift linkage, reverse the removal procedure, then adjust it as described in the following Steps.

ADJUSTMENT

▶ **Refer to illustration 5.9**

7   Loosen the nut securing the shift rod swivel to the equalizer lever.

8   Place the column selector lever in the Neutral position gate (don't use the indicator to find the position).

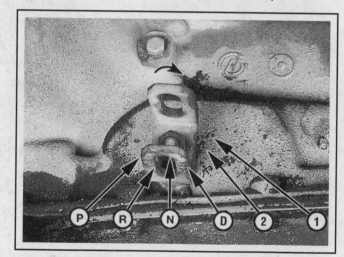

5.9 Shift positions on the transmission shift lever

9   Put the transmission shift lever in the Neutral position by turning the bottom of the lever to the forward (Park) position, then back to the second detent (two clicks) (see illustration).

10   Hold the rod tightly in the swivel and tighten the swivel nut securely.

11   Shift the selector lever into the Park position and check the adjustment with the engine running and your foot on the brake. The column selector lever must go into all of the positions easily.

12   Make sure the engine starts in the Park and Neutral positions only. If the engine cranks over in any other positions, adjust the Neutral safety switch as outlined in Section 9.

13   Align the indicator needle if necessary.

6   Transmission output shaft seal - replacement

▶ **Refer to illustrations 6.3 and 6.5**

1   Raise the vehicle and support it securely on jackstands.
2   Remove the driveshaft (see Chapter 8).

3   Using a seal removal tool or a long screwdriver, pry the old seal from the end of the transmission (see illustration).

4   Compare the new seal with the old to make sure they are the same.

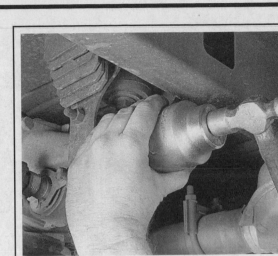

6.3 Use a seal removal tool or a long screwdriver to carefully pry the seal out of the end of the transmission

5   Drive the new seal into position using a large socket or a piece of pipe which is the same diameter as the seal (see illustration).
6   Once in position, coat the lips of the new seal with automatic

6.5 A large socket works well for installing the seal - the socket should contact the outer edge of the seal

transmission fluid.
7   Reinstall the various components in the reverse order of removal, referring to the necessary Chapters as needed.

7   Transmission - removal and installation

REMOVAL

▶ **Refer to illustrations 7.14, 7.15a, 7.15b, 7.18 and 7.22**

1   Disconnect the cable from the negative terminal of the battery.

❊❊ CAUTION:

On models equipped with an anti-theft audio system, disable the anti-theft feature before performing any operation that requires disconnecting the battery or disrupting power to the stereo (see the Anti-theft audio system procedures at the front of this manual).

2   Remove the engine cover (see Chapter 11) and the transmission fluid dipstick.
3   On carbureted and TBI engines, remove the air cleaner assembly (see Chapter 4).
4   Raise the vehicle and support it securely on jackstands.
5   Drain the transmission fluid, then reinstall the pan (see Chapter 1).
6   On 700-R4 and older 4L60 models, disconnect the TV cable from the transmission (see Section 4) and place it out of the way. Plug the cable hole to prevent leakage and contamination.
7   On 1995 and earlier models, disconnect the shift linkage from the transmission (see Section 5).
8   On 1996 and later models, disconnect the shift cable from the transmission (see Section 12).
9   On models equipped with a speedometer cable, disconnect it from the transmission and route it out of the way (see Section 10).
10   Clearly label and disconnect all electrical harness connectors from the transmission.
11   Remove the driveshaft (see Chapter 8).
12   Remove any exhaust components which will interfere with trans-

7.14 Remove the four converter housing cover bolts and carefully work the cover toward the left side of the vehicle - it's a tight fit, but it will come out!

mission removal (see Chapter 4).
13   If equipped, remove the engine-to-transmission struts or transmission braces.
14   Remove the converter housing lower cover (see illustration) to access the torque-converter-to-driveplate bolts. To access to each bolt, rotate the crankshaft clockwise using the crankshaft pulley bolt.
➡**Note: On 1996 and later models, you can access the torque-converter bolts through the starter motor opening, without removing the torque converter housing lower cover.**

15   Match-mark the driveplate to the torque converter and then remove the (3) torque converter-to-driveplate bolts (see illustrations).
16   Position a transmission jack under the transmission and secure the transmission to the jack. Raise the transmission slightly.

7.15a Mark the relationship of the torque converter to the driveplate

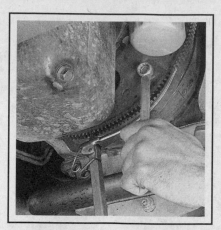

7.15b Remove the converter-to-driveplate retaining bolts (here, a flywheel wrench is being used to prevent the driveplate from turning, but a wrench on the crankshaft vibration damper bolt will also work)

7.18 Remove the angled bracket bolts on both sides of the vehicle (arrows) and the crossmember mounting bolts (being removed here with a socket and back-up wrench)

17 Unbolt the transmission mount from the transmission (see Section 8).

18 Remove the two angular brackets behind the crossmember and the crossmember mounting bolts, then remove the crossmember from the vehicle (see illustration).

19 Lower the transmission far enough to gain access to the other components. Be careful not to stretch any wires or cables.

20 Remove the transmission-to-engine bolts. Also remove the dipstick tube (it is normally secured by one of the transmission-to-engine bolts).

21 Disconnect the transmission cooler lines at the transmission. On 1995 and earlier models, use a flare-nut wrench on the tube nut and a back-up wrench on the fitting to prevent twisting the tubing. On 1996 and later models, the cooler lines are quick-connect type and secured with a spring clip, which cannot be reused. Pull back the plastic cover protecting the fitting, remove the clip and then separate the line from the fitting. On all models, cap the fittings to prevent leakage and contamination.

22 If equipped, remove the transmission damper and support (see illustration).

23 Support the engine with a floor jack and a block of wood or an engine hoist so movement will be kept to a minimum.

➡**Note: Once the transmission has been removed, the rear of the engine must remain supported at all times.**

24 Remove the starter motor and shield, if equipped (see Chapter 5).

25 Make a final check that all wires, cables, etc. are disconnected from the transmission, then separate the transmission from the engine. As this is done, the torque converter must be held back against the transmission. A special holding fixture is available or you can use a piece of bar stock across the front of the transmission for this purpose.

INSTALLATION

26 If the torque converter was removed for any reason, position it on the input shaft and gently push it in while rotating it. The converter is seated after two clunks are felt (the converter mating with the splined shafts).

27 Raise the transmission into position and push it forward onto the locating pins. Before installing the driveplate-to-converter bolts, make

7.22 The transmission damper is retained by two bolts and nuts

sure the weld nuts on the converter are flush with the driveplate and the converter rotates freely (this is another way to confirm that the converter is completely seated). Align the marks you made during removal and install the driveplate-to-converter bolts, tightening them finger tight. Once all bolts are installed, tighten them to the specified torque listed in this Chapter's Specifications.

28 Install a new oil filler tube O-ring in the case bore.

29 Install the various components in the reverse order of removal, referring to the proper Chapters where necessary for additional information.

30 After all components are installed, adjust the shift linkage (see Section 5), throttle valve cable (see Section 3), or shift cable (see Section 12) as applicable.

31 On 1996 and later models, the cooler lines at the radiator are quick-connect type and secured with a spring clip, which should not be reused. Begin assembly by installing the new spring clips so the spring ears engage in the retaining slots and are visible at all 3 places. Make sure the clips move freely once they're installed. Next, push the line into the fitting until you hear it "click" into place. When the line and fitting are correctly mated, the yellow band around the line should

NOT be visible. Confirm a secure connection by trying to pull it apart. Finally, slide the plastic cover over the fitting.

32 If the transmission has been drained and a new torque converter installed or the fluid has been drained from the existing one, a slightly different fluid filling procedure is required.

- a) Add approximately 5-1/2 quarts of transmission fluid through the filler tube.
- b) Start the engine and, with the shifter in Park, depress the accelerator enough to get the engine operating at a fast idle speed - do not race the engine.
- c) With your foot firmly on the brake pedal, move the shifter through each range.
- d) After 1 to 3 minutes of running at idle, check the fluid level with the shifter in Park (engine still running).
- e) Add enough fluid to bring the level to a point between the two marks on the dipstick. Add fluid only a little at a time, waiting 1 to 3 minutes, to prevent overfilling the transmission.

8 Transmission mount - replacement

▶ **Refer to illustrations 8.3, 8.4, 8.5 and 8.6**

1   Raise the vehicle and support it securely on jackstands.
2   Support the transmission with a floor jack and a block of wood.
3   Remove the mount-to-support bracket nut and the two support bracket-to-crossmember nuts and bolts (see illustration).
4   Remove the two mount-to-transmission bolts (see illustration).

5   Remove the two upper support bracket insulators from the support bracket (see illustration).
6   Slide the support bracket and mount out from between the transmission and the crossmember (see illustration).
7   Pull the mount from the support bracket. Installation is the reverse of the removal procedure.

8.3 The transmission mount-to-support bracket nut (center) should be removed before the support bracket bolts are removed

8.4 Mount-to-transmission bolt locations (arrows)

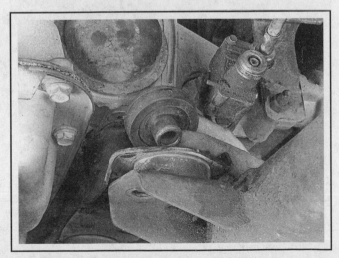

8.5 Remove the two support bracket insulators

8.6 Jack up the transmission and slide the bracket and mount out far enough to allow mount removal

9 Park/Neutral switch - adjustment and replacement

→Note: Checking the Park/Neutral switch is covered in Chapter 1.

1995 AND EARLIER MODELS

Adjustment

▶ Refer to illustration 9.2

1   The Park/Neutral switch should allow the engine to start only in Park or Neutral. If it starts in any other gear or does not start in both Park and Neutral, adjustment is required.

2   The Park/Neutral switch on these models is located inside the vehicle at the bottom of the steering column (see illustration).

3   Apply the parking brake and place the gear selector on the steering column in the Neutral (N) position.

4   Grasp the switch and rotate it in a clockwise direction as far as it will go (to the Low "L" position).

5   Now, place the gear selector in the Park (P) position. Moving the gear selector to Park will ratchet the switch to the correct position. Check switch operation (see Chapter 1). Replace the switch if it does not operate correctly after adjustment.

Replacement

6   Apply the parking brake and place the gear selector on the steering column in the Neutral (N) position.

7   Disconnect the electrical connectors from the Park/Neutral switch.

8   Spread the tangs on the housing and pull the switch out of the steering column.

9   Before installing the switch, position the tang on the switch with the hole in the shift tube.

10 Install the switch onto the steering column and press down on the front of the switch until it snaps into position.

11 Connect the electrical connectors to the switch.

12 Adjust the switch as described above.

1996 AND LATER MODELS

→Note: A special switch adjustment tool is required to properly perform this procedure unless a new switch is being installed. A new switch comes with a "positive assurance bracket" to align it, eliminating the need for the special tool.

Replacement and adjustment

▶ Refer to illustration 9.16

13 Apply the parking brake and place the gear selector in the Neutral (N) position.

14 Disconnect the cable from the negative terminal of the battery.

✻✻ CAUTION:

On models equipped with an anti-theft audio system, disable the anti-theft feature before performing any operation that requires disconnecting the battery or disrupting power to the stereo (see the Anti-theft audio system procedures at the front of this manual).

15 Raise the vehicle and support it securely on jackstands.

16 Remove the nut from the manual lever shaft and remove the lever from the shaft (see illustration).

17 Disconnect the electrical connector from the Park/Neutral switch.

18 Remove the switch retaining bolts and remove the switch from the transmission making sure the end of the shaft is free of burrs. Use a file to deburr the shaft as required.

19 Install the special adjustment tool on the switch per the manufacturer's instructions. Make sure the two detents on the switch (where the manual shaft is inserted through the switch) fit into the two lower tabs on the tool. Rotate the tool until its upper locator pin is aligned with the locator on the switch.

20 Align the switch hub flats with the flats on the manual shaft.

21 Install the switch onto the transmission and secure with the mounting bolts. Tighten the bolts to the torque listed in this Chapter's Specifications.

22 Remove the special adjustment tool from the switch.

23 Install the manual shift lever onto the transmission manual shaft. Install the nut and tighten it to the torque listed in this Chapter's Specifications.

24 The remaining steps are the reverse of removal. Check the Park/Neutral switch operation (see Chapter 1). If necessary, adjust the shift cable (see Section 12).

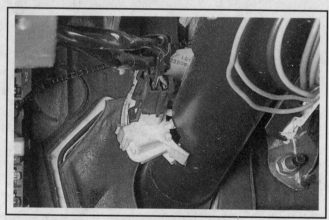

9.2 The Neutral start switch is located on the lower steering column mast on 1995 and earlier models

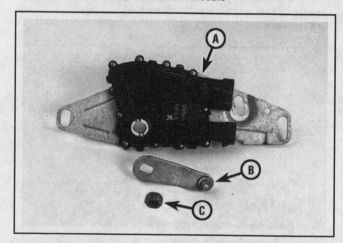

9.16 Installation details of the Park/Neutral switch used on 1996 and later models

A   Park/Neutral switch   C   Nut
B   Manual Lever

10 Speedometer driven gear - removal and installation

▶ **Refer to illustrations 10.2 and 10.4**

➡ **Note: 1991 and later models do not have a speedometer cable or driven gear. The speedometer is driven electronically, using data the vehicle speed sensor (VSS) provides to the PCM.**

1   Raise the vehicle and support it securely on jackstands.

2   Unscrew the speedometer cable from the sleeve and remove the cable housing retaining bolt from the crossmember (see illustration).

3   Remove the sleeve retainer screw and retainer. Pull the sleeve and seal from the bore.

4   Slide the driven gear out of the sleeve (see illustration). Inspect the gear for missing or damaged teeth, replacing it if necessary.

5   Installation is the reverse of the removal procedure. Be sure to install a new seal and check the transmission fluid level.

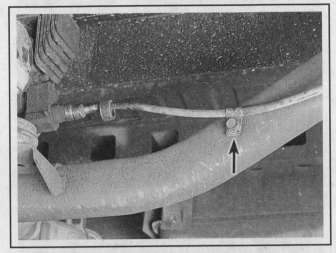

10.2 Unscrew the speedometer cable from the sleeve and remove the cable housing retaining bolt (arrow), then pull the cable out

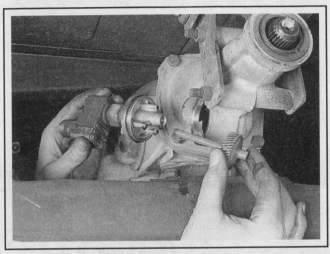

10.4 Slide the driven gear out of the sleeve and inspect it for damage

11 Vehicle speed sensor - removal and installation

GENERAL DESCRIPTION

1   The Vehicle Speed Sensor (VSS) is used by the Powertrain Control Module (PCM) to determine vehicle speed and control the Torque Converter Clutch (TCC). On 1992 and earlier models, the VSS was located behind the instrument cluster. On 1993 and later models, it is mounted on the transmission rear extension housing. Due to the complexity of this circuit, diagnosis of the VSS should be performed by a dealer service department or other qualified repair shop.

REPLACEMENT

1992 and earlier models

▶ **Refer to illustration 11.3**

2   Remove the instrument cluster (see Chapter 12).

3   Remove the VSS retaining screw (see illustration) and withdraw the VSS from the instrument cluster.

4   Installation is the reverse of removal.

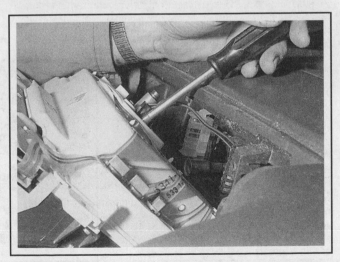

11.3 On 1992 and earlier models, the VSS is located behind the instrument cluster - the VSS electrical connector is secured by a retaining screw

1993 and later models

▶ Refer to illustration 11.6

5   Raise the rear of the vehicle and support it securely on jackstands.

6   Disconnect the electrical connector from the VSS (see illustration).

7   Position a drain pan under the sensor as required to catch any transmission fluid that may leak out.

8   Remove the retaining bolt and withdraw the sensor from the transmission.

9   Installation is the reverse of removal. Inspect the O-ring seal for damage and replace if necessary, first lubricating it with a small amount of clear transmission fluid.

10 After installation, check the transmission fluid level and add some if required (see Chapter 1).

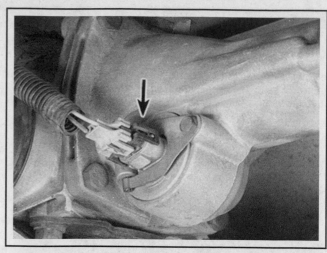

11.6 On 1993 and later models the VSS (arrow) is located on the transmission rear extension housing

12 Shift cable (1996 and later models) - removal, installation and adjustment

REMOVAL

▶ Refer to illustrations 12.4, 12.8, 12.9, 12.10, 12.12, 12.13a and 12.13b

1   Disconnect the cable from the negative terminal of the battery.

✳✳ CAUTION:

On models equipped with an anti-theft audio system, disable the anti-theft feature before performing any operation that requires disconnecting the battery or disrupting power to the stereo (see the Anti-theft audio system procedures at the front of this manual).

2   Apply the parking brake and block the wheels so the vehicle cannot roll in either direction. Place the gear selector in the Neutral (N) position.

3   Raise the vehicle and place it securely on jackstands.

12.4 Use a screwdriver to pry the shift cable from the transmission control lever

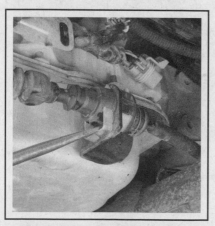

12.8 Remove the U-clip from the shift cable housing at the bracket

12.9 Using pliers, compress the cable housing retaining tabs and withdraw the cable from the bracket

12.10 From under the vehicle, push the shift cable grommet out of the floorpan and into the vehicle

12.12 Use a screwdriver to pry the shift cable end (arrow) from the shift lever at the steering column

12.13a At the steering column, remove the U-clip from the cable housing at the bracket

12.13b Using pliers, compress the locking tabs and push the cable through the bracket

4 Working under the vehicle, pry the cable end from the transmission control lever stud (see illustration).

5 Place the transmission in the Neutral (N) position. To locate Neutral, push the cable shift lever forward (towards the front of the vehicle) until it stops, then pull it back two detents. Make sure the gear selector on the steering column is in the Neutral position as well.

6 Remove the cable from the clip on the transmission oil pan.

7 Remove the three cable clips securing the cable to the body.

8 Remove the metal U-clip from the shift cable housing (see illustration).

9 Compress the locking tabs and pull the cable out of the bracket (see illustration).

10 Push the cable grommet out of the floorpan (see illustration).

11 Working inside the vehicle, remove the driver's side knee bolster, and if applicable, the lower steering column cover (see Chapter 11).

12 Pry the cable end from the shift lever at the steering column (see illustration).

13 Remove the metal U-clip from the cable housing at the bracket on the steering column, then compress the locking tabs and push the cable through the bracket (see illustrations).

14 Working under the vehicle, remove the shift cable.

INSTALLATION

15 Installation is the reverse of removal with the following additions: Before attaching the steering column cable end, make sure the cable moves freely and does not bind. Before securing the cable to the vehicle with the retaining clips or attaching the cable to the transmission control lever stud, perform the shift cable adjustment procedure below.

ADJUSTMENT

▶ **Refer to illustration 12.21**

16 Apply the parking brake and block the wheels so the vehicle cannot roll in either direction. Place the gear selector in the Neutral (N) position.

17 Raise the vehicle and place it securely on jackstands.

18 Working under the vehicle, pry the cable end from the transmission control lever stud (see illustration 12.4).

19 Remove any retaining clips securing the cable to the underside of the vehicle.

20 Place the transmission in the Neutral (N) position. To locate Neu-

12.21 Shift cable body-core adjuster (white colored) locking tab (arrow)

tral, push the cable shift lever forward (towards the front of the vehicle) until it stops, then pull it back two detents. Make sure the gear selector on the steering column inside the vehicle is in the Neutral position as well.

21 While pulling the cable tight, pry the WHITE colored body-core adjuster locking tab out from its locked position (see illustration).

22 Ensure the spring-loaded body-core adjuster moves freely by pulling on the cable. The cable should move forward when pulled, and then return to the fully rearward position freely under spring load.

23 Make sure the gear selector on the steering column and the shift lever on the transmission are still in Neutral. Reposition if necessary.

24 With the transmission and gear selector in Neutral, install the cable end onto the transmission manual control lever.

25 Push the WHITE colored locking tab into the body-core until it snaps. If necessary, wiggle the body-core while pushing in the locking tab.

26 Route the cable as required and install the cable retaining clips.

27 Check the adjustment by making sure all gears can be selected and the engine can only start in the Park or Neutral positions. If necessary, readjust the shift cable and/or the Park/Neutral switch (see Section 9).

28 Place the gear selector in Park and lower the vehicle. Test drive to check for proper operation.

13 Torque Converter Clutch (TCC)

GENERAL DESCRIPTION

1   The Powertrain Control Module (PCM) energizes a solenoid-operated valve in the transmission to mechanically "lock-up" the engine driveplate to the output shaft of the transmission through the torque converter. This reduces slippage losses in the converter, reducing emissions and increases fuel economy.

2   For the TCC to operate properly, two conditions must be met:

a) *The engine must be at operating temperature.*

b) *The vehicle must be traveling at a speed necessary to raise the hydraulic pressure in the transmission to the correct level.*

3   After the converter clutch applies, the PCM uses the information from the TPS to release the clutch when the vehicle is accelerating or decelerating at a certain rate.

4   The brake switch is also used by the PCM to deactivate the TCC when the brakes are applied.

5   The transmission on the V6 engine uses a 4th gear switch to send a signal to the PCM, letting it know it's ready for lock-up. However, the transmission does not have to be in high gear in order for the PCM to lock up the torque converter. Transmissions using gear select switches can be identified by three wires leading out of the TCC harness connector.

6   The transmission also uses a 4-3 pulse switch to open the TCC solenoid during a downshift.

7   A 3rd gear switch is placed in series on the battery side of the TCC solenoid to prevent TCC application until the transmission is in 3rd gear.

CHECK

8   If the converter clutch is applied at all times, the engine will stall immediately, just like a manual transmission left in gear.

9   If the converter clutch does not apply, fuel economy will be lower than expected.

➡ **Note: If the VSS fails (see Section 4), the TCC will not engage.**

10   A TCC-equipped transmission has different operating characteristics than a transmission without TCC. If you detect a "chuggle" or "surge" condition, perform the following check to see if the TCC is working properly.

11   If the vehicle is not equipped with a tachometer, connect one to the engine per the manufacturer's instructions. The tachometer must be placed inside the vehicle so it can be visible while driving the vehicle.

12   Drive the vehicle until it reaches operating temperature and then maintain a speed of approximately 50 to 55 mph.

13   Once the vehicle speed and engine rpm has stabilized, while observing the tachometer and maintaining throttle position, lightly touch the brake pedal. You should feel a slight bumpy sensation, indicating the TCC is releasing and the engine rpm should increase slightly.

14   Release the brake and see if you can feel the torque converter lock-up again and a slight decrease in engine rpm should be observed.

15   If you can't feel the TCC locking up or unlocking, have the transmission diagnosed by a dealer service department or other qualified transmission repair shop.

Specifications

Torque specifications	Ft-lbs
Park/Neutral switch (1996 and later models only)	
Retaining bolts	17 to 21
Manual lever nut	17 to 21
Torque converter-to-driveplate bolts	
1985	39
1986 on	46
Transmission-to-engine bolts	
Four-cylinder engine	55
V6 engine	35
Converter housing cover bolts	10

Section

Reference to other Chapters

8

CLUTCH AND DRIVETRAIN

1 General information

The information in this Chapter deals with the components from the rear of the engine to the rear wheels, except for the transmission, which is dealt with in the previous Chapter. For the purposes of this Chapter, these components are grouped into three categories; clutch, driveshaft and rear axle. Separate Sections within this Chapter offer general descriptions and checking procedures for each of these three groups.

Since nearly all the procedures covered in this Chapter involve working under the vehicle, make sure it's securely supported on sturdy jackstands or on a hoist where the vehicle can be easily raised and lowered.

2 Clutch - description and check

♦ **Refer to illustration 2.1**

1   All models equipped with a manual transmission feature a single dry plate, diaphragm spring-type clutch (see illustration). The actuation is through a hydraulic system.

2   When the clutch pedal is depressed, hydraulic fluid (under pressure from the clutch master cylinder) flows into the slave cylinder. Because the slave cylinder is connected to the clutch fork, the fork moves the release bearing into contact with the pressure plate release fingers, disengaging the clutch plate.

3   The hydraulic system locates the clutch pedal and provides clutch adjustment automatically, so no adjustment of the linkage or pedal is required.

4   Terminology can be a problem regarding the clutch components because common names have in some cases changed from that used by the manufacturer. For example, the driven plate is also called the clutch plate or disc, the clutch release bearing is sometimes called a throwout bearing, the slave cylinder is sometimes called the operating cylinder.

5   Other than to replace components with obvious damage, some preliminary checks should be performed to diagnose a clutch system failure.

   a) The first check should be of the fluid level in the clutch master cylinder. If the fluid level is low, add fluid as necessary and re-test. If the master cylinder runs dry, or if any of the hydraulic components are serviced, bleed the hydraulic system as described in Section 8.

   b) To check clutch spin down time, run the engine at normal idle speed with the transmission in Neutral (clutch pedal up - engaged). Disengage the clutch (pedal down), wait nine seconds and shift the transmission into Reverse. No grinding noise should be heard. A grinding noise would indicate component failure in the pressure plate assembly or the clutch disc.

   c) To check for complete clutch release, run the engine (with the brake on to prevent movement) and hold the clutch pedal approximately 1/2-inch from the floor mat. Shift the transmission between 1st gear and Reverse several times. If the shift is not smooth, component failure is indicated. Measure the slave cylinder pushrod travel. With the clutch pedal completely depressed the slave cylinder pushrod should extend 1.03-inches (26.16 mm). If the pushrod will not extend this far check the fluid level in the clutch master cylinder.

   d) Visually inspect the clutch pedal bushing at the top of the clutch pedal to make sure there is no sticking or excessive wear.

   e) Under the vehicle, check that the clutch fork is solidly mounted on the ball stud.

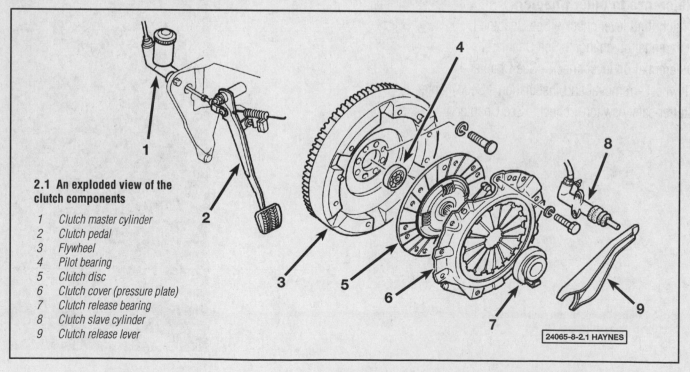

2.1 An exploded view of the clutch components

1   Clutch master cylinder
2   Clutch pedal
3   Flywheel
4   Pilot bearing
5   Clutch disc
6   Clutch cover (pressure plate)
7   Clutch release bearing
8   Clutch slave cylinder
9   Clutch release lever

24065-8-2.1 HAYNES

3   Clutch components - removal, inspection and installation

❊❊ WARNING:

Dust produced by clutch wear and deposited on clutch components contains asbestos, which is hazardous to your health. DO NOT blow it out with compressed air and DO NOT inhale it. DO NOT use gasoline or petroleum-based solvents to remove the dust. Brake system cleaner should be used to flush the dust into a drain pan. After the clutch components are wiped clean with a rag, dispose of the contaminated rags and cleaner in a covered container.

REMOVAL

▶ **Refer to illustration 3.7**

1   Access to the clutch components is normally accomplished by removing the transmission, leaving the engine in the vehicle. If, of course, the engine is being removed for major overhaul, then the opportunity should always be taken to check the clutch for wear and replace worn components as necessary. The following procedures assume that the engine will stay in place.

2   Referring to Chapter 7 Part A, remove the transmission from the vehicle. Support the engine while the transmission is out. Preferably, an engine hoist should be used to support it from above. However, if a jack is used underneath the engine, make sure a piece of wood is used between the jack and oil pan to spread the load.

❊❊ CAUTION:

The pickup for the oil pump is very close to the bottom of the oil pan. If the pan is bent or distorted in any way, engine oil starvation could occur.

3   Remove the slave cylinder (see Section 6).

4   Remove the bellhousing-to-engine bolts and then detach the housing. It may have to be gently pried off the alignment dowels with a screwdriver or pry bar.

5   The clutch fork and release bearing can remain attached to the housing for the time being.

6   To support the clutch plate during removal, install a clutch alignment tool through the clutch plate hub.

7   Carefully inspect the flywheel and pressure plate for indexing marks. The marks are usually an X, an O or a white letter. If they cannot be found, scribe marks yourself so the pressure plate and the flywheel will be in the same alignment during installation (see illustration).

8   Turning each bolt only 1/2-turn at a time, slowly loosen the pressure plate-to-flywheel bolts. Work in a diagonal pattern and loosen each bolt a little at a time until all spring pressure is relieved. Then hold the pressure plate securely and completely remove the bolts, followed by the pressure plate and clutch plate.

INSPECTION

▶ **Refer to illustrations 3.12 and 3.14**

9   Ordinarily, when a problem occurs in the clutch, it can be attributed to wear of the clutch driven plate assembly (clutch plate). However, all components should be inspected at this time.

10   Inspect the flywheel for cracks, heat checking, grooves or other signs of obvious defects. If the imperfections are slight, a machine shop can machine the surface flat and smooth, which is highly recommended regardless of the surface appearance. Refer to Chapter 2 for the flywheel removal and installation procedure.

11   Inspect the pilot bearing (see Section 5).

12   Inspect the lining on the clutch plate. There should be at least 1/16-inch of lining above the rivet heads. Check for loose rivets, distortion, cracks, broken springs and other obvious damage (see illustration). As mentioned above, ordinarily the clutch plate is replaced as a matter of course, so if in doubt about the condition, replace it with a new one.

13   Ordinarily, the release bearing is also replaced along with the clutch plate (see Section 4).

14   Check the machined surfaces of the pressure plate (see illustration). If the surface is grooved or otherwise damaged, take it to a

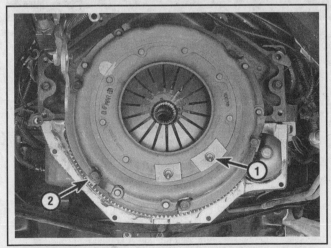

3.7   After removal of the transmission, this will be the view of the clutch components

1   *Pressure plate*   2   *Flywheel*

3.12   The clutch plate

1   *Lining* - *this will wear down in use*
2   *Rivets* - *these secure the lining and will damage the flywheel or pressure plate if allowed to contact the surfaces*
3   *Markings* - *"Flywheel side" or something similar*

3.14 The machined face of the pressure plate must be inspected for score marks and other damage - if damage is slight, a machine shop can make the surface smooth again

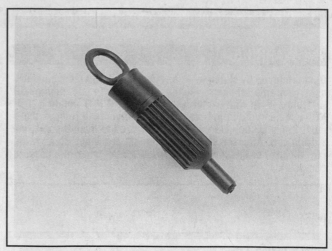

3.16 A clutch alignment tool can be purchased at most auto parts stores and eliminates all guesswork when centering the clutch plate in the pressure plate

machine shop for possible machining or replacement. Also check for obvious damage, distortion, cracking, etc. Light glazing can be removed with medium grit emery cloth. If a new pressure plate is indicated, new or factory-rebuilt units are available.

INSTALLATION

▶ Refer to illustration 3.16

15 Before installation, carefully wipe the flywheel and pressure plate machined surfaces clean. It's important that no oil or grease is on these surfaces or the lining of the clutch plate. Handle these parts only with clean hands.

16 Position the clutch plate and pressure plate with the clutch held in place with an alignment tool (see illustration). Make sure it's installed properly (most replacement clutch plates will be marked "flywheel side" or something similar - if not marked, install the clutch with the damper

springs toward the transmission).

17 Tighten the pressure plate-to-flywheel bolts only finger tight, working around the pressure plate.

18 Center the clutch plate by inserting the alignment tool through the splined hub and into the pilot bearing in the crankshaft. Tighten the pressure plate-to-flywheel bolts a little at a time, working in a criss-cross pattern to prevent distorting the cover. After all of the bolts are snug, tighten them to the specified torque. Remove the alignment tool.

19 Using high temperature grease, lubricate the inner groove of the release bearing (refer to Section 4). Also place grease on the fork fingers.

20 Install the clutch release bearing as described in Section 4.

21 Install the bellhousing and tighten the bolts to the proper torque specification.

22 Install the transmission, slave cylinder and all components removed previously, tightening all fasteners to the proper torque specifications.

4 Clutch release bearing - removal and installation

▶ Refer to illustrations 4.6a, 4.6b and 4.7

REMOVAL

1   Disconnect the negative cable from the battery.

✳✳ CAUTION:

On models equipped with an anti-theft audio system, disable the anti-theft feature before performing any operation that requires disconnecting the battery or disrupting power to the stereo (see the Anti-theft audio system procedures at the front of this manual).

2   Remove the transmission (see Chapter 7).
3   Remove the bellhousing (see Section 3).
4   Remove the clutch release fork from the ball stud.
5   Hold the center of the bearing and spin the outer portion. If the

bearing doesn't turn smoothly or if it's noisy, replace it with a new one. Wipe the bearing with a clean rag and inspect it for damage, wear and cracks. Don't immerse the bearing in solvent - it's sealed for life and to do so would ruin it.

INSTALLATION

6   Lubricate the clutch fork ends where they contact the bearing lightly with white lithium base grease. Pack the inner diameter of the bearing with the same grease (see illustrations).

7   Install the release bearing on the clutch fork so that both of the fork tabs fit into the bearing recess (see illustration).

8   Lubricate the clutch release fork ball socket with high temperature grease and push the fork onto the ball stud until it's firmly seated.

9   Install the bellhousing and tighten the bolts to the specified torque.

10   The remainder of the installation is the reverse of the removal procedure, tightening all bolts to the specified torque.

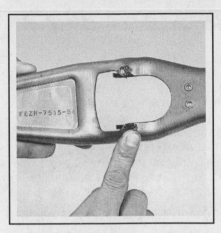

4.6a Lubricate the friction surfaces of the release lever

4.6b Lubricate the inner groove and the thrust side (where the release fingers contact it) of the release bearing with high-temperature grease

4.7 When installing the release bearing, make sure the fingers and the tabs fit into the bearing recess

5   Pilot bearing - removal, inspection and installation

▶ **Refer to illustrations 5.5 and 5.6**

1   The clutch pilot bearing is an oil impregnated type bearing which is pressed into the rear of the crankshaft. Its primary purpose is to support the front of the transmission input shaft. The pilot bearing should be inspected whenever the clutch components are removed from the engine. Due to its inaccessibility, if you are in doubt as to its condition, replace it with a new one.

➡**Note: If the engine has been removed from the vehicle, disregard the following steps which do not apply.**

2   Remove the transmission (refer to Chapter 7 Part A).

3   Remove the clutch components (see Section 3).

4   Using a clean rag, wipe the bearing clean and inspect for any excessive wear, scoring or obvious damage. A flashlight will be helpful to direct light into the recess.

5   Removal can be accomplished with a special puller and slide hammer attachment (see illustration). Remove the bearing with the puller and clean the crankshaft recess.

6   To install the new bearing, lubricate the outside surface with oil then drive it into the recess with a driver and hammer (see illustration).

7   Install the clutch components, transmission and all other components removed to gain access to the pilot bearing.

5.5 Removing the pilot bearing with a slide hammer and internal puller attachment

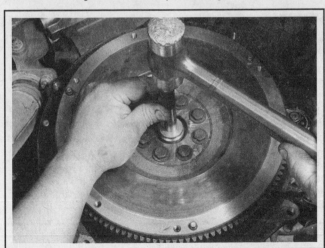

5.6 Install the pilot bearing with a properly sized driver or socket

6 Clutch slave cylinder - removal, overhaul and installation

➡Note: Before beginning this procedure, contact local parts stores and dealer service departments concerning the purchase of a rebuild kit or a new slave cylinder. Availability and cost of the necessary parts may dictate whether the cylinder is rebuilt or replaced with a new one. If it's decided to rebuild the cylinder, inspect the bore as described in Step 9 before purchasing parts.

REMOVAL

1 Disconnect the negative cable from the battery. Place the cable out of the way so it cannot accidentally come into contact with the terminal, which would again allow current to flow.

✳ CAUTION:

On models equipped with an anti-theft audio system, disable the anti-theft feature before performing any operation that requires disconnecting the battery or disrupting power to the stereo (see the Anti-theft audio system procedures at the front of this manual).

2 Raise the vehicle and support it securely on jackstands.
3 Disconnect the hydraulic line at the slave cylinder. Have a small can and rags handy, as some fluid will be spilled as the line is removed.
4 With your fingers, work the rubber dust boot out of the clutch housing, then remove the two mounting nuts.
5 Remove the slave cylinder.

OVERHAUL

6 Slide the shield off the slave cylinder and remove the pushrod and the rubber dust cover.
7 Remove the retaining clip with a small screwdriver.
8 Tap the cylinder on a block of wood to eject the plunger and seal. Also remove the spring from inside the cylinder.
9 Carefully inspect the bore of the cylinder. Check for deep scratches, score marks and ridges. The bore must be smooth to the touch. If any imperfections are found, the slave cylinder must be replaced with a new one.
10 Using the new parts in the rebuild kit, assemble the components using plenty of fresh brake fluid for lubrication. Note the installed direction of the spring and the seal. The seal lip must face toward the hydraulic fluid line.

INSTALLATION

11 Previous to installation, the system must be bled of all air.
12 Connect the hydraulic line to the slave cylinder. Tighten the connection.
13 Fill the clutch master cylinder with brake fluid (conforming to DOT 3 specifications).
14 Bleed the system as described in Section 8.
15 Install the slave cylinder on the clutch housing, working the rubber dust boot into place and tightening the nuts to the specified torque. Make sure the pushrod is seated in the release fork pocket.
16 Lower the vehicle and connect the negative battery cable.

7 Clutch master cylinder - removal, overhaul and installation

➡Note: Before beginning this procedure, contact local parts stores and dealer service departments concerning the purchase of a rebuild kit or a new master cylinder. Availability and cost of the necessary parts may dictate whether the cylinder is rebuilt or replaced with a new one. If it's decided to rebuild the cylinder, inspect the bore as described in Step 12 before purchasing parts.

REMOVAL

1 Disconnect the negative cable from the battery. Place the cable out of the way so it cannot accidentally come in contact with the negative terminal of the battery, as this would once again allow power into the electrical system of the vehicle.

✳✳ CAUTION:

On models equipped with an anti-theft audio system, disable the anti-theft feature before performing any operation that requires disconnecting the battery or disrupting power to the stereo (see the Anti-theft audio system procedures at the front of this manual).

2 Under the dashboard, disconnect the pushrod from the top of the clutch pedal. It's held in place with a spring clip.
3 Disconnect the hydraulic line at the clutch master cylinder. If available, use a flare nut wrench on the fitting, which will prevent the fitting from being rounded off. Have rags handy as some fluid will be lost as the line is removed.

✳✳ CAUTION:

Don't allow brake fluid to come into contact with paint as it will damage the finish.

4 Remove the two bolts which secure the master cylinder to the engine firewall. Remove the master cylinder, again being careful not to spill any of the fluid.

OVERHAUL

5 Remove the reservoir cap and drain all fluid from the master cylinder. Pry the reservoir from the master cylinder body.
6 Pull back the dust cover on the pushrod and remove the snapring.
7 Remove the retaining washer and the pushrod from the cylinder.
8 Tap the master cylinder on a block of wood to eject the plunger assembly from inside the bore.
9 Separate the spring from the plunger.
10 Remove the spring support, seal and shim from the pushrod.
11 Carefully remove the seals from the plunger. Note carefully which

way the seal lips face so the new ones can be reinstalled the same way.

12 Inspect the bore of the master cylinder for deep scratches, score marks and ridges. The surface must be smooth to the touch. If the bore isn't perfectly smooth, the master cylinder must be replaced with a new or factory rebuilt unit.

13 If the cylinder will be rebuilt, use the new parts contained in the rebuild kit and follow any specific instructions which may have accompanied the rebuild kit.

14 Attach the plunger seal to the plunger.

15 Assemble the shim, spring support and spring on the other end of the plunger.

16 Lubricate the bore of the cylinder and the seals with plenty of fresh brake fluid (DOT 3).

17 Carefully guide the plunger assembly into the bore, being careful not to damage the seals. Make sure the spring end is installed first, with the pushrod end of the plunger closest to the opening.

18 Position the pushrod and retaining washer in the bore, compress the spring and install a new snap-ring.

19 Apply a liberal amount of rubber grease or equivalent to the inside of the dust cover and attach it to the master cylinder.

INSTALLATION

20 Install the master cylinder on the firewall, installing the two mounting nuts finger tight.

21 Connect the hydraulic line to the master cylinder, moving the cylinder slightly as necessary to thread the fitting properly into the bore. Don't cross-thread the fitting as it's installed.

22 Tighten the two mounting nuts to the torque listed in this Chapter's Specifications.

23 Inside the vehicle, connect the pushrod to the clutch pedal.

24 Fill the clutch master cylinder reservoir with brake fluid conforming to DOT 3 specifications and bleed the clutch system as outlined in Section 8.

8   Clutch hydraulic system - bleeding

▶ **Refer to illustration 8.5**

1   The hydraulic system should be bled to remove all air whenever any part of the system has been removed or if the fluid level has fallen so low that air has been drawn into the master cylinder. The procedure is very similar to bleeding a brake system.

2   Fill the master cylinder with new brake fluid conforming to DOT 3 specifications.

❋❋ CAUTION:

Don't re-use any of the fluid coming from the system during the bleeding operation. Also, don't use fluid which has been inside an open container for an extended period of time.

3   Raise the vehicle and place it securely on jackstands to gain access to the slave cylinder, which is located on the left side of the clutch housing.

4   With your fingers, pry the rubber dust cover out of the clutch cover and then remove the two slave cylinder mounting bolts.

5   Remove the dust cap which fits over the bleeder valve and push a length of plastic hose over the valve. Place the other end of the hose in a clear container with about two inches of brake fluid. The hose end must be in the fluid at the bottom of the container. Hold the slave cylinder up at a 45° angle with the bleeder valve at its highest point (see illustration).

6   Have an assistant depress the clutch pedal and hold it. Open the bleeder valve on the slave cylinder, allowing fluid to flow through the hose. Close the bleeder valve when the flow of bubbles or old fluid

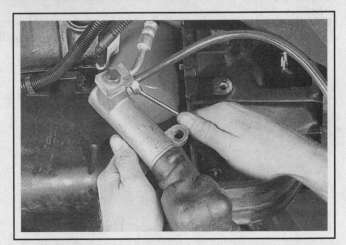

8.5 When bleeding the hydraulic clutch system, hold the slave cylinder as shown in order to get the bleed screw at the highest position

ceases. Once closed, have your assistant release the pedal.

7   Continue this process until all air is evacuated from the system, indicated by a full, solid stream of fluid being ejected from the bleeder valve each time and no air bubbles in the hose or container. Keep a close watch on the fluid level inside the master cylinder; if the level drops too low, air will be sucked back into the system and the process will have to be started all over again.

8   Install the slave cylinder and lower the vehicle. Check carefully for proper operation before placing the vehicle in normal service.

9   Clutch pedal - removal and installation

1   Disconnect the negative cable from the battery. Place the cable out of the way so it cannot accidentally come in contact with the negative terminal of the battery, as this would once again allow power into the electrical system of the vehicle.

❋❋ CAUTION:

On models equipped with an anti-theft audio system, disable the anti-theft feature before performing any operation that requires disconnecting the battery or disrupting power to the stereo (see the Anti-theft audio system procedures at the front of this manual).

2   Disconnect and remove the starter safety switch (see Section 10).

3   Disconnect the pushrod which goes into the master cylinder. It's held by a pin and spring clip.

4   Remove the pedal pivot pin retainer clip and slide the pedal to the left to remove it. Insert a long screwdriver or rod through the opposite side of the bracket to hold the brake pedal in place while the clutch pedal is removed.

5   Wipe clean all parts; however, don't use cleaning solvent on the bushings. Replace all worn parts with new ones.

6   Installation is the reverse of removal. Check that the starter safety switch allows the vehicle to be started only with the clutch pedal fully depressed.

10   Starter safety switch - removal, installation and adjustment

1   Disconnect the cable from the negative terminal of the battery.

✳✳ CAUTION:

On models equipped with an anti-theft audio system, disable the anti-theft feature before performing any operation that requires disconnecting the battery or disrupting power to the stereo (see the Anti-theft audio system procedures at the front of this manual).

2   Unplug the electrical connector from the switch.

3   Remove the mounting screw and rotate the switch down, dislodging the shaft from the clutch pedal.

4   To install the switch, reverse the removal procedure.

5   To adjust the switch, move the slider to the rear of the switch shaft. Depress the clutch pedal to the floor and hold it there, then move the slider down the shaft to contact the switch.

➥**Note: All carpets and floor mats must be in place to achieve an accurate adjustment.**

11   Driveshaft and universal joints - description and check

1   The driveshaft is a tube running between the transmission and the rear end. Universal joints are located at either end of the driveshaft and permit power to be transmitted to the rear wheels at varying angles.

2   The driveshaft features a splined yoke at the front, which slips into the extension housing of the transmission. This arrangement allows the driveshaft to slide back-and-forth within the transmission as the vehicle is in operation.

3   An oil seal is used to prevent leakage of fluid at this point and to keep dirt and contaminants from entering the transmission. If leakage is evident at the front of the driveshaft, replace the oil seal referring to the procedures in Chapter 7.

4   The driveshaft assembly requires very little service. The universal joints are lubricated for life and must be replaced if problems develop. The driveshaft must be removed from the vehicle for this procedure.

5   Since the driveshaft is a balanced unit, it's important that no undercoating, mud, etc. be allowed to stay on it. When the vehicle is raised for service it's a good idea to clean the driveshaft and inspect it for any obvious damage. Also check that the small weights used to originally balance the driveshaft are in place and securely attached. Whenever the driveshaft is removed it's important that it be reinstalled in the same relative position to preserve this balance.

6   Problems with the driveshaft are usually indicated by a noise or vibration while driving the vehicle. A road test should verify if the problem is the driveshaft or another vehicle component:

a)   On an open road, free of traffic, drive the vehicle and note the engine speed (rpm) at which the problem is most evident.

b)   With this noted, drive the vehicle again, this time manually keeping the transmission in 1st, then 2nd, then 3rd gear ranges and running the engine up to the engine speed noted.

c)   If the noise or vibration occurs at the same engine speed regardless of which gear the transmission is in, the driveshaft is not at fault because the speed of the driveshaft varies in each gear.

d)   If the noise or vibration decreased or was eliminated, visually inspect the driveshaft for damage, material on the shaft which would effect balance, missing weights and damaged universal joints. Another possibility for this condition would be tires which are out-of-balance.

7   To check for worn universal joints:

a)   On an open road, free of traffic, drive the vehicle slowly until the transmission is in High gear. Let off on the accelerator, allowing the vehicle to coast, then accelerate. A clunking or knocking noise will indicate worn universal joints.

b)   Drive the vehicle at a speed of about 10 to 15 mph and then place the transmission in Neutral, allowing the vehicle to coast. Listen for abnormal driveline noises.

c)   Raise the vehicle and support it securely on jackstands. With the transmission in Neutral, manually turn the driveshaft, watching the universal joints for excessive play.

12   Driveshaft - removal and installation

▸ **Refer to illustration 12.3**

REMOVAL

1   Disconnect the negative cable from the battery. Place the cable out of the way so it cannot accidentally come in contact with the negative terminal of the battery, as this would once again allow power into the electrical system of the vehicle.

✳✳ CAUTION:

On models equipped with an anti-theft audio system, disable the anti-theft feature before performing any operation that requires disconnecting the battery or disrupting power to the stereo (see the Anti-theft audio system procedures at the front of this manual).

2   Raise the vehicle on a hoist or support it securely on jackstands. Place the transmission in Neutral with the parking brake off.

3   Place paint marks on the driveshaft and the differential flange in line with each other (see illustration). This is to make sure the driveshaft is reinstalled in the same position to preserve the balance.

4   Remove the rear universal joint strap bolts and the straps. Turn the driveshaft (or tires) as necessary to bring the bolts into the most accessible position.

5   Tape the bearing caps to the spider to prevent the caps from coming off during removal.

6   Lower the rear of the driveshaft and then slide the front out of the transmission. Do not pry on the universal joint or yoke. Avoid making a sharp angle between the joint and yoke, which could break the joint. Inspect the joint for burrs, which can damage the transmission seal, and the yoke splines for signs of wear. If servicing the driveshaft, don't clamp the tube in a vise, because this could deform it and affect its balance.

7   To prevent loss of fluid and protect against contamination while the driveshaft is out, wrap a plastic bag over the transmission housing and hold it in place with a rubber band.

INSTALLATION

8   Remove the plastic bag on the transmission and wipe the area clean. Inspect the oil seal carefully. Procedures for replacement of this seal can be found in Chapter 7.

9   Slide the front of the driveshaft into the transmission.

12.3 Before removing the driveshaft, mark the relationship of the driveshaft yoke to the differential flange - to prevent the driveshaft from turning when loosening the strap bolts, insert a screwdriver through the yoke

10 Raise the rear of the driveshaft into position, checking to be sure that the marks are in perfect alignment. If not, turn the rear wheels to match the pinion flange and the driveshaft.

11 Remove the tape securing the bearing caps and install the straps and bolts. Tighten the bolts to the specified torque.

13 Universal joints - replacement

▶ **Refer to illustrations 13.2, 13.4 and 13.9**

➡**Note: A press or large vise will be required for this procedure. It may be advisable to take the driveshaft to a local dealer or machine shop where the universal joints can be replaced for you, normally at a reasonable charge.**

1   Remove the driveshaft as discussed in the previous Section.

2   Using a small pair of pliers, remove the snap-rings from the spider (see illustration). Original front universal joints utilize injected plastic retainers instead of snap-rings. During reassembly, use the snap-rings supplied with the new universal joint.

3   Supporting the driveshaft, place it in position on either an arbor press or on a workbench equipped with a vise.

4   Place a piece of pipe or a large socket with the same inside diameter over one of the bearing caps. Position a socket which is of slightly smaller diameter than the cap on the opposite bearing cap (see illustration) and use the vise or press to force the cap out (inside the pipe or large socket), stopping just before it comes completely out of the yoke. Use the vise or large pliers to work the cap the rest of the way out.

SNAP RING   BEARING CAP   NEEDLE BEARINGS   SEAL   UNIVERSAL JOINT

24010-8-13.2 HAYNES

13.2 Exploded view of the universal joint components

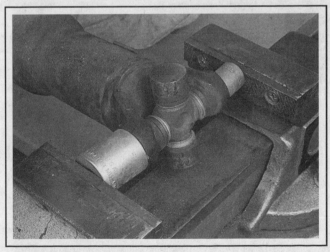

13.4 To press the universal joint out of the driveshaft, set it up in a vise with the small socket (on the right) pushing the joint and bearing cap into the large socket

5   Transfer the sockets to the other side and press the opposite bearing cap out in the same manner.

6   Pack the new universal joint bearings with grease. Ordinarily, specific instructions for lubrication will be included with the universal joint servicing kit and should be followed carefully.

7   Position the spider in the yoke and partially install one bearing cap in the yoke.

8   Start the spider into the bearing cap and then partially install the other cap. Align the spider and press the bearing caps into position, being careful not to damage the dust seals.

9   Install the snap-rings. If difficulty is encountered in seating the snap-rings, strike the driveshaft yoke sharply with a hammer. This will spring the yoke ears slightly and allow the snap-rings to seat in the groove (see illustration).

10 Follow the same procedures to replace the rear universal joint, noting that only one-half of the spider needs to be pressed out, since the other two ends are held to the pinion flange with straps and bolts.

11 Install the driveshaft and all components previously removed.

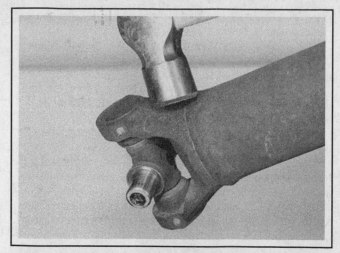

13.9 To relieve stress produced by pressing the bearing caps into the yokes, strike the yoke in the area shown

14 Rear axle - description and check

▶ **Refer to illustration 14.3**

DESCRIPTION

1   The rear axle assembly is a hypoid, semi-floating type (the centerline of the pinion gear is below the centerline of the ring gear). The differential carrier is a casting with a pressed steel cover and the axle tubes are made of steel, pressed and welded into the carrier.

2   An optional locking rear axle is also available. This differential allows for normal differential operation until one wheel loses traction. The unit utilizes multi-disc clutch packs and a speed sensitive engagement mechanism which locks both axleshafts together, applying equal rotational power to both wheels.

3   In order to undertake certain operations, particularly replacement of the axleshafts, it's important to know the axle identification number. It's located on the front face of the right side axle tube (see illustration).

CHECK

4   Many times a fault is suspected in the rear axle area when, in fact, the problem lies elsewhere. For this reason, a thorough check should be performed before assuming a rear axle problem.

5   The following noises are those commonly associated with rear axle diagnosis procedures:

   a) *Road noise is often mistaken for mechanical faults. Driving the vehicle on different surfaces will show whether the road surface is the cause of the noise. Road noise will remain the same if the vehicle is under power or coasting.*

   b) *Tire noise is sometimes mistaken for mechanical problems. Tires which are worn or low on pressure are particularly susceptible to emitting vibrations and noises. Tire noise will remain about the same during varying driving situations, where rear axle noise will change during coasting, acceleration, etc.*

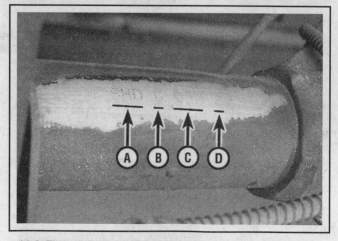

14.3 The rear axle identification number is near the top of the right axle tube, close to the differential

A	Axle code	C	Day built
B	Axle type	D	Shift

   c) *Engine and transmission noise can be deceiving because it will travel along the driveline. To isolate engine and transmission noises, make a note of the engine speed at which the noise is most pronounced. Stop the vehicle and place the transmission in Neutral and run the engine to the same speed. If the noise is the same, the rear axle is not at fault.*

6   Overhaul and general repair of the rear axle is beyond the scope of the home mechanic due to the many special tools and critical measurements required. Thus, the procedures listed here will involve axleshaft removal and installation, axleshaft oil seal replacement, axleshaft bearing replacement and removal of the entire unit for repair or replacement.

15 Axleshaft - removal and installation

▶ **Refer to illustrations 15.3a, 15.3b, 15.4 and 15.5**

1   Raise the rear of the vehicle, support it securely and remove the wheel and brake drum (refer to Chapter 9).

2   Remove the pressed steel cover from the differential carrier and allow the oil to drain into a container.

3   Remove the lock bolt from the differential pinion shaft. Remove the pinion shaft (see illustrations).

4   Push the outer (flanged) end of the axleshaft in and remove the C-lock from the inner end of the shaft (see illustration).

5   Withdraw the axleshaft, taking care not to damage the oil seal in the end of the axle housing as the splined end of the axleshaft passes through it (see illustration).

6   Installation is the reverse of removal. Tighten the lock bolt to the specified torque.

7   Always use a new cover gasket and tighten the cover bolts to the specified torque.

8   Refill the axle with the correct quantity and grade of lubricant (see Chapter 1).

15.3a Remove the pinion shaft lock bolt . . .

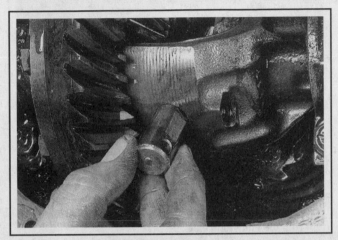

15.3b . . . then carefully remove the pinion shaft from the differential carrier (don't turn the wheels or the carrier after the shaft has been removed, or the spider gears may fall out)

15.4 Push the axle flange in, then remove the C-lock from the inner end of the axleshaft

15.5 Carefully pull the axleshaft from the housing to avoid damaging the seal

16 Axleshaft oil seal - replacement

♦ **Refer to illustrations 16.2 and 16.3**

1   Remove the axleshaft as described in the preceding Section.

2   Pry the old oil seal out of the end of the axle housing, using a large screwdriver or the inner end of the axleshaft itself as a lever (see illustration).

3   Using a large socket as a seal driver, tap the seal into position so that the lips are facing in and the metal face is visible from the end of the axle housing (see illustration). When correctly installed, the face of the oil seal should be flush with the end of the axle housing. Lubricate the lips of the seal with gear oil.

4   Installation of the axleshaft is described in the preceding Section.

16.2 The axleshaft oil seal can sometimes be pried out with the end of the axle

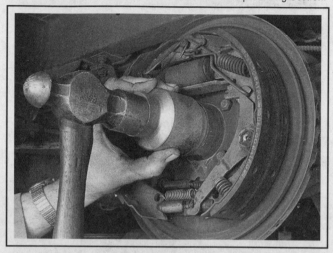

16.3 A large socket can be used to install the new seal squarely

17 Axleshaft bearing - replacement

♦ **Refer to illustrations 17.2 and 17.4**

1   Remove the axleshaft (refer to Section 15) and the oil seal (refer to Section 16).

2   A bearing puller will be required or a tool which will engage behind the bearing will have to be fabricated (see illustration).

3   Attach a slide hammer and pull the bearing from the axle housing.

4   Clean out the bearing recess and drive in the new bearing using a special bearing driver, which is available in most auto parts stores (see illustration).

❋❋ **CAUTION:**

Failure to use this tool could result in bearing damage. Lubricate the new bearing with gear lubricant. Make sure that the bearing is tapped into the full depth of its recess and that the numbers on the bearing are visible from the outer end of the housing.

5   Discard the old oil seal and install a new one, then install the axleshaft.

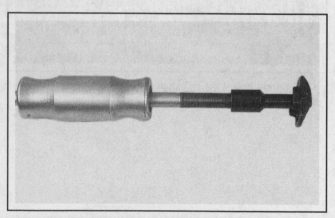

17.2 A slide hammer with a special bearing puller attachment is required to pull the axleshaft bearing from the axle housing

17.4 A special bearing driver is needed to install the axleshaft bearing without damaging it

18 Rear axle assembly - removal and installation

▶ Refer to illustrations 18.6 and 18.10

REMOVAL

1   Loosen the rear wheel lug nuts, raise the rear of the vehicle and support it securely on jackstands positioned under the frame. Remove the wheels.

2   Remove the brake drums (refer to Chapter 9 if any difficulty is encountered).

3   Disconnect the vent hose from the top of the axle housing.

4   Remove the lower shock absorber nuts and bolts and separate the shock absorbers from their mounts.

5   Mark the driveshaft yoke and pinion flange relationship and remove the four universal joint strap bolts (refer to Section 12). Slide the driveshaft as far forward as possible and suspend it with a piece of wire.

6   Disconnect the brake lines at the junction block bolted to the differential cover and unbolt the brake hose from the support bracket (see illustration). Plug the holes in the fitting to prevent excessive fluid loss or hydraulic system contamination.

7   Disconnect the parking brake cables at the equalizer (see Chapter 9).

8   Remove the rear stabilizer bar, if so equipped (refer to Chapter 10).

9   Support the rear axle assembly with a jack positioned under the differential housing.

10 Remove the anchor plate nuts and U-bolt nuts and remove the anchor plates from the top of the leaf springs (see illustration). Use caution while doing this, as the axle will be balancing on the floor jack after the anchor plates have been removed.

11 Carefully lower the rear axle assembly to the floor.

INSTALLATION

12 Raise the rear axle assembly into place.

13 Install the U-bolts and anchor plates and tighten the nuts to the specified torque.

➡**Note: If the vehicle is equipped with a stabilizer bar, it should be installed before tightening any of the bolts.**

14 Connect the brake hose junction block to the support bracket, install the two hydraulic lines in the junction block and tighten the tube nuts securely.

15 Connect the driveshaft to the pinion flange, lining up the marks made during removal. Tighten the strap bolts to the specified torque.

16 Connect the parking brake cables at the equalizer.

17 Install the shock absorbers and tighten the nuts securely.

18 Install the vent hose on the differential housing.

19 Install the brake drums and rear wheels. Tighten the lug nuts to the specified torque.

20 Check the rear axle lubricant level (see Chapter 1).

21 Bleed the brakes as outlined in Chapter 9.

22 Adjust the parking brake if necessary, following the procedure outlined in Chapter 9.

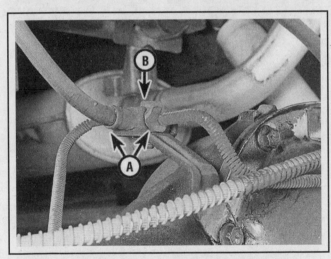

18.6 Loosen the two tube nuts (A) and separate the hydraulic lines from the junction block, then remove the bolt (B) securing the hose to the support bracket

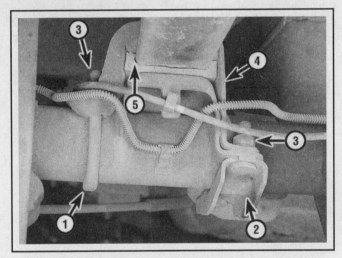

18.10 Typical rear axle assembly mounting components (right side shown)

1	U-bolt	4	Anchor plate
2	U-bolt bracket	5	Rubber insulator
3	Anchor plate nuts		

Specifications

Clutch

Fluid type	See Chapter 1
Clutch disc runout limit	0.020 inch
Slave cylinder pushrod travel	1.03 inch

Torque specifications	Ft-lbs
Pressure plate-to-flywheel bolts	
four-cylinder engine	18
V6 engine	30
Bellhousing-to-engine bolts	
four-cylinder engine	46
V6 engine (carburetor equipped)	46
V6 engine (TBI equipped)	55
Clutch master cylinder nuts	13
Clutch slave cylinder nuts	13

Driveshaft

Torque specifications	Ft-lbs
Universal joint strap bolts	
1985	12 to 17
1986 on	27

Rear axle

Torque specifications	Ft-lbs
Differential cover bolts	20 to 23
Pinion shaft lock bolt	
1985 through 2000	25
2001 and later	19
U-bolt nuts	
1985 through 2000	
Inner	41
Outer	48
2001 and later	74

Section

Reference to other Chapters

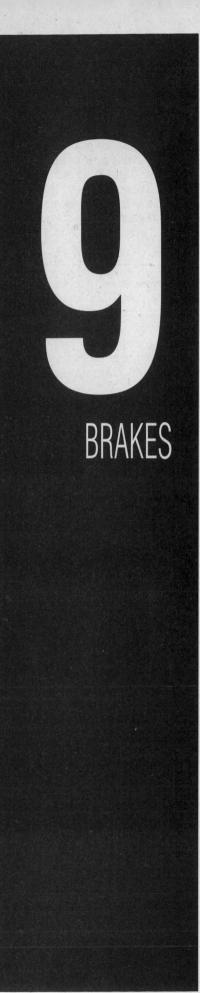

9

BRAKES

The vehicles covered by this manual are equipped with hydraulically operated front and rear brake systems. The front brakes are disc type, and the rear brakes are drum type on earlier models and disc type on later models. Both the front and rear brakes are self adjusting. The disc brakes automatically compensate for pad wear, while the drum brakes incorporate an adjustment mechanism that is activated as the brakes are applied when the vehicle is driven in reverse.

HYDRAULIC SYSTEM

The hydraulic system consists of separate front and rear circuits. The master cylinder has separate reservoirs for the two circuits and in the event of a leak or failure in one hydraulic circuit, the other circuit will remain operative. A visual warning of circuit failure or air in the system is given by a warning light activated by displacement of the piston in the pressure differential switch portion of the combination valve from its normal "in balance" position.

COMBINATION VALVE

A combination valve, located in the engine compartment below the master cylinder, consists of three sections providing the following functions: The metering section limits pressure to the front brakes until a predetermined front input pressure is reached and until the rear brakes are activated. There is no restriction at inlet pressures below 3 psi, allowing pressure equalization during non-braking periods. The proportioning section proportions outlet pressure to the rear brakes after a predetermined rear input pressure has been reached, preventing early rear wheel lock-up under heavy brake loads. The valve is also designed to assure full pressure to one brake system should the other system fail. The pressure differential warning switch incorporated into the combination valve is designed to continuously compare the front and rear brake pressure

from the master cylinder and energize the dash warning light in the event of either a front or rear brake system failure. The design of the switch and valve are such that the switch will stay in the "warning" position once a failure has occurred. The only way to turn the light off is to repair the cause of the failure and apply a brake pedal force of 450 psi.

POWER BRAKE BOOSTER

Two very different types of power brake assist systems are used on these models. Both types of boosters are mounted on the firewall and have the brake master cylinder attached to it.

From 1985 until 1993, all power assist applied to the brakes came from a traditional vacuum powered brake booster. Vacuum from the engine is applied to a diaphragm inside the booster when the brake pedal is depressed, providing additional force to actuate the master cylinder.

Beginning on some models in 1993 and all later models, a new type of power brake booster replaced the vacuum type booster, which uses hydraulic pressure produced by the power steering pump to provide the brake power assist. The hydraulic brake booster unit is connected to the power steering pump by flexible hoses and uses a common fluid reservoir.

PARKING BRAKE

The parking brake operates the rear brakes only, through cable actuation. It's activated by a pedal mounted on the left side kick panel.

SERVICE

After completing any operation involving disassembly of any part of the brake system, always test drive the vehicle to check for proper

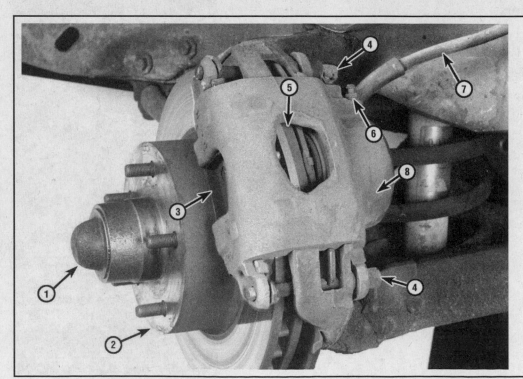

A typical disc brake assembly

1   Dust cap
2   Front brake rotor
3   Outer brake pad
4   Caliper mounting bolts
5   Inner brake pad
6   Bleed screw
7   Brake hose
8   Caliper

Typical drum brake components

1 Hold-down pins
2 Backing plate
3 Parking brake lever
4 Secondary shoe
5 Shoe guide
6 Parking brake strut
7 Actuator lever
8 Actuator link
9 Return spring
10 Return spring
11 Hold-down spring
12 Lever pivot
13 Lever return spring
14 Strut spring
15 Adjusting screw assembly
16 Adjusting screw spring
17 Primary shoe

braking performance before resuming normal driving. When testing the brakes, perform the tests on a clean, dry flat surface. Conditions other than these can lead to inaccurate test results.

Test the brakes at various speeds with both light and heavy pedal pressure. The vehicle should stop evenly without pulling to one side

or the other. Avoid locking the brakes because this slides the tires and diminishes braking efficiency and control of the vehicle.

Tires, vehicle load and front-end alignment are factors which also affect braking performance.

2 Disc brake pads - replacement

▸ Refer to illustrations 2.5a and 2.5b

✳✳ WARNING:

Disc brake pads must be replaced on both front wheels at the same time - never replace the pads on only one wheel. Also, the dust created by the brake system is harmful to your health. Never blow it out with compressed air and don't inhale any of it. An approved filtering mask should be worn when working on the brakes. Do not, under any circumstances, use petroleum-based solvents to clean brake parts. Use brake cleaner or denatured alcohol only!

➡Note: On later models, which have disc brakes front and rear, the rear calipers have a single piston, whereas the front calipers have dual pistons. Earlier front calipers had a single piston, as reflected in the procedures and illustrations in this Chapter, which are typical.

1 Remove the covers from the brake fluid reservoir.
2 Loosen the wheel lug nuts, raise the front or rear of the vehicle and support it securely on jackstands.
3 Remove the wheels. Work on one brake assembly at a time, using the assembled brake for reference if necessary.
4 Inspect the rotor carefully as outlined in Section 4. If machining

2.5a Before disassembling the brake, wash it thoroughly with brake system cleaner and allow it to dry. Position a drain pan under the brake to catch the residue. DO NOT use compressed air to blow off the brake dust!

is necessary, follow the information in that Section to remove the rotor, at which time the pads can be removed from the calipers as well.

5 Clean the brake with brake system cleaner (see illustration). Push

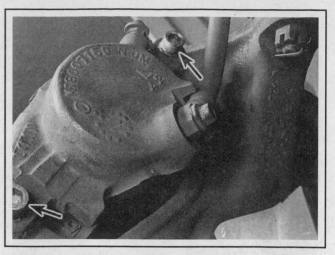

2.5b Using a large C-clamp, push the pistons back into the caliper bore - note that one end of the clamp is on the flat area near the brake hose fitting and the other end (screw end) is pressing against the outer brake pad

2.6a Remove the two caliper-to-steering knuckle mounting bolts (arrows) (this will require the use of an Allen head socket wrench)

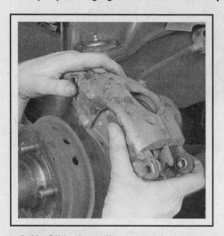

2.6b Slide the caliper up and off the rotor

2.6c Pull the inner pad straight out, disengaging the retainer spring from the caliper piston

2.6d Transfer the retainer spring from the old inner pad to the new one - hook the end of the spring in the hole at the top of the pad, then insert the two prongs of the spring into the slot on the pad backing plate

the piston(s) back into its/their bore(s). If necessary, a C-clamp can be used, but a flat bar will usually do the job (see illustration). (You may need to reposition the C-clamp to bottom the pistons.) As the piston is depressed to the bottom of the caliper bore, the fluid in the master cylinder will rise. Make sure that it doesn't overflow. If necessary, siphon off some of the fluid.

2.6e Lubricate the caliper mounting ears with multi-purpose grease

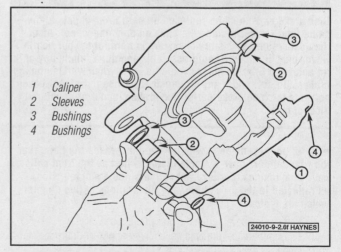

1 Caliper
2 Sleeves
3 Bushings
4 Bushings

24010-9-2.6f HAYNES

2.6f Push the mounting bolt sleeves out of the bores, remove the old bushings and install the new ones supplied with the brake pads

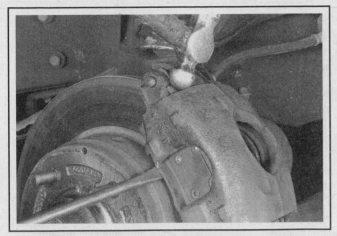

2.6g Slide the caliper assembly over the rotor, install the mounting bolts, then insert a screwdriver between the rotor and outer brake pad, pry up and strike the pad ears with a hammer to eliminate all play between the pad and caliper

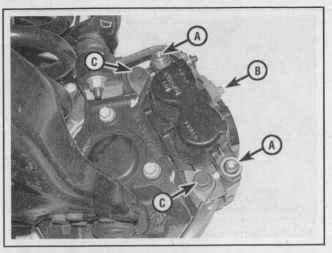

2.7a Front caliper mounting details (2003 and later models)

A *Caliper mounting bolts*   C *Caliper mounting bracket*
B *Brake hose inlet fitting bolt*   *bolts*

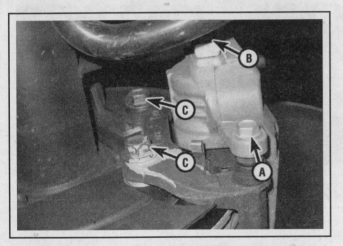

2.7b Rear caliper mounting details (2003 and later models)

A *Caliper mounting bolt*   C *Caliper mounting bracket*
   (upper bolt not visible)   *bolts*
B *Brake hose inlet fitting bolt*

2002 AND EARLIER MODELS

▸ **Refer to illustrations 2.6a through 2.6g**

6 Follow the accompanying photos, beginning with illustration 2.6a, for the actual pad replacement procedure. Be sure to stay in order and read the caption under each illustration. Once finished, proceed to Step 8.

2003 AND LATER MODELS

▸ **Refer to illustrations 2.7a through 2.7k**

➡**Note: This procedure applies to the front and rear brake pads.**

7 Follow the accompanying photo sequence for the actual pad replacement procedure (see illustrations 2.7a through 2.7k). Be sure to stay in order and read the caption under each illustration. Once finished, proceed to the next Step.

2.7c If you're replacing the front brake pads, remove the lower mounting bolt and pivot the caliper up, supporting it in this position. If you're replacing the rear pads, hold the upper caliper slide pin with an open-end wrench, remove the upper mounting bolt with another wrench, then pivot the caliper down for access to the brake pads

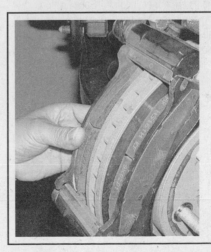

2.7d Remove the inner brake pad

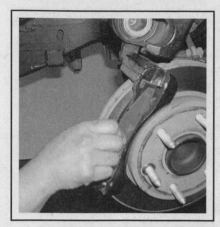

2.7e Remove the outer brake pad

2.7f Remove the upper and lower pad retainers from the caliper mounting bracket; if they are cracked or distorted, replace them

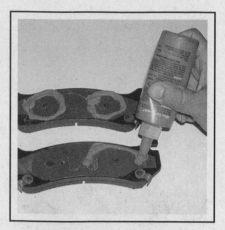

2.7g Apply anti-squeal compound to the back of both pads (let the compound "set up" a few minutes before installing them)

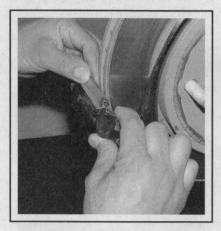

2.7h Install the upper and lower pad retainers on the caliper mounting bracket

2.7i Install the inner brake pad . . .

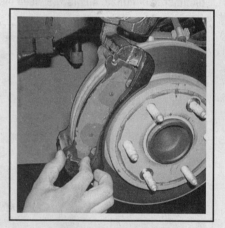

2.7j . . . and the outer brake pad

ALL MODELS

8   When reinstalling the caliper, be sure to tighten the mounting bolts to the torque listed in this Chapter's Specifications. Tighten the wheel lug nuts to the torque listed in the Chapter 1 Specifications.

9   After the job has been completed, firmly depress the brake pedal a few times to bring the pads into contact with the disc. Check the level of the brake fluid, adding some if necessary (see Chapter 1). Check the operation of the brakes carefully before placing the vehicle into normal service.

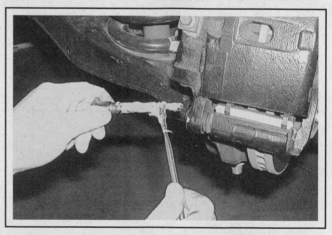

2.7k If you're replacing the front pads, inspect the caliper mounting bolt for scoring and corrosion, then lubricate it with high-temperature brake grease (if it was dry, pivot the caliper up again, slide the upper mounting bolt out of the bracket and lubricate it, too). If you're replacing the rear pads, check the condition of the slide pin and the rubber boot, then lubricate the pin with high-temperature brake grease (if it was dry, do the same thing to the lower slide pin)

3   Disc brake caliper - removal, overhaul and installation

※ WARNING:

Dust created by the brake system is harmful to your health. Never blow it out with compressed air and don't inhale any of it. An approved filtering mask should be worn when working on the brakes. Do not, under any circumstances, use petroleum-based solvents to clean brake parts. Use brake cleaner or denatured alcohol only!

➡️**Note 1:** If an overhaul is indicated (usually because of fluid leakage) explore all options before beginning the job. New and factory rebuilt calipers are available on an exchange basis, which makes this job quite easy. If it's decided to rebuild the calipers, make sure that a rebuild kit is available before proceeding. Always rebuild the calipers on both front or rear wheels - never rebuild just one of them.

➡️**Note 2:** On later models, which have disc brakes front and rear, the rear calipers have a single piston, whereas the front calipers have dual pistons. Earlier front calipers had a single piston, as reflected in the procedures and illustrations in this chapter, which are typical.

REMOVAL

♦ **Refer to illustration 3.4**

1   Remove the cover from the brake fluid reservoir. If the fluid level is higher than the midpoint, remove and discard enough fluid to bring the level to approximately the midpoint.

2   Loosen the wheel lug nuts, raise the front or rear of the vehicle and support it securely on jackstands. Remove the wheels.

3   Bottom the piston in the caliper bore (see illustration 2.5b).

4   Clean any dirt or debris from the end of the brake hose. Remove the brake hose inlet fitting bolt and detach the hose (see illustration). Have a rag handy to catch spilled fluid and wrap a plastic bag tightly around the end of the hose to prevent fluid loss and contamination.

➡️**Note:** If you are simply removing the caliper for access to other components, leave the brake hose connected and suspend the caliper with a length of wire - don't let it hang by the hose.

5   Remove the two mounting bolts and detach the caliper from the vehicle (refer to Section 2 if necessary).

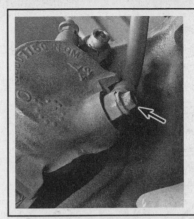

3.4 It's easier to remove the brake hose inlet fitting bolt (arrow) before removing the caliper mounting bolts

※ CAUTION:

Do not step on the brake pedal with the brake system disassembled. This may cause damage.

OVERHAUL

♦ **Refer to illustrations 3.8, 3.9, 3.10, 3.11, 3.15, 3.16, 3.17 and 3.18**

6   Refer to Section 2 and remove the brake pads from the caliper.

7   Clean the exterior of the caliper with brake cleaner or denatured alcohol. Never use gasoline, kerosene or petroleum-based cleaning solvents. Place the caliper on a clean workbench.

8   Position a wooden block or several shop rags in the caliper as a cushion, then use compressed air to remove the piston from the caliper (see illustration). Use only enough air pressure to ease the piston out of the bore. If the piston is blown out, even with the cushion in place, it may be damaged.

※ WARNING:

Never place your fingers in front of the piston in an attempt to catch or protect it when applying compressed air, as serious injury could occur.

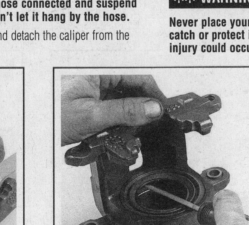

3.8 With the caliper padded to catch the piston, use compressed air to force the piston out of the bore - make sure your hands or fingers are not between the piston and caliper!

3.9 When prying the dust boot out of the caliper, be very careful not to scratch the bore surface

3.10 Remove the piston seal with a wooden or plastic tool to avoid scratching the bore and seal groove

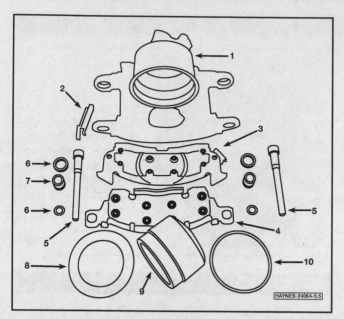

3.11 An exploded view of the disc brake caliper components (2002 and earlier models)

1	Caliper housing	6	Bushing
2	Pad retainer	7	Sleeve
3	Inner pad	8	Dust boot
4	Outer pad	9	Piston
5	Mounting bolt	10	Piston seal

9   Carefully pry the dust boot out of the caliper bore (see illustration).

10 Using a wood or plastic tool, remove the piston seal from the groove in the caliper bore (see illustration). Metal tools may cause bore damage.

11 Remove the caliper bleeder screw, then remove and discard the sleeves and bushings from the caliper ears. Discard all rubber parts (see illustration).

12 Clean the remaining parts with brake system cleaner or denatured alcohol then blow them dry with compressed air. (Some compressed air sources contain minute droplets of oil from the compressor and should

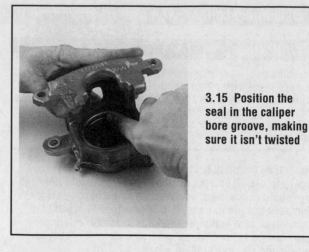

3.15 Position the seal in the caliper bore groove, making sure it isn't twisted

not be used here.) Also blow out the passages in the caliper housing and bleeder valve.

13 Carefully examine the piston for nicks and burrs and loss of plating. If surface defects are present, the parts must be replaced.

14 Check the caliper bore in a similar way. Light polishing with crocus cloth is permissible to remove light corrosion and stains. Discard the mounting bolts if they're corroded or damaged.

15 When assembling, lubricate the piston bores and seal with clean brake fluid. Position the seal in the caliper bore groove (see illustration).

16 Lubricate the piston with clean brake fluid, then install a new boot in the piston groove with the fold toward the open end of the piston (see illustration).

17 Insert the piston squarely in the caliper bore, then apply force to bottom the piston in the bore (see illustration).

18 Position the dust boot in the caliper counterbore, then use a drift to drive it into position (see illustration). Make sure the boot is recessed evenly below the caliper face. Release any trapped air under the piston boot by lifting the edge next to the piston with a small screwdriver. Clean the outside of the caliper boot with denatured alcohol.

19 Install the bleeder screw.

20 Install new bushings in the mounting bolt holes and fill the area between the bushings with the silicone grease supplied in the rebuild kit. Push the sleeves into the mounting bolt holes.

3.16 Install the new dust boot in the piston groove (note that the folds are at the open end of the piston)

3.17 Install the piston squarely in the caliper bore then bottom it by pushing down evenly

3.18 Seat the boot in the counterbore (a seal driver is being used in this photo, but a drift punch will work if care is exercised)

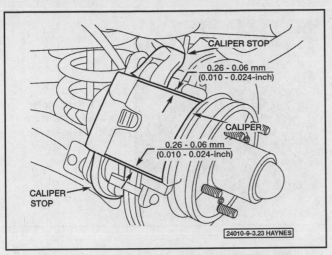

3.23 Measure the gap between the top and bottom of the caliper and the steering knuckle, then add the measurements together - the total should be within the specified limit

INSTALLATION

▶ **Refer to illustration 3.23**

21 Inspect the mounting bolts for excessive corrosion.

22 If necessary, push the piston back into its bore using a C-clamp or flat bar, as in Section 2, Step 5. Place the caliper in position over the rotor and mounting bracket, install the bolts using Loctite, and tighten them to the specified torque.

23 Check to make sure the total clearance between the caliper and the bracket stops is between 0.010 and 0.024 inch (see illustration).

24 Install the brake hose and inlet fitting bolt, using new copper washers, then tighten the bolt to the specified torque. Be sure to bleed the brakes (see Section 10) if a brake hose was disconnected.

25 Install the wheels and lower the vehicle.

26 After the job has been completed, firmly depress the brake pedal a few times to bring the pads into contact with the rotor.

4   Brake rotor (disc) - inspection, removal and installation

INSPECTION

▶ **Refer to illustrations 4.3, 4.4a, 4.4b and 4.5**

1 Loosen the wheel lug nuts, raise the vehicle and support it securely on jackstands. Remove the wheel.

2 Remove the brake caliper as outlined in Section 3. It's not necessary to disconnect the brake hose. After removing the caliper bolts, suspend the caliper out of the way with a piece of wire. Don't let the caliper hang by the hose and don't stretch or twist the hose.

3 Visually check the rotor surface for score marks and other damage. Light scratches and shallow grooves are normal after use and may not always be detrimental to brake operation, but deep score marks - over 0.015-inch (0.38 mm) - require rotor removal and refinishing by an automotive machine shop. Be sure to check both sides of the rotor (see illustration). If pulsating has been noticed during application of the brakes, suspect rotor runout.

4 To check rotor runout, place a dial indicator at a point about 1/2-inch from the outer edge of the rotor (see illustration). Set the indicator to zero and turn the rotor. The indicator reading should not exceed the specified allowable runout limit. If it does, the rotor should be refinished by an automotive machine shop.

➡**Note: Professionals recommend resurfacing of brake rotors regardless of the dial indicator reading (to produce a smooth, flat surface, that will eliminate brake pedal pulsations and other undesirable symptoms related to questionable rotors). At the very least, if you elect not to have the rotors resurfaced, deglaze the brake pad surface with medium-grit emery cloth (use a swirling motion to ensure a non-directional finish) (see illustration).**

5 The rotor must not be machined to a thickness less than the specified minimum refinish thickness. The minimum wear (or discard) thickness is cast into the inside of the rotor. It should not be confused with the minimum refinish thickness. The rotor thickness can be checked with a micrometer (see illustration).

4.3 Check the rotor for deep grooves and score marks (be sure to inspect both sides of the rotor)

4.4a Check rotor runout with a dial indicator - if the reading exceeds the maximum allowable runout, the rotor will have to be resurfaced or replaced

4.4b Using a swirling motion, remove the glaze from the rotor with medium-grit emery cloth

REMOVAL

2002 and earlier models

6 Refer to Chapter 1, Section 36, for the rotor removal procedure (it's part of the hub and comes off when the hub is removed).

2003 and later models

7 Remove the caliper mounting bracket (see illustrations 2.7a and 2.7b).

8 Slide the rotor off of the hub. If it sticks, strike it with a rubber mallet or soft-faced hammer to break it loose.

INSTALLATION

9 If you're working on a 2002 or earlier model, install the rotor and hub assembly and adjust the wheel bearings (see Chapter 1).

10 If you're working on a 2003 or later model, place the rotor onto the hub and install the caliper mounting bracket, tightening the bolts to the torque listed in this Chapter's Specifications. Install the brake pads in the mounting bracket.

11 Install the caliper, tightening the bolts to the torque listed in this Chapter's Specifications.

4.5 A micrometer is used to measure rotor thickness

12 Install the wheel, then lower the vehicle to the ground. Tighten the lug nuts to the torque listed in the Chapter 1 Specifications.

13 Depress the brake pedal a few times to bring the brake pads into contact with the rotor. Bleeding of the system will not be necessary unless the brake hose was disconnected from the caliper.

14 Check the operation of the brake system carefully before placing the vehicle into normal service.

5 Rear brake shoes - inspection and replacement

▶ Refer to illustrations 5.4a through 5.4v

✳✳ WARNING:

Drum brake shoes must be replaced on both wheels at the same time - never replace the shoes on only one wheel. Also, the dust created by the brake system is harmful to your health. Never blow it out with compressed air and don't inhale any of it. An approved filtering mask should be worn when working on the brakes. Do not, under any circumstances, use petroleum-based solvents to clean brake parts. Use brake cleaner or denatured alcohol only!

✳✳ CAUTION:

Whenever the brake shoes are replaced, the retractor and holddown springs should also be replaced. Due to the continuous heating/cooling cycle that the springs are subjected to, they lose their tension over a period of time and may allow the shoes to drag on the drum and wear at a much faster rate than normal. When replacing the rear brake shoes, use only high quality nationally recognized brand-name parts.

1 Loosen the wheel lug nuts, raise the rear of the vehicle and support it securely on jackstands. Block the front wheels to keep the vehicle from rolling.

2 Release the parking brake.

3 Remove the wheel.

➡Note: All four rear brake shoes must be replaced at the same time, but to avoid mixing up parts, work on only one brake assembly at a time.

4 Follow the accompanying photos (illustrations 5.4a through 5.4v) for the inspection and replacement of the brake shoes. Be sure to stay in order and read the caption under each illustration.

➡Note: If the brake drum cannot be easily pulled off the axle

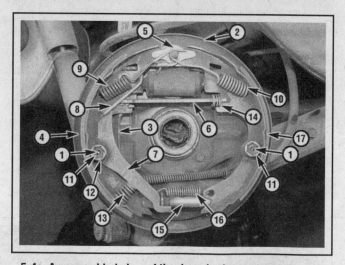

5.4a An assembled view of the drum brake components

1	Hold-down pins	10	Return spring
2	Backing plate	11	Hold-down spring
3	Parking brake lever	12	Lever pivot
4	Secondary shoe	13	Lever return spring
5	Shoe guide	14	Strut spring
6	Parking brake strut	15	Adjusting screw assembly
7	Actuator lever	16	Adjusting screw spring
8	Actuator link	17	Primary shoe
9	Return spring		

and shoe assembly, make sure that the parking brake is completely released, then squirt some penetrating oil around the center hub area. Allow the oil to soak in and try to pull the drum off. If the drum still cannot be pulled off, the brake shoes will have to be retracted. This is accomplished by first removing the

5.4b Remove the shoe return springs - the spring tool shown here is available at most auto parts stores and makes this job much easier and safer

5.4c Pull the bottom of the actuator lever toward the secondary brake shoe, compressing the lever return spring- the actuator link can now be removed

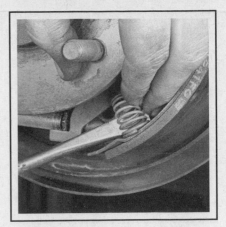

5.4d Pry the actuator lever spring out with a large screwdriver

5.4e Slide the parking brake strut out from between the axle flange and primary shoe

5.4f Remove the hold down springs and pins - the hold down spring tool shown here is available at most auto parts stores

5.4g Remove the actuator lever and pivot - be careful not to let the pivot fall out of the lever

plug from the backing plate with a hammer and chisel. With the plug removed, pull the lever off the adjusting screw wheel with one small screwdriver while turning the adjusting wheel with another small screwdriver, moving the shoes away from the drum. The drum should now come off.

5 Before reinstalling the drum it should be checked for cracks, score marks, deep scratches and hard spots, which will appear as small

5.4h Spread the top of the shoes apart and slide the assembly around the axle

5.4l Unhook the parking brake lever from the secondary shoe

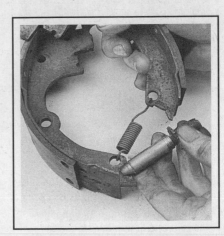

5.4j Spread the bottom of the shoes apart and remove the adjusting screw assembly

5.4k Clean the adjusting screw with solvent, dry it off and lubricate the threads and end with high-temperature brake grease, then reinstall the adjusting screw assembly between the new brake shoes

5.4l Lubricate the shoe contact points on the backing plate with high-temperature brake grease

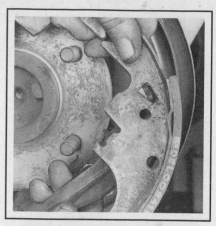

5.4m Insert the parking brake lever into the opening in the secondary brake shoe

5.4n Spread the shoes apart and slide them into position on the backing plate

5.4o Install the hold down-pin and spring through the backing plate and primary shoe

5.4p Insert the lever pivot into the actuator lever, place the lever over the secondary shoe hold down pin and install the hold down spring

5.4q Guide the parking brake strut behind the axle flange and engage the rear end of it in the slot on the parking brake lever - spread the shoes enough to allow the other end to seat against the primary shoe

5.4r Place the shoe guide over the anchor pin

5.4s Hook the lower end of the actuator link to the actuator lever, then loop the top end over the anchor pin

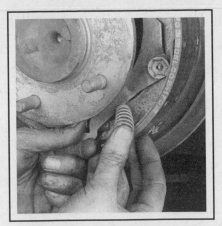

5.4t Install the lever return spring over the tab on the actuator lever, then push the spring up onto the brake shoe

5.4u Install the primary and secondary shoe return springs

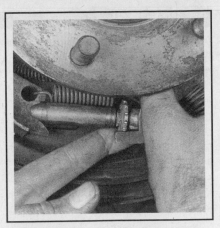

5.4v Pull out on the actuator lever to disengage it from the adjusting screw wheel, turn the wheel to adjust the shoes in or out as necessary - the brake drum should slide over the shoes and turn with a very slight amount of drag

discolored areas. If the hard spots cannot be removed with fine emery cloth or if any of the other conditions listed above exist, the drum must be taken to an automotive machine shop to have it turned.

➡Note: Professionals recommend resurfacing the drums whenever a brake job is done. Resurfacing will eliminate the possibility of out-of-round drums. If the drums are worn so much that they can't be resurfaced without exceeding the maximum allowable diameter (stamped into the drum), then new ones will be required. At the very least, if you elect not to have the drums

resurfaced, remove the glazing from the surface with medium-grit emery cloth using a swirling motion.

6   Install the brake drum on the axle flange.
7   Mount the wheel, install the lug nuts, then lower the vehicle.
8   Make a number of forward and reverse stops to adjust the brakes until satisfactory pedal action is obtained.

6   Wheel cylinder - removal, overhaul and installation

➡Note: If an overhaul is indicated (usually because of fluid leakage or sticky operation) explore all options before beginning the job. New wheel cylinders are available, which makes this job quite easy. If it's decided to rebuild the wheel cylinder, make sure that a rebuild kit is available before proceeding. Never overhaul only one wheel cylinder - always rebuild both of them at the same time.

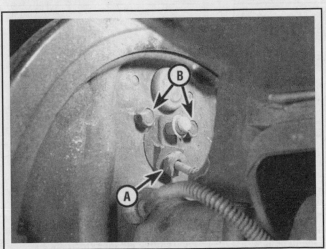

6.4 Completely loosen the brake line fitting (A) then remove the two wheel cylinder mounting bolts (B)

REMOVAL

▶ **Refer to illustration 6.4**

1   Raise the rear of the vehicle and support it securely on jackstands. Block the front wheels to keep the vehicle from rolling.
2   Remove the brake shoe assembly (Section 5).
3   Remove all dirt and foreign material from around the wheel cylinder.
4   Disconnect the brake line (see illustration). Don't pull the brake line away from the wheel cylinder.
5   Remove the wheel cylinder mounting bolts.
6   Detach the wheel cylinder from the brake backing plate and place it on a clean workbench. Immediately plug the brake line to prevent fluid loss and contamination.

OVERHAUL

▶ **Refer to illustration 6.7**

7   Remove the bleeder valve, seals, pistons, boots and spring assembly from the wheel cylinder body (see illustration).
8   Clean the wheel cylinder with brake fluid, denatured alcohol or brake system cleaner.

※※ WARNING:

Do not, under any circumstances, use petroleum based solvents to clean brake parts!

9   Use compressed air to remove excess fluid from the wheel cylinder and to blow out the passages.

10 Check the cylinder bore for corrosion and score marks. Crocus cloth can be used to remove light corrosion and stains, but the cylinder must be replaced with a new one if the defects cannot be removed easily, or if the bore is scored.

11 Lubricate the new seals with brake fluid.

12 Assemble the brake cylinder components. Make sure the seal lips face in.

INSTALLATION

13 Place the wheel cylinder in position and install the bolts.

14 Connect the brake line and install the brake shoe assembly.

15 Bleed the brakes (see Section 10).

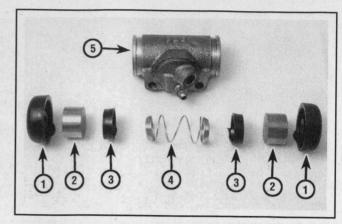

6.7 Wheel cylinder components - exploded view

1	Dust boot	4	Cup return spring
2	Piston	5	Wheel cylinder body
3	Piston cup		

7   Master cylinder - removal, overhaul and installation

➡Note: Before deciding to overhaul the master cylinder, investigate the availability and cost of a new or factory rebuilt unit and also the availability of a rebuild kit.

REMOVAL

▶ Refer to illustrations 7.2, 7.4, 7.7, 7.8 and 7.9

1   Place rags under the brake line fittings and prepare caps or plastic bags to cover the ends of the lines once they are disconnected.

※※ CAUTION:

Brake fluid will damage paint. Cover all body parts and be careful not to spill fluid during this procedure.

2   Loosen the tube nuts at the ends of the brake lines where they enter the master cylinder. To prevent rounding off the flats on these nuts, a flare-nut wrench, which wraps around the nut, should be used (see illustration).

3   Pull the brake lines away from the master cylinder slightly and plug the ends to prevent contamination.

4   Remove the two master cylinder mounting nuts disconnect the pushrod from the brake pedal (non-power brakes), move the bracket retaining the combination valve forward slightly, taking care not to bend the hydraulic lines running to the combination valve, and remove the master cylinder from the vehicle (see illustration).

5   Remove the reservoir covers and reservoir diaphragms, then discard any fluid remaining in the reservoir.

6   Mount the master cylinder in a vise. Be sure to line the vise jaws with blocks of wood to prevent damage to the cylinder body.

7.2 Disconnect the brake lines from the master cylinder - a flare-nut wrench should be used

7.4 Pull the combination valve forward, being careful not to bend or kink the lines, then slide the master cylinder off the mounting studs

7.7 Push the primary piston in and remove the lock-ring with a screwdriver

7.8 Pull the primary piston and spring assembly out of the bore

7.9 To remove the secondary piston, tap the cylinder against a block of wood

7.14 Lay the reservoir face down on the bench, with the secondary reservoir propped up on a block of wood - push the master cylinder straight down over the reservoir tubes using a rocking motion

7.15a Pry the secondary piston spring retainer off with a small screwdriver, then remove the seal

7 Remove the primary piston lock-ring by depressing the piston and prying the ring out with a screwdriver (see illustration).

8 Remove the primary piston assembly from the cylinder bore (see illustration).

9 Remove the secondary piston assembly from the cylinder bore. It may be necessary to remove the master cylinder from the vise and invert it, carefully tapping it against a block of wood to expel the piston (see illustration).

10 Pry the reservoir from the cylinder body with a screwdriver.

Remove the grommets.

11 Do not attempt to remove the quick take-up valve from the master cylinder body - it's not serviceable.

OVERHAUL

♦ **Refer to illustrations 7.14, 7.15a, 7.15b, 7.15c, 7.15d, 7.16 and 7.17**

12 Inspect the cylinder bore for corrosion and damage. If any corrosion or damage is found, replace the master cylinder body with a new one, as abrasives cannot be used on the bore.

13 Lubricate the new reservoir grommets with silicone lubricant and press them into the master cylinder body. Make sure they're properly seated.

14 Lay the reservoir on a hard surface and press the master cylinder body onto the reservoir, using a rocking motion (see illustration).

15 Remove the old seals from the secondary piston assembly and install the new secondary seals with the lips facing away from each other (see illustrations). The lip on the primary seal must face in (see illustration).

7.15b Remove the secondary seals from the piston (some only have one seal)

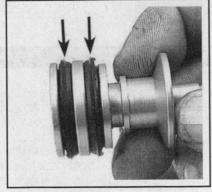

7.15c Install the secondary seals with the lips facing away from each other (on the single seal design, the seal lip should face away from the center of the piston)

7.15d Install a new primary seal on the secondary piston with the seal lip facing in the direction shown

7.16 Install a new spring retainer over the end of the secondary piston and push it into place with a socket

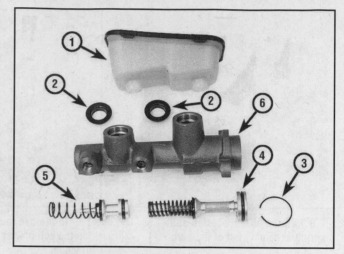

7.17 Master cylinder components - exploded view

1 Reservoir assembly (cover, diaphragm and reservoir body)
2 Grommets
3 Lock-ring
4 Primary piston and spring
5 Secondary piston and spring
6 Master cylinder body

16 Attach the spring retainer to the secondary piston assembly (see illustration).

17 Lubricate the cylinder bore with clean brake fluid and install the spring and secondary piston assembly (see illustration).

18 Install the primary piston assembly in the cylinder bore, depress it and install the lock-ring.

19 Inspect the reservoir cover and dia-phragm for cracks and deformation. Replace any damaged parts with new ones and attach the diaphragm to the cover.

➡**Note: Whenever the master cylinder is removed, the complete hydraulic system must be bled. The time required to bleed the system can be reduced if the master cylinder is filled with fluid and bench bled (refer to Steps 21 through 25) before the master cylinder is installed on the vehicle.**

20 Insert threaded plugs of the correct size into the cylinder outlet holes and fill the reservoirs with brake fluid. The master cylinder should be supported in such a manner that brake fluid will not spill during the bench bleeding procedure.

21 Loosen one plug at a time and push the piston assembly into the bore to force air from the master cylinder. To prevent air from being drawn back into the cylinder, the appropriate plug must be replaced before allowing the piston to return to its original position.

22 Stroke the piston three or four times for each outlet to ensure that all air has been expelled.

23 Since high pressure is not involved in the bench bleeding pro-cedure, an alternative to the removal and replacement of the plugs with each stroke of the piston assembly is available. Before pushing in on

the piston assembly, remove one of the plugs completely. Before releas-ing the piston, however, instead of replacing the plug, simply put your finger tightly over the hole to keep air from being drawn back into the master cylinder. Wait several seconds for the brake fluid to be drawn from the reservoir to the piston bore, then repeat the procedure. When you push down on the piston it will force your finger off the hole, allow-ing the air inside to be expelled. When only brake fluid is being ejected from the hole, replace the plug and go on to the other port.

24 Refill the master cylinder reservoirs and install the diaphragm and cover assembly.

➡**Note: The reservoirs should only be filled to the top of the res-ervoir divider to prevent overflowing when the cover is installed.**

INSTALLATION

25 Carefully install the master cylinder by reversing the removal steps, then bleed the brakes (refer to Section 10).

---

8   Combination valve - check and replacement

CHECK

1 Disconnect the wire from the pressure differential switch.

➡**Note: When unplugging the connector, squeeze the side lock releases, moving the inside tabs away from the switch, then pull up. Pliers may be used as an aid if necessary.**

2 Using a jumper wire, connect the switch wire to a good ground, such as the engine block.

3 Turn the ignition key to the On position. The warning light in the instrument panel should light.

4 If the warning light does not light, either the bulb is burned out or the electrical circuit is defective. Replace the bulb (refer to Chapter 10)

or repair the electrical circuit as necessary.

5 When the warning light functions correctly, turn the ignition switch off, disconnect the jumper wire and reconnect the wire to the switch terminal.

6 Make sure the master cylinder reservoirs are full, then attach a bleeder hose to one of the rear wheel bleeder valves and immerse the other end of the hose in a container partially filled with clean brake fluid.

7 Turn the ignition switch on.

8 Open the bleeder valve while a helper applies moderate pressure to the brake pedal. The brake warning light on the instrument panel should light.

9   Close the bleeder valve before the helper releases the brake pedal.

10 Reapply the brake pedal with moderate to heavy pressure. The brake warning light should go out.

11 Attach the bleeder hose to one of the front brake bleeder valves and repeat Steps 8 through 10. The warning light should react in the same manner as in Steps 8 and 10.

12 Turn the ignition switch off.

13 If the warning light did not come on in Steps 8 and 11, but does light when a jumper is connected to ground, the warning light switch portion of the combination valve is defective and the combination valve must be replaced with a new one since the components of the combination valve are not individually serviceable.

REPLACEMENT

14 Place a container under the combination valve and protect all painted surfaces with newspapers or rags.

15 Disconnect the hydraulic lines at the combination valve, then plug the lines to prevent further loss of fluid and to protect the lines from contamination.

16 Disconnect the electrical connector from the pressure differential switch.

17 Remove the bolt holding the valve to the mounting bracket and remove the valve from the vehicle.

18 Installation is the reverse of the removal procedure.

19 Bleed the entire brake system.

---

9   Brake hoses and lines - inspection and replacement

INSPECTION

1   About every six months, with the vehicle raised and supported securely on jackstands, the rubber hoses which connect the steel brake lines with the front and rear brake assemblies should be inspected for cracks, chafing of the outer cover, leaks, blisters and other damage. These are important and vulnerable parts of the brake system and inspection should be complete. A light and mirror will be helpful for a thorough check. If a hose exhibits any of the above conditions, replace it with a new one.

REPLACEMENT

Front brake hose

▶ **Refer to illustration 9.2**

2   Using a back-up wrench, disconnect the brake line from the hose fitting, being careful not to bend the frame bracket or brake line (see illustration).

3   Use a pair of pliers to remove the U-clip from the female fitting at the bracket, then detach the hose from the bracket.

4   At the caliper end of the hose, remove the bolt from the fitting block, then remove the hose and the copper washers on either side of

the fitting block.

5   When installing the hose, always use new copper washers on either side of the fitting block and lubricate all bolt threads with clean brake fluid before installation.

6   With the fitting flange engaged with the caliper locating ledge, attach the hose to the caliper.

7   Without twisting the hose, install the female fitting in the hose bracket. It will fit the bracket in only one position.

8   Install the U-clip retaining the female fitting to the frame bracket.

9   Using a back-up wrench, attach the brake line to the hose fitting.

10 When the brake hose installation is complete, there should be no kinks in the hose. Make sure the hose doesn't contact any part of the suspension. Check this by turning the wheels to the extreme left and right positions. If the hose makes contact, remove the hose and correct the installation as necessary.

Rear brake hose

11 Using a back-up wrench, disconnect the hose at the frame bracket, being careful not to bend the bracket or steel lines.

12 Remove the U-clip with a pair of pliers and separate the female fitting from the bracket.

13 Disconnect the two hydraulic lines at the junction block, then unbolt and remove the hose.

14 Without twisting the hose, install the female end of the hose in the frame bracket. It will fit the bracket in only one position. Bolt the junction block to the support bracket and connect the lines, tightening them securely.

15 Install the U-clips retaining the female end to the bracket.

16 Using a back-up wrench, attach the steel line fittings to the female fittings. Again, be careful not to bend the bracket or steel line.

17 Make sure the hose installation did not loosen the frame bracket. Tighten the bracket if necessary.

18 Fill the master cylinder reservoir and bleed the system (refer to Section 10).

Metal brake lines

19 When replacing brake lines be sure to use the correct parts. Don't use copper tubing for any brake system components. Purchase steel brake lines from a dealer or auto parts store.

20 Prefabricated brake line, with the tube ends already flared and fittings installed, is available at auto parts stores and dealers. These lines are also bent to the proper shapes.

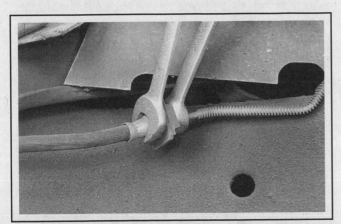

9.2 Place a wrench on the hose fitting to prevent it from turning and disconnect the line with a flare-nut wrench

21 If prefabricated lines are not available, obtain the recommended steel tubing and fittings to match the line to be replaced. Determine the correct length by measuring the old brake line (a piece of string can usually be used for this) and cut the new tubing to length, allowing about 1/2-inch extra for flaring the ends.

22 Install the fitting over the cut tubing and flare the ends of the line with a flaring tool.

23 If necessary, carefully bend the line to the proper shape. A tube bender is recommended for this.

✳✳ WARNING:

Do not crimp or damage the line.

24 When installing the new line make sure it's securely supported in the brackets and has plenty of clearance between moving or hot components.

25 After installation, check the master cylinder fluid level and add fluid as necessary. Bleed the brake system as outlined in the next Section and test the brakes carefully before driving the vehicle in traffic.

10 Brake system bleeding

▶ **Refer to illustration 10.8**

✳✳ WARNING:

Wear eye protection when bleeding the brake system. If the fluid comes in contact with your eyes, immediately rinse them with water and seek medical attention.

➡ **Note: Bleeding the hydraulic system is necessary to remove any air that manages to find its way into the system when it's been opened during removal and installation of a hose, line, caliper or master cylinder.**

1 It will probably be necessary to bleed the system at all four brakes if air has entered the system due to low fluid level, or if the brake lines have been disconnected at the master cylinder.

2 If a brake line was disconnected only at a wheel, then only that caliper or wheel cylinder must be bled.

3 If a brake line is disconnected at a fitting located between the master cylinder and any of the brakes, that part of the system served by the disconnected line must be bled.

4 Remove any residual vacuum from the brake power booster by applying the brake several times with the engine off.

5 Remove the master cylinder reservoir cover and fill the reservoir with brake fluid. Reinstall the cover.

➡ **Note: Check the fluid level often during the bleeding operation and add fluid as necessary to prevent the fluid level from falling low enough to allow air bubbles into the master cylinder.**

6 Have an assistant on hand, as well as a supply of new brake fluid, an empty clear plastic container, a length of clear tubing to fit over the bleeder valve and a wrench to open and close the bleeder valve.

7 Beginning at the right rear wheel, loosen the bleeder valve slightly, then tighten it to a point where it is snug but can still be loosened quickly and easily.

8 Place one end of the tubing over the bleeder valve and submerge the other end in brake fluid in the container (see illustration).

9 Have the assistant pump the brakes slowly a few times to get pressure in the system, then hold the pedal firmly depressed.

10 While the pedal is held depressed, open the bleeder valve just enough to allow a flow of fluid to leave the valve. Watch for air bubbles

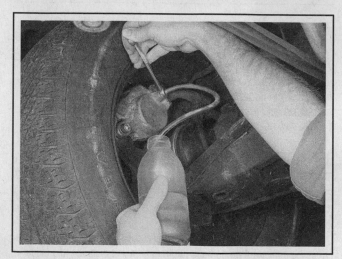

10.8 When bleeding the brakes, a hose is connected to the bleeder valve and then submerged in brake fluid - air will be seen as bubbles in the container and the hose

to exit the submerged end of the tube. When the fluid flow slows after a couple of seconds, close the valve and have your assistant release the pedal.

11 Repeat Steps 9 and 10 until no more air is seen leaving the tube, then tighten the bleeder valve and proceed to the left rear wheel, the right front wheel and the left front wheel, in that order, and perform the same procedure. Be sure to check the fluid in the master cylinder reservoir frequently.

12 Never use old brake fluid. It contains moisture which will deteriorate the brake system components.

13 Refill the master cylinder with fluid at the end of the operation.

14 Check the operation of the brakes. The pedal should feel solid when depressed, with no sponginess. If necessary, repeat the entire process.

✳✳ WARNING:

Do not operate the vehicle if you are in doubt about the effectiveness of the brake system.

11 Parking brake - adjustment

2002 AND EARLIER MODELS

▶ **Refer to illustration 11.4**

1 Apply the parking brake lever exactly two ratchet clicks.
2 Raise the vehicle and support it securely on jackstands.
3 Before adjusting, make sure the equalizer nut groove is lubricated liberally with multi-purpose grease.
4 Tighten the adjusting nut (see illustration) until the left rear wheel can just be turned backwards with two hands, but locks when forward motion is attempted.
5 Release the parking brake lever and make sure the rear wheels turn freely in both directions with no drag.
6 Lower the vehicle.

2003 AND LATER MODELS

7 The parking brake cable is self-adjusting, but if the parking brake pedal travel is excessive or won't hold the vehicle on an incline, the

11.4 To adjust the parking brake cable, turn the adjusting nut on the equalizer while preventing the cable from turning with a pair of locking pliers clamped to the end of the adjuster rod

parking brake shoes may need to be adjusted or replaced (see Section 18).

12 Parking brake cables - replacement

REAR CABLES

▶ **Refer to illustrations 12.4 and 12.5**

1 Loosen the wheel lug nuts, raise the rear of the vehicle and support it securely on jackstands. Remove the wheel(s).
2 Loosen the equalizer nut to slacken the cables, then disconnect the cable to be replaced from the equalizer.
3 Remove the brake drum from the axle flange. Refer to Section 5 if any difficulty is encountered.
4 Remove the brake shoe assembly far enough to disconnect the cable end from the parking brake lever (see illustration).
5 Depress the tangs on the cable housing retainer and push the housing and cable through the backing plate (see illustration).
6 To install the cable, reverse the removal procedure and adjust the

cable as described in the preceding Section.

FRONT CABLE

▶ **Refer to illustration 12.9**

7 Raise the vehicle and support it securely on jackstands.
8 Loosen the equalizer assembly to provide slack in the cable.
9 Disconnect the front cable from the cable joiner near the equalizer (see illustration).
10 Disconnect the cable from the pedal assembly.
11 Free the cable from the routing clips and push the cable and grommet through the firewall.
12 To install the cable, reverse the removal procedure and adjust the cable as described in the preceding Section.

12.4 To disconnect the cable end from the parking brake lever, pull back on the return spring and maneuver the cable out of the slot in the lever

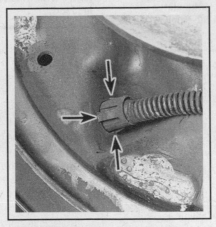

12.5 Depress the retention tangs (arrows) to free the cable and housing from the backing plate

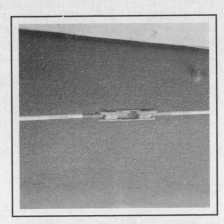

12.9 The rear end of the front cable and the front end of the rear are attached at this connector, which is located just forward of the equalizer assembly

13 Parking brake pedal - removal and installation

▶ **Refer to illustration 13.3**

1   Disconnect the cable from the negative terminal of the battery.

�֎ CAUTION:

On models equipped with an anti-theft audio system, disable the anti-theft feature before performing any operation that requires disconnecting the battery or disrupting power to the stereo (see the Anti-theft audio system procedures at the front of this manual).

2   On 1991 and earlier models, refer to Chapter 11 and remove the instrument panel and dash assembly.

3   Disconnect the release rod or cable from the pedal assembly (see illustration).

4   Unplug the electrical connector from the parking brake switch.

5   Remove the three pedal mechanism mounting bolts.

6   Disconnect the parking brake cable then depress the tangs on the cable housing, pushing the housing through the pedal bracket.

7   Remove the pedal assembly.

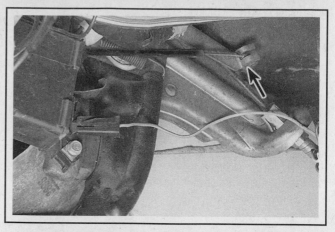

13.3   Remove the release cable from the parking brake release arm (arrow)

8   To install the pedal assembly, reverse the removal procedure and adjust the parking brake cable as outlined in Section 11.

14 Brake pedal - removal and installation

➡**Note: On vehicles equipped with a manual transmission, the clutch pedal must be removed first. Refer to Chapter 8 for the clutch pedal removal and installation procedure.**

1992 AND EARLIER MODELS

Removal

1   Disconnect the cable from the negative battery terminal.

✻ CAUTION:

On models equipped with an anti-theft audio system, disable the anti-theft feature before performing any operation that requires disconnecting the battery or disrupting power to the stereo (see the Anti-theft audio system procedures at the front of this manual).

2   Remove the driver's side knee bolster (see Chapter 11).

3   Remove the master cylinder pushrod retaining clip from the actuator lever and slide the pushrod and washers off the pin.

4   Remove the actuator lever pinch bolt.

5   Slide the brake pedal retaining clip off the pivot pin and pull the pedal from the mounting bracket. Note how the bushings, actuator lever and spacers are situated.

Installation

6   Insert the pedal pivot bushings into the mounting bracket, position the actuator lever and spacer between the two bushings and slide the brake pedal pivot pin through them. Install the retaining clip.

7   Install the actuator lever pinch bolt and tighten it securely.

8   Attach the master cylinder pushrod to the actuator lever (with a washer on each side of it) and install the clip. Insert the actuator to

brake pedal bolt and tighten the nut securely.

9   The remaining installation steps are the reverse of removal.

1993 AND LATER MODELS

Removal

10   Disconnect the cable from the negative terminal of the battery.

✻ CAUTION:

On models equipped with an anti-theft audio system, disable the anti-theft feature before performing any operation that requires disconnecting the battery or disrupting power to the stereo (see the Anti-theft audio system procedures at the front of this manual).

11   If necessary, remove the driver's side knee bolster (see Chapter 11).

12   Remove the brake light switch and master cylinder pushrod from the brake pedal.

13   Note the position of the bushings and spacers, then remove the nut and pivot bolt located at the top of the pedal arm.

14   Withdraw the pedal arm from the mounting bracket and retrieve the spacers and bushings.

15   Inspect the pedal arm, pivot bolt, bushings and spacers for excessive wear or damage and replace if necessary.

Installation

16   Install the bushings and spacers onto the pedal arm in their proper locations.

17   Position the pedal arm in the mounting bracket and secure with the pivot bolt and nut. Tighten the nut securely.

18   The remaining installation steps are the reverse of removal.

15 Power brake booster - inspection and replacement

VACUUM BOOSTER (1992 AND EARLIER MODELS)

Inspection

1   The vacuum power brake booster unit requires no special maintenance or service besides inspecting the vacuum line hose, grommets and connections for tightness and deterioration. Replace parts as necessary.

2   Repair of the booster unit requires special tools and is outside the scope of the home mechanic. If a problem develops with the booster unit, it is recommended that a new or rebuilt unit be installed.

Replacement

▶ **Refer to illustration 15.6**

3   Working in the engine compartment, remove the nuts securing the master cylinder to the booster and carefully pull the master cylinder, along with the ABS RWAL control module and isolation/dump valve (if equipped) forward until it clears the mounting studs. Be careful not to bend or kink the brake lines.

4   Disconnect the vacuum hose from the booster.

5   Working inside the vehicle, remove the driver's side knee bolster (see Chapter 11).

6   Disconnect the pushrod from the brake pedal, then remove the 4 nuts securing the booster unit to the firewall (see illustration).

7   Carefully withdraw the booster unit from the firewall and remove it from the engine compartment.

8   Installation is the reverse of removal. Tighten the booster unit and master cylinder mounting nuts to the torque listed in this Chapter's Specifications.

9   Test the operation of the braking system before placing the vehicle in normal service.

HYDRAULIC BOOSTER (1993 AND LATER MODELS)

Inspection

10   Beginning in 1993, a hydraulically powered brake booster replaced the traditional vacuum powered brake booster. It mounts in the conventional way, between the firewall and the brake master cylinder. Instead of using vacuum to boost power to the brakes, this system uses hydraulic pressure produced from the power steering pump.

11   During heavy braking, a hissing noise can be heard coming from the booster unit especially when the vehicle is motionless. This is normal. Also, upon quickly releasing the brake pedal after a hard braking incident, a clunk, clatter or clicking noises will be heard. This is also within the normal operating parameters of the system.

12   The hydraulic power brake booster unit requires no special maintenance or service. In order for the booster to function properly, the following conditions must be acceptable:

 a) The drivebelt must be in good condition and correctly installed on the power steering pump pulley (see Chapter 1).

 b) The power steering fluid level must be correct (see Chapter 1).

 c) The pressure lines must be free of kinks and the connections must be tight.

 d) The booster unit must be free of leaks.

15.6 Remove the four brake booster-to-firewall mounting nuts (arrows)

 e) The power steering system must be free of air bubbles (see Chapter 10).

13   Repair of the booster unit requires special tools and is outside the scope of the home mechanic. If a problem develops with the booster unit, it is recommended that a new or rebuilt unit be installed.

Replacement

14   Working in the engine compartment, remove the nuts securing the master cylinder to the booster and carefully pull the master cylinder, along with the ABS RWAL control module and isolation/dump valve (if equipped) forward until it clears the mounting studs. Be careful not to bend or kink the brake lines.

15   Some power steering fluid will be lost during this operation, so place a drain pan under the vehicle.

16   Disconnect the power steering fluid reservoir supply hose from the booster unit and plug it immediately to prevent leakage. Then place it out of the way.

17   Disconnect the pressure lines from the booster and cap them immediately to prevent leakage and contamination.

18   Working inside the vehicle, remove the driver's side knee bolster if necessary (see Chapter 11).

19   Disconnect the pushrod from the brake pedal.

20   Remove the 4 nuts securing the booster unit to the firewall (see illustration 15.6).

21   Carefully withdraw the booster unit from the firewall and remove it from the engine compartment.

22   Installation is the reverse of removal. Tighten the booster unit and master cylinder mounting nuts, and hose fittings to the torque listed in this Chapter's Specifications.

23   Start the vehicle and add power steering fluid as required (see Chapter 1) and check for leaks.

24   Bleed the power steering system (see Chapter 10).

25   Test the operation of the braking system before placing the vehicle in normal service.

16 Brake light switch - removal, installation and adjustment

1993 AND EARLIER MODELS

Removal

▶ **Refer to illustration 16.4**

1 The brake light switch is mounted on a bracket at the top of the brake pedal.
2 Disconnect the cable from the negative terminal of the battery.

✳✳ CAUTION:

On models equipped with an anti-theft audio system, disable the anti-theft feature before performing any operation that requires disconnecting the battery or disrupting power to the stereo (see the Anti-theft audio system procedures at the front of this manual).

3 Remove the driver's side knee bolster (see Chapter 11).
4 Locate the switch at the top of the brake pedal (see illustration). If equipped with cruise control, there will be another switch very similar in appearance. The brake light switch is the one on the left of the bracket.
5 Disconnect the wire harnesses at the brake light switch.
6 Depress the brake pedal and pull the switch out of the clip. The switch appears to be threaded, but it's designed to be pushed into and out of the clip and not turned.

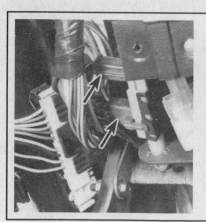

16.4 Unplug the brake light switch electrical connectors (arrows) and pull the switch out of the mounting bracket (1993 and earlier models)

Installation and adjustment

7 With the brake pedal depressed, push the new switch into the clip. Note that audible clicks will be heard as this is done.
8 Pull the brake pedal all the way to the rear against the pedal stop until the clicking sounds can no longer be heard. This action will automatically move the switch the proper amount and no further adjustment will be required.
➡**Note: Do not apply excessive force during this adjustment procedure, as power booster damage may result.**

9 Connect the wires at the switch and the battery. Have an assistant check that the rear brake lights are functioning properly.

1994 AND LATER MODELS

Removal

10 The brake light switch is mounted on the brake pedal.
11 Disconnect the cable from the negative terminal of the battery.

✳✳ CAUTION:

On models equipped with an anti-theft audio system, disable the anti-theft feature before performing any operation that requires disconnecting the battery or disrupting power to the stereo (see the Anti-theft audio system procedures at the front of this manual).

12 Remove the driver's side knee bolster if necessary (see Chapter 11).
13 Remove the retainer securing the brake switch and master cylinder pushrod to the brake pedal.
14 Detach the brake light switch and master cylinder pushrod from the brake pedal.
15 Disconnect the electrical connector from the switch and remove it from the vehicle.

Installation and adjustment

16 Place the brake light switch over the master cylinder pushrod and install them onto the brake pedal. Secure with the retaining clip.
17 Connect the harness electrical connector to the brake light switch.
18 The remaining installation steps are the reverse of removal.
19 This type of switch is self adjusting, no adjustment is required.
20 Check brake light operation.

17 Anti-lock brake system - general information

▶ **Refer to illustrations 17.4a and 17.4b**

DESCRIPTION

The Anti-lock brake system is designed to maintain vehicle maneuverability, directional stability and optimum deceleration under severe braking conditions on most road surfaces. It does so by monitoring the rotational speed of the wheels and controlling the brake line pressure during braking. This prevents the wheels from locking up prematurely.

Two types of systems are used: Rear Wheel Anti-Lock (RWAL) and Four Wheel Anti-Lock (4WAL). RWAL only controls lockup on the rear wheels, whereas 4WAL prevents lockup on all four wheels.

COMPONENTS

Actuator assembly

The actuator assembly includes the master cylinder and a control valve which consists of a dump valve and an isolation valve. The valve operates by changing the brake fluid pressure in response to signals from the control unit.

Control unit

The control unit for the anti-lock brakes is called the Electro-Hydraulic Control Unit (EHCU) on 4WAL systems and Control module on

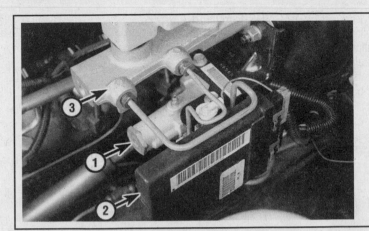

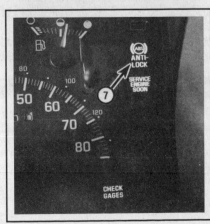

17.4a RWAL system components

1   Isolation/dump valve
2   Control module
3   Master cylinder

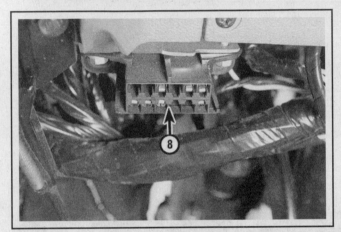

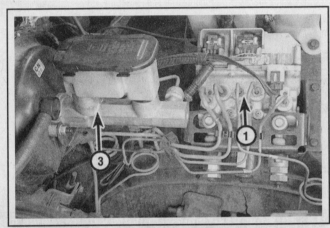

17.4b Four-wheel anti-lock (4WAL) brake system diagram

1   4WAL EHCU valve
2   Combination valve
3   Master cylinder
4   Brake pedal switch
5   Parking brake switch
6   Wheel speed sensors
7   Warning light
8   Assembly Line Diagnostic Link (ALDL)

RWAL systems. The unit is mounted in the engine compartment below the master cylinder and is the "brain" for the system (see illustrations). The function of the control unit is to accept and process information received from the speed sensor(s) and brake light switch to control the hydraulic line pressure, avoiding wheel lockup. The control unit also constantly monitors the system, even under normal driving conditions, to find faults within the system.

If a problem develops within the system, the BRAKE (RWAL system) or ANTI-LOCK (4WAL system) warning light will glow on the dashboard. A diagnostic code will also be stored, which, when retrieved by a service technician, will indicate the problem area or component.

Speed sensor

On 4WAL systems, each wheel has a speed sensor. On RWAL systems, a rear wheel speed sensor is located in the transmission extension housing on 2WD models and in the transfer case on All-Wheel Drive models. The speed sensor(s) sends a signal to the control unit indicating wheel rotational speed.

Brake light switch

The brake light switch signals the control unit when the driver steps on the brake pedal. Without this signal the anti-lock system won't activate.

DIAGNOSIS AND REPAIR

If the BRAKE or ANTI-LOCK warning light on the dashboard comes on and stays on, make sure the parking brake is released and there's no problem with the brake hydraulic system. If neither of these is the cause, the anti-lock system is probably malfunctioning. Although special test procedures are necessary to properly diagnose the system, the home mechanic can perform a few preliminary checks before taking the vehicle to a dealer service department.

a) *Make sure the brakes, calipers and wheel cylinders are in good condition.*
b) *Check the electrical connectors at the control unit.*
c) *Check the fuses.*
d) *Follow the wiring harness to the speed sensor(s) and brake light switch and make sure all connections are secure and the wiring isn't damaged.*

If the above preliminary checks don't rectify the problem, the vehicle should be diagnosed by a dealer service department or other qualified repair shop.

---

18 Parking brake shoes (models with rear disc brakes) - replacement

▶ **Refer to illustrations 18.3, 18.4 and 18.6**

✳ WARNING:

The dust created by the brake system is harmful to your health. Never blow it out with compressed air and don't inhale any of it. An approved filtering mask should be worn when working on the brakes. Do not, under any circumstances, use petroleum-based solvents to clean brake parts. Use brake system cleaner only!

1   Loosen the rear wheel lug nuts, raise the rear of the vehicle and support it securely on jackstands. Release the parking brake. Block the front wheels to prevent the vehicle from rolling, then remove the rear wheels.

2   Remove the brake caliper (see Section 3), mounting bracket and the brake disc (see Section 4). From underneath the vehicle, pull the equalizer toward the rear of the vehicle to expose the inner parking brake cable. Clamp the exposed cable with padded locking pliers to hold the cable extended, which will relieve tension on the rear cables.

✳ CAUTION:

Don't nick or cut the plastic anti-corrosion coating on the cable.

3   Wash the brake assembly with brake system cleaner. Slide the shoe downward until it clears the retaining clip (see illustration), then slide the shoe up and off of the actuator.

4   Lift one end of the shoe over the axle flange, then work the shoe over the flange and remove it (see illustration).

18.3 Slide the parking brake shoe downward, out of the clip

5   Before installing the new shoe, turn the adjuster screw star wheel in, then make sure the slots in the adjusting screw and the tappet are parallel with the backing plate.

6   To install the shoe, reverse the removal procedure. Slide the shoe under the retaining clip, then make sure the ends of the shoe seat properly in the slots in the adjuster screw and tappet (see illustration).

7   When installing the new shoe and lining assembly, turn the adjuster screw until the shoe lining just drags on the braking surface inside the disc. Then remove the disc and back-off the adjuster screw

18.4 Lift one end of the parking brake shoe over the axle flange, then "wind" the rest of the shoe over the flange

18.6 Make sure the ends of the shoe seat in the adjuster screw slot (A) and the tappet slot (B); C is the adjuster screw star wheel

until the shoe lining doesn't drag when the disc is installed and turned. The actual clearance between the lining surface of the shoe and the braking surface inside the disc should be 0.026-inch.

8   Installation is otherwise the reverse of the removal procedure. Be sure to tighten the caliper bracket bolts and the caliper mounting bolts to the torque listed in this Chapter's Specifications, and the wheel lug nuts to the torque listed in the Chapter 1 Specifications.

Specifications

Brake fluid type	See Chapter 1

Disc brakes

Minimum brake pad thickness	See Chapter 1
Rotor discard thickness	Cast into disc
Lateral runout	
2000 and earlier	0.002 inch maximum
2001 and later	0.001 maximum
Rotor thickness variation (parallelism)	0.0005 inch
Caliper-to-knuckle clearance (clearance on each end added together for total)	0.010 to 0.024 inch

Drum brakes

Minimum brake lining thickness	See Chapter 1
Drum discard diameter	Cast into drum
Out-of-round	0.006 inch maximum
Taper	0.003 inch maximum

Torque specifications

	Ft-lbs
ABS wheel speed sensor mounting bolt	156 in-lbs
Master cylinder mounting nuts	
1985 through 1994	21
1995 and later	27
Power booster mounting nuts	
Vacuum type	21
Hydraulic type	27
Caliper mounting bolts	
1985 through 1994	37
1995 through 2002	
Front	38
2003 and later	
Front	80
Rear	31
Caliper mounting bracket bolts (2003 and later models)	
Front	129
Rear	148
Wheel cylinder mounting bolts	13
Brake hose-to-caliper inlet fitting bolt	29 to 32
Wheel lug nuts	See Chapter 1

Section

Reference to other Chapters

10

STEERING AND SUSPENSION SYSTEMS

1.1 Front suspension components - typical

1	*Stabilizer bar*	5	*Steering gear*	9	*Shock absorber*	12	*Lower suspension balljoint*
2	*Relay rod*	6	*Idler arm*	10	*Steering knuckle*	13	*Lower control arm*
3	*Connecting rod*	7	*Upper control arm*	11	*Coil spring*	14	*Tie-rod*
4	*Pitman arm*	8	*Upper suspension balljoint*				

1.2 Rear suspension components (single-leaf spring type shown)

1	*Composite leaf spring*	3	*Rear axle tube*	5	*U-bolt*
2	*Shock absorber*	4	*Lower plate*	6	*Anchor plate*

1   General information

♦ **Refer to illustrations 1.1 and 1.2**

❊❊ WARNING 1:

Whenever any of the suspension or steering fasteners are loosened or removed they must be inspected and if necessary, replaced with new ones of the same part number or of original equipment quality and design. Torque specifications must be followed for proper reassembly and component retention. Never attempt to heat, straighten or weld any suspension or steering component. Instead, replace any bent or damaged part with a new one.

❊❊ WARNING 2:

Most 1993 and later models are equipped with airbags. Always disable the airbag system before working in the vicinity of the impact sensors, steering column or instrument panel to avoid the possibility of accidental deployment of the airbag(s), which could cause personal injury (see Chapter 12 for the airbag disabling procedure). The yellow wires and connectors routed through the instrument panel are for this system. Do not use electrical test equipment on these yellow wires or tamper with them in any way while working under the instrument panel.

1   The front suspension is independent, allowing each wheel to compensate for road surface changes without appreciably affecting the other (see illustration). Each wheel is connected to the frame by a spindle, balljoint, lower control arms and a shock absorber/strut assembly positioned vertically between the spindle and frame. Each side uses a coil spring mounted between the lower control arm and the frame. Body side roll is controlled by a stabilizer bar.

2   The rear suspension consists of a solid axle, leaf springs, shock absorbers and a stabilizer bar (see illustration). The front of each leaf spring is attached to the frame by a hanger assembly, while the rear is connected by a moveable shackle. Two different types of leaf springs are used on these models. On 1985 to 1994 models, a constant-rate composite material single-leaf spring was installed. On 1995 and later models, a more traditional steel multi-leaf type spring is used.

3   The steering system consists of the steering wheel, steering column, an articulated intermediate shaft, the steering gear which connect the steering gear to the spindles, power steering pump and tie rods.

2   Front stabilizer bar - removal and installation

♦ **Refer to illustrations 2.2 and 2.3**

REMOVAL

1   Raise the vehicle and support it securely on jackstands. Apply the parking brake.

2   Remove the stabilizer bar-to-lower control arm nuts, and bolts, noting how the spacers, washers and bushings are positioned (see illustration).

3   Remove the stabilizer bar bracket bolts and detach the bar from the vehicle (see illustration).

4   Pull the brackets off the stabilizer bar and inspect the bushings for cracks, hardening and other signs of deterioration. If the bushings are damaged, replace them.

INSTALLATION

5   Position the stabilizer bar bushings on the bar with the slits facing the front of the vehicle.

➡**Note: The offset in the bar must face down.**

6   Push the brackets over the bushings and raise the bar up to the frame. Install the bracket bolts but don't tighten them completely at this time.

7   Install the stabilizer bar-to-lower control arm bolts, washers, spacers and rubber bushings and tighten the nuts securely.

8   Tighten the bracket bolts securely.

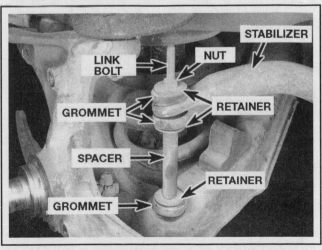

2.2   Note the positions of the bushings, spacers and washers before removing the stabilizer bar-to-control arm nut/bolt

2.3   Remove the stabilizer bar bracket bolts - the nuts are welded to the frame, so there's no need to put a wrench on them to prevent them from turning

3 Front shock absorbers - removal and installation

▶ **Refer to illustration 3.3**

REMOVAL

1 Loosen the wheel lug nuts, raise the vehicle and support it securely on jackstands. Apply the parking brake. Remove the wheel.

2 Remove the upper shock absorber stem nut. Use an open end wrench to keep the stem from turning. If the nut won't loosen because of rust, squirt some penetrating oil on the stem threads and allow it to soak in for awhile. It may be necessary to keep the stem from turning with a pair of locking pliers, since the flats provided for a wrench are quite small.

3 Remove the two lower shock mount bolts (see illustration) and

pull the shock absorber out through the bottom of the lower control arm. Remove the washers and the rubber grommets from the top of the shock absorber.

INSTALLATION

4 Extend the new shock absorber as far as possible. Position a new washer and rubber grommet on the stem and guide the shock up through the coil spring and into the upper mount.

5 Install the upper rubber grommet and washer and wiggle the stem back-and-forth to ensure that the grommets are centered in the mount. Tighten the stem nut securely.

6 Install the lower mounting bolts and tighten them securely.

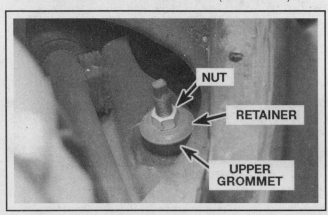
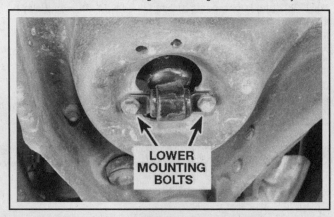

3.3 The bottom of the shock absorber is bolted to the lower control arm

4 Balljoints - replacement

▶ **Refer to illustrations 4.4, 4.5 and 4.8**

UPPER BALLJOINT

1 Loosen the wheel lug nuts, raise the vehicle and support it securely on jackstands. Apply the parking brake. Remove the wheel.

2 Place a jack or a jackstand under the lower control arm.

➡**Note: The jack or jackstand must remain under the control arm during removal and installation of the balljoint to hold the spring and control arm in position.**

Unbolt the bracket for the wheel speed sensor harness (if equipped) from the upper control arm.

3 Remove the cotter pin from the balljoint stud and back off the nut two turns.

4 Separate the balljoint from the steering knuckle with a special tool, available at most auto parts stores, to press the balljoint out of the steering knuckle (see illustration). An equivalent tool can be fabricated from a large bolt, nut, washer and socket. Countersink the center of the bolt head with a large drill bit to prevent it from slipping off the balljoint stud. Install the tool as shown in the illustration, hold the bolt head with a wrench and tighten the nut against the washer until the balljoint pops out. Notice that the balljoint nut hasn't been completely removed. Support the steering knuckle to prevent it from damaging the brake hose.

5 Using a 1/8-inch drill bit, drill a 1/4-inch deep hole in the center

of each rivet head (see illustration).

6 Using a 1/2-inch diameter drill bit, drill off the rivet heads.

7 Use a punch to knock the rivet shanks out, then remove the balljoint from the control arm.

8 Position the new balljoint on the control arm and install the bolts and nuts supplied in the kit (see illustration). Tighten the replacement balljoint-to-upper control arm bolts to the torque listed in this Chapter's Specifications. Remove the steering knuckle support.

9 Insert the balljoint stud into the steering knuckle and install the nut. Tighten the upper balljoint-to-steering knuckle nut to the torque listed in this Chapter's Specifications.

10 Install a new cotter pin, tightening the nut slightly if necessary to align a slot in the nut with the hole in the balljoint stud. Reinstall the bracket for the wheel speed sensor harness (if equipped).

11 Install the balljoint grease fitting and fill the joint with grease.

12 Install the wheel, tightening the lug nuts to the specified torque.

13 Drive the vehicle to an alignment shop to have the front end alignment checked and, if necessary, adjusted.

LOWER BALLJOINT

14 The lower balljoint is a press fit in the lower control arm and requires special tools to remove and replace it. Refer to Section 6, remove the lower control arm and take it to an automotive machine shop to have the old balljoint pressed out and the new balljoint pressed in.

4.4 A special tool, available at most auto parts stores, is required to push the balljoints out of the steering knuckle (an alternative tool can be fabricated from a large bolt, nut, washer and socket)

4.5 Drill into the balljoint rivets with a 1/8-inch drill bit, then use a 1/2-inch drill bit to cut the rivet heads off - be careful not to enlarge the holes in the control arm

4.8 You don't need to rivet the new upper balljoint - install it with the nuts and bolts included in the kit

5   Control arms - removal and installation

▶ **Refer to illustration 5.4**

UPPER CONTROL ARM

Removal

1   Loosen the wheel lug nuts, raise the front of the vehicle and support it securely on jackstands. Apply the parking brake. Remove the wheel.

2   Support the lower control arm with a jack or jackstand. The support point must be as close to the balljoint as possible to give maximum leverage on the lower control arm. Unbolt the bracket for the wheel speed sensor harness (if equipped) from the upper control arm.

3   Disconnect the upper balljoint from the steering knuckle (refer to Section 4).

➡**Note: DO NOT use a "pickle fork" type balljoint separator - it may damage the balljoint seals.**

Support the steering knuckle to prevent it from damaging the brake hose.

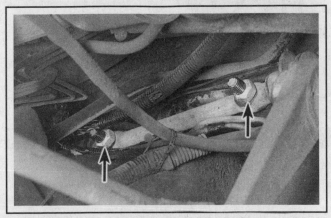

5.4 Upper control arm mounting nuts (arrows) (note the positions of the alignment shims and return them to their original positions)

4   Remove the control arm-to-frame nuts and bolts, recording the position of any alignment shims. They must be reinstalled in the same location to maintain wheel alignment (see illustration).

5   Detach the control arm from the vehicle.

➡**Note: The control arm bushings are pressed into place and require special tools for removal and installation. If the bushings must be replaced, take the control arm to a dealer service department or an automotive machine shop to have the old bushings pressed out and the new ones pressed in.**

Installation

6   Position the control arm on the frame and install the bolts and nuts. Install any alignment shims that were removed. Tighten the nuts to the specified torque.

7   Insert the balljoint stud into the steering knuckle and tighten the nut to the specified torque. Install a new cotter pin, tightening the nut slightly, if necessary, to align a slot in the nut with the hole in the balljoint stud.

8   Install the wheel and lug nuts and lower the vehicle. Tighten the lug nuts to the specified torque.

9   Drive the vehicle to an alignment shop to have the front end alignment checked and, if necessary, adjusted.

LOWER CONTROL ARM

Removal

10 Loosen the wheel lug nuts, raise the vehicle and support it securely on jackstands. Apply the parking brake. Remove the wheel.

11 Disconnect the stabilizer bar from the lower control arm (Section 2).

12 Remove the coil spring as described in Section 8.

13 Remove the cotter pin and back off the lower control arm balljoint stud nut two turns. Break the balljoint loose from the steering knuckle and remove the nut (see illustration 4.4).

14 Remove the control arm from the vehicle.

➡Note: The control arm bushings are pressed into place and require special tools for removal and installation. If the bushings must be replaced, take the control arm to a dealer service department or an automotive machine shop to have the old bushings pressed out and the new ones pressed in.

INSTALLATION

15 Insert the balljoint stud into the steering knuckle, tighten the nut to the specified torque and install a new cotter pin. If necessary, tighten the nut slightly to align a slot in the nut with the hole in the balljoint stud.

16 Install the coil spring (Section 8) and the lower control arm pivot bolts and nuts, but don't completely tighten them at this time.

17 Connect the stabilizer bar to the lower control arm.

18 Install the wheel and lug nuts, lower the vehicle and tighten the lug nuts to the specified torque listed in the Chapter 1 Specifications.

---

6 Hub and bearing assembly (front) - removal and installation

▶ Refer to illustrations 6.3 and 6.4

✳✳ WARNING:

The dust created by the brake system is harmful to your health. Never blow it out with compressed air and don't inhale any of it. Do not, under any circumstances, use petroleum-based solvents to clean brake parts. Use brake system cleaner only.

➡Note: The hub and bearing assembly is sealed-for-life. If worn or damaged, it must be replaced as a unit.

REMOVAL

1 Loosen the front wheel lug nuts, raise the vehicle and support it securely on jackstands. Remove the wheel.

2 Remove the brake caliper and hang it out of the way with a piece of wire, then remove the caliper mounting bracket (see Chapter 9). Pull the disc off the hub.

3 Remove the wheel speed sensor from the hub (see illustration).

4 Working from the back side of the steering knuckle, remove the hub retaining bolts from the steering knuckle (see illustration). Remove the disc shield.

5 Remove the hub from the steering knuckle. The hub assembly should come right out of the steering knuckle, but if it doesn't, tap it from side-to-side to free it.

INSTALLATION

6 Clean the mating surfaces on the steering knuckle, bearing flange and knuckle bore.

7 Position the disc shield on the steering knuckle, then insert the hub and bearing assembly into the steering knuckle. Install the bolts, tightening them to the torque listed in this Chapter's Specifications.

8 Insert the ABS wheel speed sensor into its hole in the hub, tightening the bolt to the torque listed in the Chapter 9 Specifications.

9 Install the brake disc, caliper mounting bracket and caliper (see Chapter 9).

10 On 4WD models, install the hub nut and tighten it to the torque listed in the Chapter 8 Specifications. Prevent the axle from turning by inserting a screwdriver through the caliper and into a disc cooling vane. Install the hub cover.

11 Install the wheel, lower the vehicle and tighten the lug nuts to the torque listed in the Chapter 1 Specifications.

6.3 The ABS wheel speed sensor is retained to the hub by one screw; remove the screw and pull the sensor straight out (don't pry on it)

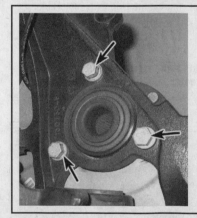

6.4 Remove the bolts that secure the hub and bearing assembly to the knuckle

---

7 Steering knuckle - removal and installation

REMOVAL

1 Loosen the wheel lug nuts, raise the vehicle and support it securely on jackstands placed under the frame. Apply the parking brake. Remove the wheel.

2 Remove the brake caliper and place it on top of the upper control arm (see Chapter 9 if necessary).

3 Remove the brake rotor (see Chapter 9).

4 Remove the splash shield from the steering knuckle.

➡Note: On models equipped with ABS, disconnect the wheel speed sensor harness connector.

5 Separate the tie-rod end from the steering arm (see Section 17).

6   If the steering knuckle must be replaced, remove the dust seal from the spindle by prying it off with a screwdriver. If it's damaged, replace it with a new one.

7   Position a floor jack under the lower control arm balljoint and raise it slightly to take the spring pressure off the suspension stop. The jack must remain in this position throughout the entire procedure.

8   Remove the cotter pins from the upper and lower balljoint studs and back off the nuts two turns each.

9   Break the balljoints loose from the steering knuckle (see illustration 4.4).

➡ **Note: A pickle fork type balljoint separator may damage the balljoint seals.**

10   Remove the nuts from the balljoint studs, separate the control arms from the steering knuckle and remove the knuckle from the vehicle.

INSTALLATION

11   Place the knuckle between the upper and lower control arms and insert the balljoint studs into the knuckle, beginning with the lower balljoint. Install the nuts and tighten them to the specified torque. Install new cotter pins, tightening the nuts slightly to align the slots in the nuts with the holes in the balljoint studs, if necessary.

12   Install the splash shield. If equipped, connect the wheel speed sensor harness connector.

13   Connect the tie-rod end to the steering arm and tighten the nut to the specified torque. Be sure to use a new cotter pin.

14   Install the brake rotor (see Chapter 9).

15   Install the brake caliper (see Chapter 9).

16   Install the wheel and lug nuts. Lower the vehicle to the ground and tighten the nuts to the torque listed in the Chapter 1 Specifications.

8   Coil spring - removal and installation

REMOVAL

▶ **Refer to illustrations 8.5 and 8.7**

1   Loosen the front wheel lug nuts, raise the vehicle and place it securely on jackstands. Remove the wheel.

2   Remove the shock absorber (see Section 3).

3   Remove the stabilizer bar link bolt (see Section 2).

4   Disconnect the outer tie-rod end from the steering knuckle (see Section 17).

5   Install a suitable internal type spring compressor in accordance with the tool manufacturer's instructions (see illustration). Compress the spring enough to relieve all pressure from the spring seats (but don't compress it any more than necessary, or it could be ruined). When you can wiggle the spring, it's compressed enough. (You can buy a suitable spring compressor at most auto parts stores or rent one from a tool rental yard.)

6   Support the lower control arm with a floor jack.

7   Remove the control arm pivot bolts and nuts (see illustration).

8   Pull the lower control arm down and to the rear, then guide the compressed coil spring out. Do not use force.

9   If the coil spring is being replaced, carefully unscrew the spring compressor.

INSTALLATION

10   Inspect the upper and lower spring insulators. If either insulator is cracked or excessively worn, replace it. Inspect the coil spring for chips in the corrosion protection coating. If the coating has been chipped or damaged, replace the spring.

11   If the coil spring is being replaced, install the spring compressor and compress the spring.

12   With the lower spring insulator in place, position the spring on the lower control arm with the flat end of the spring facing up and the tapered end facing down. The identification tape should be toward the lower control arm. Make sure the tapered end seats on the lower control arm with the lower end of the spring seated in the lowest part of the spring seat. The end of the spring must cover all or part of one of the drain holes in the lower control arm, but the other hole must not be covered. The top spring insulator should also be in place.

13   Put the floor jack under the lower control arm and raise the arm into position in the frame. Install the control arm pivot bolts and nuts. Tighten the nuts until they're snug but don't torque them yet.

8.5 A typical aftermarket internal type spring compressor: The hooked arms grip the upper coils of the spring, the plate is inserted between the lower coils, and when the nut on the threaded rod is turned, the spring is compressed

8.7 Remove the control arm mounting bolts (arrow indicates the front pivot bolt)

14 Remove the spring compressor.
15 Reattach the outer tie-rod end to the steering knuckle (see Section 17).
16 Reattach the stabilizer bar link to the lower control arm (see Section 2).
17 Install the shock absorber (see Section 3).

18 Position the floor jack under the lower control arm balljoint and raise the arm to simulate normal ride height. Tighten the lower control arm pivot bolt nuts to the torque listed in this Chapter's Specifications.
19 Install the wheel, remove the jackstands and lower the vehicle. Tighten the wheel lug nuts to the torque listed in the Chapter 1 Specifications.

9 Rear stabilizer bar - removal and installation

➡**Note: Not all models are equipped with a rear stabilizer bar.**

1 Raise the rear of the vehicle and support it securely on jackstands. Block the front wheels to keep the vehicle from rolling.
2 Remove the upper link-to-frame bracket bolts.
3 Remove the nuts and clamps from the anchor block studs and detach the stabilizer bar from the vehicle.

4 Inspect all of the rubber bushings and grommets for cracks, hardening and general deterioration, replacing any faulty components as necessary.
5 Installation is the reverse of the removal procedure. Tighten the bracket bolts, link bolts and clamp nuts to the torque listed in this Chapter's Specifications.

10 Rear shock absorbers - removal and installation

♦ **Refer to illustrations 10.2a and 10.2b**

1 Raise the rear of the vehicle and support it securely on jackstands. Block the front wheels to keep the vehicle from rolling.

2 Remove the shock absorber lower mounting nut and bolt. On the right side shock absorber, remove the parking brake cable bracket (see illustrations).
3 Remove the shock absorber upper mounting nut and slide the shock absorber off the bolt that protrudes through the frame.
4 Installation is the reverse of the removal procedure.

10.2a The left side lower shock absorber mount (if the vehicle is equipped with gas-filled shock absorbers it may be necessary to raise the shock with a jack to insert the bolt through the mounting hole)

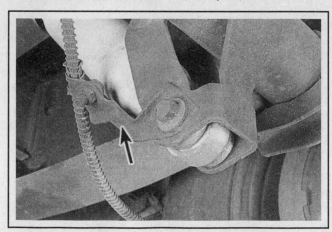

10.2b The right side lower shock absorber mounting bolt also holds the right parking brake cable bracket

11 Leaf spring - removal and installation

✳✳ CAUTION:

The single-leaf springs are constructed of a fiberglass composite material. Be extremely careful not to scratch or gouge the spring when working near it, as spring failure may occur during vehicle operation.

REMOVAL

♦ **Refer to illustrations 11.2, 11.5, 11.6, 11.8a and 11.8b**

1 Raise the rear of the vehicle and support it securely on jackstands placed underneath the frame. Block the front wheels to keep the vehicle from rolling.

11.2 On earlier models, the bumper bracket must be removed to allow the lower shackle bolt to be pulled out

11.5 Remove the U-bolt nuts and the lower plate nuts (arrows) . . .

11.6 . . . then pry the anchor plate off the spring and axle

11.8a The spring retainer is fastened to the forward spring hanger with four nuts (arrows) (single-leaf model)

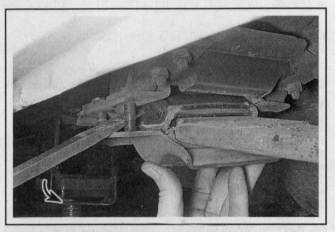

11.8b Carefully pry the spring retainer off (pry between the retainer and hanger, not against the spring) (single-leaf model)

2   On earlier models, remove the bumper bracket (see illustration). On later models, this is unnecessary.

3   Remove the shock absorber lower mounting bolt (Section 10).

4   Support the rear axle assembly with a floor jack positioned under the axle tube on the side being worked on.

5   Remove the U-bolt nuts and the lower plate nuts (see illustration).

➡**Note: If the vehicle is equipped with a rear stabilizer bar, disconnect it from the anchor plate (see Section 9).**

6   Remove the anchor plate (see illustration), then lower the jack just enough to relieve spring pressure on the axle. Make sure the axle doesn't stretch the brake hose.

7   Loosen, but don't remove the shackle nuts.

8   On single-leaf spring models, remove the 4 spring retainer nuts and carefully pry the retainer from the hanger (see illustrations).

9   On multi-leaf spring models, remove the spring-to-hanger nut and bolt.

10 Allow the spring to hang down and remove the lower spring-to-shackle nut, washers and bolt. Detach the spring from the vehicle.

INSTALLATION

▶ **Refer to illustrations 11.13 and 11.15**

11 Insert the rear of the spring into the shackle and install the bolt, washers and nut. Don't tighten the nut at this time.

12 Place the other end of the spring into the spring hanger and install the retainer, washers and nuts. Tighten the nuts to the specified torque.

13 Raise the axle to the spring, ensuring that the locating lug on the spring engages with the pocket on the axle (see illustration).

14 Install the anchor plate over the top of the spring. Apply rubber

11.13 When raising the axle into position, the lug on the spring must engage with the pocket in the axle

lubricant to the leaf spring insulator, to aid installation. Position the U-bolt and lower plate and install the nuts and bolts, tightening them to the specified torque.

15 Raise the rear axle until the top of the axle is 5.9-inches (150 mm) (5.5-inches for 1997 and later models) from the bottom of the frame (see illustration), then tighten the shackle nuts to the specified torque. This adjusts the rear suspension trim height.

16 Install the shock absorber.

17 If equipped, attach the rear stabilizer bar to the anchor plate (see Section 9).

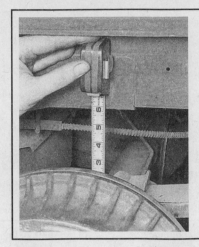

11.15 Before tightening the shackle nuts, raise the axle until the distance between the axle and the frame is 5.9-inches

12 Steering wheel - removal and installation

▶ Refer to illustrations 12.2, 12.3, 12.4, 12.5a, 12.5b, 12.6a, 12.6b, 12.6c, 12.7, 12.8 and 12.9

❋❋ WARNING:

Most 1993 and later models are equipped with airbags. Always disable the airbag system before working in the vicinity of the impact sensors, steering column or instrument panel to avoid the possibility of accidental deployment of the airbag(s), which could cause personal injury (see Chapter 12 for the airbag disabling procedure). The yellow wires and connectors routed through the instrument panel are for this system. Do not use electrical test equipment on these yellow wires or tamper with them in any way while working under the instrument panel.

❋❋ CAUTION:

If the vehicle is equipped with a Delco Loc II or Theftlock audio system, make sure you have the correct activation code before disconnecting the battery.

1 Park the vehicle with the wheels pointing straight ahead. Disconnect the cable from the negative terminal of the battery. On airbag-equipped models, disable the airbag system, as described in the Airbag Section in Chapter 12.

2 On models without an airbag, the horn pad is attached to the steering wheel with various combinations of clips and/or screws (see illustration).

3 On 1995 and earlier airbag-equipped models, remove the four screws from the backside of the steering wheel (the side facing the dash) (see illustration).

4 On 1996 and later models, turn the steering wheel 90-degrees to gain access to the holes in the backside (the side facing the dash) of the steering wheel. Insert a screwdriver into the hole for each of the four spring clips (see illustration) and push the spring aside to release the pin. There are four pins and four springs.

5 On airbag-equipped models, lift the airbag module carefully away from the steering wheel and disconnect the yellow airbag electrical connector. This is a two-part disconnection, as there is a plastic clip that must be removed before the connector can be disconnected (see illustrations). Remove the module.

❋❋ WARNING:

When carrying the airbag module, keep the driver's side of it away from your body, and, when you place it on the bench, have the driver's side facing up.

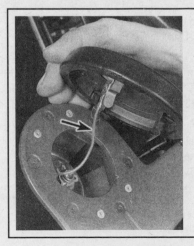

12.2 On models without airbags, the horn pad is removed by gripping it firmly and pulling it from the steering wheel - detach the horn wire (arrow)

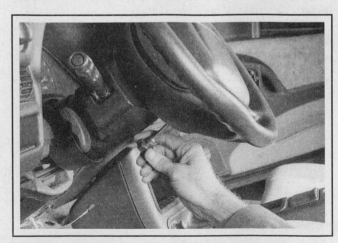

12.3 On 1993 through 1995 airbag-equipped models, unscrew the four airbag module screws from the backside of the steering wheel

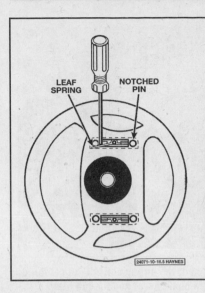

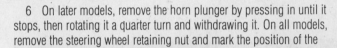

12.4 To detach the airbag module from the steering wheel on 1996 and later models, rotate the wheel 90-degrees, insert a screwdriver into each of the four holes in the backside of the steering wheel and pry each leaf spring aside to release it from its notched pin (there are four of them)

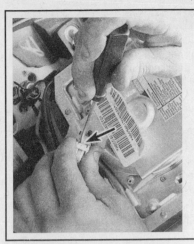

12.5a Use a small screwdriver to release the plastic locking clip (arrow) from the airbag module connector, then . . .

steering wheel to the shaft, if marks don't already exist or don't line up (see illustrations).

7   Use a puller to detach the steering wheel from the shaft (see illustration). Don't hammer on the shaft to dislodge the wheel.

8   Remove the puller and disconnect the horn and ground connectors (see illustration). Remove the steering wheel.

6   On later models, remove the horn plunger by pressing in until it stops, then rotating it a quarter turn and withdrawing it. On all models, remove the steering wheel retaining nut and mark the position of the

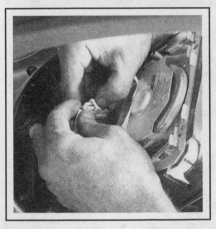

12.5b . . . squeeze the tab and separate the airbag connector

12.6a A pair of snap-ring pliers can be used to remove the safety clip (if equipped) from the shaft

12.6b Remove the steering wheel nut with a deep socket

12.6c Check to be sure that there are alignment marks on the steering shaft and steering wheel (arrow) - if there aren't any or if they don't match up, scribe or paint a line from the shaft to the steering wheel

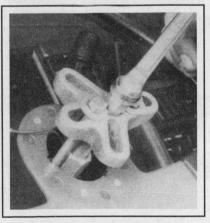

12.7 Remove the wheel from the shaft with a puller - DO NOT hammer on the shaft!

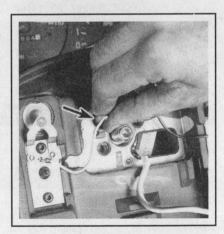

12.8 Disconnect the horn and ground wire connectors

9   Installation is the reverse of removal. On airbag-equipped models, before the steering wheel is installed, make sure the SIR coil is centered (see illustration). If it isn't, see the Airbag Section in Chapter 12 for the centering procedure. Connect the airbag connector to the back of the airbag module just as it was before steering wheel removal, i.e. with the plastic locking device in place. Be sure to tighten the steering wheel nut and airbag screws to the torque listed in this Chapter's Specifications.

12.9 When properly aligned, the airbag coil will be centered with the marks aligned (in circle here) and the tab fitted between the projections on the top of the steering column

13 Intermediate shaft - removal and installation

▶ **Refer to illustration 13.2**

1   Turn the front wheels to the straight-ahead locked position. On 1996 and later models, unbolt the underhood electrical (relay/fuse) center and place it aside. Then unbolt the intermediate shaft boot from the cowl.
2   Using white paint, place alignment marks on the upper universal joint, the steering shaft, the lower universal joint and the steering gear input shaft (see illustration). On later models, this will require sliding the boot over the upper or lower universal joint to expose the joint and shaft.
3   Remove the upper and lower universal joint pinch bolts.
4   On earlier models, remove the steering gear-to-frame mounting bolts (Section 14) and lower the steering gear enough to allow interme-

13.2 Mark the relationship of the intermediate shaft to the steering shaft and also to the steering gear input shaft

diate shaft removal.
5   Pry the intermediate shaft off the steering shaft with a large screwdriver, then pull the shaft from the steering gearbox in the same manner.
6   Installation is the reverse of the removal procedure. Be sure to align the marks and tighten the pinch bolts to the specified torque.

14 Steering gear - removal and installation

▶ **Refer to illustrations 14.2, 14.4 and 14.5**

REMOVAL

1   Turn the front wheels to the straight-ahead locked position. Apply the parking brake. Raise the front of the vehicle and support it securely on jackstands. On later models, remove the lower fan shroud.
2   Place a drain pan under the steering gear (power steering only). Remove the power steering pressure and return lines and cap the ends to prevent excessive fluid loss and contamination (see illustration).
3   Mark the relationship of the lower intermediate shaft universal joint to the steering gear input shaft. Remove the lower intermediate

14.2 The power steering pressure and return line fittings (arrows) can be reached from under the vehicle

14.4 Place alignment marks on the Pitman arm and the shaft, then remove the nut and washer

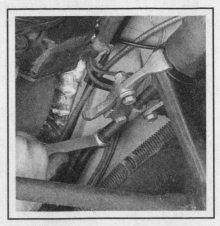

14.5 Use a puller to remove the Pitman arm from the Pitman shaft

shaft pinch bolt (see Section 13).

4 Mark the relationship of the Pitman arm to the Pitman shaft so it can be installed in the same position (see illustration). Remove the nut and washer.

5 Remove the Pitman arm from the shaft with a two-jaw puller (see illustration).

6 Support the steering gear and remove the steering gear-to-frame mounting bolts. Lower the unit, separate the intermediate shaft from the steering gear input shaft and remove the steering gear from the vehicle.

INSTALLATION

7 Raise the steering gear into position and connect the intermediate

shaft, aligning the marks.

8 Install the mounting bolts and washers and tighten them to the specified torque.

9 Slide the Pitman arm onto the Pitman shaft, ensuring that the marks are aligned. Install the washer and nut and tighten the nut to the specified torque.

10 Install the lower intermediate shaft pinch bolt and tighten it to the specified torque.

11 Connect the power steering pressure and return hoses to the steering gear and fill the power steering pump reservoir with the recommended fluid (Chapter 1).

12 Lower the vehicle and bleed the steering system as outlined in Section 16.

15 Power steering pump - removal and installation

➡Note: On models with serpentine drivebelts, special power steering pump pulley puller and installation tools (available at auto parts stores) are required to properly perform this procedure.

REMOVAL

◆ Refer to illustrations 15.8, 15.9 and 15.10

1 Disconnect the cable from the negative terminal of the battery.

❊❊ CAUTION:

On models equipped with an anti-theft audio system, disable the anti-theft feature before performing any operation that requires disconnecting the battery or disrupting power to the stereo (see the Anti-theft audio system procedures at the front of this manual).

2 Place a drain pan under the power steering pump to catch the fluid that will leak out.

3 On 1992 and later VIN W and X models, remove the air cleaner assembly (see Chapter 4).

4 On 1993 and later models, remove the hood latch mechanism and place it out of the way (see Chapter 11).

15.8 On 1997 and later models equipped with speed dependent steering, disconnect the wiring harness connector from the steering assist control solenoid at the bottom of the power steering pump (arrow)

5 On 1993 and later models, remove the upper and lower fan shroud (see Chapter 3).

6 Remove the drivebelt from the power steering pump pulley (see Chapter 1).

15.9 On models with a serpentine drivebelt the power steering pump pulley must be removed to access the pump mounting bolts. Use a puller like this one to remove the pulley

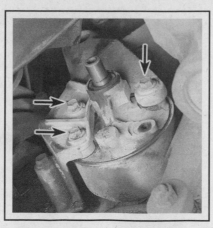

15.10 Remove the power steering pump mounting bolts (serpentine belt model shown)

15.16 Using an installation tool (or a fabricated tool, as shown) press the pulley onto the pump shaft until it's flush with the shaft, but no further

7 On 1990 and earlier V6 engines, remove the brace at the rear of the pump.

8 On 1997 and later models equipped with speed dependent steering, disconnect the wiring harness from the steering assist control solenoid (see illustration).

9 On models equipped with a single serpentine drivebelt, use a puller to remove the pulley from the power steering pump (see illustration).

➡ **Note: This operation is not normally required on multi-belted engines unless the old pulley must be transferred to a new pump.**

10 Remove the pump mounting bolts (see illustration). On later models, where the pump has a rear mounting bracket, remove the nut. (It may be necessary to raise the vehicle and support it securely on jackstands to do this, after which the vehicle can be lowered.) Move the pump as necessary to access the hose fittings.

➡ **Note 1: If the hose fittings are accessible, it's much easier to loosen them while the pump is still mounted.**

➡ **Note 2: On some models it will be easier to unbolt the pump from its mounting bracket, while on other models it will be easier to remove the pump and mounting bracket as an assembly.**

11 Disconnect the hoses from the pump and immediately plug the hose ends and pump fittings to prevent fluid leakage and contamination.

12 Remove the pump from the vehicle, taking care not to spill power steering fluid on any painted surfaces.

INSTALLATION

▶ **Refer to illustration 15.16**

13 If the pump and mounting bracket were removed as an assembly, position the pump/bracket assembly on the block, connect the hoses and tighten the bolts/nut finger-tight.

14 If the pump was removed separately, position the pump on the block, connect the hoses, install the bolts/nut and tighten them securely.

15 If not already done, connect the hoses to the power steering pump and tighten them securely.

16 On models equipped with a serpentine drivebelt, install the drivebelt pulley onto the pump using a special tool designed for the purpose. An appropriate tool can be fabricated from a long bolt, nut, washer and a socket of the same diameter as the pulley hub (see illustration). Press the pulley onto the shaft until the front of the hub is flush with the shaft, but no further. After installation, rotate the pulley to make sure it isn't rubbing on anything.

17 Install the drivebelt (see Chapter 1) and, on multi-belted engines, adjust the belt tension (see Chapter 1).

18 The remaining installation steps are the reverse of removal.

19 After installation is complete, fill the power steering fluid reservoir with the recommended fluid (see *Recommended lubricants and fluids* in Chapter 1) and bleed the power steering system (see Section 16).

16 Power steering system - bleeding

1 Following any operation in which the power steering fluid lines have been disconnected, the power steering system must be bled to remove all air and obtain proper steering performance.

2 With the front wheels in the straight ahead position, check the power steering fluid level and, if low, add fluid until it reaches the Cold mark on the dipstick.

3 Start the engine and allow it to run at fast idle. Recheck the fluid level and add more if necessary to reach the Cold mark on the dipstick.

4 Bleed the system by turning the wheels from side-to-side, without hitting the stops. This will work the air out of the system. Keep the reservoir full of fluid as this is done.

5 When the air is worked out of the system, return the wheels to the straight ahead position and leave the vehicle running for several more minutes before shutting it off.

6 Road test the vehicle to be sure the steering system is functioning normally and noise free.

7 Recheck the fluid level to be sure it is up to the Hot mark on the dipstick while the engine is at normal operating temperature. Add fluid if necessary (see Chapter 1).

17 Steering linkage - inspection, removal and installation

✳✳ WARNING:

Whenever any of the suspension or steering fasteners are loosened or removed they must be inspected and if necessary, replaced with new ones of the same part number or of original equipment quality and design. Torque specifications must be followed for proper reassembly and component retention. Never attempt to heat, straighten or weld any suspension or steering component. Instead, replace any bent or damaged part with a new one.

✳✳ CAUTION:

DO NOT use a "pickle fork" type balljoint separator - it may damage the balljoint seals.

➡Note: Many of the nuts used to fasten the suspension components are of the "prevailing torque" type, which create a "thread interface" between the nut and bolt. The factory recommends not reusing these nuts unless they and the bolt are clean, rust-free, and can achieve the specified torque.

INSPECTION

▶ Refer to illustrations 17.1 and 17.5

1 The steering linkage connects the steering gear to the front wheels and keeps the wheels in proper relation to each other (see illustration). The linkage consists of the Pitman arm, fastened to the steering gear shaft, which moves the relay rod back-and-forth through the connecting rod. The relay rod is supported on each end by frame-mounted idler arms. The back-and-forth motion of the relay rod is transmitted to the steering knuckles through a pair of tie-rod assemblies. Each tie-rod is made up of an inner and outer tie-rod end, a threaded adjuster tube and two clamps.

2 Set the wheels in the straight ahead position and lock the steering wheel.

3 Raise one side of the vehicle until the tire is approximately 1-inch off the ground.

4 Mount a dial indicator with the needle resting on the outside edge of the wheel. Grasp the front and rear of the tire and using light pressure, wiggle the wheel back-and-forth and note the dial indicator reading. The gauge reading should be less than 0.108-inch. If the play in the steering system is more than specified, inspect each steering linkage pivot point and ball stud for looseness and replace parts if necessary.

5 Raise the vehicle and support it on jackstands. Push up, then pull down on the relay rod end of the idler arm, exerting a force of approximately 25 pounds each way. Measure the total distance the end of the arm travels (see illustration). If the play is greater than 1/4-inch, replace the idler arm.

6 Check for torn ball stud boots, frozen joints and bent or damaged linkage components.

REMOVAL AND INSTALLATION

▶ Refer to illustrations 17.9, 17.11, 17.13 and 17.15

Tie-rod

7 Loosen the wheel lug nuts, raise the vehicle and support it securely on jackstands. Apply the parking brake. Remove the wheel.

8 Remove the cotter pin and loosen, but do not remove, the castellated nut from the ball stud.

9 Using a two jaw puller, separate the tie-rod end from the steering knuckle (see illustration). Remove the castellated nut and pull the tie-rod end from the knuckle.

10 Remove the nut securing the inner tie-rod end to the relay rod. Separate the inner tie-rod end from the relay rod in the same manner as in Step 9.

11 If the inner or outer tie-rod end must be replaced, measure the distance from the end of the adjuster tube to the center of the ball stud and record it (see illustration). Loosen the adjuster tube clamp and unscrew the tie-rod end.

12 Lubricate the threaded portion of the tie-rod end with chassis grease. Screw the new tie-rod end into the adjuster tube and adjust

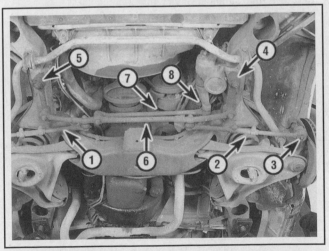

17.1 An exploded view of the steering system components

1	Inner tie-rod	4	Idler arm	7	Power steering
2	Adjuster tube	5	Idler arm		pump
3	Outer tie-rod	6	Relay rod	8	Pitman arm

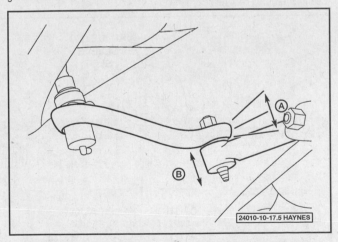

24010-10-17.5 HAYNES

17.5 To check for play in the idler arms, apply approximately 25 lbs. of force up and down (B) on the idler arm - if the total idler arm movement (A) is greater than 1/4-inch, replace the idler arm

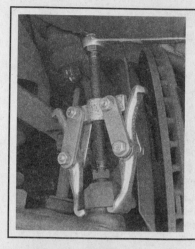

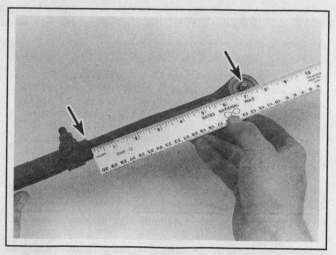

17.9 Using a puller, separate the tie-rod end from the steering knuckle; notice that the nut hasn't been completely removed - this will prevent violent separation of the parts!

the distance from the tube to the ball stud to the previously measured dimension. The number of threads showing on the inner and outer tie-rod ends should be equal within three threads. Don't tighten the clamp yet.

13 To install the tie-rod, insert the inner tie-rod end ball stud into the relay rod until it's seated. Install the nut and tighten it to the specified torque. If the ball stud spins when attempting to tighten the nut, force it into the tapered hole with a large pair of pliers (see illustration).

14 Connect the outer tie-rod end to the steering knuckle and install the castellated nut. Tighten the nut to the specified torque and install a new cotter pin. If necessary, tighten the nut slightly to align a slot in the nut with the hole in the ball stud.

15 Tighten the clamp nuts. The center of the bolt should be nearly horizontal and the adjuster tube slot must not line up with the gap in the clamps (see illustration).

16 Install the wheel and lug nuts, lower the vehicle and tighten the lug nuts to the specified torque. Drive the vehicle to an alignment shop to have the front end alignment checked and, if necessary, adjusted.

Idler arm

17 Raise the vehicle and support it securely on jackstands. Apply the parking brake.

18 Loosen but do not remove the idler arm-to-relay rod nut.

19 Separate the idler arm from the relay rod with a two jaw puller (see illustration 17.9). Remove the nut.

17.11 Measure the distance from the adjuster tube to the ball stud centerline so the new tie-rod end can be set to this dimension

17.13 It may be necessary to force the ball stud into the hole to keep it from spinning when tightening the nut

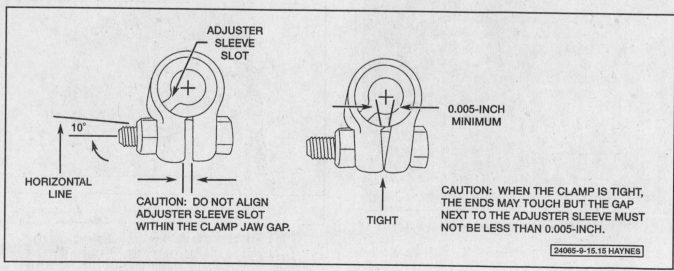

ADJUSTER SLEEVE SLOT

10°

HORIZONTAL LINE

CAUTION: DO NOT ALIGN ADJUSTER SLEEVE SLOT WITHIN THE CLAMP JAW GAP.

0.005-INCH MINIMUM

TIGHT

CAUTION: WHEN THE CLAMP IS TIGHT, THE ENDS MAY TOUCH BUT THE GAP NEXT TO THE ADJUSTER SLEEVE MUST NOT BE LESS THAN 0.005-INCH.

24065-9-15.15 HAYNES

17.15 Note these guidelines when installing the tie-rod ends

20 Remove the idler arm-to-frame bolts.

21 To install the idler arm, position it on the frame and install the bolts, tightening them to the specified torque.

22 Insert the idler arm ball stud into the relay rod and install the nut. Tighten the nut to the specified torque. If the ball stud spins when attempting to tighten the nut, force it into the tapered hole with a large pair of pliers.

Relay rod

23 Raise the vehicle and support it securely on jackstands. Apply the parking brake.

24 Separate the two inner tie-rod ends from the relay rod.

25 Separate the connecting rod from the relay rod.

26 Separate both idler arms from the relay rod.

27 Installation is the reverse of the removal procedure. If the ball studs spin when attempting to tighten the nuts, force them into the tapered holes with a large pair of pliers. Be sure to tighten all of the nuts to the specified torque.

Connecting rod

28 Raise the front of the vehicle and support it securely on jackstands. Apply the parking brake.

29 Loosen, but do not remove, the nut securing the connecting rod ball stud to
the relay rod. Separate the joint with a two jaw puller then remove the nut.

30 Separate the connecting rod from the Pitman arm.

31 Installation is the reverse of the removal procedure. If the ball studs spin when attempting to tighten the nuts, force them into the tapered holes with a large pair of pliers. Be sure to tighten all of the nuts to the specified torque.

Pitman arm

32 Refer to Section 14 of this Chapter for the Pitman arm removal procedure.

18 Wheels and tires - general information

▶ **Refer to illustration 18.1**

All vehicles covered by this manual are equipped with metric-sized fiberglass or steel belted radial tires (see illustration). Use of other size or type of tires may affect the ride and handling of the vehicle. Don't mix different types of tires, such as radials and bias belted, on the same vehicle as handling may be seriously affected. It's recommended that tires be replaced in pairs on the same axle, but if only one tire is being replaced, be sure it's the same size, structure and tread design as the other.

Because tire pressure has a substantial effect on handling and wear, the pressure on all tires should be checked at least once a month or before any extended trips (see Chapter 1).

Wheels must be replaced if they are bent, dented, leak air, have elongated bolt holes, are heavily rusted, out of vertical symmetry or if the lug nuts won't stay tight. Wheel repairs that use welding or peening are not recommended.

Tire and wheel balance is important in the overall handling, braking and performance of the vehicle. Unbalanced wheels can adversely affect handling and ride characteristics as well as tire life. Whenever a tire is installed on a wheel, the tire and wheel should be balanced by a shop with the proper equipment.

All vehicles covered by this manual are equipped with a compact spare tire, which is designed to save space as well as being easier to handle due to its lighter weight. The spare tire pressure should be checked at least once a month, and maintained at 60 psi. The compact spare tire and wheel are designed for use with each other only, and neither the tire nor the wheel should be coupled with other types or size of wheels and tires. Because the compact spare is designed as a temporary replacement for an out-of-service standard wheel and tire, the compact spare should be used on the vehicle only until the standard wheel and tire are repaired or replaced. Continuous use of the compact

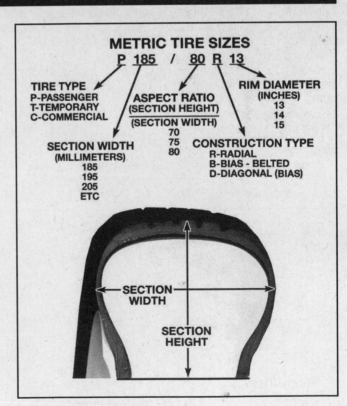

18.1 Metric tire size code

spare at speeds of over 50 mph is not recommended. In addition, the expected tread life of the compact spare is only 3000 miles.

19 Front end alignment - general information

▶ **Refer to illustration 19.1**

A front end alignment refers to the adjustments made to the front wheels so they are in proper angular relationship to the suspension and the ground. Front wheels that are out of proper alignment not only affect steering control, but also increase tire wear. The front end adjustments normally required are camber, caster and toe-in (see illustration).

Getting the proper front wheel alignment is a very exacting process, one in which complicated and expensive machines are necessary to perform the job properly. Because of this, you should have a technician with the proper equipment perform these tasks. We will, however, use this space to give you a basic idea of what is involved with front end alignment so you can better understand the process and deal intelligently with the shop that does the work.

Toe-in is the turning in of the front wheels. The purpose of a toe specification is to ensure parallel rolling of the front wheels. In a vehicle with zero toe-in, the distance between the front edges of the wheels will be the same as the distance between the rear edges of the wheels. The actual amount of toe-in is normally only a fraction of an inch. Toe-in adjustment is controlled by the tie-rod end position on the inner tie-rod. Incorrect toe-in will cause the tires to wear improperly by making them scrub against the road surface.

Camber is the tilting of the front wheels from the vertical when viewed from the front of the vehicle. When the wheels tilt out at the top, the camber is said to be positive (+). When the wheels tilt in at the top the camber is negative (-). The amount of tilt is measured in degrees from the vertical and this measurement is called the camber angle. This angle affects the amount of tire tread which contacts the road and compensates for changes in the suspension geometry when the vehicle is cornering or travelling over an undulating surface.

Caster is the tilting of the front steering axis from the vertical. A tilt toward the rear is positive caster and a tilt toward the front is negative caster.

Caster is adjusted by moving shims from one end of the upper control arm mount to the other.

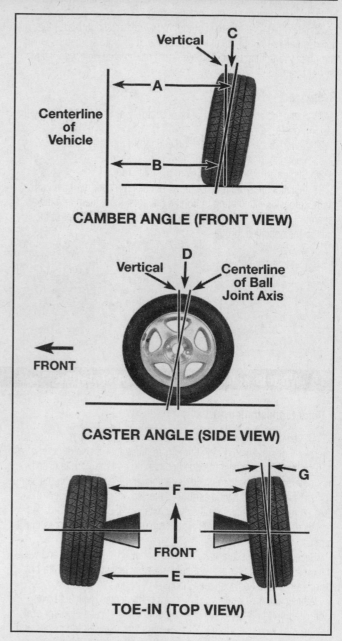

CAMBER ANGLE (FRONT VIEW)

CASTER ANGLE (SIDE VIEW)

TOE-IN (TOP VIEW)

19.1 Front end alignment details

A minus B = C (degrees camber)
D = caster (measured in degrees)
E minus F = toe-in (measured in inches)
G = toe-in (expressed in degrees)

Specifications

Torque specifications	Ft-lbs (unless otherwise indicated)

Front suspension

Upper control arm-to-frame nuts	
1985 through 1987	66
1988 through 1995	75
1996 and later	81
Lower control arm pivot bolt nuts	
1985 through 1992	96
1993 through 1995	20
1996 and later	66
Replacement balljoint-to-upper control arm bolts	
1985 through 1988	8
1989	17
1990	22
1991	17
1992 and later	22
Hub and bearing assembly-to-steering knuckle bolts	
(2003 and later models)	133
Lower balljoint-to-steering knuckle nut	
1985 through 1987	81
1988 on	90
Upper balljoint-to-steering knuckle nut	
1985 through 1987	52
1988 on	65
2003 on	61

Rear suspension

Single-leaf spring (1985 to 1994 models)	
Spring retainer-to-hanger assembly nuts	18 to 28
Hanger assembly-to-frame nuts	81
Shackle nuts (all through 1988)	81
Shackle-to-spring nuts (1989 on)	103
Shackle-to-frame nuts (1989 on)	81
U-bolt-to-anchor plate nuts	48
Lower plate-to-anchor plate nuts	
1985 through 1992	114
1993 and 1994	52
Multi-leaf spring (1995 and later models)	
Anchor plate nuts	
1994 and 1995	52
1996 through 2002	41
2003	74
Bumper nut	
1994 and 1995	18
1996 and later	33
Hanger assembly-to-frame and spring nuts	
1994 and 1995	81
1996 and later	74

Torque specifications (continued) Ft-lbs (unless otherwise indicated)

Rear suspension (continued)

Shackle-to-frame nut/bolt (1996 and later only)	103
Shackle-to-spring nut	
1994 and 1995	81
1996 and later	74
Shock absorber nuts	75
Stabilizer link	
Lower nuts	14
Upper nuts	33
Stabilizer shaft clamp nuts (1996 and later only)	44

Steering

Airbag module screws (1993 through 1995 only)	
1993 and 1994	27 in-lbs
1995	70 in-lbs
Steering wheel-to-steering shaft nut	30
Intermediate shaft pinch bolts	30
Steering gear-to-frame bolts	
1985	70
1986 on	55
Tie-rod end-to-steering knuckle nut	30 to 35
Tie-rod adjuster tube clamp nuts	
1985 through 1996	156 in-lbs
1997 and later	18
Tie-rod-to-relay rod	
1985 through 1987	66
1988 on	35
Idler arm-to-relay rod	
1985 through 1987	66
1988 through 1996	35
1997 and later	40
Idler arm-to-frame	
1985 through 1987	33 to 37
1988 through 1996	52
1997 and later	77
Connecting rod-to-relay rod	
1985 through 1987	66
1988 on	35
Connecting rod-to-Pitman arm	
1985 through 1987	66
1988 through 1996	35
1997 and later	48
Pitman arm-to-steering gear	185
Wheel lug nuts	See Chapter 1

Section

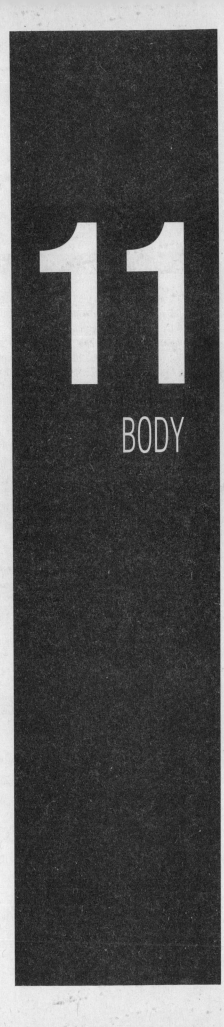

11

BODY

1   General information

The vehicles covered in this manual have a separate frame and body.

As with other parts of the vehicle, proper maintenance of body components plays an important part in preserving the vehicle's market value. It's far less costly to handle small problems before they grow into larger ones. Information in this Chapter will tell you all you need to know to keep seals sealing, body panels aligned and general appearance up to par.

Major body components which are particularly vulnerable in accidents are removable. These include the hood, fenders, grille and doors. It's often cheaper and less time consuming to replace an entire panel than it is to attempt a restoration of the old one. However, this must be decided on a case-by-case basis.

2   Body and frame - maintenance

1   The condition of your vehicle's body is very important, because it's on this that the second hand value will mainly depend. It's much more difficult to repair a neglected or damaged body than it is to repair mechanical components. The hidden areas of the body, such as the fender wells, the frame, and the engine compartment, are equally important, although they obviously don't require as frequent attention as the rest of the body.

2   Once a year, or every 12,000 miles, it's a good idea to have the underside of the body and the frame steam cleaned. All traces of dirt and oil will be removed and the underside can then be inspected carefully for rust, damaged brake lines, frayed electrical wiring, damaged cables and other problems. The front suspension components should be greased after this job is done.

3   At the same time, clean the engine and the engine compartment with either a steam cleaner or a water soluble degreaser.

4   The fender wells should be given particular attention, as undercoating can peel away and stones and dirt thrown up by the tires can cause the paint to chip and flake, allowing rust to set in. If rust is found, clean down to the bare metal and apply an anti-rust paint.

5   The body should be washed as needed. Wet the vehicle thoroughly to soften the dirt, then wash it down with a soft sponge and plenty of clean soapy water. If the surplus dirt is not washed off very carefully, it will in time wear down the paint.

6   Spots of tar or asphalt coating thrown up from the road should be removed with a cloth soaked in solvent.

7   Once every six months, give the body and chrome trim a thorough waxing. If a chrome cleaner is used to remove rust from any of the vehicle's plated parts, remember that the cleaner also removes part of the chrome, so use it sparingly.

3   Upholstery and carpets - maintenance

1   Every three months remove the carpets or mats and clean the interior of the vehicle (more frequently if necessary). Vacuum the upholstery and carpets to remove loose dirt and dust.

2   If the upholstery is soiled, apply upholstery cleaner with a damp sponge and wipe it off with a clean, dry cloth.

4   Body repair - minor damage

See photo sequence

REPAIR OF MINOR SCRATCHES

1   If the scratch is superficial and does not penetrate to the metal of the body, repair is very simple. Lightly rub the scratched area with a fine rubbing compound to remove loose paint and built up wax. Rinse the area with clean water.

2   Apply touch-up paint to the scratch, using a small brush. Continue to apply thin layers of paint until the surface of the paint in the scratch is level with the surrounding paint. Allow the new paint at least two weeks to harden, then blend it into the surrounding paint by rubbing with a very fine rubbing compound. Finally, apply a coat of wax to the scratch area.

3   If the scratch has penetrated the paint and exposed the metal of the body, causing the metal to rust, a different repair technique is required. Remove all loose rust from the bottom of the scratch with a pocket knife, then apply rust inhibiting paint to prevent the formation of rust in the future. Using a rubber or nylon applicator, coat the scratched area with glaze-type filler. If required, the filler can be mixed with thinner to provide a very thin paste, which is ideal for filling narrow scratches. Before the glaze filler in the scratch hardens, wrap a piece of smooth cotton cloth around the tip of a finger. Dip the cloth in thinner and then quickly wipe it along the surface of the scratch. This will ensure that the surface of the filler is slightly hollow. The scratch can now be painted over as described earlier in this section.

REPAIR OF DENTS

4   When repairing dents, the first job is to pull the dent out until the affected area is as close as possible to its original shape. There is no point in trying to restore the original shape completely as the metal in

the damaged area will have stretched on impact and cannot be restored to its original contours. It is better to bring the level of the dent up to a point which is about 1/8-inch below the level of the surrounding metal. In cases where the dent is very shallow, it is not worth trying to pull it out at all.

5 If the back side of the dent is accessible, it can be hammered out gently from behind using a soft-face hammer. While doing this, hold a block of wood firmly against the opposite side of the metal to absorb the hammer blows and prevent the metal from being stretched.

6 If the dent is in a section of the body which has double layers, or some other factor makes it inaccessible from behind, a different technique is required. Drill several small holes through the metal inside the damaged area, particularly in the deeper sections. Screw long, self tapping screws into the holes just enough for them to get a good grip in the metal. Now the dent can be pulled out by pulling on the protruding heads of the screws with locking pliers.

7 The next stage of repair is the removal of paint from the damaged area and from an inch or so of the surrounding metal. This is easily done with a wire brush or sanding disk in a drill motor, although it can be done just as effectively by hand with sandpaper. To complete the preparation for filling, score the surface of the bare metal with a screwdriver or the tang of a file or drill small holes in the affected area. This will provide a good grip for the filler material. To complete the repair, see the Section on filling and painting.

REPAIR OF RUST HOLES OR GASHES

8 Remove all paint from the affected area and from an inch or so of the surrounding metal using a sanding disk or wire brush mounted in a drill motor. If these are not available, a few sheets of sandpaper will do the job just as effectively.

9 With the paint removed, you will be able to determine the severity of the corrosion and decide whether to replace the whole panel, if possible, or repair the affected area. New body panels are not as expensive as most people think and it is often quicker to install a new panel than to repair large areas of rust.

10 Remove all trim pieces from the affected area except those which will act as a guide to the original shape of the damaged body, such as headlight shells, etc. Using metal snips or a hacksaw blade, remove all loose metal and any other metal that is badly affected by rust. Hammer the edges of the hole inward to create a slight depression for the filler material.

11 Wire brush the affected area to remove the powdery rust from the surface of the metal. If the back of the rusted area is accessible, treat it with rust-inhibiting paint.

12 Before filling is done, block the hole in some way. This can be done with sheet metal riveted or screwed into place, or by stuffing the hole with wire mesh.

13 Once the hole is blocked off, the affected area can be filled and painted. See the following sub-section on filling and painting.

FILLING AND PAINTING

14 Many types of body fillers are available, but generally speaking, body repair kits which contain filler paste and a tube of resin hardener are best for this type of repair work. A wide, flexible plastic or nylon applicator will be necessary for imparting a smooth and contoured finish to the surface of the filler material. Mix up a small amount of filler

on a clean piece of wood or cardboard (use the hardener sparingly). Follow the manufacturer's instructions on the package, otherwise the filler will set incorrectly.

15 Using the applicator, apply the filler paste to the prepared area. Draw the applicator across the surface of the filler to achieve the desired contour and to level the filler surface. As soon as a contour that approximates the original one is achieved, stop working the paste. If you continue, the paste will begin to stick to the applicator. Continue to add thin layers of paste at 20-minute intervals until the level of the filler is just above the surrounding metal.

16 Once the filler has hardened, the excess can be removed with a body file. From then on, progressively finer grades of sandpaper should be used, starting with a 180-grit paper and finishing with 600-grit wet-or-dry paper. Always wrap the sandpaper around a flat rubber or wooden block, otherwise the surface of the filler will not be completely flat. During the sanding of the filler surface, the wet-or-dry paper should be periodically rinsed in water. This will ensure that a very smooth finish is produced in the final stage.

17 At this point, the repair area should be surrounded by a ring of bare metal, which in turn should be encircled by the finely feathered edge of good paint. Rinse the repair area with clean water until all of the dust produced by the sanding operation is gone.

18 Spray the entire area with a light coat of primer. This will reveal any imperfections in the surface of the filler. Repair the imperfections with fresh filler paste or glaze filler and once more smooth the surface with sandpaper. Repeat this spray-and-repair procedure until you are satisfied that the surface of the filler and the feathered edge of the paint are perfect. Rinse the area with clean water and allow it to dry completely.

19 The repair area is now ready for painting. Spray painting must be carried out in a warm, dry, windless and dust free atmosphere. These conditions can be created if you have access to a large indoor work area, but if you are forced to work in the open, you will have to pick the day very carefully.

If you are working indoors, dousing the floor in the work area with water will help settle the dust which would otherwise be in the air. If the repair area is confined to one body panel, mask off the surrounding panels. This will help minimize the effects of a slight mismatch in paint color. Trim pieces such as chrome strips, door handles, etc., will also need to be masked off or removed. Use masking tape and several thicknesses of newspaper for the masking operations.

20 Before spraying, shake the paint can thoroughly, then spray a test area until the spray painting technique is mastered. Cover the repair area with a thick coat of primer. The thickness should be built up using several thin layers of primer rather than one thick one. Using 600-grit wet-or-dry sandpaper, rub down the surface of the primer until it is very smooth. While doing this, the work area should be thoroughly rinsed with water and the wet-or-dry sandpaper periodically rinsed as well. Allow the primer to dry before spraying additional coats.

21 Spray on the top coat, again building up the thickness by using several thin layers of paint. Begin spraying in the center of the repair area and then, using a circular motion, work out until the whole repair area and about two inches of the surrounding original paint is covered. Remove all masking material 10 to 15 minutes after spraying on the final coat of paint. Allow the new paint at least two weeks to harden, then use a very fine rubbing compound to blend the edges of the new paint into the existing paint. Finally, apply a coat of wax.

These photos illustrate a method of repairing simple dents. They are intended to supplement Body repair - minor damage in this Chapter and should not be used as the sole instructions for body repair on these vehicles.

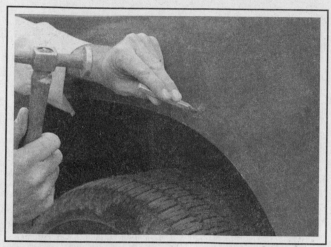

1   If you can't access the backside of the body panel to hammer out the dent, pull it out with a slide-hammer-type dent puller. In the deepest portion of the dent or along the crease line, drill or punch hole(s) at least one inch apart . . .

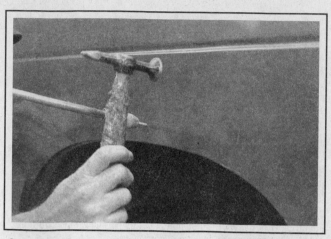

2   . . . then screw the slide-hammer into the hole and operate it. Tap with a hammer near the edge of the dent to help 'pop' the metal back to its original shape. When you're finished, the dent area should be close to its original contour and about 1/8-inch below the surface of the surrounding metal

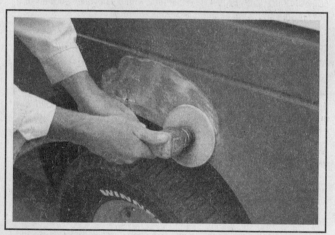

3   Using coarse-grit sandpaper, remove the paint down to the bare metal. Hand sanding works fine, but the disc sander shown here makes the job faster. Use finer (about 320-grit) sandpaper to feather-edge the paint at least one inch around the dent area

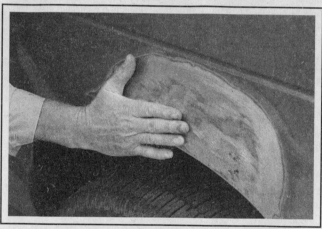

4   When the paint is removed, touch will probably be more helpful than sight for telling if the metal is straight. Hammer down the high spots or raise the low spots as necessary. Clean the repair area with wax/silicone remover

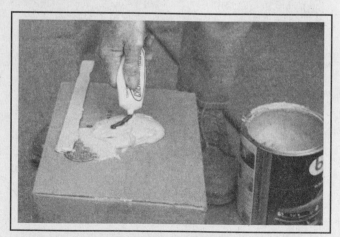

5   Following label instructions, mix up a batch of plastic filler and hardener. The ratio of filler to hardener is critical, and, if you mix it incorrectly, it will either not cure properly or cure too quickly (you won't have time to file and sand it into shape)

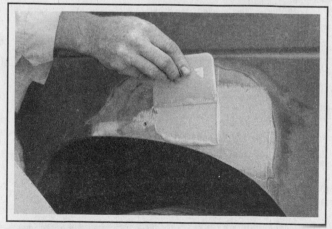

6   Working quickly so the filler doesn't harden, use a plastic applicator to press the body filler firmly into the metal, assuring it bonds completely. Work the filler until it matches the original contour and is slightly above the surrounding metal

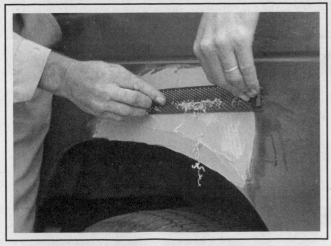

7 Let the filler harden until you can just dent it with your fingernail. Use a body file or Surform tool (shown here) to rough-shape the filler

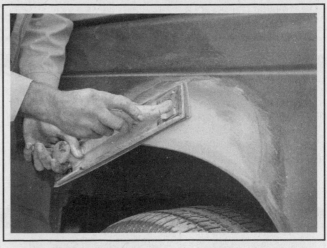

8 Use coarse-grit sandpaper and a sanding board or block to work the filler down until it's smooth and even. Work down to finer grits of sandpaper - always using a board or block - ending up with 360 or 400 grit

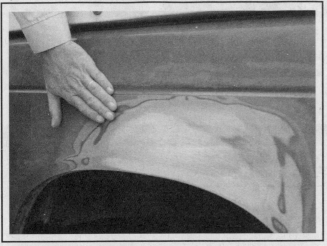

9 You shouldn't be able to feel any ridge at the transition from the filler to the bare metal or from the bare metal to the old paint. As soon as the repair is flat and uniform, remove the dust and mask off the adjacent panels or trim pieces

10 Apply several layers of primer to the area. Don't spray the primer on too heavy, so it sags or runs, and make sure each coat is dry before you spray on the next one. A professional-type spray gun is being used here, but aerosol spray primer is available inexpensively from auto parts stores

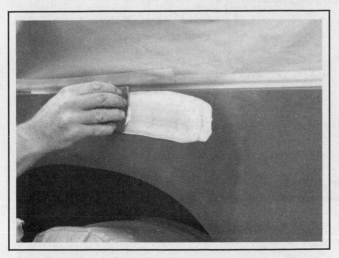

11 The primer will help reveal imperfections or scratches. Fill these with glazing compound. Follow the label instructions and sand it with 360 or 400-grit sandpaper until it's smooth. Repeat the glazing, sanding and respraying until the primer reveals a perfectly smooth surface

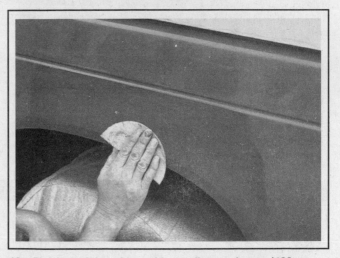

12 Finish sand the primer with very fine sandpaper (400 or 600-grlt) to remove the primer overspray. Clean the area with water and allow it to dry. Use a tack rag to remove any dust, then apply the finish coat. Don't attempt to rub out or wax the repair area until the paint has dried completely (at least two weeks)

5 Body and frame repair - major damage

1 Major damage must be repaired by an auto body/frame repair shop with the necessary welding and hydraulic straightening equipment.
2 If the damage has been serious, it's vital that the frame be checked for proper alignment or the vehicle's handling characteristics may be adversely affected. Other problems, such as excessive tire wear and wear in the driveline and steering may occur.

3 Due to the fact that all of the major body components (hood, fenders, etc.) are separate and replaceable units, any seriously damaged components should be replaced rather than repaired. Sometimes these components can be found in a wrecking yard that specializes in used vehicle components, often at considerable savings over the cost of new parts.

6 Hinges and locks - maintenance

Every 3000 miles or three months, the door and hood hinges should be lubricated with a few drops of oil. Lubricate the locks with

graphite spray. The door striker plates should also be given a thin coat of white lithium-base grease to reduce wear and ensure free movement.

7 Windshield and fixed glass - removal and installation

1 Replacement of the windshield and fixed glass requires the use of special fast-setting adhesive/caulk materials. These operations should be left to a dealer service department or a shop specializing in glass work.

2 Windshield mounted rear view mirror support removal is also best left to experts, as the bond to the glass also requires special tools and adhesives.

8 Hood - removal, installation and adjustment

→Note: The hood is somewhat awkward to remove and install - at least two people should perform this procedure.

REMOVAL

1 Use blankets or pads to cover the cowl area of the body and both fenders. This will protect the body and paint as the hood is lifted off.
2 Open the hood and support it on the prop rod.
3 Scribe alignment marks around the bolt heads and hinges to aid alignment during installation (a permanent-type felt-tip marker also will work for this).
4 Have an assistant support one side of the hood while you support the other. Simultaneously remove the hinge-to-hood bolts.
5 Lift off the hood.

INSTALLATION

6 Installation is the reverse of removal. Align the marks around the hinges and bolts (one side at a time) and then check for proper clearance. Readjust as necessary (see below).

ADJUSTMENT

7 Adjustment of the hood is done by moving the hood in relation to the hinge plate after loosening the bolts. The hood must be aligned so there is a 0.16-inch gap (approximate) to the front fenders and flush with the top surface.
8 Scribe or trace a line around the entire hinge plate so you can judge the amount of movement.
9 Loosen the bolts and move the hood into correct alignment. Move it only a little at a time. Tighten the hinge bolts and carefully lower the hood to check the alignment.
10 Adjust the hood bumpers on the radiator support so the hood is flush with the fenders when closed.
11 The hood latch assembly can also be adjusted up-and-down and side-to-side after loosening the nuts. Make sure you place alignment marks around the hood latch assembly before loosening the mounting nuts.
12 The hood latch assembly, as well as the hinges, should be periodically lubricated with white lithium-base grease to prevent sticking and wear.

9 Hood release latch and cable - removal and installation

LATCH

Removal
▶ Refer to illustration 9.2

1 Before removing the hood latch, use a felt tip pen or scribe to

mark the location of the hood latch to the radiator support. Trace the outline of the hood latch base so it can be reinstalled in the same position.
2 Use a screwdriver to pry the cable housing from the latch mechanism and then detach the cable end (see illustration).
3 Remove the 2 bolts securing the latch mechanism to the radiator support.

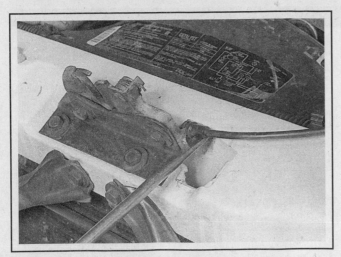

9.2 Use a screwdriver to pry the cable out of the clip

Installation

▶ **Refer to illustration 9.4**

4   Installation is the reverse of removal (see illustration). Align the scribe marks at installation. Tighten the latch mounting bolts securely.

CABLE

Removal

5   Use a screwdriver to pry the cable housing from the latch mechanism and then detach the cable end (see illustration 9.2).
6   Follow the cable to the grommet and pry it out of the firewall.
7   Securely fasten a suitable length of string (or wire) to the cable end. The string (or wire) will be used to pull the cable back through for installation.
8   On 1995 and earlier models, working inside the vehicle, locate the handle locking tabs on the backside of the driver's side knee bolster, compress the tabs and withdraw the handle assembly from the knee bolster. Then carefully pull the cable through the knee bolster and

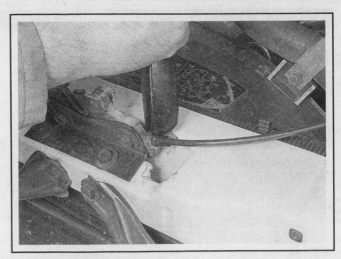

9.4 Use a screwdriver handle to seat the cable in the clip

firewall - do not let the guide string (or wire) become detached. Once the cable is free, detach the guide string (or wire).
9   On 1996 and later models, remove the driver's side knee bolster (see Section 34) and left cowl side vent cover (see Section 32). Remove the 2 bolts attaching the cable handle to the body panel. Then carefully pull the cable through firewall - do not let the guide string (or wire) become detached. Once the cable is free, detach the guide string (or wire).

Installation

10 To install the cable, fasten the guide string (or wire) onto the end, route it into position and install the end into the latch mechanism. Make sure the grommet is securely installed into the firewall. If necessary, apply sealant to ensure a good seal.
11 On 1995 and earlier models, press the handle into the knee bolster until it locks in place.
12 On 1996 and later models, install the handle assembly on the body panel, install the kick panel and driver's side knee bolster (see Section 34).

10   Cowl ventilator grille - removal and installation

▶ **Refer to illustration 10.3**

1   Remove the windshield wiper arms (see Chapter 12).
2   If applicable, remove the radio antenna (see Chapter 12 if necessary).
3   Remove the retaining screws with a Torx bit (see illustration).
4   Lift the grille off.
5   To install the grille, place it in position and install the screws. Tighten the screws securely.
6   Install the wiper arms and antenna (if applicable).

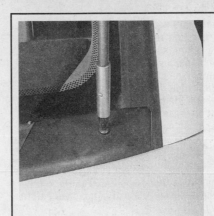

10.3 A special Torx bit will be required to remove the cowl ventilator screws

11 Engine cover - removal and installation

REMOVAL

▶ **Refer to illustrations 11.1a, 11.1b and 11.4**

1   On 1989 and earlier models, working in the engine compartment, use a long handled screwdriver to loosen the engine cover upper retaining screws located on the firewall (see illustrations).

➡**Note: The screws are captured in the brackets and cannot be removed.**

2   Working inside the vehicle, place the front seats in their fully rearward positions.

3   Remove the instrument panel lower extension (see Section 30).

4   If equipped, disengage the support rods from the engine cover studs (see illustration).

5   If equipped, remove the floor ducting from the engine cover.

6   On 1990 and later models, remove the instrument panel support braces from each side of the engine cover.

7   On 1990 and later models, loosen the engine cover upper retaining screws.

➡**Note: The screws are captured in the brackets and cannot be removed.**

8   Unlock the two latches securing the bottom of the engine cover and carefully remove it from the vehicle. Be careful when removing the engine cover: the captured fasteners can easily scratch interior surfaces. Also the underside of the engine cover may be oily and dirty.

INSTALLATION

9   Before installing the engine cover, inspect the rubber seal for damage and replace if necessary. If this seal is damaged or does not seat correctly, water and/or harmful vapors may enter the passenger compartment.

10 Make sure the front seats are in their fully rearward positions.

11 Maneuver the engine cover into position. Make sure the seal is seated properly and nothing is caught between the cover and its mating surfaces.

12 Tighten the engine cover upper screws and then lock the bottom latches.

13 The remaining installation steps are the reverse of removal.

11.1a Use a large screwdriver to unscrew the right side engine cover screw (arrow). . .

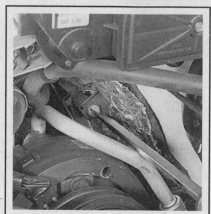

11.1b . . . and left side engine cover screw (arrow) in the engine compartment

11.4 Detach the support rods from the studs

12 Fender - removal and installation

✳✳ WARNING:

Most 1993 and later models are equipped with airbags. Always disable the airbag system before working in the vicinity of the impact sensors, steering column or instrument panel to avoid the possibility of accidental deployment of the airbag(s), which could cause personal injury (see Chapter 12 for the airbag disabling procedure). The yellow wires and connectors routed through the instrument panel are for this system. Do not use electrical test equipment on these yellow wires or tamper with them in any way while working under the instrument panel.

✳✳ CAUTION:

If the vehicle is equipped with a Delco Loc II or Theftlock audio system, make sure you have the correct activation code before disconnecting the battery.

REMOVAL

▶ **Refer to illustration 12.7**

1   On models with sealed-beam headlights, remove the headlight bezel (see Section 17).

2   Remove the front bumper (see Section 18) and the wheelhouse panel extension (see Section 16).

3   On later models, remove the two screws holding the windshield frame filler panel (the small triangular piece at the lower side corner of the windshield, above the fender). On all models, remove the cowl ventilator grille (see Section 10).

4   With the aid of another person, remove the hood hinge-to-fender bolts and remove the hood from the vehicle.

5   On 1994 and earlier models, remove the front wheel house panel extension (see Section 16).

12.7 Remove the fender mounting bolts (arrows) - typical. The rear fender bolts (two right-most arrows) are accessed with the door open

13 Radiator grille - removal and installation

❋❋ WARNING:

Most 1993 and later models are equipped with airbags. Always disable the airbag system before working in the vicinity of the impact sensors, steering column or instrument panel to avoid the possibility of accidental deployment of the airbag(s), which could cause personal injury (see Chapter 12 for the airbag disabling procedure). The yellow wires and connectors routed through the instrument panel are for this system. Do not use electrical test equipment on these yellow wires or tamper with them in any way while working under the instrument panel.

❋❋ CAUTION:

If the vehicle is equipped with a Delco Loc II or Theftlock audio system, make sure you have the correct activation code before disconnecting the battery.

REMOVAL

▶ **Refer to illustration 13.5**

1 On models equipped with sealed beam headlights, remove the headlight bezels (see Section 17).

2 On models equipped with composite headlights, remove the turn signal lamp assemblies (see Chapter 12).

14 Front end panel - removal and installation

❋❋ WARNING:

Most 1993 and later models are equipped with airbags. Always disable the airbag system before working in the vicinity of the impact sensors, steering column or instrument panel to avoid the possibility of accidental deployment of the airbag(s), which could cause personal injury (see Chapter 12 for the airbag disabling procedure). The yellow wires and connectors routed through the instrument panel are for this system. Do not use electrical test equipment on these yellow wires or tamper with them in any way while working under the instrument panel.

6 On 1995 and later models, remove the front rocker panel body cladding.

7 Remove the fender mounting bolts (see illustration) and carefully remove the fender from the vehicle to avoid damaging painted surfaces.

INSTALLATION

8 Carefully place the fender on the vehicle and install the bolts hand-tight. Maneuver the fender to obtain the best alignment and tighten the retaining bolts.

9 With the aid of another person, install the hood and secure with the hinge pin bolts. Check the alignment of the hood and adjust if necessary (see Section 8).

10 The remaining installation steps are the reverse of removal.

13.5 Remove the grille mounting screws (arrows) - typical

3 On models equipped with sealed beam headlights, remove the plastic splash panel under the front bumper.

4 On models equipped with sealed beam headlights, working under the vehicle, remove the four bolts securing the bottom of the grille to the front end panel (see illustration 13.5).

5 Remove the bolts securing the grille to the radiator support (see illustration).

INSTALLATION

6 Installation is the reverse of removal.

1 Remove the front bumper (see Section 18).

2 Remove the radiator grille (see Section 13).

3 Remove the front end panel mounting bolts and lower the panel from the vehicle.

4 To install the panel, place it in position and install the bolts. Tighten the bolts securely.

5 Install the grille and front bumper.

15 Front air deflector - removal and installation

1 Remove the air deflector-to-bumper and fender bolts.
2 Detach the air deflector.

3 Installation is the reverse of removal.

16 Front wheelhouse panel extension - removal and installation

▶ **Refer to illustrations 16.2a and 16.2b**

1 Remove the panel extension mounting bolts.
2 Release the retainers fastening the panel extension to the fender by prying out the center of each retainer with wire cutters (don't cut the center piece off, just pull it out) (see illustration). Once the center has

been pulled out, the retainer can be removed (see illustration).
3 Remove the panel extension from the fender.
4 To install the extension, place it in position and install the retainers. Once the extension and retainers are in place, press the center of each retainer in to lock them in place.
5 Install the bolts.

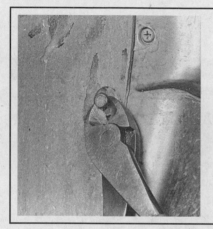

16.2a Pry the center of the plastic retainer out with wire cutters - DO NOT cut it off!

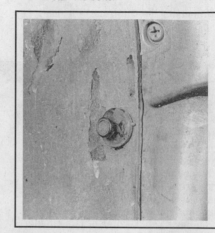

16.2b With the center pulled out, the retainer can be removed

17 Headlight bezel (sealed beam headlights only) - removal and installation

▶ **Refer to illustrations 17.1 and 17.2**

1 Remove the four bezel retaining screws (see illustration). A special Torx screwdriver will be needed for this.

2 Rotate the bezel, disconnect the parking light and turn signal bulbs on the back side by turning the bulb holders a half turn and remove it from the vehicle (see illustration).
3 Installation is the reverse of removal.

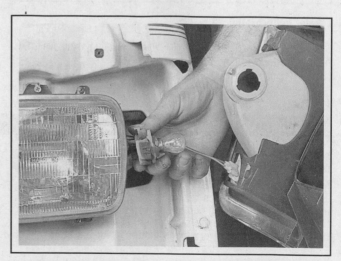

17.1 The Torx screws retaining the headlight bezel are accessible after raising the hood

17.2 Rotate the bulb holders a half turn to detach them from the bezel

18 Bumpers - removal and installation

✳✳ WARNING:

Most 1993 and later models are equipped with airbags. Always disable the airbag system before working in the vicinity of the impact sensors, steering column or instrument panel to avoid the possibility of accidental deployment of the airbag(s), which could cause personal injury (see Chapter 12 for the airbag disabling procedure). The yellow wires and connectors routed through the instrument panel are for this system. Do not use electrical test equipment on these yellow wires or tamper with them in any way while working under the instrument panel.

⟨⟩ CAUTION:

If the vehicle is equipped with a Delco Loc II or Theftlock audio system, make sure you have the correct activation code before disconnecting the battery.

FRONT BUMPER

Removal

1   On 1994 and earlier models, remove the front air deflector (see Section 15).

2   On 1995 and later models, remove the radiator grille (see Section 13).

3   On 1995 and later models, remove the two fascia-to-brace bolts, one under each lower corner of the fascia. (The nuts are welded to the brace and are not removable). Remove the fascia-to-fender bolts, one under each upper corner of the fascia. Remove the fascia-to-bumper push-in fasteners. Remove the fascia from the bumper.

4   Remove the brace-to-frame bolts (if equipped) and the bracket-to-frame bolts and separate the bumper from the vehicle.

Installation

5   Installation is the reverse of removal. Tighten the bumper mounting bolts to the torque listed in this Chapter's Specifications.

REAR BUMPER

Removal

6   On 1995 and later models, remove the fascia from the bumper.

➡ **Note: Push-in fasteners are used (under the fascia).**

7   Remove the brace-to-frame and the bracket-to-frame bolts/nuts and separate the bumper from the vehicle. (The brackets attach the bumpers to the bottom of the frame rails; the braces are the curved pieces at the outside ends of the bumper.)

Installation

8   Installation is the reverse of removal. Tighten the bumper mounting bolts to the torque listed in this Chapter's Specifications.

19 Swing-out window and latch - removal and installation

▶ **Refer to illustrations 19.2 and 19.3**

1   Open the window.

2   Remove the window latch-to-body screws (see illustration).

3   While an assistant supports the window glass, remove the hinge screws (see illustration). Lift the window from the vehicle.

4   If a new window is being installed, transfer the latch and other hardware to the new glass.

5   Place the glass in position, have an assistant hold it in place and install the hinge screws. Install the latch and make sure the window closes securely.

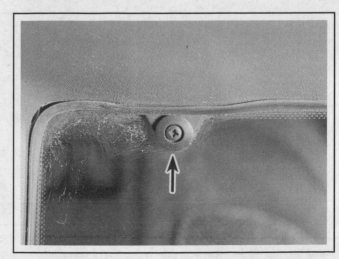

19.2 Use a Phillips screwdriver to remove the latch mounting screws

19.3 Remove the hinge screws (arrow) and carefully lift the window from the vehicle

20 Door trim panel - removal and installation

FRONT DOOR

1995 and earlier models

Removal

▶ **Refer to illustrations 20.3, 20.4, 20.5 and 20.7**

1   On models equipped with power windows, disconnect the cable from the negative terminal of the battery.

> ※※ **CAUTION:**
>
> **On models equipped with an anti-theft audio system, disable the anti-theft feature before performing any operation that requires disconnecting the battery or disrupting power to the stereo (see the Anti-theft audio system procedures at the front of this manual).**

2   Remove the two screws securing the armrest and remove it from the vehicle.

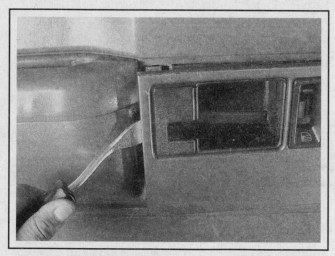

20.3 The door handle cover simply pries off on most models - be sure it clears the door handle as it's removed

3   Carefully pry off the door handle cover (see illustration) and, if equipped, disconnect the wiring harness connectors.

4   On models with manual windows, remove the window crank handle using a special tool (see illustration) which is available at most auto parts stores. If the tool isn't available, press the door panel in and use a piece of hooked wire to pull off the retaining spring. Another method is to take a small piece of cloth and work it back-and-forth behind the crank handle until the retainer is dislodged.

5   Remove the four retaining screws (two are inside the armrest recess) and then pry around the outer edge of the door panel using a screwdriver or putty knife to disengage the push-in fasteners (see illustration).

6   Lift the door panel away from the door, pry it off the door frame at the window seal and remove it from the vehicle.

7   If access to the inner door components is required, carefully peel away the watershield, taking care not to tear it (see illustration). The watershield can easily be reinstalled by pressing it back into place.

1996 and later models

Removal

▶ **Refer to illustrations 20.12, 20.13, 20.14, 20.15a and 20.15b**

8   On models equipped with power windows, disconnect the cable from the negative terminal of the battery.

> ※※ **CAUTION:**
>
> **On models equipped with an anti-theft audio system, disable the anti-theft feature before performing any operation that requires disconnecting the battery or disrupting power to the stereo (see the Anti-theft audio system procedures at the front of this manual).**

9   Remove the 2 screws from the bottom of the armrest and then pivot it up and remove it from the door trim panel.

10 If equipped with manual windows, remove the crank handle as described in Step 4.

11 Carefully pry the top corner of the trim panel covering the exterior mirror mounting nuts from the door and then pivot it down to remove it.

20.4 A special tool can be used to pull the clip off the manual regulator crank handle, but a hooked piece of wire works just as well

20.5 Insert a screwdriver behind the door panel and pry out very carefully to disengage the retainers

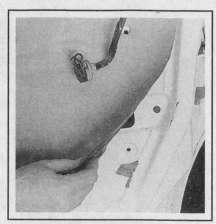

20.7 Peel the watershield back carefully - try not to tear it!

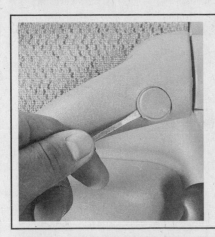

20.12 Use a small screwdriver to pry out the trim caps from the pull handle

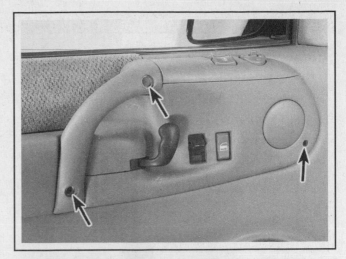

20.13 Pull handle and accessory switch mount plate bolt locations (arrows)

16 If equipped, remove the door courtesy lamp and disconnect the wiring harness.

17 Pry around the outer edge of the door panel using a screwdriver to disengage the push-in fasteners.

18 Lift the door panel away from the door, pry it off the door frame at the window seal and remove it from the vehicle.

19 If access to the inner door components is required, carefully peel away the watershield, taking care not to tear it (see illustration 20.7). The watershield can easily be reinstalled by pressing it back into place.

Installation

20 Prior to installing the door panel, remove any push-in fasteners that may have remained in the door and place them into position on the door panel.

21 Fit the door panel into position on the door and press around the outer edges to engage the push-in fasteners.

22 Install the retaining screws as required.

23 To install the window crank on manual windows, first determine the proper position of the handle by comparing it to the opposite side door, install the retaining clip onto the crank handle, place the plastic washer on the regulator shaft and then push the crank handle onto the shaft until it snaps into place.

24 The remaining installation steps are the reverse of removal.

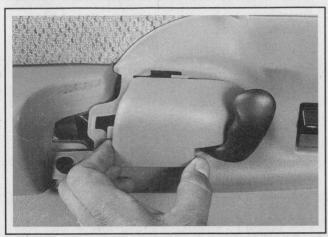

20.14 Slide the door latch handle trim cover forward and remove it from the door

12 Pry out the 2 trim caps from the door pull handle (see illustration).

13 Remove the bolts from the pull handle and accessory switch mount plate (see illustration).

14 Remove the door pull handle, then slide the door latch handle trim cover forward and remove it from the door panel (see illustration).

15 Remove the accessory switch mount plate and disconnect the electrical connectors from the switches (see illustrations).

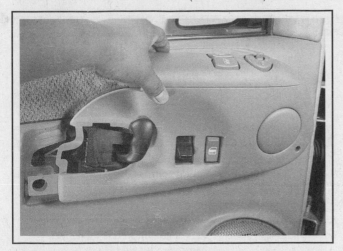

20.15a Removing the accessory switch mount plate

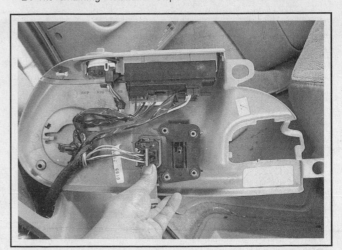

20.15b Disconnect the electrical connectors from the back side of the accessory switch mount plate

SIDE DOOR

▶ **Refer to illustration 20.25**

25 Remove the retaining screws and lift off the handle cover (see illustration).

26 Remove the retaining screws and then pry around the outer edge of the trim panel with a large screwdriver to disengage the retainers.

27 Pry off the garnish molding.

28 Lift off the trim panel.

29 To install the trim panel, press it into place, making sure the retainers are fully seated.

30 Install the handle cover and screws.

31 Place the garnish molding in position and press it in until the retainers engage.

REAR DOOR

▶ **Refer to illustrations 20.32a, 20.32b and 20.33**

32 Remove the screws, pry around the edges of the panel to disengage the retainers and lower the door garnish panel to remove it (see illustrations).

33 Remove the screws and detach the trim panel from the door

20.25 Remove the side door handle cover retaining screws (arrows)

(see illustration).

34 Place the trim panel in position and install the screws.

35 Raise the garnish panel, engage the pins at the top in the holes in the door and press it into place until the retainers engage. Install the screws.

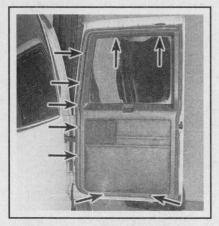

20.32a Remove the screws around the perimeter of the rear door trim panel (arrows)

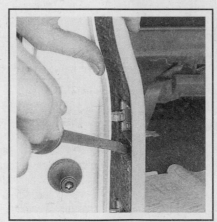

20.32b Pry the retainers out of the door very carefully with a screwdriver

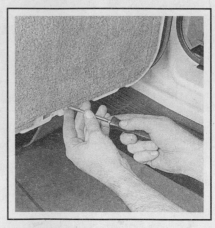

20.33 Remove the lower trim panel screws

21 Front door - removal and installation

REMOVAL

▶ **Refer to illustration 21.4**

➡Note: A special door hinge spring compressor tool is required to properly perform this procedure.

1 Disconnect the cable from the negative terminal of the battery.

❊❊ CAUTION:

On models equipped with an anti-theft audio system, disable the anti-theft feature before performing any operation that requires disconnecting the battery or disrupting power to the stereo (see the Anti-theft audio system procedures at the front of this manual).

2 If the door has power features, remove the door trim panel (see Section 20), and disconnect the wiring harness from the door.

3 Compress the hinge spring using a special tool or cover the spring with heavy cloth before removal.

❊❊ WARNING:

The door spring will be released when the door is removed and could fly out of the door, causing personal injury, if it's not properly held in place.

4 Compress the clip on the lower hinge pin and remove the pin by pulling it out with locking pliers while tapping on the end with a soft-face hammer (see illustration). The clip will ride up on the pin as it's removed and then fall out.

5   Insert a bolt into the lower hinge pin hole and then remove the upper hinge pin.
6   Remove the bolt and lift the door off.

INSTALLATION

7   To install the door, place it in position and install the bolt in the lower hinge pin hole.
8   Assemble the upper hinge pin using a new hinge pin clip.
9   Remove the bolt and install the lower hinge pin, using a new clip.
10 Install the spring.
11 If necessary, install the door trim panel and connect the wiring harness.
12 Connect the cable to the negative terminal of the battery.

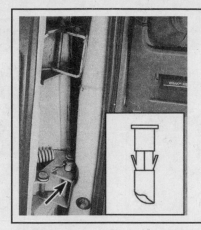

21.4 Compress the clips on the lower hinge pin (arrow)

22 Sliding door - removal, installation and adjustment

REMOVAL

▶ **Refer to illustrations 22.3a, 22.3b, 22.4, 22.5, 22.7 and 22.9**

1   Remove the door trim panel and garnish molding.
2   Remove the right tail light lens (see Chapter 12) and the rear stowage compartment for access to the track cover bolts.
3   Remove the bolts and lift off the track cover (see illustrations).
4   Remove the upper roller bracket cover (see illustration).
5   Outline the lower bracket bolt heads (see illustration).
6   Remove the upper roller bracket Torx head screws.
7   With an assistant supporting the weight of the door, disconnect the latch cable from the lower bracket (see illustration).
8   Remove the lower bracket.
9   With the help of an assistant, roll the door to the end of the track, rotate the roller out and lift the door off (see illustration).

INSTALLATION

10 To install the door, engage the roller with the track and move the door forward until the latch cable can be connected to the lower roller.
11 Install the lower roller bolts in the marked positions. Tighten the bolts securely.
12 Install the upper roller screws and the covers.

ADJUSTMENT

13 The door can be adjusted up-and-down and in-and-out slightly to provide an even fit in the door opening. Adjustment points are:
   1) *The roller at the lower rear edge of the door - its bracket bolts can be loosened and the roller moved.*
   2) *The roller track along the middle of the door.*
   3) *The latch striker at the front edge of the door.*

Front edge up-and-down

14 Open the door part way and loosen the lower front catch and the roller track assembly bolts.
15 Move the front edge up-or-down and then tighten the bolts securely. Make sure the door is not moved beyond the point where the door alignment pins no longer engage.

Rear edge up-and-down

16 Loosen the roller track bolts.
17 Open the door and adjust the striker up-or-down. The roller track bolts must be loosened before any adjustment to the striker bolt is made.
18 Close the door and check the rear edge height.
19 If the height is correct, move the roller track up to meet the roller, tighten the bolts and recheck the height, adjusting as necessary.

22.3a Once the rear storage compartment has been removed, remove the door track cover screws (arrows)

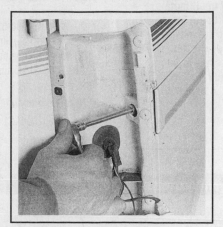

22.3b Remove the rubber plug and use a ratchet and socket extension to remove the rear track cover bolt

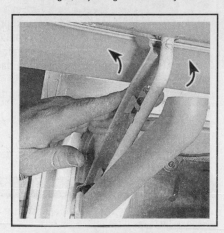

22.4 Remove the upper roller bracket covers by unsnapping them in the direction shown (arrows)

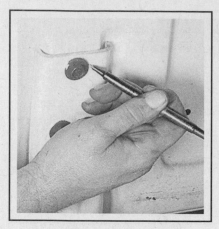

22.5 Mark the location of the lower bracket bolts with a scribe or pencil

22.7 Use a small screwdriver to disconnect the latch cable from the lower bracket

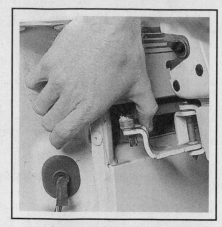

22.9 Rotate the roller out of the track and lift the door off

Front and rear edge in-and-out

20 Loosen the rear latch striker bolt and move it in-or-out as required to achieve the proper gap.

21 Install the track cover, tail light lens, the rear stowage compartment and the side door garnish molding and trim panel.

23 Rear door and liftgate - removal and installation

1 Disconnect the negative cable at the battery. Place the cable out of the way so it cannot accidentally come in contact with the negative terminal of the battery, as this would once again allow power into the electrical system of the vehicle.

✳✳ CAUTION:

On models equipped with an anti-theft audio system, disable the anti-theft feature before performing any operation that requires disconnecting the battery or disrupting power to the stereo (see the Anti-theft audio system procedures at the front of this manual).

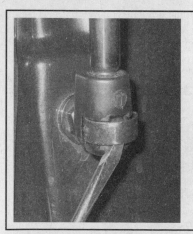

23.9 Detach either end of the liftgate support strut by pulling out the retainer clip with a small screwdriver, then pull the strut from the ballstud

2 On right side door, remove the trim panel and unplug the electrical harnesses.

3 Remove the check strap.

4 With an assistant supporting the weight of the door, drive out the hinge pins with a hammer and pin punch. Detach the door from the vehicle.

5 Have an assistant hold the door in place, insert the hinge pins and drive them into place with a hammer.

6 Installation is the reverse of removal.

LIFTGATE

▶ **Refer to illustration 23.9**

7 Remove the molding from around the liftgate window.

8 Disconnect the wiring harness from the wiper motor, power lock actuator and the glass defogger, then free the harness clips from the liftgate

9 With the help of an assistant, support the liftgate and remove the support struts (see Illustration).

10 Mark the position of each hinge to the body.

11 While supporting the liftgate, remove the hinge-to-body nuts and remove the liftgate.

➡ **Note: Two assistants would be helpful at this time.**

12 Installation is the reverse of removal.

24 Door lock assembly - removal and installation

FRONT DOOR

▶ **Refer to illustration 24.2**

1 Remove the door trim panel and watershield (see Section 20).

2 Disconnect the lock cylinder rods, remove the screws and detach the assembly from the door (see illustration).

3 Installation is the reverse of removal.

SIDE DOOR

▶ **Refer to illustration 24.5**

4   Remove the door trim panel (see Section 20).
5   Disconnect the outside handle and locking rods, remove the retaining screws and detach the lock from the door (see illustration).
6   Installation is the reverse of removal.

REAR DOOR

7   Remove the door trim panel (see Section 20). Disconnect the locking and lock cylinder rods, remove the screws and detach the lock.
8   Installation is the reverse of removal.

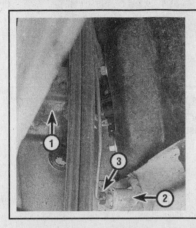

24.2 Front door lock assembly details

1   Lock assembly
2   Lock cylinder
3   Retaining clip

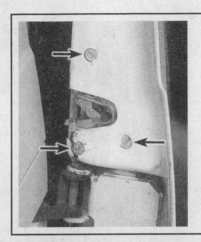

24.5 Remove the lock assembly mounting screws (arrows)

25 Door lock striker - removal and installation

▶ **Refer to illustration 25.2**

1   Mark the position of the striker bolt with a pencil.
2   Use a large Torx driver to unscrew the striker bolt (see illustration).
3   To install the striker, reverse the removal procedure and line the bolt up with the marks made during removal. Tighten the bolt to the specified torque.

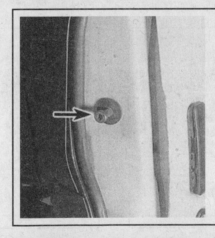

25.2 Use a Torx bit to remove the striker bolt (arrow)

26 Door handle - removal and installation

OUTSIDE HANDLE

▶ **Refer to illustrations 26.2, 26.3 and 26.4**

1   On front doors, remove the door trim panel and watershield (see Section 20). Remove the door handle mounting nuts. Disconnect the lock rod clips and remove the handle from the door. Installation is the reverse of removal.
2   On the side door, remove the door trim panel. Disconnect the lock rod clips, remove the mounting nuts and detach the handle from the door (see illustration). Installation is the reverse of removal.
3   On rear doors, remove the access cover (see illustration).
4   Push the clip off with a small screwdriver and disconnect the rods (see illustration).
5   Remove the mounting nuts and detach the handle assembly from the door.
6   Installation is the reverse of removal.

26.2 Remove the outside door handle retaining nuts (arrows)

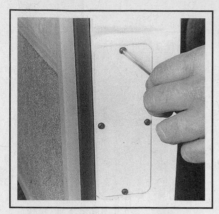

26.3 The rear door handle access cover is held in place by four screws

26.4 Push the rod clip off in the direction shown (arrow)

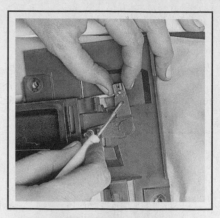

26.7a Insert a small screwdriver through the hole in the clip to disconnect it from the rod

INSIDE HANDLE

▶ **Refer to illustrations 26.7a, 26.7b, 26.7c, 26.8, 26.9a and 26.9b**

7   On front doors, remove the door trim panel and door handle mounting bolts. Disconnect the door handle lock rod clip by inserting a screwdriver through the hole in the handle arm (see illustration). Disengage the rod and lower the door handle assembly from the door (see illustrations). Installation is the reverse of removal.

8   On some models, the inside handle is riveted to the door panel. Drill out the center of the rivets with a 3/16-inch drill bit, disengage the lock rods and lift the handle from the door (see illustration). Install the handle using 1/4-inch diameter by 1/2-inch long bolts, nuts and lock washers.

9   On the side door, remove the door trim panel. Disconnect the rod at the handle (see illustration). Remove the mounting screws and detach the handle (see illustration). Installation is the reverse of removal.

26.7b Lift the assembly away from the door . . .

26.7c . . . while guiding the rod out

26.8 Some inside handles are riveted to the door panel

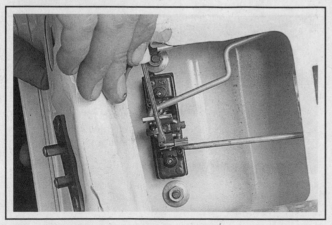

26.9a Push the clip down with a small screwdriver and disengage the rod

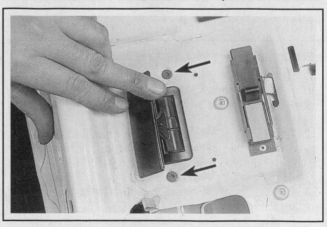

26.9b Side door inside handle mounting screws (arrows)

27 Front door window regulator - removal and installation

REMOVAL

1   Disconnect the negative cable at the battery. Place the cable out of the way so it cannot accidentally come in contact with the negative terminal of the battery, as this would once again allow power into the electrical system of the vehicle.

❋❋ CAUTION:

On models equipped with an anti-theft audio system, disable the anti-theft feature before performing any operation that requires disconnecting the battery or disrupting power to the stereo (see the Anti-theft audio system procedures at the front of this manual).

2   Remove the door trim panel and watershield (see Section 20).
3   Remove the armrest bracket.
4   Secure the window glass in the up position with strong adhesive tape fastened to the glass and wrapped over the frame.
5   Punch out the center of the rivets securing the window regulator to the door and drill out the rivets with a 3/16-inch drill bit.
6   Push the regulator into the door, move it forward and then to the rear to disconnect the arms from the glass and fold the arms together. Lift the regulator from the door through the access hole. On power window regulators unplug the electrical connector.

7   If the power regulator motor is being replaced, install a self-tapping sheet metal screw through the sector gear and backing plate in the provided hole, to lock it.

❋❋ WARNING:

The regulator arms are under considerable tension from the counterbalance spring and could cause injury if the motor is removed from the regulator without first locking the sector gear in place. Drill out the rivets and remove the motor from the regulator.

Place the new motor in position and secure it with new rivets. Remove the metal sheet screw and lubricate the gear teeth with multipurpose grease.

INSTALLATION

8   Place the regulator in position in the door and connect the arm rollers to the sash and regulator rail.
9   Align the regulator holes with the holes in the door and secure it with 1/4-inch diameter by 1/2-inch long bolts with nuts and lock washers. On power regulators, plug in the electrical connector.
10 Install the armrest bracket, watershield and door trim panel.
11 Remove the tape from the glass and connect the battery.

28 Front door window glass - removal and installation

1   Remove the door trim panel and watershield (see Section 20).
2   Remove the lower glass run channel and the other window sealing strip.

3   Remove the window regulator (see Section 27).
4   Slide the glass up and out of the door to remove it.
5   Installation is the reverse of removal.

29 Exterior mirror - removal and installation

◗ **Refer to illustration 29.4**

1   On 1995 and earlier models, remove the door trim panel and peel back the watershield in the area of the mirror (see Section 20).
2   Carefully pry the top corner of the trim panel covering the exterior mirror mounting nuts from the door and then pivot it down to remove it.
3   On power mirrors, unplug the electrical connector.
4   Remove the nuts or screws and detach the mirror (see illustration).
5   Installation is the reverse of removal.

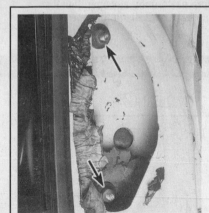

29.4 With the door trim panel removed, remove the mirror mounting nuts (arrows) and remove the mirror

30 Instrument panel lower extension - removal and installation

REMOVAL

▶ **Refer to illustrations 30.2a and 30.2b**

1   Disconnect the cable from the negative terminal of the battery.

✳✳ CAUTION:

On models equipped with an anti-theft audio system, disable the anti-theft feature before performing any operation that requires disconnecting the battery or disrupting power to the stereo (see the Anti-theft audio system procedures at the front of this manual).

2   Remove the lower extension mounting bolts and nuts (see illustrations).

3   Pull the lower extension away from the engine cover just enough to disconnect the electrical connectors and remove it from the vehicle.

INSTALLATION

4   Installation is the reverse of removal.

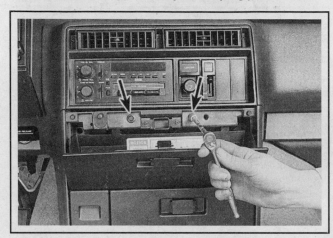

30.2a On 1989 and earlier models, the instrument panel lower extension upper retaining bolts are located inside the glove box (arrows)

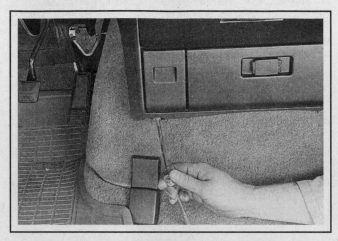

30.2b On all models, the lower extension is secured to the engine cover at the bottom by nuts on each side

31 Roof console - removal and installation

1   Disconnect the negative cable at the battery. Place the cable out of the way so it cannot accidentally come in contact with the negative terminal of the battery, as this would once again allow power into the electrical system of the vehicle.

✳✳ CAUTION:

On models equipped with an anti-theft audio system, disable the anti-theft feature before performing any operation that requires disconnecting the battery or disrupting power to the stereo (see the Anti-theft audio system procedures at the front of this manual).

2   Remove the mounting bolts.

3   Slide the roof console forward for access to the electrical harness. Unplug the electrical connector and lower the console from the vehicle.

4   To install the console, raise it into place, plug in the connector, insert the console into the slots in the roof and install the mounting bolts.

32 Interior trim panels - removal and installation

FRONT DOOR GARNISH MOLDING

1   Remove the mounting screw (some models), pry at the molding retainers to disengage the molding and remove it.

2   To install the molding, press it into place until the retainers seat and install the screw.

FRONT DOOR HINGE PILLAR GARNISH MOLDING

3   Remove the mounting screw and lift the molding off.

4   Installation is the reverse of removal.

COWL SIDE VENT COVER

5   On 1996 and later models, remove the respective knee bolster (see Section 34).

6   Remove the mounting screw and pry the cover from the body panel. Installation is the reverse of removal.

SLIDING DOOR PILLAR MOLDINGS

7   Remove the screws and detach the moldings.

8   Installation is the reverse of removal.

BODY SIDE TRIM PANELS

9 Pry around the outer edge of the trim panel until the retainers are all detached.

10 Prior to installation make sure any retainers that were pulled from the panels during removal are reinstalled in the clips. Place the panel in position and press it into place until the retainers are seated.

ACCESSORY TRIM PANEL (1985 TO 1995 MODELS ONLY)

▶ **Refer to illustration 32.12**

11 Remove the instrument panel lower extension (see Section 30).

12 Remove the retaining screws (see illustration) and withdraw the accessory trim panel from the dash. On 1993 to 1995 models, label and disconnect the electrical connectors from the accessory switches.

13 Installation is the reverse of removal.

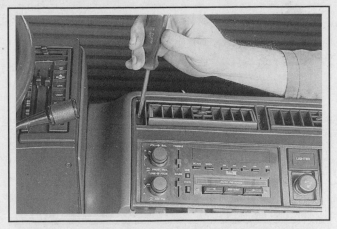

32.12 The accessory trim panel retaining screws on 1992 and earlier models are located on each side of the vent openings, as shown; on 1993 and later models, the screws are located beneath the center air vents

33 Seats - removal and installation

FRONT SEATS

▶ **Refer to illustrations 33.1 and 33.3**

1 Use a screwdriver to pry off the seatbelt guide (see illustration).

2 On models equipped with power seats, disconnect the wiring harness connector.

3 Remove the 4 retaining nuts, lift the seat off the studs and withdraw it from the vehicle (see illustration).

REAR SEATS

▶ **Refer to illustration 33.5**

4 Lift the lever and tilt the seatback forward.

5 Lift the latches in the rear seat legs and rotate the seat forward and out of the floor mounts (see illustration).

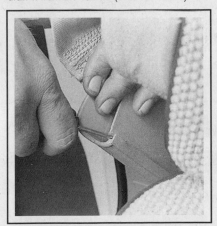

33.1 Pry the seatbelt guide off with a screwdriver

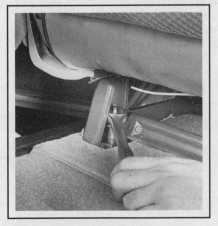

33.3 Use a socket wrench to remove the front seat mounting nuts

33.5 Hold the latch up and rotate the rear seat forward

34 Knee bolsters - removal and installation

REMOVAL

▶ **Refer to illustrations 34.4a and 34.4b**

1 On 1995 and earlier models, remove the instrument panel lower extension (see Section 30).

2 On 1995 and earlier models, in order to completely remove the driver's side knee bolster from the vehicle, the hood release cable must be removed (see Section 9).

3 If you are removing the driver's side knee bolster, disconnect the parking brake release rod or cable from the pedal assembly (see Chapter 9 if necessary) and withdraw it from the knee bolster. If you are removing the passenger's side knee bolster on later models, remove the instrument panel lower extension (see Section 30) if not already done in Step 1.

4 Remove the 3 mounting screws and remove the knee bolster from the vehicle (see illustrations).

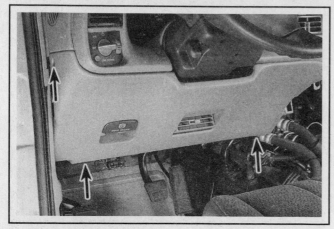

34.4a Driver's side knee bolster screw locations (arrows) (1996 and later models)

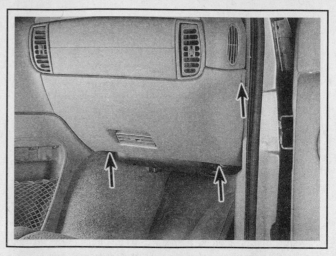

34.4b Passenger's side knee bolster screw locations (arrows) (1996 and later models)

INSTALLATION

5 Installation is the reverse of removal.

35 Steering column cover - removal and installation

✳✳ WARNING:

Most 1993 and later models are equipped with airbags. Always disable the airbag system before working in the vicinity of the impact sensors, steering column or instrument panel to avoid the possibility of accidental deployment of the airbag(s), which could cause personal injury (see Chapter 12 for the airbag disabling procedure). The yellow wires and connectors routed through the instrument panel are for this system. Do not use electrical test equipment on these yellow wires or tamper with them in any way while working under the instrument panel.

1 Disconnect the negative battery cable. On airbag-equipped models, disable the airbag system (see Chapter 12).

2 Remove the steering wheel (see Chapter 10).

3 Remove the tilt lever by pulling it straight out.

4 Remove the two retaining screws and the lower half of the steering column trim cover.

5 Remove the Torx-head screw and the upper half of the trim cover.

6 Installation is the reverse of removal. When replacing the tilt lever, push it straight back into the slot.

Specifications

Torque specifications

	Ft-lbs
Front bumper bolts	
1985 through 1995	26
1996 and later	
Front bumper-to-bracket bolts	43
Front bracket-to-frame bolts	41
Rear bumper-to-bracket bolts/nuts	26
Rear bracket-to-frame bolts	41
Rear bumper-to-brace nut	26
Rear brace-to-frame bolt	26
Lock striker-to-body-pillar bolt (hinged doors)	49
Lock striker-to-body-pillar bolt (sliding door front)	49
Lock striker bolt (sliding door rear)	18
Front seat nuts	
1985 through 1995	26
1996 and later	32

12 CHASSIS ELECTRICAL SYSTEM

Section

Reference to other Chapters

Battery check and maintenance - See Chapter 1
Battery - removal and installation - See Chapter 5
Brake light switch - removal, installation and adjustment - See Chapter 9
Park/Neutral switch - replacement - See Chapter 7
Starter/Neutral safety switch check - See Chapter 1
Starter safety switch - removal, installation and adjustment - See Chapter 8
Wiper blade inspection and replacement - See Chapter 1

1 General information

The electrical system is a 12-volt, negative ground type. Power for the lights and all electrical accessories is supplied by a lead/acid-type battery which is charged by the alternator.

This Chapter covers repair and service procedures for the various electrical components not associated with the engine. Information on the battery, alternator, distributor and starter motor can be found in Chapter 5.

It should be noted that when portions of the electrical system are serviced, the negative battery cable should be disconnected from the battery to prevent electrical shorts and/or fires.

✳✳ CAUTION:

On models equipped with an anti-theft audio system, disable the anti-theft feature before performing any operation that requires disconnecting the battery or disrupting power to the stereo (see the Anti-theft audio system procedures at the front of this manual).

2 Electrical troubleshooting - general information

A typical electrical circuit consists of an electrical component, any switches, relays, motors, fuses, fusible links or circuit breakers related to that component and the wiring and connectors that link the component to both the battery and the chassis. To help you pinpoint an electrical circuit problem, wiring diagrams are included at the end of this book.

Before tackling any troublesome electrical circuit, first study the appropriate wiring diagrams to get a complete understanding of what makes up that individual circuit. Trouble spots, for instance, can often be narrowed down by noting if other components related to the circuit are operating properly. If several components or circuits fail at one time, chances are the problem is in a fuse or ground connection, because several circuits are often routed through the same fuse and ground connections.

Electrical problems usually stem from simple causes, such as loose or corroded connections, a blown fuse, a melted fusible link or a bad relay. Visually inspect the condition of all fuses, wires and connections in a problem circuit before troubleshooting it.

If testing instruments are going to be utilized, use the diagrams to plan ahead of time where you will make the necessary connections in order to accurately pinpoint the trouble spot.

The basic tools needed for electrical troubleshooting include a circuit tester or voltmeter (a 12-volt bulb with a set of test leads can also be used), a continuity tester, which includes a bulb, battery and set of test leads, and a jumper wire, preferably with a circuit breaker incorporated, which can be used to bypass electrical components. Before attempting to locate a problem with test instruments, use the wiring diagrams(s) to decide where to make the connections.

VOLTAGE CHECKS

Voltage checks should be performed if a circuit is not functioning properly. Connect one lead of a circuit tester to either the negative battery terminal or a known good ground. Connect the other lead to a connector in the circuit being tested, preferably nearest to the battery or fuse. If the bulb of the tester lights, voltage is present, which means that the part of the circuit between the connector and the battery is problem free. Continue checking the rest of the circuit in the same fashion. When you reach a point at which no voltage is present, the problem lies between that point and the last test point with voltage. Most of the time the problem can be traced to a loose connection.

➡Note: Keep in mind that some circuits receive voltage only when the ignition key is in the Accessory or Run position.

FINDING A SHORT

One method of finding shorts in a circuit is to remove the fuse and connect a test light or voltmeter in its place to the fuse terminals. There should be no voltage present in the circuit. Move the wiring harness from side-to-side while watching the test light. If the bulb goes on, there is a short to ground somewhere in that area, probably where the insulation has rubbed through. The same test can be performed on each component in the circuit, even a switch.

GROUND CHECK

Perform a ground test to check whether a component is properly grounded. Disconnect the battery and connect one lead of a selfpowered test light, known as a continuity tester, to a known good ground. Connect the other lead to the wire or ground connection being tested. If the bulb goes on, the ground is good. If the bulb does not go on, the ground is not good.

CONTINUITY CHECK

A continuity check is done to determine if there are any breaks in a circuit - if it is passing electricity properly. With the circuit off (no power in the circuit), a self-powered continuity tester can be used to check the circuit. Connect the test leads to both ends of the circuit (or to the "power" end and a good ground), and if the test light comes on the circuit is passing current properly. If the light doesn't come on, there is a break somewhere in the circuit. The same procedure can be used to test a switch, by connecting the continuity tester to the power in and power out sides of the switch. With the switch turned On, the test light should come on.

FINDING AN OPEN CIRCUIT

When diagnosing for possible open circuits, it is often difficult to locate them by sight because oxidation or terminal misalignment are hidden by the connectors. Merely wiggling a connector on a sensor or in the wiring harness may correct the open circuit condition. Remember this when an open circuit is indicated when troubleshooting a circuit. Intermittent problems may also be caused by oxidized or loose connections.

Electrical troubleshooting is simple if you keep in mind that all elec-trical circuits are basically electricity running from the battery, through the wires, switches, relays, fuses and fusible links to each electrical component (light bulb, motor, etc.) and to ground, from which it is passed back to the battery. Any electrical problem is an interruption in the flow of electricity to and from the battery.

3   Fuses - general information

▶ **Refer to illustrations 3.1a, 3.1b, 3.2 and 3.4**

1   The electrical circuits of the vehicle are protected by a combina-tion of fuses, circuit breakers and fusible links. The fuse block is located under the instrument panel on the driver's side (see illustrations).

2   On 1996 and later models, a fuse and relay block called the Underhood Electrical Center is located inside the engine compartment and mounted on the firewall on the driver's side. The fuse and relay positions are shown on the cover (see illustration).

3   Each of the fuses is designed to protect a specific circuit, and the various circuits are identified on the fuse panel itself.

4   These models use miniaturized fuses. These compact fuses, with blade terminal design, allow fingertip removal and replacement. If an electrical component fails, always check the fuse first. The easiest way to check fuses is with a test light. Check for power at the exposed ter-minal tips of each fuse. If power is present on one side of the fuse but not the other, the fuse is blown. A blown fuse can also be confirmed by visually inspecting it (see illustration).

5   Be sure to replace blown fuses with the correct type. Fuses of different ratings are physically interchangeable, but only fuses of the proper rating should be used. Replacing a fuse with one of a higher or lower value than specified is not recommended. Each electrical circuit needs a specific amount of protection. The amperage value of each fuse is molded into the fuse body.

3.1a The fuse block is located under the instrument panel to the left of the steering column

3.1b The fuse block on 1996 and later models has a flip-open cover with the fuse and circuit breaker identification on it. The turn signal and hazard flasher unit (arrow) is located to the left of the fuse block

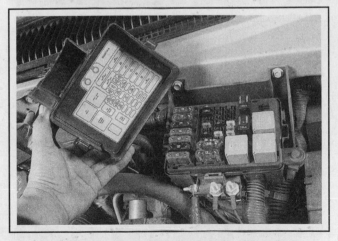

3.2 On 1996 and later models, the Underhood Electrical Center is located inside the engine compartment mounted on the firewall on the driver's side - Fuse and relay circuit identification is located on the underside of the cover

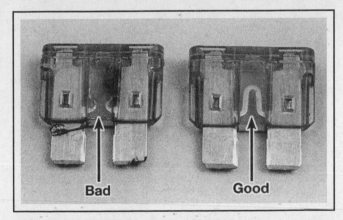

3.4 When a fuse blows, the element between the terminals melts - the fuse on the left is blown, the fuse on the right is good

※※ **CAUTION:**

Never bypass a fuse with pieces of metal or foil. Serious damage to the electrical system could result.

6   If the replacement fuse immediately fails, don't replace it again until the cause of the problem is isolated and corrected. In most cases, this will be a short circuit in the wiring caused by a broken or deteriorated wire.

4   Fusible links - general information

Some circuits are protected by fusible links. The links are used in circuits which are not ordinarily fused, such as the ignition circuit.

Although the fusible links appear to be a heavier gauge than the wire they are protecting, the appearance is due to the thick insulation. All fusible links are four wire gauges smaller than the wire they are designed to protect. The location of the fusible links on your particular vehicle may be determined by referring to the wiring diagrams at the end of this book.

Fusible links cannot be repaired, but a new link of the same size wire can be put in its place. The procedure is as follows:

a) *Disconnect the negative cable from the battery.*

※※ **CAUTION:**

On models equipped with an anti-theft audio system, disable the anti-theft feature before performing any operation that requires disconnecting the battery or disrupting power to the stereo (see the Anti-theft audio system procedures at the front of this manual).

b) *Disconnect the fusible link from the wiring harness.*
c) *Cut the damaged fusible link out of the wiring just behind the connector.*
d) *Strip the insulation back approximately 1/2-inch.*
e) *Position the connector on the new fusible link and crimp it into place.*
f) *Use rosin core solder at each end of the new link to obtain a good solder joint.*
g) *Use plenty of electrical tape around the soldered joint. No wires should be exposed.*
h) *Connect the battery ground cable. Test the circuit for proper operation.*

5   Circuit breakers - general information

Circuit breakers, which are located in the main fuse block, protect accessories such as power windows, power door locks and the rear window defogger.

The headlight wiring is also protected by a circuit breaker. An electrical overload in the system will cause the lights to go off and come on or, in some cases, to remain off. If this happens, check the headlight circuit immediately. The circuit breaker will function normally once the overload condition is corrected. Refer to the wiring diagrams at the end of this book for the location of the circuit breakers in your vehicle.

6   Turn signal and hazard flasher - replacement

▸ **Refer to illustration 6.2**

1   The turn signal and hazard flasher units on these models are located under the instrument panel, near the fuse block.
2   On 1995 and earlier models, separate turn signal and hazard flasher units are used; the hazard flasher unit is located above the fuse block and the turn signal flasher is attached to the trim panel below the hood release handle with a retaining clip (see illustration).
3   On 1996 and later models, a single flasher unit is used to control both systems (see illustration 3.1b).
4   Remove the driver's side knee bolster (see Chapter 11).
5   Locate the flasher unit. If the flasher units are not labeled, activate one of the two systems and you will be able to hear and feel the flasher click on-and-off as it operates. To replace a flasher unit, simply unplug the old one and install the new one in its place.

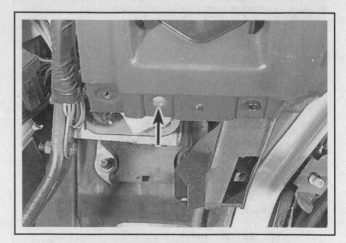

6.2 The turn signal flasher on 1995 and earlier models is located on the trim panel below the hood release handle (arrow)

7 Multi-function switch lever - replacement

✳✳ WARNING:

Most 1993 and later models are equipped with airbags. Always disable the airbag system before working in the vicinity of the impact sensors, steering column or instrument panel to avoid the possibility of accidental deployment of the airbag(s), which could cause personal injury (see Section 20). The yellow wires and connectors routed through the instrument panel are for this system. Do not use electrical test equipment on these yellow wires or tamper with them in any way while working under the instrument panel.

➡Note: The multi-function lever on 1996 and later models is part of the multi-function switch, which is replaced by following the procedure in Section 8.

1992 AND EARLIER MODELS

1 Detach the cable from the negative terminal of the battery.
2 Put the shift lever in Low and place the turn signal lever in the right turn position.
3 On tilt-column models, place the column in the Up position.

4 Locate the multi-function lever electrical connector at the base of the steering column and unplug it.
5 Attach a length of wire to the pigtail to pull the pigtail back through during installation.
6 Pull the lever straight out of the steering column.
7 Pull the multi-function lever pigtail lead up through the steering column. Detach the wire from the pigtail lead and attach it to the lead of the new multi-function lever.
8 Carefully thread the new connector and lead back through steering column.
9 Installation is the reverse of removal.

1993 THROUGH 1995 MODELS

10 Place the lever in the center position (middle detent - turn signals off).
11 Carefully pry the side cover from the steering column.
12 Disconnect the lever wiring harness at the connector near the lever.
13 Pull the lever straight out of the turn signal switch.
14 Installation is the reverse of removal.

8 Steering column switches and ignition key lock cylinder - replacement

▶ Refer to illustrations 8.3, 8.4, 8.5, 8.6, 8.7, 8.12, 8.13 and 8.14

✳✳ WARNING:

Most 1993 and later models are equipped with airbags. Always disable the airbag system before working in the vicinity of the impact sensors, steering column or instrument panel to avoid the possibility of accidental deployment of the airbag(s), which could cause personal injury (see Section 20). The yellow wires and connectors routed through the instrument panel are for this system. Do not use electrical test equipment on these yellow wires or tamper with them in any way while working under the instrument panel.

➡Note 1: A special lock plate spring compressor is required to properly perform this procedure.

➡Note 2: If you have a tilt wheel, place it in the straight position.

1 Disconnect the cable from the negative terminal of the battery and, on models equipped with airbags, disable the airbag (see Section 20).

✳✳ CAUTION:

On models equipped with an anti-theft audio system, disable the anti-theft feature before performing any operation that requires disconnecting the battery or disrupting power to the stereo (see the Anti-theft audio system procedures at the front of this manual).

2 Remove the steering wheel (see Chapter 10). Also remove the column trim plate and lock plate cover (if equipped).

8.3 To release the SIR coil from the steering shaft, remove this snap-ring

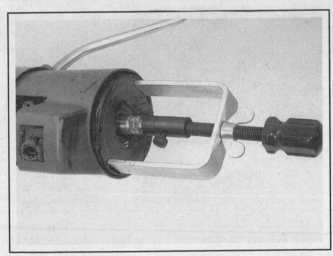

8.4 A special tool is required to depress the steering shaft lockplate so the retaining ring can be removed

8.5 Remove the turn signal canceling cam and spring (arrows)

8.6 Remove the screw and hazard warning knob

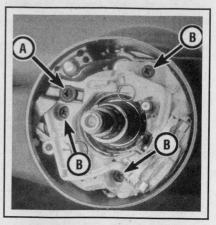

8.7 Place the turn signal switch in the right turn position and remove the lever arm screw (A) and the 3 switch mounting screws (B)

3   On airbag-equipped models, remove the airbag coil snap-ring and remove the coil assembly; let the coil hang by the wiring harness (see illustration). Remove the wave washer.

4   On all models, compress the lock plate with a special tool and pry off the retaining ring (see illustration). Remove the lock plate.

1995 AND EARLIER MODELS

5   Remove the turn signal canceling cam and spring (see illustration).

6   Use a Phillips head screwdriver to remove the hazard warning knob (see illustration).

7   Place the turn signal lever on right turn, remove the turn signal lever arm screw and the 3 switch mounting screws (see illustration).

8   Remove the multi-function switch lever (see Section 7).

9   Remove the driver's side knee bolster (see Chapter 11).

10   Working on the lower portion of the steering column, locate the turn signal harness connector (mounted lengthwise on the column) and detach the connectors from the steering column. If the turn signal switch assembly is to be replaced, disconnect the wiring harness connectors.

11   Pull the turn signal switch assembly out of the steering column just enough to clear the shaft and place it out of the way. If the switch is to be replaced, carefully pull the switch and wiring harness from the steering column. If only the turn signal switch is being replaced, proceed to Step 14.

12   Using needle nose pliers, remove the buzzer switch (see illustration).

➡**Note: Don't lose the small retainer clip that holds the buzzer switch in place. The clip must be reinstalled in its original position.**

13   Remove the lock cylinder retaining screw (see illustration), turn the key to the ON position and withdraw it from the steering column.

14   Installation is basically the reverse of removal, with the following additions:

a   *If the turn signal switch assembly was removed, feed the new switch wiring harness down into the steering column, attach it to the connector and secure it onto the steering column.*

b   *The airbag coil should be centered as shown (see illustration). If it is not centered, center it as described in Section 20.*

8.12 Carefully remove the buzzer switch and clip using needle nose pliers

8.13 Remove the lock cylinder retaining screw (arrow), rotate the key to the On position and pull it out of the steering column

8.14 When properly installed, the airbag coil will be centered with the marks aligned (circle) and the tab fitted between the projections on the top of the steering column (arrow)

1996 AND LATER MODELS

Combination switch

➡Note: The combination switch includes the headlight, dimmer, turn signal, cruise control and hazard switches.

15 Remove the steering column upper and lower covers (see Chapter 11). To do this on earlier models, it may be necessary to lower the steering column. This can be done by first removing the driver's side knee bolster (see Chapter 11), then unscrewing the two nuts that secure the steering column to the bottom of the dash (see illustration 9.5).

16 Cut the cable ties that secure the switch wiring harness to the steering column. Disconnect the electrical connector for the switch that's located on the column.

17 Remove the two screws and separate the combination switch from the steering column.

18 Carefully pull the switch assembly and wiring harness pigtail from the steering column.

19 Installation is the reverse of removal. Use new cable ties to secure the wiring harness to the steering column.

Key lock cylinder

20 Remove the steering column upper and lower covers, as described in Step 15.

21 Hold the ignition key in the Start position.

22 Push a 1/16-inch Allen wrench through the access hole directly above the lock cylinder and press down firmly.

23 Return the key to the Run position and pull the lock cylinder out of the steering column assembly.

24 Installation is the reverse of removal.

Ignition switch

25 Remove the steering column upper and lower covers, as described in Step 15.

26 Remove the alarm switch from the lock assembly by gently prying its retaining clip with a small, flat-bladed screwdriver, then rotating the alarm switch 1/4-turn.

27 Remove the two screws (if applicable) and pull the ignition switch from the steering column assembly.

28 Trace the wiring harness from the switch to the firewall, cutting the cable ties that secure it.

29 Disconnect the electrical connector at the firewall and remove the switch.

30 Installation is the reverse of removal. Be sure to secure the wiring harness to the column with new cable ties.

9 Ignition switch (1995 and earlier models) - replacement

▶ **Refer to illustrations 9.4, 9.5 and 9.7**

➡Note: Ignition switch replacement on 1996 and later models is covered in Section 8.

1 Disconnect the cable from the negative terminal of the battery.

❈❈ CAUTION:

On models equipped with an anti-theft audio system, disable the anti-theft feature before performing any operation that requires disconnecting the battery or disrupting power to the stereo (see the Anti-theft audio system procedures at the front of this manual).

2 Place the ignition switch in the LOCK position. If the key lock cylinder has been removed, pull the actuating rod up until a definite stop can be felt and then move it back down one detent.

3 Remove the driver's side knee bolster (see Chapter 11).

4 On models equipped with an analog (not electronic) instrument cluster, remove the retaining screw and disconnect the shift indicator cable from the steering column shroud (see illustration).

5 Remove the two nuts securing the steering column to the instrument panel (see illustration) and lower it down onto the driver's seat.

6 Remove the two bolts securing the switch to the steering column, withdraw the actuating rod and remove it from the vehicle.

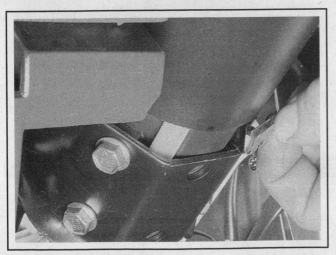

9.4 The instrument cluster shift indicator cable is attached to the steering column shroud

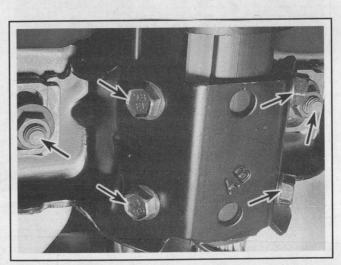

9.5 While supporting the steering column, remove the two nuts (arrows) and lower it from the instrument panel

7   Prior to installation, verify the ignition switch itself is in the LOCK position (see illustration).

➡**Note: If a new switch is being installed, it will be pinned in the LOCK position with a plastic pin; be sure to remove it after tightening the switch mounting bolts.**

8   Insert the actuator rod and position the switch on the steering column and secure it with the retaining bolts. After installation, check for proper switch operation and, if necessary, reposition the switch to achieve proper operation.

9   The remaining installation steps are the reverse of removal. Tighten the steering column support nuts securely.

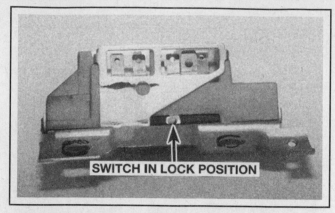

9.7 The ignition switch must be in the Lock position before installation

10 Headlight switch - replacement

▸ Refer to illustrations 10.3a and 10.3b

✳✳ WARNING:

Most 1993 and later models are equipped with airbags. Always disable the airbag system before working in the vicinity of the impact sensors, steering column or instrument panel to avoid the possibility of accidental deployment of the airbag(s), which could cause personal injury (see Chapter 12 for the airbag disabling procedure). The yellow wires and connectors routed through the instrument panel are for this system. Do not use electrical test equipment on these yellow wires or tamper with them in any way while working under the instrument panel.

1   Detach the cable from the negative terminal of the battery.

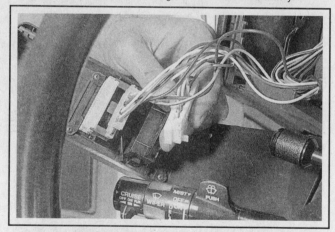

10.3a Unplug the connectors from the headlight switch . . .

✳✳ CAUTION:

On models equipped with an anti-theft audio system, disable the anti-theft feature before performing any operation that requires disconnecting the battery or disrupting power to the stereo (see the Anti-theft audio system procedures at the front of this manual).

2   Remove the instrument cluster trim plate (see Section 17).

3   Unplug the electrical connectors. Remove the screws (see illustrations) or, on later models, release the retaining tabs.

4   Installation is the reverse of removal.

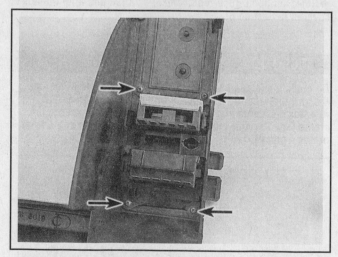

10.3b . . . and remove the screws (arrows)

11 Headlight - replacement

SEALED BEAM

▸ Refer to illustrations 11.3a, 11.3b and 11.4

1   Make sure the headlight switch is in the OFF position.

2   On earlier models, remove the headlight bezel (see Chapter 11).

3   Remove the screws and detach the headlight retaining ring (see illustrations).

4   Unplug the headlight electrical connector and remove the headlight (see illustration).

5   Installation is the reverse of removal.

6   Adjust the headlights (see Section 12).

11.3a Remove only the headlight retaining screws, not the adjusting screws . . .

11.3b . . . and lift off the retaining ring

HALOGEN BULB-TYPE (COMPOSITE)

> ✻✻ **WARNING:**
>
> **Halogen gas-filled bulbs are under pressure and may shatter if the surface is scratched or the bulb dropped. Wear eye protection and handle the bulbs carefully, grasping only the base whenever possible. Do not touch the surface of the bulb with your fingers because the oil from your skin could cause it to overheat and fail prematurely. If you touch the bulb surface, clean it with rubbing alcohol.**

7   Make sure the headlight switch is in the OFF position.

8   Remove the side lens retaining bolt located at the corner of the fender and withdraw it from the fender.

9   Remove the 3 mounting screws and remove the headlight assembly from the core support.

10   Disconnect the bulb electrical connections.

11   Grasp the bulb holder and rotate it counterclockwise to remove it from the headlight assembly.

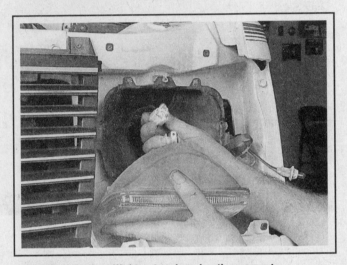

11.4 Pull the headlight out and unplug the connector

12   Installation is the reverse of removal. If necessary, adjust the headlights (see Section 12).

12   Headlights - adjustment

▶ **Refer to illustrations 12.2 and 12.4**

1   It's important that the headlights be aimed correctly. If adjusted incorrectly they could blind an oncoming driver and cause a serious accident or seriously reduce your ability to see the road. The headlights should be checked for proper aim every 12 months and each time a new sealed beam headlight is installed or front end body work is performed.

2   Each sealed beam headlight has two spring-loaded screws, one on the top controlling vertical movement and one on the side controlling horizontal movement (see illustration).

3   On composite type headlights, the hood must be opened to access the horizontal and vertical adjuster screws, which are located on top of each headlight assembly. The adjustment screws are Torx head number T-15.

4   There are several methods of adjusting the headlights. The simplest method requires an empty wall 25 feet in front of the vehicle and a level floor (see illustration).

5   Park the vehicle on a level floor 25 feet from the wall.

6   Position masking tape vertically on the wall in reference to the vehicle centerline and the centerlines of both headlights.

7   Position a horizontal tape line in reference to the centerline of all the headlights.

➡ **Note: It may be easier to position the tape on the wall with the vehicle parked only a few inches away.**

8   Adjustment should be made with the vehicle sitting level, the gas tank half-full and no unusually heavy load in the vehicle.

9   Starting with the Low beam adjustment, position the high intensity zone two inches below the horizontal line and two inches to the right of the headlight vertical line. Adjustment is made by turning the top adjusting screw clockwise to raise the beam and counterclockwise to lower the beam. The adjusting screw on the side should be used in the same manner to move the beam left-or-right.

10   With the High beams on, the high intensity zone should be vertically centered with the exact center just below the horizontal line.

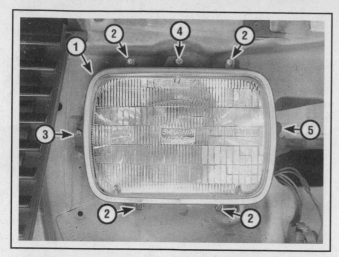

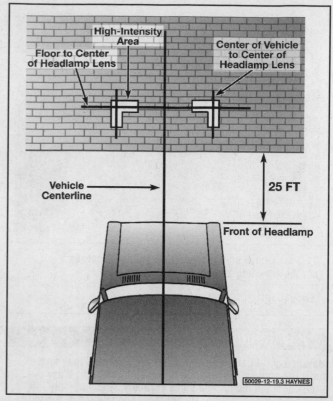

12.2 Headlight ring and adjustment screw locations (sealed beam type)

1   Headlight retaining ring
2   Headlight retaining ring screws
3   Right headlight horizontal adjusting screw
4   Vertical adjusting screw
5   Left headlight horizontal adjusting screw

➡️Note: It may not be possible to position the headlight aim exactly for both High and Low beams. If a compromise must be made, keep in mind that the Low beams are the most used and have the greatest effect on driver safety.

12.4 Headlight aiming details

13 Bulb replacement

1   Detach the cable from the negative terminal of the battery before attempting to replace any of the following bulbs.

❊❊ CAUTION:

On models equipped with an anti-theft audio system, disable the anti-theft feature before performing any operation that requires disconnecting the battery or disrupting power to the stereo (see the Anti-theft audio system procedures at the front of this manual).

FRONT TURN SIGNAL/SIDE MARKER LIGHTS

2   On 1994 and earlier models, remove the headlight bezel (see Chapter 11).

3   On 1995 and later models, remove the 2 mounting screws and pivot the light assembly away from the vehicle.

4   Turn the bulb holders counterclockwise and remove them from the lens assembly.

5   Installation is the reverse of removal.

13.6a Open the rear door, remove the taillight lens screws . . .

13.6b . . . and rotate the lens out of the tabs in the body

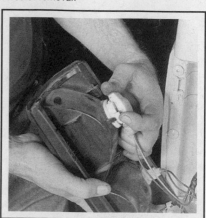

13.7 Turn the bulb holder counterclockwise and withdraw it from the lens assembly

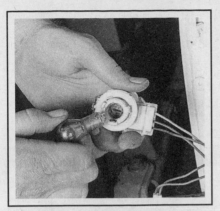

13.8 Push in and turn the bulb to remove it from the holder

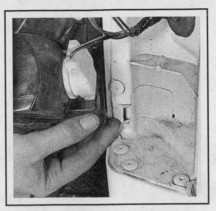

13.9 Fit the taillight lens into the tabs in the body and rotate it into position

13.18 Turn the instrument cluster bulb, then lift it out

REAR TURN SIGNAL AND BACK-UP LIGHTS

▶ **Refer to illustrations 13.6a, 13.6b, 13.7, 13.8 and 13.9**

6 Open the rear doors, remove the screws along the inner edge and rotate the tail light lens out of the tabs in the body (see illustrations).

7 Rotate the bulb holders counterclockwise to remove them from the lens assembly (see illustration).

8 Push the bulb in and turn it counterclockwise to remove it from the holder (see illustration).

9 Installation is the reverse of removal, taking care to align the lens tabs with the body tab recesses (see illustration). If the holder contains grease, don't remove the grease.

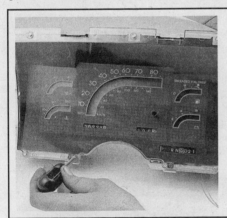

13.19a Remove the cluster front case cover with a nut driver . . .

LICENSE PLATE LIGHT

10 Remove the two bolts and lower the license plate light assembly.

11 Pull the bulb from the light assembly and press a new one into place.

DOME LIGHT

12 Pry off the plastic dome light lens with a small screwdriver.

13 Pull the bulb straight out.

14 Installation is the reverse of removal.

INSTRUMENT CLUSTER LIGHTS

▶ **Refer to illustrations 13.18, 13.19a, 13.19b, 13.19c and 13.19d**

15 Remove the instrument cluster (see Section 16).

16 Turn the instrument cluster upside down and lay it down on a clean work surface or shop rag.

17 Turn the bulb holder counterclockwise and pull it out of the back of the case.

18 Pull the bulb out of the bulb holder (see illustration).

➡**Note: On some models, the instrument cluster bulb and holder are inseparable and must be replaced as an assembly.**

19 On 1992 and earlier models, to remove the upper bulbs, remove the case front and rear covers, unplug the circuit board and pull

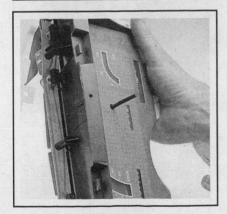

13.19b . . . lift off the back cover . . .

13.19c . . . and unplug the circuit board/bulb holder

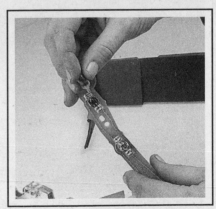

13.19d The bulbs are replaced by pulling them straight out

the bulbs from the holders (see illustrations).

20 Installation is the reverse of removal.

CENTER HIGH-MOUNTED STOP LAMP (CHMSL)

21 The CHMSL is not serviceable and must be replaced as an assembly when bulb failure occurs.

22 On all models except 1996 through 1998, remove the 2 mounting screws, lift up the CHMSL and disconnect the electrical connector.

23 On 1996 through 1998 models, remove the rear door frame garnish molding (see Chapter 11), then locate and disconnect the electrical harness connector under the headliner. Remove the 2 mounting screws and withdraw the CHMSL from the vehicle.

24 Installation is the reverse of removal.

14 Radio and speakers - removal and installation

✳✳ WARNING:

Most 1993 and later models are equipped with airbags. Always disable the airbag system before working in the vicinity of the impact sensors, steering column or instrument panel to avoid the possibility of accidental deployment of the airbag(s), which could cause personal injury (see Section 20). The yellow wires and connectors routed through the instrument panel are for this system. Do not use electrical test equipment on these yellow wires or tamper with them in any way while working under the instrument panel.

1 Disconnect the cable from the negative terminal of the battery prior to performing any of these operations.

✳✳ CAUTION:

On models equipped with an anti-theft audio system, disable the anti-theft feature before performing any operation that requires disconnecting the battery or disrupting power to the stereo (see the Anti-theft audio system procedures at the front of this manual).

RADIO

▶ Refer to illustrations 14.6, 14.7a, 14.7b and 14.8

2 Remove the instrument panel lower extension (see Chapter 11).

3 On 1995 and earlier models, remove the accessory trim panel (see Chapter 11).

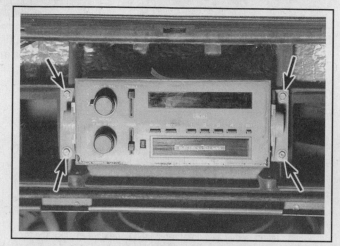

14.6 Remove the radio mounting screws (arrows) (1995 and earlier models)

4 On 1996 and later models, remove the driver's side knee bolster (see Chapter 11).

5 On 1996 and later models, remove the instrument panel cluster trim plate (see Section 16).

6 On 1995 and earlier models, remove the radio mounting bolts and withdraw the radio from the dash (see illustration).

7 On 1996 and later models, the radio unit slides into its mount and is secured by retaining clips located on each side of the unit. Depress the retaining clips and pull the radio out of the dash (see illustrations).

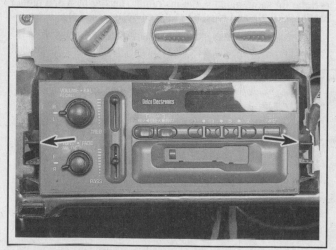

14.7a On 1996 and later models, the radio slides into the mount and is secured on each side by retaining clips (arrows)

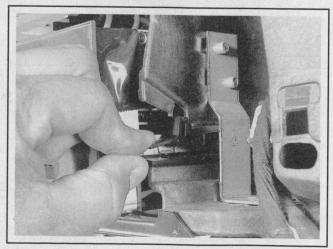

14.7b Depress the retaining clips and pull the radio out of the instrument panel

14.8 Typical radio electrical connections and antenna lead (arrows)

8 Disconnect the antenna lead and wiring harness connectors from the rear of the radio and remove it from the vehicle (see illustration).

9 Installation is the reverse of removal.

REMOTE CASSETTE PLAYER (1996 AND LATER MODELS)

10 Carefully pry the instrument panel lower extension trim plate (and cup holder if equipped) from the lower extension.

11 Depress the retaining clips and pull the cassette player from the instrument panel lower extension. Disconnect the wiring harness connectors from the rear of the unit and remove it from the vehicle.

12 Installation is the reverse of removal.

SPEAKERS

Dashboard

13 Using a screwdriver, carefully pry off the speaker grille.

14 Remove the mounting screws, withdraw the speaker, disconnect the speaker wires and remove it from the vehicle.

15 Installation is the reverse of removal.

Front door (1996 and later models)

Lower speaker

16 Remove the door trim panel (see Chapter 11).

17 Remove the 4 speaker mounting screws, withdraw the speaker, disconnect the speaker wires and remove it from the vehicle.

18 Installation is the reverse of removal.

Upper speaker

19 Using a screwdriver, pry out the 2 trim caps covering the door pull handle mounting screws (see Chapter 11, if necessary).

20 Remove the bolts from the pull handle and accessory switch mount plate.

21 Remove the door pull handle, then slide the door latch handle trim cover forward and remove it from the door panel.

22 Detach the accessory switch mount plate from the door trim panel.

14.24a Carefully pry the rear door speaker grille off with a small screwdriver . . .

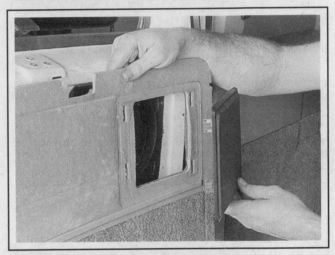

14.24b . . . and remove the door trim panel

23 Disconnect the speaker wires and remove the speaker by rotating it as necessary to align the mounting tabs with the slots in the speaker frame.

Rear door

▶ **Refer to illustrations 14.24a and 14.24b**

24 Remove the speaker grille and the door trim panel (see illustrations).

25 Remove the mounting screws, withdraw the speaker, disconnect the speaker wires and remove it from the vehicle.

➡**Note: On some models the speaker may be secured with rivets. To remove the rivets, drill off the rivet head using the appropriate size drill bit and then punch it out of the hole with a punch and hammer.**

26 Installation is the reverse of removal.

Pillar speakers

27 Using a screwdriver, carefully pry off the speaker grille.

28 Remove the mounting screws, withdraw the speaker, disconnect the speaker wires and remove it from the vehicle.

29 Installation is the reverse of removal.

15 Radio antenna - removal and installation

1 Unscrew the old antenna mast.
2 Install the new antenna and tighten it securely.

16 Instrument cluster - removal and installation

▶ **Refer to illustrations 16.9. 16.10a and 16.10b**

✳✳ **WARNING:**

Most 1993 and later models are equipped with airbags. Always disable the airbag system before working in the vicinity of the impact sensors, steering column or instrument panel to avoid the possibility of accidental deployment of the airbag(s), which could cause personal injury (see Section 20). The yellow wires and connectors routed through the instrument panel are for this system. Do not use electrical test equipment on these yellow wires or tamper with them in any way while working under the instrument panel.

✳✳ **CAUTION:**

The instrument cluster on some models is an Electro-Static Discharge (ESD) sensitive electronic device, meaning a static electricity discharge from your body could possibly damage internal electrical components. Avoid touching the electrical terminals unless absolutely necessary.

1 Disconnect the cable from the negative terminal of the battery prior to performing any of these operations.

✳✳ **CAUTION:**

On models equipped with an anti-theft audio system, disable the anti-theft feature before performing any operation that requires disconnecting the battery or disrupting power to the stereo (see the Anti-theft audio system procedures at the front of this manual).

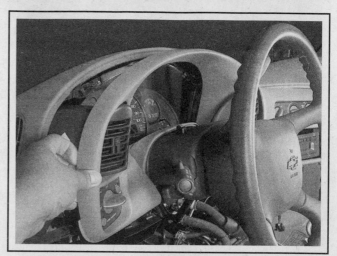

16.9 The instrument trim panel on 1996 and later models uses self-locking fasteners - carefully pry it away with a screwdriver and pull it off by hand

2 On 1992 and earlier models, remove the instrument panel lower and upper trim panel mounting screws. Pull the upper trim panel away from the dashboard, disconnect the electrical connectors and remove it from the vehicle.
3 On 1993 and later models, remove the driver's side knee bolster (see Chapter 11).
4 If applicable, disconnect the gear position indicator cable from the steering column (see illustration 9.4).
5 If equipped, place the tilt steering column in the lowest position.

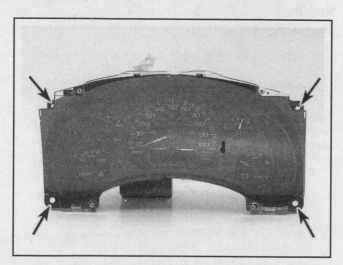

16.10a Instrument cluster mounting screw locations (arrows) - 1998 model shown, others similar

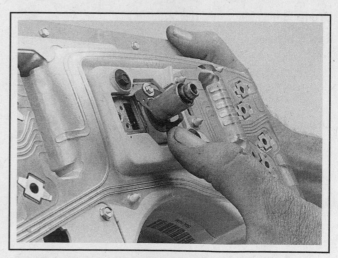

16.10b On 1992 and earlier models, the vehicle speed sensor electrical connector is secured by a screw

6   Apply the parking brake and block the wheels. Turn the ignition key ON and place the gear selector in "1" or "L" as applicable.

➡**Note: On models equipped with brake-transmission interlock, power must be restored to accomplish this operation. After positioning the transmission gear selector in "1" or "L" as applicable, disconnect the cable from the negative battery terminal.**

7   On 1993 to 1995 models, remove the air conditioner and heater controls from the instrument panel (see Chapter 3).

8   On 1993 to 1995 models, remove the instrument trim panel mounting screws, pull it away from the dashboard, disconnect the electrical connectors and remove it from the vehicle.

9   On 1996 and later models, using a screwdriver, carefully pry the instrument cluster trim plate from the dashboard, disconnect the electrical connectors and remove it from the vehicle (see illustration).

10   Remove the four instrument cluster retaining screws, pull it away from the dashboard just enough to disconnect the electrical connectors and speedometer cable (if applicable), then remove it from the vehicle (see illustrations).

11   Installation is the reverse of removal.

17 Speedometer cable - replacement

✳✳ WARNING:

Most 1993 and later models are equipped with airbags. Always disable the airbag system before working in the vicinity of the impact sensors, steering column or instrument panel to avoid the possibility of accidental deployment of the airbag(s), which could cause personal injury (see Section 20). The yellow wires and connectors routed through the instrument panel are for this system. Do not use electrical test equipment on these yellow wires or tamper with them in any way while working under the instrument panel.

➡**Note: 1991 and later models do not have a speedometer cable. The speedometer is driven electronically, using data the vehicle speed sensor (VSS) provides to the PCM.**

1   Detach the negative cable from the battery.

✳✳ CAUTION:

On models equipped with an anti-theft audio system, disable the anti-theft feature before performing any operation that requires disconnecting the battery or disrupting power to the stereo (see the Anti-theft audio system procedures at the front of this manual).

2   Disconnect the speedometer cable at the transmission.

3   Detach the cable from the fasteners in the engine compartment and pull it up enough to provide slack for disconnecting the cable from the speedometer.

4   Remove the instrument cluster screws and disconnect the speedometer cable from the back of the cluster.

5   Remove the cable from the vehicle.

6   Installation is the reverse of removal.

18 Windshield wiper arm - removal and installation

◆ **Refer to illustrations 18.1 and 18.2**

1   Use a screwdriver to pry the release lever away from the windshield wiper arm (see illustration).

2   Lift the wiper arm off, disconnect the washer fluid tube and remove the arm from the vehicle (see illustration).

3   Installation is the reverse of removal.

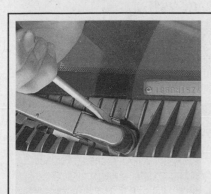

18.1 Pry the lever away from the wiper arm to release it

18.2 Pull the washer fluid connector off the tube

19 Windshield wiper motor - removal and installation

▶ **Refer to illustrations 19.3 and 19.5**

1 Remove the wiper arms (see Section 18).
2 Remove the cowl ventilator grille (see Chapter 11).
3 Remove the plastic wiper motor cavity grille (see illustration).

4 Remove the nut from the wiper motor and disconnect the wiper actuating arms.
5 Unplug the electrical connector, remove the mounting bolts and lift the wiper motor from the engine compartment (see illustration).
6 Installation is the reverse of removal.

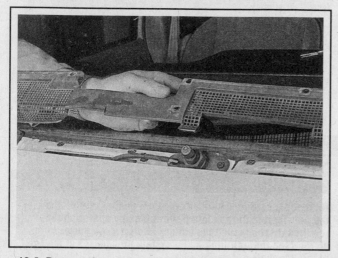

19.3 Remove the retainers and lift the grille out

19.5 Unplug the connector and remove the wiper motor bolts

20 Airbag - general information

DESCRIPTION

1 Most 1993 and later models are equipped with a Supplemental Inflatable Restraint (SIR) system, more commonly known as airbags. This SIR system is designed to protect the driver and, depending on year and model, also the passenger, from serious injury in the event of a head-on or frontal collision up to 30 degrees of the centerline of the vehicle. The driver's side airbag is located in the center of the steering wheel and, if equipped, the passenger airbag is mounted on the dashboard directly in front of the passenger seat. Besides the airbag(s), the system includes two forward discriminating (impact) sensors mounted on each frame rail near the front, an arming sensor which is centrally located under the vehicle and a Diagnostic Energy Reserve Module (DERM) (1995 and earlier models) or a Sensing and Diagnostic Module (SDM) (1996 and later models) located inside the passenger compartment. The DERM or SDM contains the circuitry that supplies power to the airbag modules in the event of a collision, even if battery power is disconnected. On 1995 and earlier models, the DERM stores enough voltage to deploy the airbags for 2 minutes after power has been disconnected. On 1996 and later models, the SDM can deploy the airbags for up to 10 minutes after power is lost.

2 The system is equipped with a self-diagnostic function. Each time the ignition key is turned ON, or the vehicle is started, the AIRBAG warning light in the instrument panel will flash seven times as the system checks itself. If the system is functioning properly, the light will flash seven times and then go out and remain off. The DERM or SDM constantly monitors the system and will illuminate the AIRBAG warning light if any problems are detected. If the AIRBAG warning light stays on after starting the engine or illuminates while driving, a fault has been detected that must be corrected for the system to operate properly.

When a problem is indicated, first check the airbag system fuse and replace if necessary. If the fuse is OK, take the vehicle to your local dealer service department as soon as possible - your safety depends on it.

3 The airbag system uses bright-yellow wires and connectors that you will see routed through the steering column and under the dash. Do not disturb these wires (except to disable the system) and, in particular, do not perform continuity checks with a self-powered test light, as this could cause the airbags to deploy and possibly cause personal injury.

4 Whenever working in the vicinity of the system sensors, steering wheel or the passenger's side of the instrument panel (on models equipped with a passenger side airbag), always disable the system as outlined below.

DISABLING THE AIRBAG SYSTEM

▶ **Refer to illustration 20.8**

5 Turn the steering wheel to the straight ahead position, place the ignition switch in LOCK and remove the key.
6 Remove the driver's side knee bolster (see Chapter 11). If equipped with a passenger side air bag, remove the passenger side knee bolster.
7 Remove the airbag system fuse from the fuse block (see Chapter 12).
8 Locate and disconnect the bright yellow Connector Position Assurance (CPA) connector leading from the airbag module (see illustration). For the driver's side airbag, the CPA is located at the bottom of the steering column. For the passenger's side airbag, the CPA is located under the instrument panel, near the engine cover.

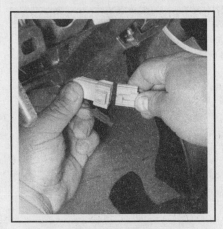

20.8 The airbag Connector Position Assurance (CPA) connector is easily recognized by its bright yellow color (driver's side shown)

20.14a To release the SIR coil from the steering shaft, remove this snap-ring . . .

20.14b . . . then remove the coil

ENABLING THE SIR SYSTEM

9   Connect the CPA connector(s) and install the knee bolster(s).

10 Install the airbag system fuse in the fuse block and connect the cable to the negative terminal of the battery.

11 Turn the ignition key On (engine OFF) and verify the AIRBAG warning light flashes seven times, then goes out. This indicates the system is functioning properly.

CENTERING THE SIR COIL

◆ Refer to illustrations 20.14a, 20.14b, 20.15 and 20.17

12 Anytime some part of the steering system is disassembled for service or replacement, the steering column should be immobilized to ensure that the SIR coil doesn't become uncentered (moved). This can occur, for example, if the steering column is sep-

arated from the steering gear, or if the centering spring is pushed down, allowing the hub to rotate while the coil is removed from the steering column. If the coil becomes accidentally uncentered, re-center it as follows before reassembling the steering system:

13 Make sure that the wheels are pointed straight ahead.

14 Remove the coil assembly snap-ring (see illustration) and remove the coil assembly (see illustration).

15 Holding the coil assembly with its bottom side facing up, depress the spring lock (see illustration) and rotate the hub in the direction of the arrow until it stops (the arrow is on the back of the coil assembly). The coil ribbon should be wound up snug against the center hub.

16 Rotate the coil hub in the opposite direction two and one half turns, then release the spring lock. The coil is now centered.

17 Install the SIR coil and secure it with the snap-ring. The tab will be at the top and the marks aligned when the coil is properly installed (see illustration).

20.15 To center the SIR coil, hold it with its underside facing up, depress the spring lock and rotate the hub in the direction of the arrow on the coil assembly until it stops

20.17 When properly installed, the airbag coil will be centered with the marks aligned (circle) and the tab fitted between the projections on the top of the steering column (arrow)

21 Horns - check and replacement

1   The horns are mounted on the radiator core support behind the radiator grille.

2   If the horns do not sound, first check the fuse (see Section 3), and replace if necessary.

3   Remove the radiator grille (see Chapter 11).

4   To check horn operation, use fused jumper wires to connect battery voltage to the horn electrical terminals. The horn should sound; if it doesn't replace it.

5   Using a voltmeter or test light, verify battery voltage is reaching the horn electrical connector when the horn pad is depressed. If no voltage is present, check the circuit and wires for obvious damage or a short.

➡Note: On models equipped with airbags, have the steering wheel horn connections checked by your local dealer service department or other qualified repair shop.

6   To replace a horn, remove the radiator grille (see Chapter 11).

7   Disconnect the electrical connector, remove the mounting bolt and remove it from the vehicle.

8   Installation is the reverse of removal.

22 Power mirror control system - description and check

1   Electric exterior rear view mirrors use two motors to move the glass; one for up-and-down, and one for left-to-right adjustments.

2   The control switch has a selector portion which energizes the left or right mirror. With the ignition On (engine OFF), roll down the windows and operate the mirror control switch through all the functions.

3   Listen carefully for the sound of the motors operating inside the mirror housing.

4   If the motors can be heard but the mirror doesn't move, there's probably a problem in the drive mechanism inside the mirror. Carefully pop out the glass and examine the drive mechanism. Repair or replace as necessary.

5   If you don't hear the motors, check the fuse (see Section 3). Replace if necessary and check the operation again.

6   If the fuse is OK, remove the mirror control switch from its mounting without disconnecting the wires from it. Turn the ignition switch ON (engine OFF) and check for battery voltage at the switch. There should be voltage at one of the terminals. If no voltage is present, check the wires and circuit for obvious damage or a short.

7   If voltage is present at the switch, check for continuity at all switch positions. If the switch doesn't have continuity in all positions, replace it.

8   Re-connect the wiring to the switch. Locate the wire going from the switch to ground. Use a jumper wire to ground this terminal and check mirror operation again. If they now operate, repair the faulty ground connection.

9   If the mirror still doesn't operate, remove the door trim panel (see Chapter 11) and, using a test light or voltmeter, check for battery voltage at the mirror electrical connector when the switch is actuated. Check operation with the ignition switch ON (engine OFF) and the mirror selector switch on the appropriate side. There should be battery voltage at one of the wires with the switch in each position (except the neutral "off" position).

10 If no voltage is present, check the wires for obvious damage or a short and repair as necessary.

➡Note: It's common for wires to break in the portion of the harness between the body and door (opening and closing the door fatigues and eventually breaks the wires).

11 If voltage is present, replace the mirror assembly (see Chapter 11).

23 Power door lock system - description and check

1   Power door lock systems are operated by bi-directional solenoids located in the doors. The lock switches have two operating positions: Lock and Unlock. These switches activate a relay which in turn connects voltage to the door lock solenoids. Depending on which way the relay is activated, it reverses polarity, allowing the two sides of the circuit to be used alternately as the feed (positive) and ground side.

2   Always check the fuse or circuit breaker first (see Section 3).

3   Operate the door lock switches in both directions (Lock and Unlock) with the engine off. Listen for the faint click of the relay operating.

4   If there's no click, check for voltage at the switches. If no voltage is present, check the wiring between the fuse panel and the switches for shorts and opens.

5   If voltage is present but no click is heard, use an ohmmeter or self-powered test light to check for continuity between the switch terminals with the switch in each position. Replace the switch if there's no continuity in either position.

6   If the switch has continuity but the relay doesn't click, check the wiring between the switch and relay for continuity. Repair the wiring if there's no continuity.

7   If the relay is receiving voltage from the switch but is not sending voltage to the solenoids, check for a bad ground at the relay case. If the relay case is grounding properly, replace the relay.

8   If all but one lock solenoid operates, remove the trim panel from the affected door (see Chapter 11) and check for voltage at the solenoid while the lock switch is operated. One of the wires should have voltage in the Lock position; the other should have voltage in the unlock position.

9   If the inoperative solenoid is receiving voltage, replace the solenoid.

10 If the inoperative solenoid isn't receiving voltage, check for an open or short in the wire between the lock solenoid and the relay.

➡Note: It's common for wires to break in the portion of the harness between the body and door (opening and closing the door fatigues and eventually breaks the wires).

24 Power window system - description and check

1 The power window system consists of the control switches, the motors, glass mechanisms (regulators), and associated wiring.

2 Power windows are wired so they can be lowered and raised from the master control switch by the driver or by remote switches located at the individual windows. Each window has a separate motor which is reversible. The position of the control switch determines the polarity and therefore the direction of operation. The system is equipped with a relay that controls current flow to the motors.

3 The power window system operates only when the ignition switch is in the ON position. In addition, these models have a window lockout switch located at the master control switch that, when depressed, deactivates the other remote window switches, giving the driver complete window control. When checking window operation, check the position of this switch first.

4 These procedures are general in nature, so if you can't find the problem using them, take the vehicle to a dealer service department or other qualified repair shop.

5 If the power windows don't work at all, check the fuse or circuit breaker.

6 Check the wiring between the switches and fuse panel for continuity. Repair the wiring, if necessary.

7 If only one window is inoperative from the master control switch, try the other control switch at the window.

➡**Note: This doesn't apply to the driver's door window.**

8 If the same window works from one switch, but not the other, check the switch for continuity. Use an ohmmeter or self-powered test light to check for continuity between the switch terminals with the switch in each position. Replace the switch if there's no continuity in either position.

9 If the switch tests OK, check for a short or open in the wiring between the affected switch and the window motor.

➡**Note: Its common for wires to break in the portion of the harness between the body and door (opening and closing the door fatigues and eventually breaks the wires).**

10 If one window is inoperative from both switches, remove the trim panel from the affected door (see Chapter 11) and check for voltage at the motor while the switch is operated.

11 If voltage is reaching the motor, disconnect the glass from the regulator (see Chapter 11). Move the window up and down by hand while checking for binding and damage. Also check for binding and damage to the regulator. If the regulator is not damaged and the window moves up and down smoothly, replace the motor (see Chapter 11, Section 27). If there's binding or damage, lubricate, repair or replace parts, as necessary.

12 If voltage isn't reaching the motor, check the wiring in the circuit for continuity between the switches and motors. Check that the relay is grounded properly and receiving voltage from the switches. Also check that the relay sends voltage to the motor when the switch is turned on. If it doesn't, replace the relay.

13 Test the windows after you are done to confirm proper repairs.

25 Wiring diagrams

Since it isn't possible to include all wiring diagrams for every year covered by this manual, the following diagrams are those that are typical and most commonly needed.

Prior to troubleshooting any circuits, check the fuse and circuit breakers (if equipped) to make sure they're in good condition. Make sure the battery is properly charged and check the cable connections (see Chapter 1).

When checking a circuit, make sure that all connectors are clean, with no broken or loose terminals. When disconnecting a connector, do not pull on the wires. Pull only on the connector housing itself.

Refer to the accompanying chart and legend for the wire color codes applicable to your vehicle.

Specifications

Bulb application	Number
Dome light	211-2
Brake/tail lights	2057
Front parking/turn signal lights	
1994 and earlier	2057
1995 and later	3157NA
Headlights	
Sealed beam	
1991 and earlier	6014
1992 and later	6052 or H6054
Composite	
Low beam	9006
High beam	9005
Side marker lights	194
Back-up lights	1156

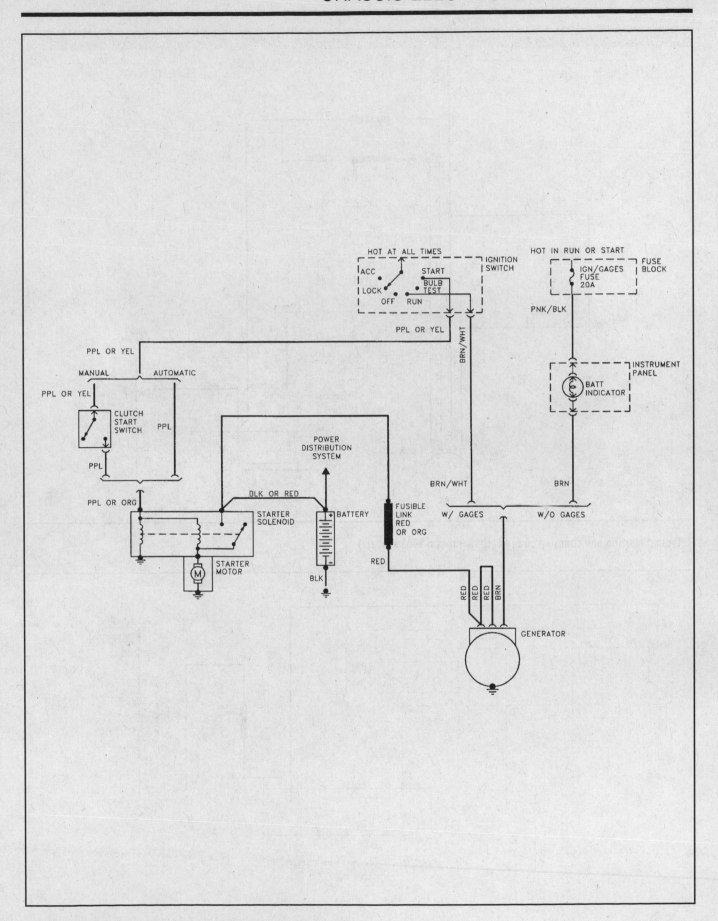

Typical starting and charging circuit (1985 through 1988 models)

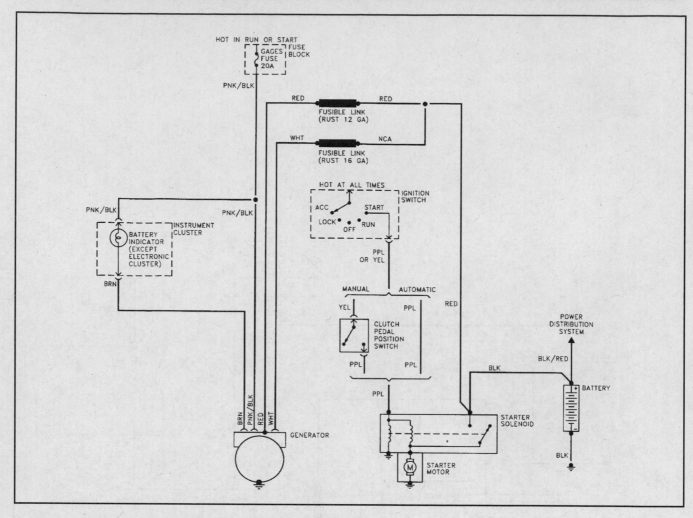

Typical starting and charging circuits (1989 through 1993 models)

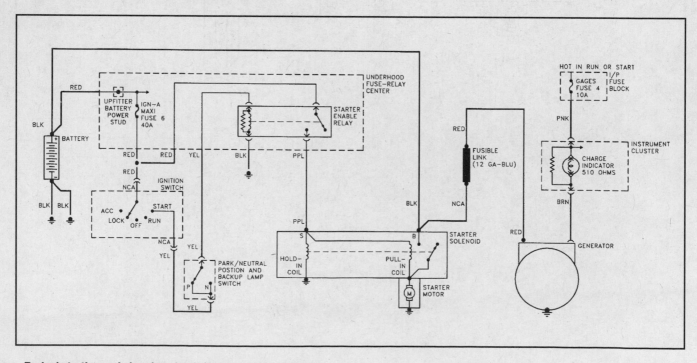

Typical starting and charging circuit (1994 and later models)

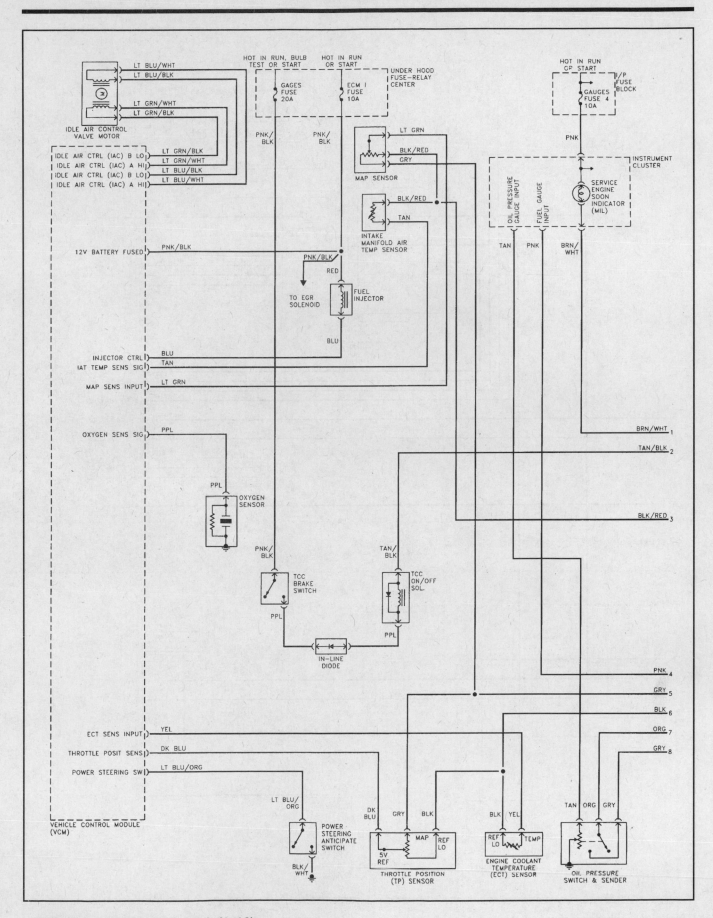

Typical four-cylinder engine controls (1 of 2)

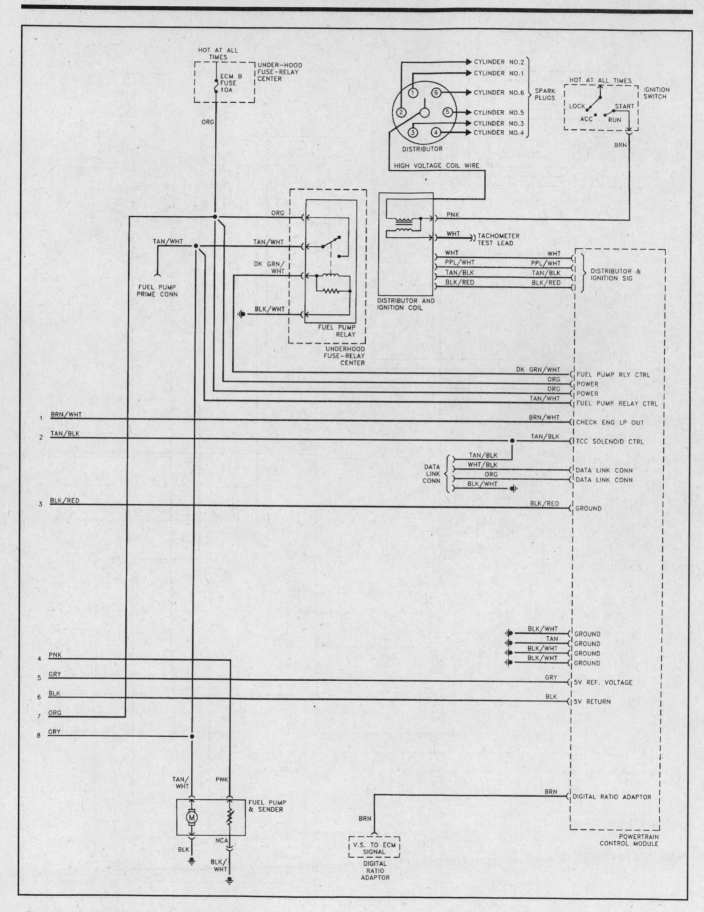

Typical four-cylinder engine controls (2 of 2)

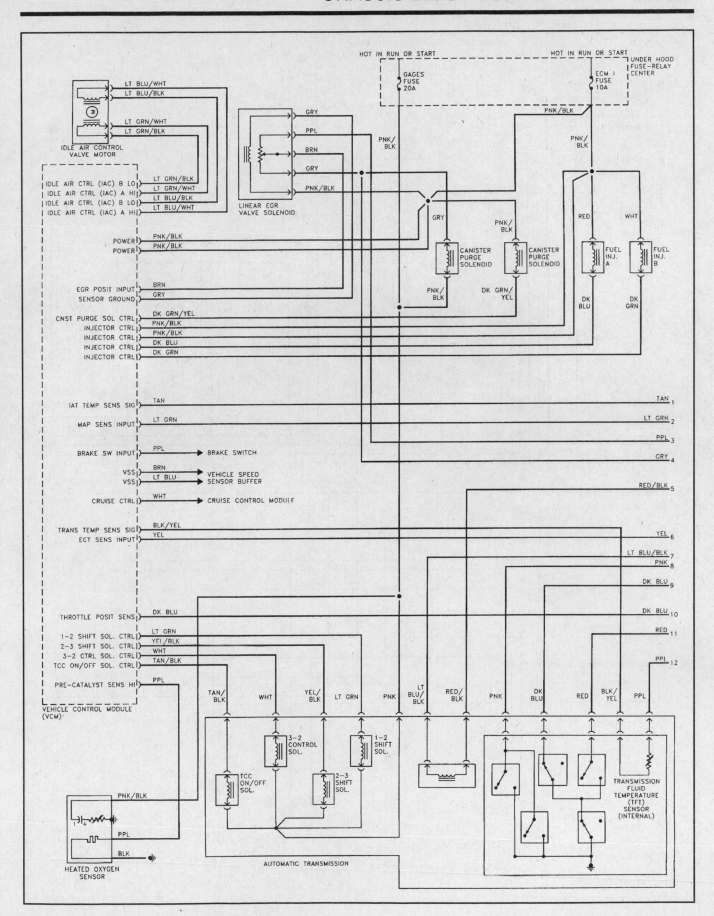

Typical 1993 and earlier V6 engine controls (1 of 3)

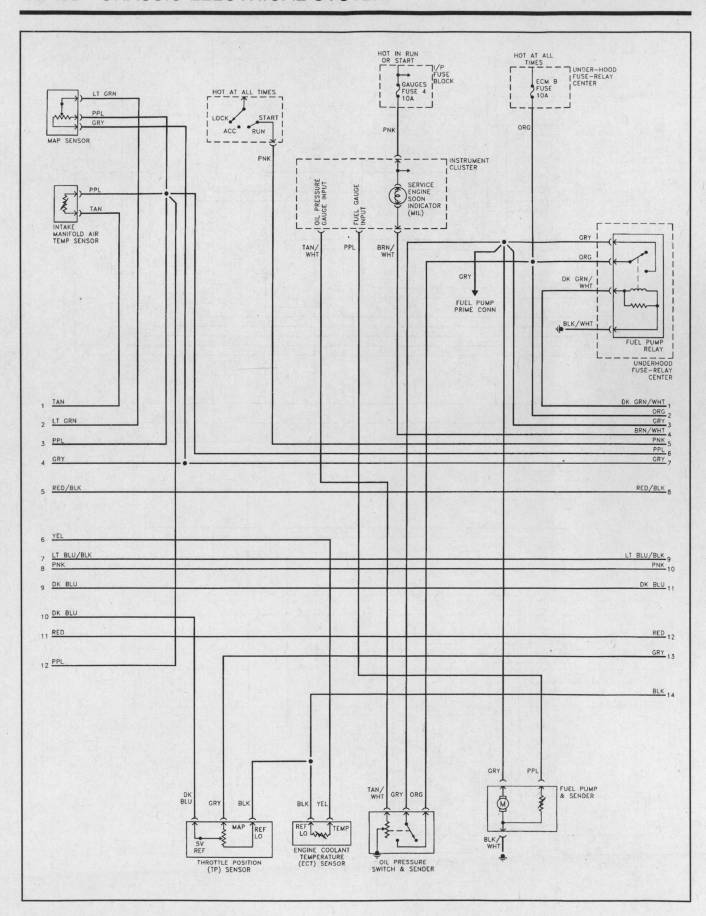

Typical 1993 and earlier V6 engine controls (2 of 3)

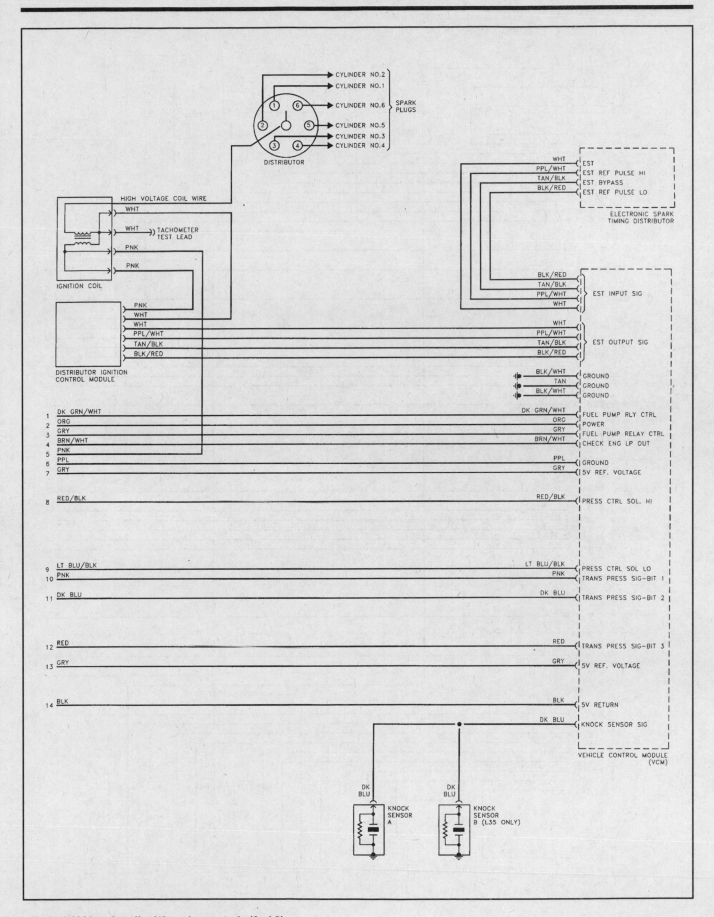

Typical 1993 and earlier V6 engine controls (3 of 3)

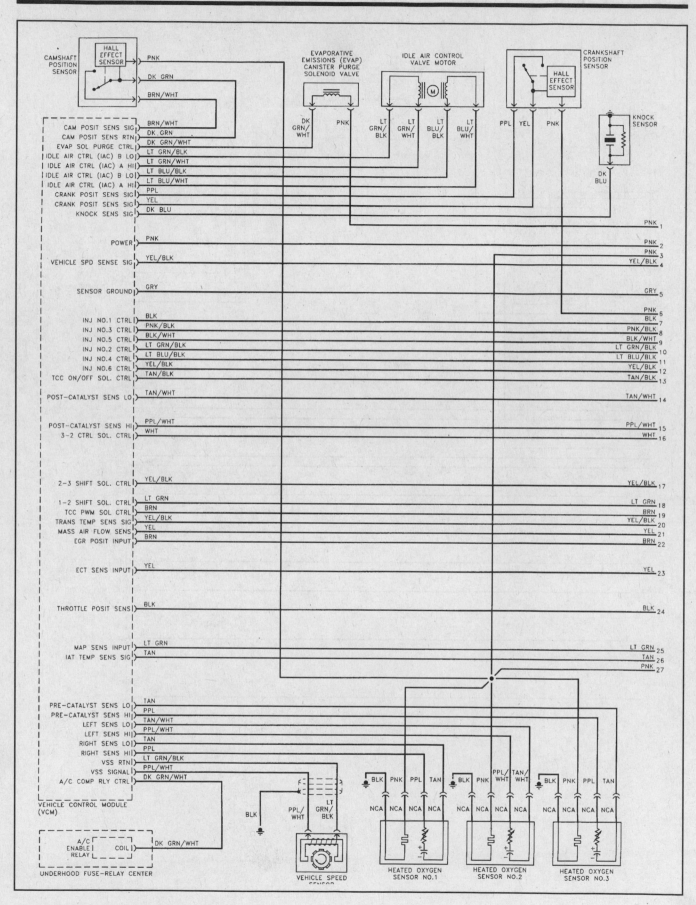

Typical 1994 and later V6 engine controls (1 of 4)

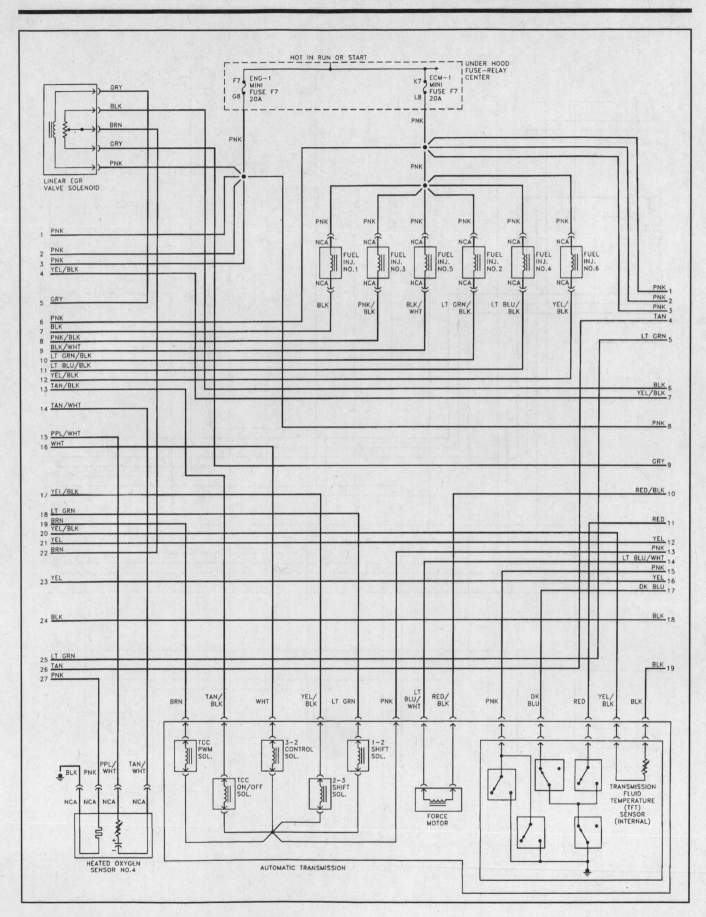

Typical 1994 and later V6 engine controls (2 of 4)

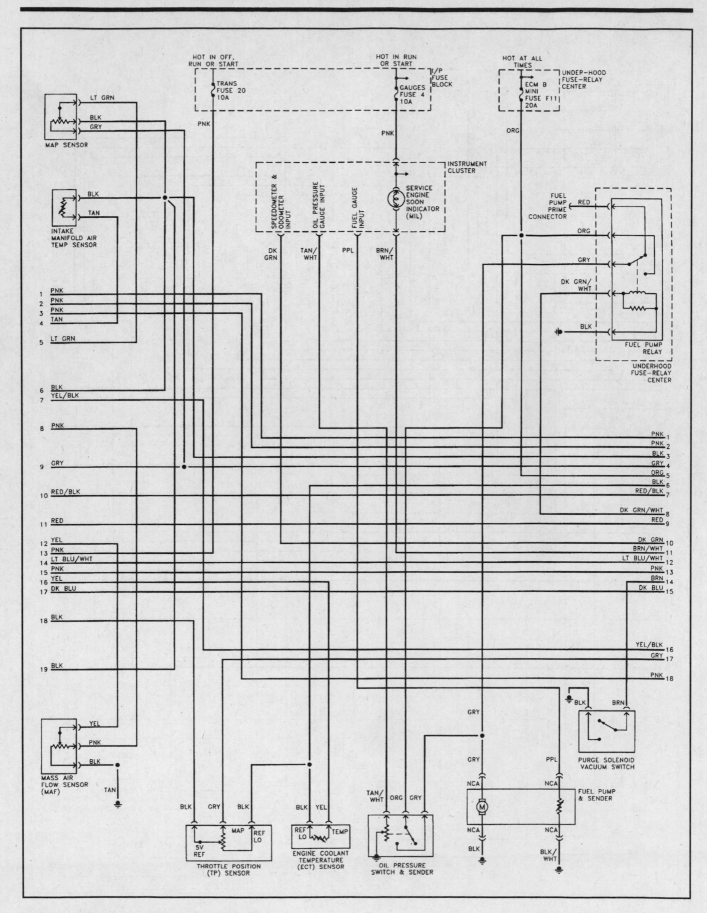

Typical 1994 and later V6 engine controls (3 of 4)

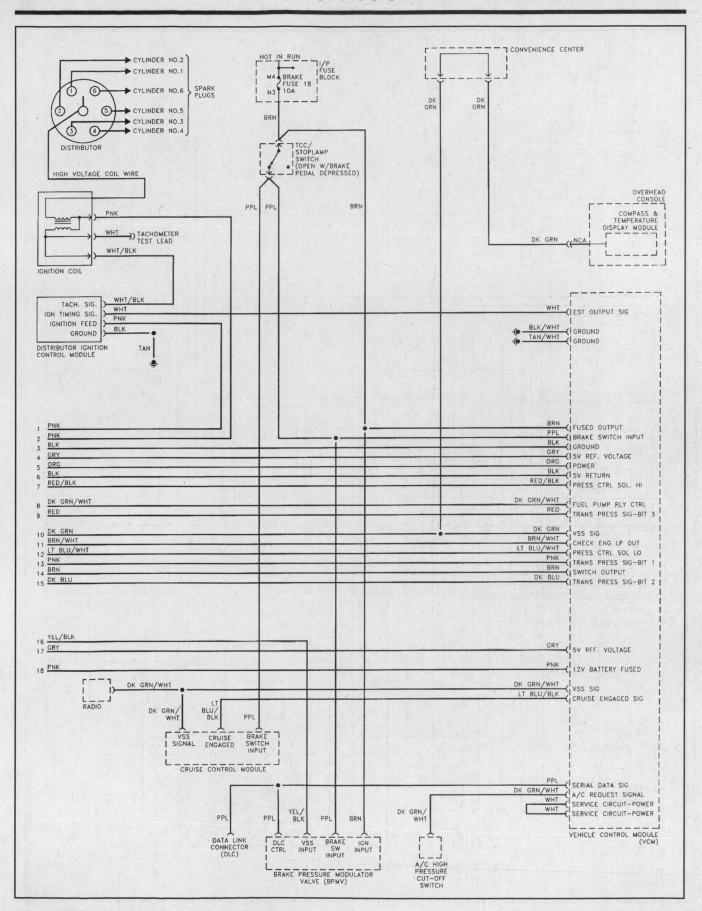

Typical 1994 and later V6 engine controls (4 of 4)

Typical exterior lighting circuit

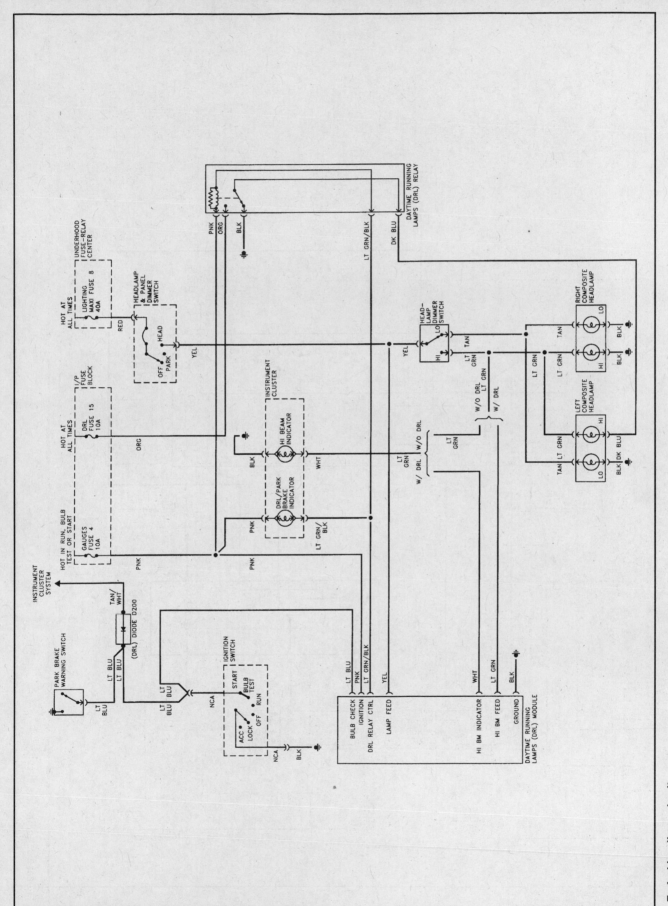

Typical headlamp circuit

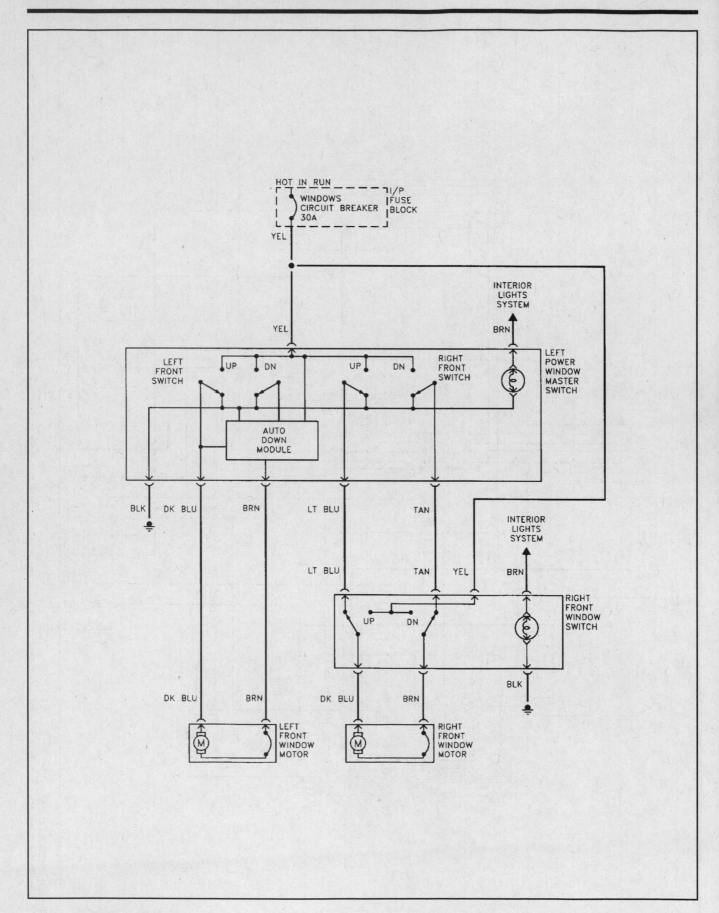

Typical power window circuit

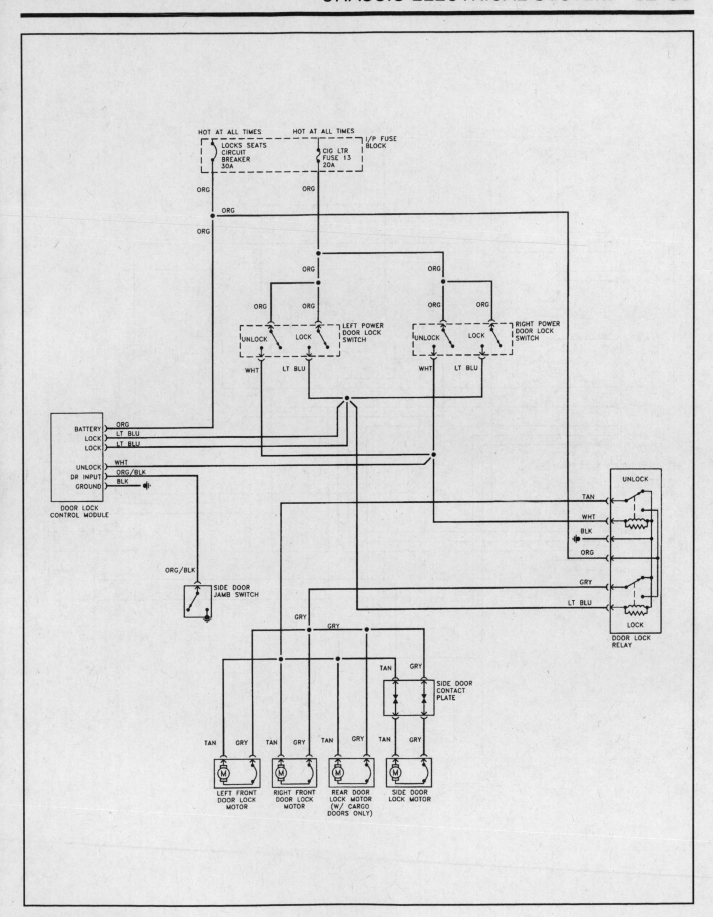

Typical power door lock circuit

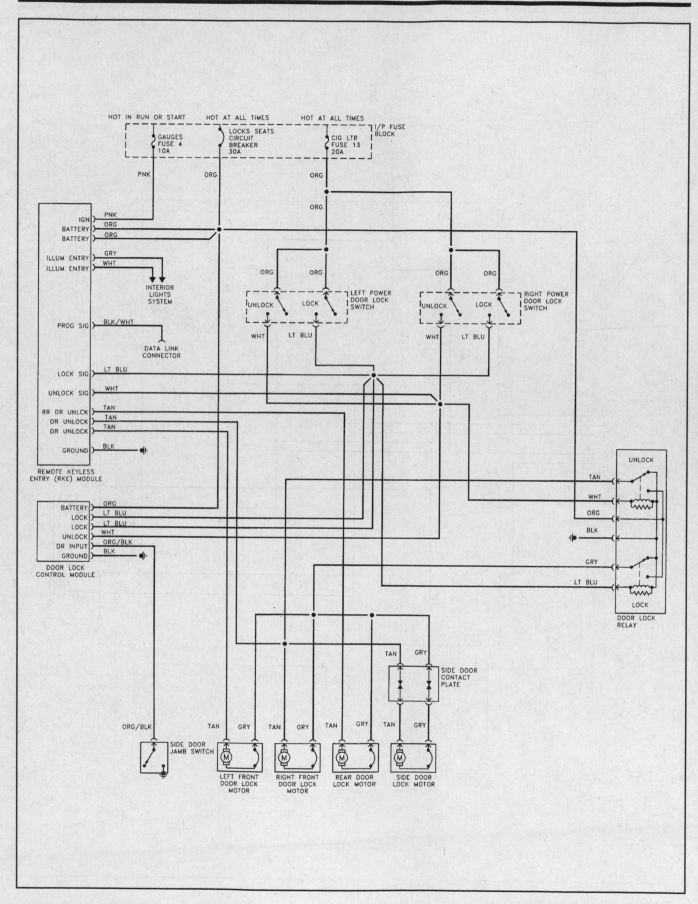

Typical keyless entry circuit

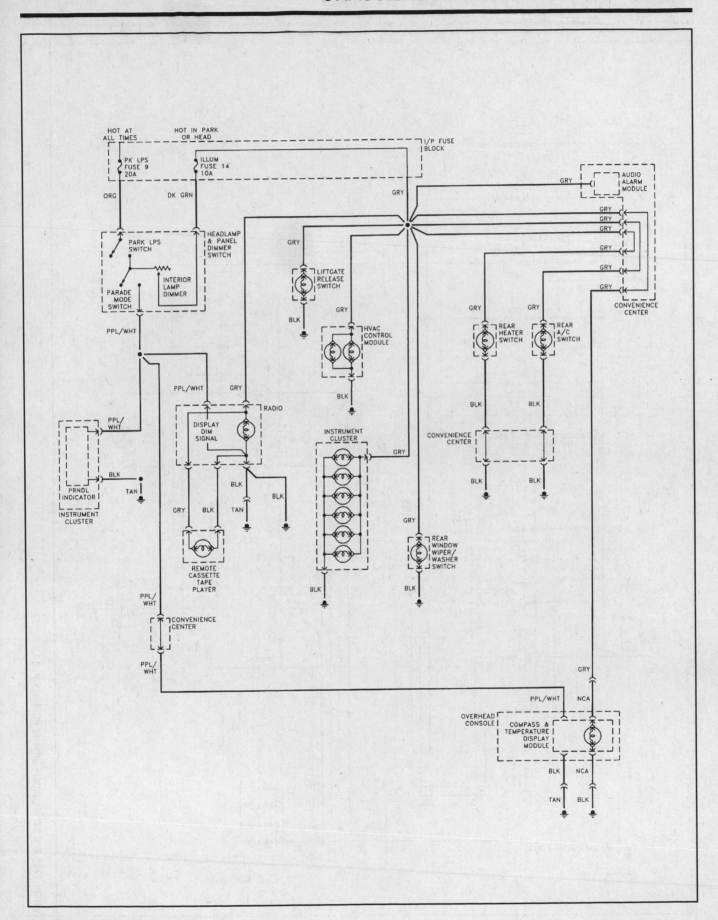

Typical instrument illumination circuit

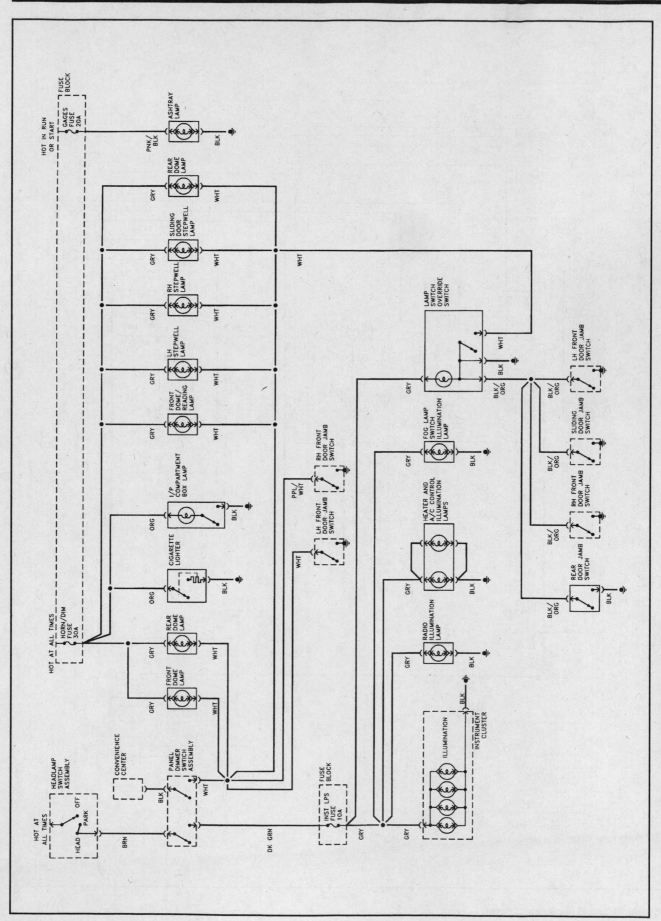

Typical 1994 and earlier model interior lighting circuit

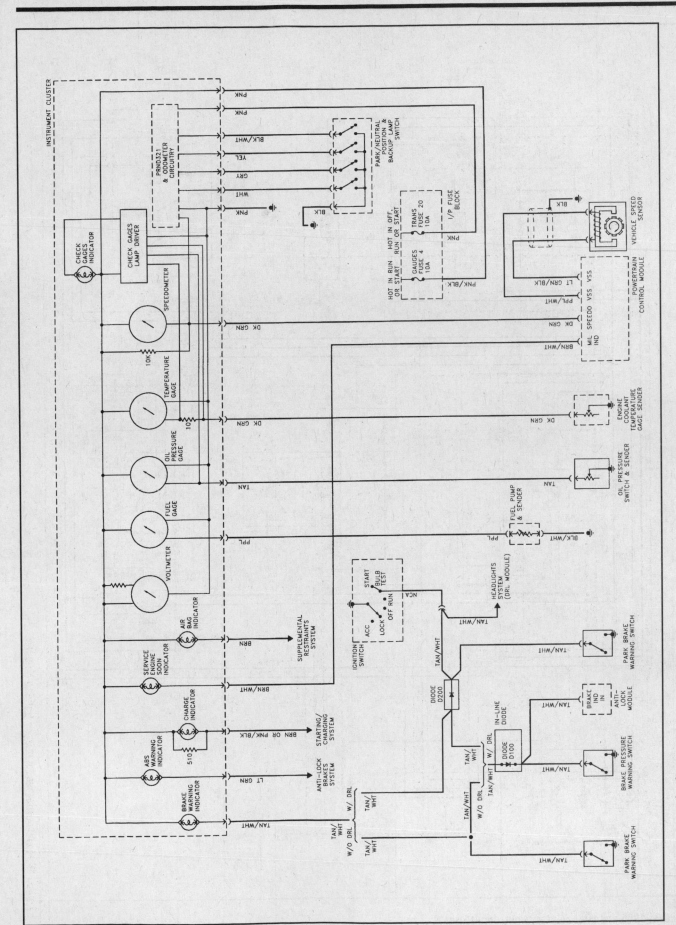

Typical warning lamp circuit

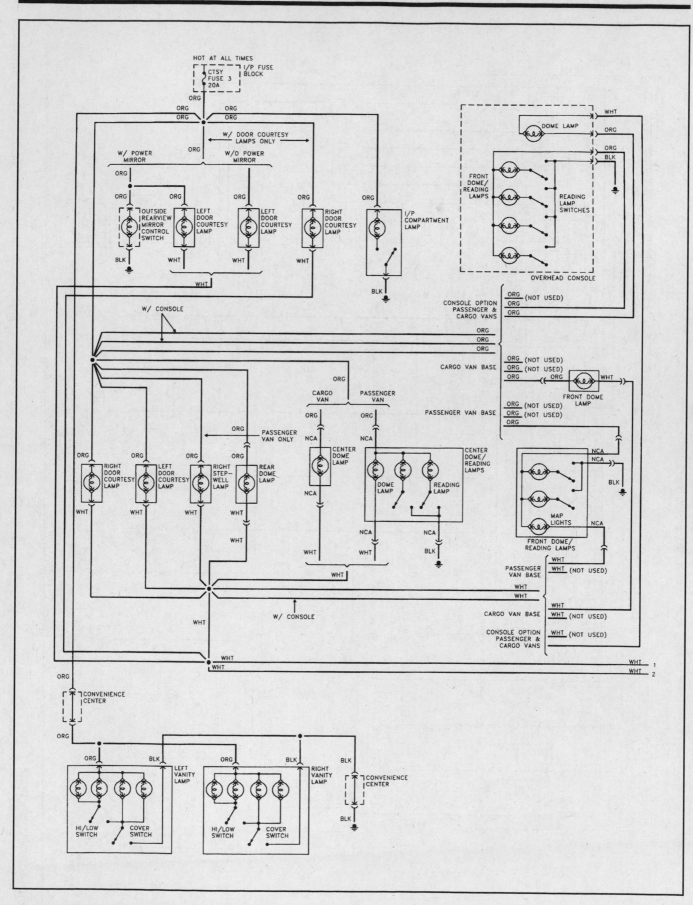

Typical 1995 and later courtesy light circuit (1 of 2)

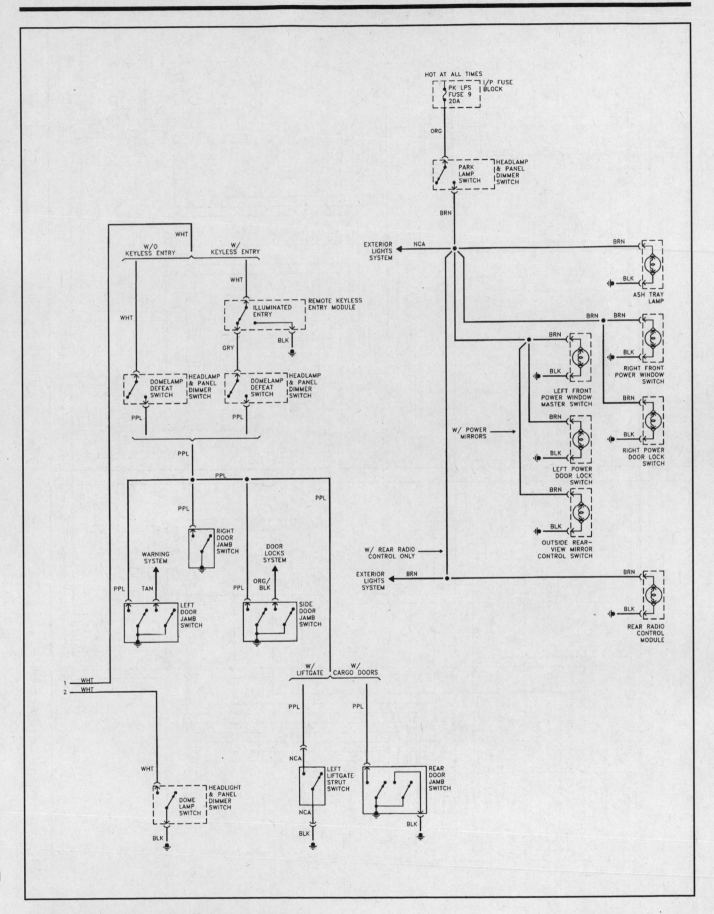

Typical 1995 and later courtesy light circuit (2 of 2)

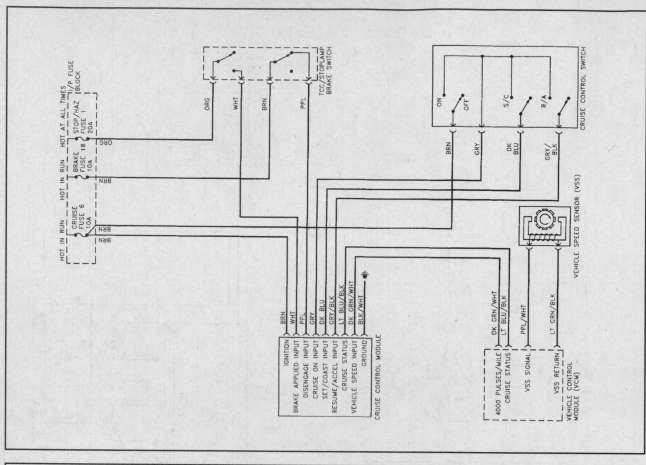

Typical 1995 and later cruise control circuit

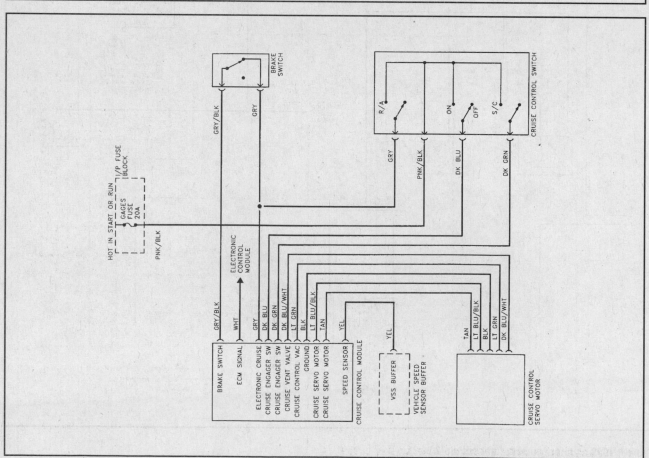

Typical 1994 and earlier model cruise control circuit

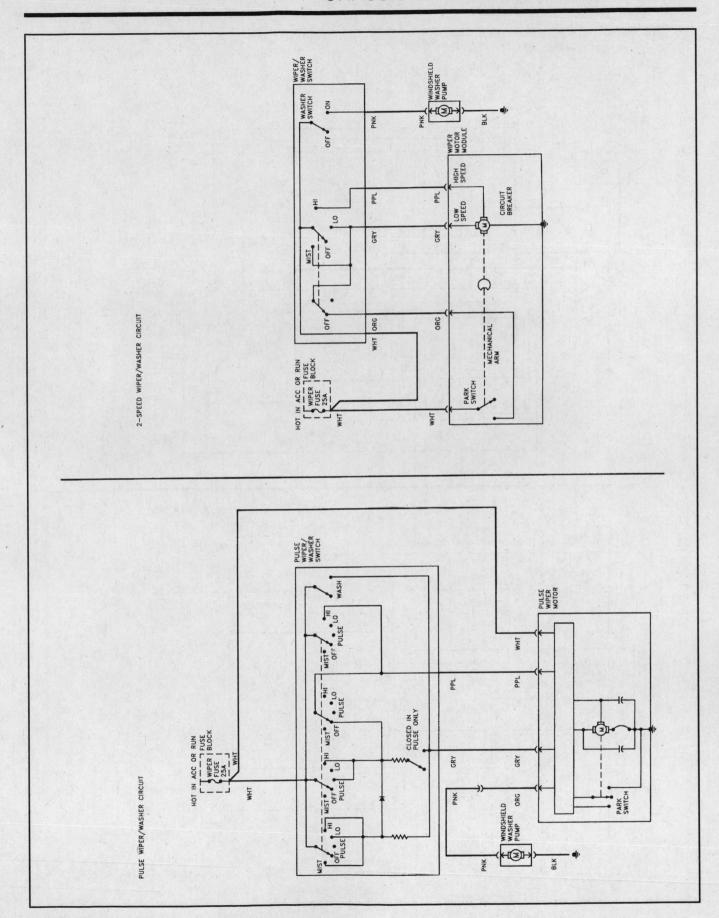

Typical 1990 and earlier model windshield wiper/washer circuit

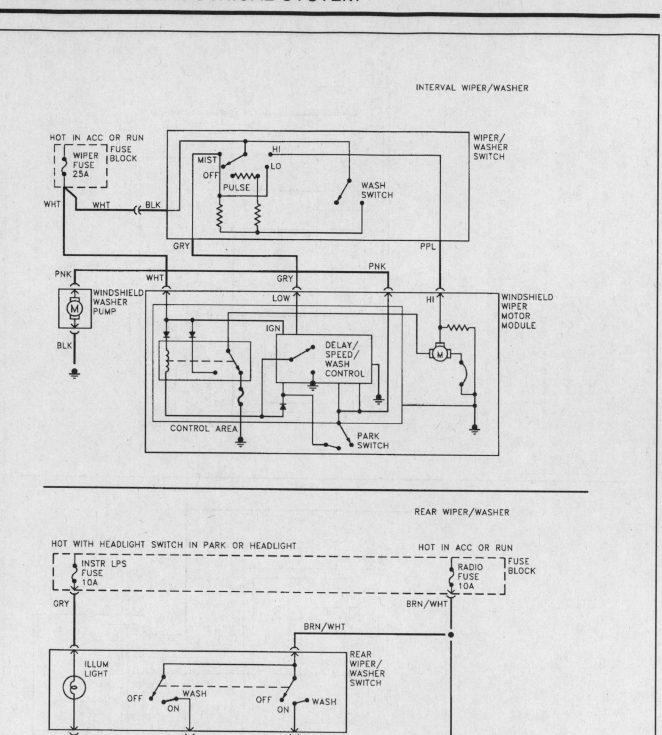

Typical 1991 through 1994 model windshield wiper/washer circuit

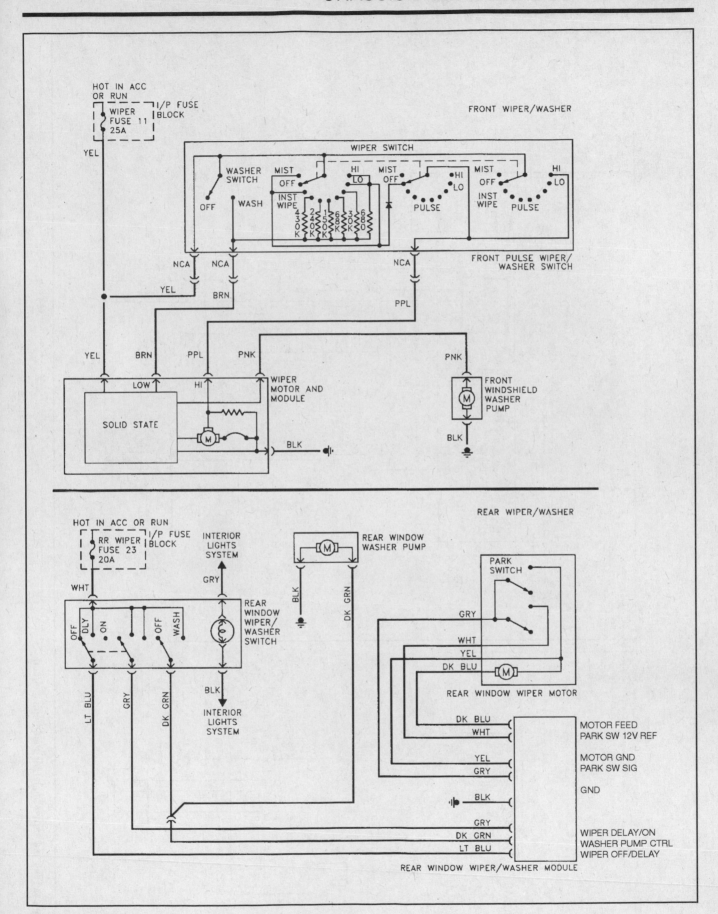

Typical 1995 and later model windshield wiper/washer circuit

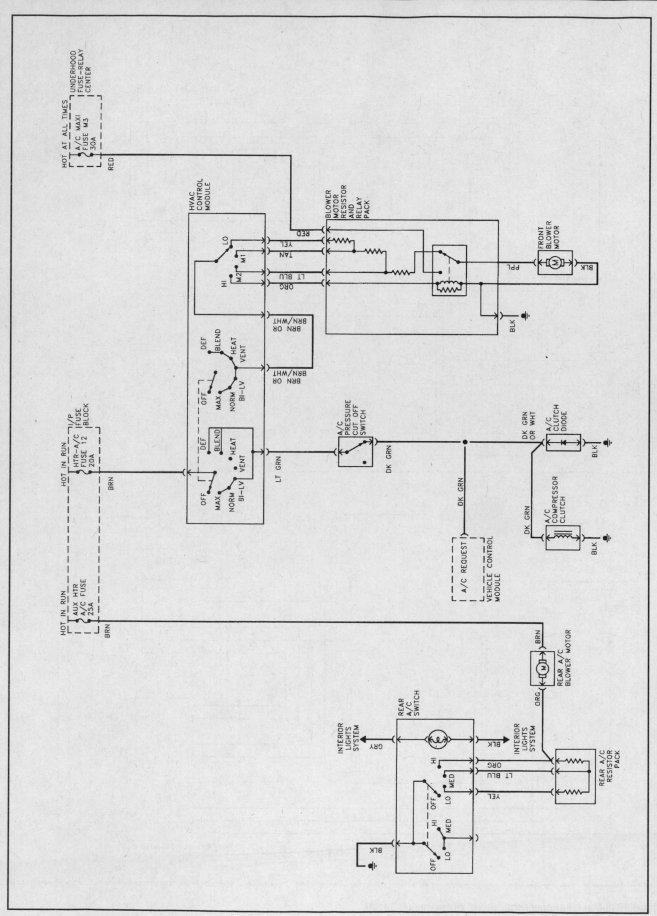

Typical 1994 and earlier model air conditioning circuit

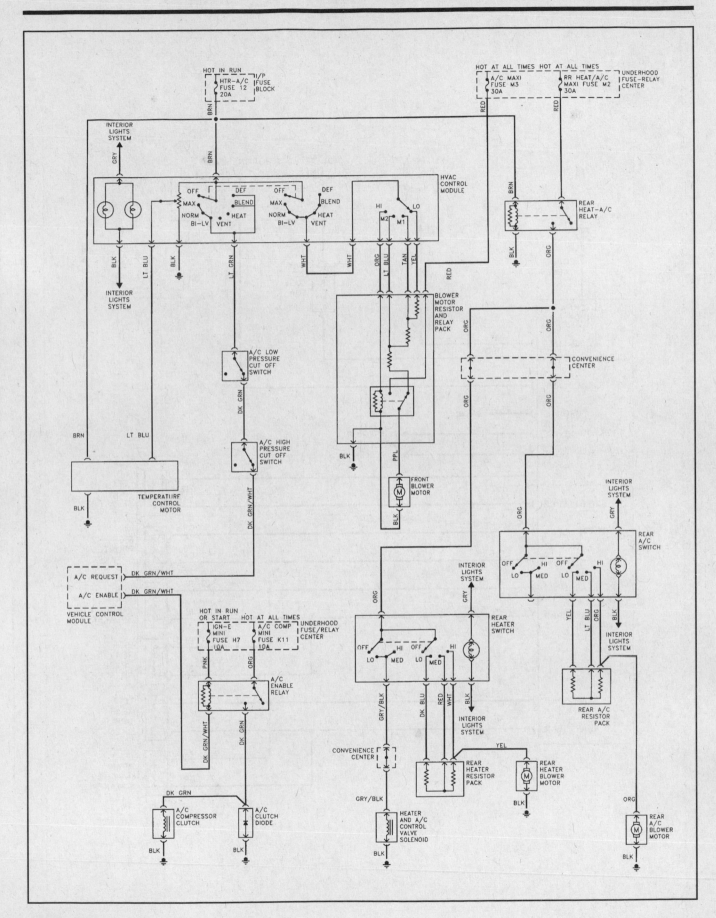

Typical 1995 and later model air conditioning circuit

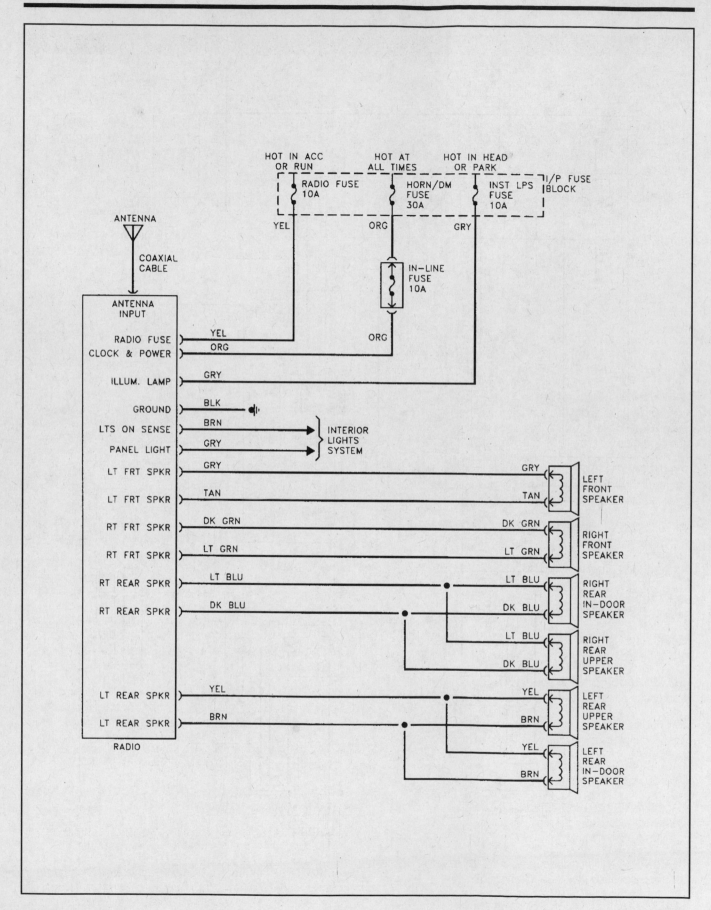

Typical radio circuit

GLOSSARY

AIR/FUEL RATIO: The ratio of air-to-gasoline by weight in the fuel mixture drawn into the engine.

AIR INJECTION: One method of reducing harmful exhaust emissions by injecting air into each of the exhaust ports of an engine. The fresh air entering the hot exhaust manifold causes any remaining fuel to be burned before it can exit the tailpipe.

ALTERNATOR: A device used for converting mechanical energy into electrical energy.

AMMETER: An instrument, calibrated in amperes, used to measure the flow of an electrical current in a circuit. Ammeters are always connected in series with the circuit being tested.

AMPERE: The rate of flow of electrical current present when one volt of electrical pressure is applied against one ohm of electrical resistance.

ANALOG COMPUTER: Any microprocessor that uses similar (analogous) electrical signals to make its calculations.

ARMATURE: A laminated, soft iron core wrapped by a wire that converts electrical energy to mechanical energy as in a motor or relay. When rotated in a magnetic field, it changes mechanical energy into electrical energy as in a generator.

ATMOSPHERIC PRESSURE: The pressure on the Earth's surface caused by the weight of the air in the atmosphere. At sea level, this pressure is 14.7 psi at 32°F (101 kPa at 0°C).

ATOMIZATION: The breaking down of a liquid into a fine mist that can be suspended in air.

AXIAL PLAY: Movement parallel to a shaft or bearing bore.

BACKFIRE: The sudden combustion of gases in the intake or exhaust system that results in a loud explosion.

BACKLASH: The clearance or play between two parts, such as meshed gears.

BACKPRESSURE: Restrictions in the exhaust system that slow the exit of exhaust gases from the combustion chamber.

BAKELITE: A heat resistant, plastic insulator material commonly used in printed circuit boards and transistorized components.

BALL BEARING: A bearing made up of hardened inner and outer races between which hardened steel balls roll.

BALLAST RESISTOR: A resistor in the primary ignition circuit that lowers voltage after the engine is started to reduce wear on ignition components.

BEARING: A friction reducing, supportive device usually located between a stationary part and a moving part.

BIMETAL TEMPERATURE SENSOR: Any sensor or switch made of two dissimilar types of metal that bend when heated or cooled due to the different expansion rates of the alloys. These types of sensors usually function as an on/off switch.

BLOWBY: Combustion gases, composed of water vapor and unburned fuel, that leak past the piston rings into the crankcase during normal engine operation. These gases are removed by the PCV system to prevent the buildup of harmful acids in the crankcase.

BRAKE PAD: A brake shoe and lining assembly used with disc brakes.

BRAKE SHOE: The backing for the brake lining. The term is, however, usually applied to the assembly of the brake backing and lining.

BUSHING: A liner, usually removable, for a bearing; an anti-friction liner used in place of a bearing.

CALIPER: A hydraulically activated device in a disc brake system, which is mounted straddling the brake rotor (disc). The caliper contains at least one piston and two brake pads. Hydraulic pressure on the piston(s) forces the pads against the rotor.

CAMSHAFT: A shaft in the engine on which are the lobes (cams) which operate the valves. The camshaft is driven by the crankshaft, via a belt, chain or gears, at one half the crankshaft speed.

CAPACITOR: A device which stores an electrical charge.

CARBON MONOXIDE (CO): A colorless, odorless gas given off as a normal byproduct of combustion. It is poisonous and extremely dangerous in confined areas, building up slowly to toxic levels without warning if adequate ventilation is not available.

CARBURETOR: A device, usually mounted on the intake manifold of an engine, which mixes the air and fuel in the proper proportion to allow even combustion.

CATALYTIC CONVERTER: A device installed in the exhaust system, like a muffler, that converts harmful byproducts of combustion into carbon dioxide and water vapor by means of a heat-producing chemical reaction.

CENTRIFUGAL ADVANCE: A mechanical method of advancing the spark timing by using flyweights in the distributor that react to centrifugal force generated by the distributor shaft rotation.

CHECK VALVE: Any one-way valve installed to permit the flow of air, fuel or vacuum in one direction only.

CHOKE: A device, usually a moveable valve, placed in the intake path of a carburetor to restrict the flow of air.

CIRCUIT: Any unbroken path through which an electrical current can flow. Also used to describe fuel flow in some instances.

CIRCUIT BREAKER: A switch which protects an electrical circuit from overload by opening the circuit when the current flow exceeds a predetermined level. Some circuit breakers must be reset manually, while most reset automatically.

COIL (IGNITION): A transformer in the ignition circuit which steps up the voltage provided to the spark plugs.

COMBINATION MANIFOLD: An assembly which includes both the intake and exhaust manifolds in one casting.

COMBINATION VALVE: A device used in some fuel systems that routes fuel vapors to a charcoal storage canister instead of venting them into the atmosphere. The valve relieves fuel tank pressure and allows fresh air into the tank as the fuel level drops to prevent a vapor lock situation.

COMPRESSION RATIO: The comparison of the total volume of the cylinder and combustion chamber with the piston at BDC and the piston at TDC.

CONDENSER: 1. An electrical device which acts to store an electrical charge, preventing voltage surges. 2. A radiator-like device in the air conditioning system in which refrigerant gas condenses into a liquid, giving off heat.

CONDUCTOR: Any material through which an electrical current can be transmitted easily.

CONTINUITY: Continuous or complete circuit. Can be checked with an ohmmeter.

COUNTERSHAFT: An intermediate shaft which is rotated by a mainshaft and transmits, in turn, that rotation to a working part.

CRANKCASE: The lower part of an engine in which the crankshaft and related parts operate.

CRANKSHAFT: The main driving shaft of an engine which receives reciprocating motion from the pistons and converts it to rotary motion.

CYLINDER: In an engine, the round hole in the engine block in which the piston(s) ride.

CYLINDER BLOCK: The main structural member of an engine in which is found the cylinders, crankshaft and other principal parts.

CYLINDER HEAD: The detachable portion of the engine, usually fastened to the top of the cylinder block and containing all or most of the combustion chambers. On overhead valve engines, it contains the valves and their operating parts. On overhead cam engines, it contains the camshaft as well.

DEAD CENTER: The extreme top or bottom of the piston stroke.

DETONATION: An unwanted explosion of the air/fuel mixture in the combustion chamber caused by excess heat and compression, advanced timing, or an overly lean mixture. Also referred to as "ping".

DIAPHRAGM: A thin, flexible wall separating two cavities, such as in a vacuum advance unit.

DIESELING: A condition in which hot spots in the combustion chamber cause the engine to run on after the key is turned off.

DIFFERENTIAL: A geared assembly which allows the transmission of motion between drive axles, giving one axle the ability to turn faster than the other.

DIODE: An electrical device that will allow current to flow in one direction only.

DISC BRAKE: A hydraulic braking assembly consisting of a brake disc, or rotor, mounted on an axle, and a caliper assembly containing, usually two brake pads which are activated by hydraulic pressure. The pads are forced against the sides of the disc, creating friction which slows the vehicle.

DISTRIBUTOR: A mechanically driven device on an engine which is responsible for electrically firing the spark plug at a predetermined point of the piston stroke.

DOWEL PIN: A pin, inserted in mating holes in two different parts allowing those parts to maintain a fixed relationship.

DRUM BRAKE: A braking system which consists of two brake shoes and one or two wheel cylinders, mounted on a fixed backing plate, and a brake drum, mounted on an axle, which revolves around the assembly.

DWELL: The rate, measured in degrees of shaft rotation, at which an electrical circuit cycles on and off.

ELECTRONIC CONTROL UNIT (ECU): Ignition module, module, amplifier or igniter. See Module for definition.

ELECTRONIC IGNITION: A system in which the timing and firing of the spark plugs is controlled by an electronic control unit, usually called a module. These systems have no points or condenser.

END-PLAY: The measured amount of axial movement in a shaft.

ENGINE: A device that converts heat into mechanical energy.

EXHAUST MANIFOLD: A set of cast passages or pipes which conduct exhaust gases from the engine.

FEELER GAUGE: A blade, usually metal, or precisely predetermined thickness, used to measure the clearance between two parts.

FIRING ORDER: The order in which combustion occurs in the cylinders of an engine. Also the order in which spark is distributed to the plugs by the distributor.

FLOODING: The presence of too much fuel in the intake manifold and combustion chamber which prevents the air/fuel mixture from firing, thereby causing a no-start situation.

FLYWHEEL: A disc shaped part bolted to the rear end of the crankshaft. Around the outer perimeter is affixed the ring gear. The starter drive engages the ring gear, turning the flywheel, which rotates the crankshaft, imparting the initial starting motion to the engine.

FOOT POUND (ft. lbs. or sometimes, ft.lb.): The amount of energy or work needed to raise an item weighing one pound, a distance of one foot.

FUSE: A protective device in a circuit which prevents circuit overload by breaking the circuit when a specific amperage is present. The device is constructed around a strip or wire of a lower amperage rating than the circuit it is designed to protect. When an amperage higher than that stamped on the fuse is present in the circuit, the strip or wire melts, opening the circuit.

GEAR RATIO: The ratio between the number of teeth on meshing gears.

GENERATOR: A device which converts mechanical energy into electrical energy.

HEAT RANGE: The measure of a spark plug's ability to dissipate heat from its firing end. The higher the heat range, the hotter the plug fires.

HUB: The center part of a wheel or gear.

HYDROCARBON (HC): Any chemical compound made up of hydrogen and carbon. A major pollutant formed by the engine as a byproduct of combustion.

HYDROMETER: An instrument used to measure the specific gravity of a solution.

INCH POUND (inch lbs.; sometimes in.lb. or in. lbs.): One twelfth of a foot pound.

INDUCTION: A means of transferring electrical energy in the form of a magnetic field. Principle used in the ignition coil to increase voltage.

INJECTOR: A device which receives metered fuel under relatively low pressure and is activated to inject the fuel into the engine under relatively high pressure at a predetermined time.

INPUT SHAFT: The shaft to which torque is applied, usually carrying the driving gear or gears.

INTAKE MANIFOLD: A casting of passages or pipes used to conduct air or a fuel/air mixture to the cylinders.

JOURNAL: The bearing surface within which a shaft operates.

KEY: A small block usually fitted in a notch between a shaft and a hub to prevent slippage of the two parts.

MANIFOLD: A casting of passages or set of pipes which connect the cylinders to an inlet or outlet source.

MANIFOLD VACUUM: Low pressure in an engine intake manifold formed just below the throttle plates. Manifold vacuum is highest at idle and drops under acceleration.

MASTER CYLINDER: The primary fluid pressurizing device in a hydraulic system. In automotive use, it is found in brake and hydraulic clutch systems and is pedal activated, either directly or, in a power brake system, through the power booster.

MODULE: Electronic control unit, amplifier or igniter of solid state or integrated design which controls the current flow in the ignition primary circuit based on input from the pick-up coil. When the module opens the primary circuit, high secondary voltage is induced in the coil.

NEEDLE BEARING: A bearing which consists of a number (usually a large number) of long, thin rollers.

OHM: (Ω) The unit used to measure the resistance of conductor-to-electrical flow. One ohm is the amount of resistance that limits current flow to one ampere in a circuit with one volt of pressure.

OHMMETER: An instrument used for measuring the resistance, in ohms, in an electrical circuit.

OUTPUT SHAFT: The shaft which transmits torque from a device, such as a transmission.

OVERDRIVE: A gear assembly which produces more shaft revolutions than that transmitted to it.

OVERHEAD CAMSHAFT (OHC): An engine configuration in which the camshaft is mounted on top of the cylinder head and operates the valve either directly or by means of rocker arms.

OVERHEAD VALVE (OHV): An engine configuration in which all of the valves are located in the cylinder head and the camshaft is located in the cylinder block. The camshaft operates the valves via lifters and pushrods.

OXIDES OF NITROGEN (NOx): Chemical compounds of nitrogen produced as a byproduct of combustion. They combine with hydrocarbons to produce smog.

OXYGEN SENSOR: Use with the feedback system to sense the presence of oxygen in the exhaust gas and signal the computer which can reference the voltage signal to an air/fuel ratio.

PINION: The smaller of two meshing gears.

PISTON RING: An open-ended ring with fits into a groove on the outer diameter of the piston. Its chief function is to form a seal between the piston and cylinder wall. Most automotive pistons have three rings: two for compression sealing; one for oil sealing.

PRELOAD: A predetermined load placed on a bearing during assembly or by adjustment.

PRIMARY CIRCUIT: the low voltage side of the ignition system which consists of the ignition switch, ballast resistor or resistance wire, bypass, coil, electronic control unit and pick-up coil as well as the connecting wires and harnesses.

PRESS FIT: The mating of two parts under pressure, due to the inner diameter of one being smaller than the outer diameter of the other, or vice versa; an interference fit.

RACE: The surface on the inner or outer ring of a bearing on which the balls, needles or rollers move.

REGULATOR: A device which maintains the amperage and/or voltage levels of a circuit at predetermined values.

RELAY: A switch which automatically opens and/or closes a circuit.

RESISTANCE: The opposition to the flow of current through a circuit or electrical device, and is measured in ohms. Resistance is equal to the voltage divided by the amperage.

RESISTOR: A device, usually made of wire, which offers a preset amount of resistance in an electrical circuit.

RING GEAR: The name given to a ring-shaped gear attached to a differential case, or affixed to a flywheel or as part of a planetary gear set.

ROLLER BEARING: A bearing made up of hardened inner and outer races between which hardened steel rollers move.

ROTOR: 1. The disc-shaped part of a disc brake assembly, upon which the brake pads bear; also called, brake disc. 2. The device mounted atop the distributor shaft, which passes current to the distributor cap tower contacts.

SECONDARY CIRCUIT: The high voltage side of the ignition system, usually above 20,000 volts. The secondary includes the ignition coil, coil wire, distributor cap and rotor, spark plug wires and spark plugs.

SENDING UNIT: A mechanical, electrical, hydraulic or electro-magnetic device which transmits information to a gauge.

SENSOR: Any device designed to measure engine operating conditions or ambient pressures and temperatures. Usually electronic in nature and designed to send a voltage signal to an on-board computer, some sensors may operate as a simple on/off switch or they may provide a variable voltage signal (like a potentiometer) as conditions or measured parameters change.

SHIM: Spacers of precise, predetermined thickness used between parts to establish a proper working relationship.

SLAVE CYLINDER: In automotive use, a device in the hydraulic clutch system which is activated by hydraulic force, disengaging the clutch.

SOLENOID: A coil used to produce a magnetic field, the effect of which is to produce work.

SPARK PLUG: A device screwed into the combustion chamber of a spark ignition engine. The basic construction is a conductive core inside of a ceramic insulator, mounted in an outer conductive base. An electrical charge from the spark plug wire travels along the conductive core and jumps a preset air gap to a grounding point or points at the end of the conductive base. The resultant spark ignites the fuel/air mixture in the combustion chamber.

SPLINES: Ridges machined or cast onto the outer diameter of a shaft or inner diameter of a bore to enable parts to mate without rotation.

TACHOMETER: A device used to measure the rotary speed of an engine, shaft, gear, etc., usually in rotations per minute.

THERMOSTAT: A valve, located in the cooling system of an engine, which is closed when cold and opens gradually in response to engine heating, controlling the temperature of the coolant and rate of coolant flow.

TOP DEAD CENTER (TDC): The point at which the piston reaches the top of its travel on the compression stroke.

TORQUE: The twisting force applied to an object.

TORQUE CONVERTER: A turbine used to transmit power from a driving member to a driven member via hydraulic action, providing changes in drive ratio and torque. In automotive use, it links the driveplate at the rear of the engine to the automatic transmission.

TRANSDUCER: A device used to change a force into an electrical signal.

TRANSISTOR: A semi-conductor component which can be actuated by a small voltage to perform an electrical switching function.

TUNE-UP: A regular maintenance function, usually associated with the replacement and adjustment of parts and components in the electrical and fuel systems of a vehicle for the purpose of attaining optimum performance.

TURBOCHARGER: An exhaust driven pump which compresses intake air and forces it into the combustion chambers at higher than atmospheric pressures. The increased air pressure allows more fuel to be burned and results in increased horsepower being produced.

VACUUM ADVANCE: A device which advances the ignition timing in response to increased engine vacuum.

VACUUM GAUGE: An instrument used to measure the presence of vacuum in a chamber.

VALVE: A device which control the pressure, direction of flow or rate of flow of a liquid or gas.

VALVE CLEARANCE: The measured gap between the end of the valve stem and the rocker arm, cam lobe or follower that activates the valve.

VISCOSITY: The rating of a liquid's internal resistance to flow.

VOLTMETER: An instrument used for measuring electrical force in units called volts. Voltmeters are always connected parallel with the circuit being tested.

WHEEL CYLINDER: Found in the automotive drum brake assembly, it is a device, actuated by hydraulic pressure, which, through internal pistons, pushes the brake shoes outward against the drums.

MASTER INDEX